Lecture Notes in Statistics 175

Edited by P. Bickel, P. Diggle, S. Fienberg, K. Krickeberg, I. Olkin, N. Wermuth, S. Zeger

Springer
Berlin
Heidelberg
New York
Hong Kong
London
Milan
Paris
Tokyo

Jürgen Groß

Linear Regression

Springer

Jürgen Groß
University of Dortmund
Department of Statistics
Vogelpothsweg 87
44221 Dortmund
Germany
gross@statistik.uni-dortmund.de

Cataloging-in-Publication Data applied for

A catalog record for this book is available from the Library of Congress.

Bibliographic information published by Die Deutsche Bibliothek
Die Deutsche Bibliothek lists this publication in the Deutsche Nationalbibliografie; detailed bibliographic data is available in the Internet at <http://dnb.ddb.de>.

ISBN 3-540-40178-4 Springer-Verlag Berlin Heidelberg New York

Springer-Verlag Berlin Heidelberg New York
a member of BertelsmannSpringer Science+Business Media GmbH

http://www.springer.de

Printed in Germany

Cover design: *design & production*, Heidelberg
Typesetting by the author using LaTeX
Printed on acid-free paper 40/3142ck-5 4 3 2 1 0

Preface

In linear regression the ordinary least squares estimator plays a central role and sometimes one may get the impression that it is the only reasonable and applicable estimator available. Nonetheless, there exists a variety of alternatives, proving useful in specific situations.

Purpose and Scope. This book aims at presenting a comprehensive survey of different *point estimation methods* in linear regression, along with the theoretical background on a *advanced courses level.* Besides its possible use as a companion for specific courses, it should be helpful for purposes of further reading, giving detailed explanations on many topics in this field. Numerical examples and graphics will aid to deepen the insight into the specifics of the presented methods.

For the purpose of self-containment, the basic theory of linear regression models and least squares is presented. The fundamentals of decision theory and matrix algebra are also included. Some prior basic knowledge, however, appears to be necessary for easy reading and understanding.

Contents. The book starts with an introduction into the basics of *decision theory* being followed by an introduction in linear regression models along with a discussion of model assumptions. *Least squares theory* is considered in detail, and the optimality of the ordinary least squares estimator within the class of unbiased linear estimators is worked out. A discussion of the concept of unbiasedness follows and it is demonstrated that under certain conditions, such as *collinearity*, it may be beneficial to use alternative biased estimators. In the sequel, *a number of various estimators* is introduced, such as restricted least squares, pretest, principal components, ridge, shrinkage, minimax, Bayes, and robust estimators. The underlying estimation principles, the non-linear variants of linear estimators, the relationships between estimators, and the possible applications are widely discussed and illustrated by numerous examples. A course on *linear admissibility* follows, giving deeper insight into the structure and construction of estimators under different states of prior knowledge about the regression parameters. Then, an outline of the process of generalization of point estimation methods to the case of *non-homogeneous and/or correlated errors* is provided. The book is concluded by a review of *regression diagnostic methods.*

Linear Versus Non-Linear Estimation. Usually, biased linear estimators depend on further unknown parameters, which for the sake of theoretical considerations are assumed to be given. For practical purposes, however, these unknowns have to be estimated, so that the resulting estimators for the regression parameters are not only biased but also non-linear. It is one of the aspects of this book to emphasize the aforementioned difference between theoretical and practical needs and to give both, theoretical results under mild assumptions as well as guidelines for the application of biased non-linear estimators.

Examples. For the purpose of elucidating the ideas behind points estimators and illustrating their behavior compared to ordinary least squares, there are several small sample size examples with simulated data included. When alternative estimators perform better in these examples, then this should definitely not be understood as some kind of evidence in favor of these estimators, but as a demonstration of their *potential.* The graphics for the examples were created with maple 6, except for the graphics in Appendix C. Computations have been done with maple 6, S-PLUS 4.5 and R 1.6.1.

Acknowledgements. For reading different earlier versions and providing a lot of helpful comments, I am grateful to Ursula Enneking, Götz Trenkler, Sven-Oliver Troschke, Barbara Bredner and Eva Brune, each of them deserving special emphasize. Of course, responsibility for remaining errors is solely due to me.

This book has been created as part of a research project on point estimators and their admissibility in linear models under the direction of Götz Trenkler and Holger Dette. Financial support was provided by the Deutsche Forschungsgemeinschaft (DFG) under grants Tr 253/3-1 and Tr 253/3-2.

Dortmund
May 2003

Jürgen Groß

Table of Contents

Part I Point Estimation and Linear Regression

Part III Miscellaneous Topics

Part I

Point Estimation and Linear Regression

1 Fundamentals

In this chapter, a brief introduction into the theory of linear regression models is given and a small numerical example is created, providing the opportunity to pose some of the central problems. In addition, because of the relevance for comparison of point estimators, an introduction into the basics of decision theory is delivered.

1.1 Linear Models

A short description of the linear regression model and its representation in matrix notation is given. Different types of linear models are introduced and the process of analysis via linear models is discussed.

1.1.1 Application of Linear Models

In a scientific investigation one may often be inclined to formulate a hypothesis about a linear relationship

$$y = \beta_1 x_1 + \cdots + \beta_p x_p$$

between a variable y on the one side and variables $x_1, \ldots, x_p$ on the other side, making is possible to explain y via $x_1, \ldots, x_p$. This can be quite useful when values of $x_1, \ldots, x_p$ are rather easy to obtain, while this is not the case for the corresponding value of y. Then the above equation may be employed for predicting the outcome of y given values $x_1, \ldots, x_p$.

Although one may be able to specify such a hypothesis for different reasons, the exact nature of the considered linear relationship will not be known, or, in other words, the parameters $\beta_1, \ldots, \beta_p$ will be unknown. Information about them can be drawn from a given set of observations of the variables y and $x_1, \ldots, x_p$.

Example 1.1. Consider for a batch of cement the heat y evolving during the hardening of the cement. Suppose that the batch consists of amounts of four main ingredients denoted by x_1, x_2, x_3, and x_4. Now, suppose that it is desired to predict the heat on the basis of the amount of one or more of the ingredients. One possible hypothesis might be

$$y = \beta_1 + \beta_2 x_1 + \beta_3 x_2 ,$$

postulating that the heat is a linear function of the two amounts of ingredients x_1 and x_2 (and a constant term). In order to obtain a formula for predicting the heat for given x_1 and x_2, information about β_1, β_2 and β_3 is required. This may be given in form of point estimates of the three parameters, using the data from Table 1.1. It provides observations of the variable y (recorded in calories per gram) and of the variables $x_1, \dots, x_4$ (recorded in addition percentages) for $n = 13$ batches. See also Appendix C for an analysis of this data with the statistical-computing environment R. □

Table 1.1. Cement data, see [49, p. 647]

Batch i	Heat y	Ingredient 1 x_1	Ingredient 2 x_2	Ingredient 3 x_3	Ingredient 4 x_4
1	78.5	7	26	6	60
2	74.3	1	29	15	52
3	104.3	11	56	8	20
4	87.6	11	31	8	47
5	95.9	7	52	6	33
6	109.2	11	55	9	22
7	102.7	3	71	17	6
8	72.5	1	31	22	44
9	93.1	2	54	18	22
10	115.9	21	47	4	26
11	83.8	1	40	23	34
12	113.3	11	66	9	12
13	109.4	10	68	8	12

In the above example we cannot expect that the heat is an exact linear function of the amount of the two ingredients x_1 and x_2. As a matter of fact, in practice there will be an *approximate* linear rather than an *exact* linear relationship between the variables of interest. One may account for this by adding a further non-observable variable, fancied as a collection of small errors. Then the hypothesis is altered to an equation of the form

$$y = \beta_1 x_1 + \cdots + \beta_p x_p + \varepsilon ,$$

letting y not only depend on $x_1, \dots, x_p$, but also on ε (the error variable), being random but non-observable.

If it is assumed that for given variables y and $x_1, \ldots, x_p$ the above hypothesis is correct, then statements about the parameters $\beta_1, \ldots, \beta_p$ can be derived, which should be as close as possible to the unknowns, but nonetheless can only be of stochastic nature. Such 'statements' may be given in form of *point estimates* based on (say n) rows of observed values

$$(y_1, x_{1,1}, \ldots, x_{1,p}), \ldots, (y_n, x_{n,1}, \ldots, x_{n,p}) \ .$$

The $y_1, \ldots, y_n$ are regarded as sample values of y, given fixed values $(x_{1,1}, \ldots, x_{1,p}), \ldots, (x_{n,1}, \ldots, x_{n,p})$. Thus y_i is the observed realization of a random variable, while the observation $x_{i,j}$ is assumed to be non-stochastic.

In the literature on linear models, there is usually no notational distinction between a realization y_i and the corresponding random variable standing behind the observation. The random variable y_i satisfies

$$y_i = \beta_1 x_{i,1} + \cdots + \beta_p x_{i,p} + \varepsilon_i \ ,$$

where ε_i is a random non-observable variable. It is obvious that if for a fixed i one has observed not only one but (infinitely) many realizations of y_i for fixed $(x_{i,1}, \ldots, x_{i,p})$, then the corresponding errors should have equalized. Hence it is assumed that the expectation of ε_i equals 0. Moreover, it is assumed that all $\varepsilon_1, \ldots, \varepsilon_n$ have the same, possibly unknown, variance σ^2. Eventually, two error variables ε_i and ε_k are assumed to be uncorrelated for $i \neq k$.

The n equations together with the assumptions about ε_i can be written in matrix notation as

$$\boldsymbol{y} = \boldsymbol{X}\boldsymbol{\beta} + \boldsymbol{\varepsilon}, \qquad \boldsymbol{\varepsilon} \sim (\boldsymbol{0}, \sigma^2 \boldsymbol{I}_n) \ ,$$

where

$$\boldsymbol{y} = \begin{pmatrix} y_1 \\ \vdots \\ \vdots \\ y_n \end{pmatrix}, \ \boldsymbol{X} = \begin{pmatrix} x_{1,1} & \cdots & x_{1,p} \\ \vdots & & \vdots \\ \vdots & & \vdots \\ x_{n,1} & \cdots & x_{n,p} \end{pmatrix}, \ \boldsymbol{\beta} = \begin{pmatrix} \beta_1 \\ \vdots \\ \beta_p \end{pmatrix}, \ \boldsymbol{\varepsilon} = \begin{pmatrix} \varepsilon_1 \\ \vdots \\ \vdots \\ \varepsilon_n \end{pmatrix} .$$

In the following, this equation will be the considered description of a linear model. The variable y will be called *dependent* variable while the variables $x_1, \ldots, x_p$ will be called *independent* variables. The vector $\boldsymbol{y}$ is the vector of sample variables (each one fancied as a random observable copy of y given a set of x-values), while $\boldsymbol{X}$ is the matrix comprising the values of the independent variables.

Example 1.2. In example 1.1 the observation vector $\boldsymbol{y}$ and the matrix $\boldsymbol{X}$ are given as

$$\boldsymbol{y} = \begin{pmatrix} 78.5 \\ 74.3 \\ 104.3 \\ 87.6 \\ 95.9 \\ 109.2 \\ 102.7 \\ 72.5 \\ 93.1 \\ 115.9 \\ 83.8 \\ 113.3 \\ 109.4 \end{pmatrix} \quad \text{and} \quad \boldsymbol{X} = \begin{pmatrix} 1 & 7 & 26 \\ 1 & 1 & 29 \\ 1 & 11 & 56 \\ 1 & 11 & 31 \\ 1 & 7 & 52 \\ 1 & 11 & 55 \\ 1 & 3 & 71 \\ 1 & 1 & 31 \\ 1 & 2 & 54 \\ 1 & 21 & 47 \\ 1 & 1 & 40 \\ 1 & 11 & 66 \\ 1 & 10 & 68 \end{pmatrix} .$$

The ordinary least squares estimator for $\boldsymbol{\beta} = (\beta_1, \beta_2, \beta_3)'$ is given by

$$\widehat{\boldsymbol{\beta}} = (\boldsymbol{X}'\boldsymbol{X})^{-1}\boldsymbol{X}'\boldsymbol{y} = \begin{pmatrix} 52.5774 \\ 1.4683 \\ 0.6623 \end{pmatrix} .$$

The estimated relationship, neglecting the error variable, is therefore

$$\begin{array}{rl} y = & \underset{(2.2862)}{52.5774} + \underset{(0.1213)}{1.4683} x_1 + \underset{(0.0459)}{0.6623} x_2 \end{array} ,$$

where the numbers in brackets are the standard errors (see also Sect. 2.4.1). This equation may be applied to predict the heat y for given x_1 and x_2. For example for $x_1 = 15$ and $x_2 = 50$ the predicted heat is 107.72. See also Appendix C for a more detailed analysis. □

1.1.2 Types of Linear Models

Depending on the type of variables $x_1, \ldots, x_p$, linear models can be classified in the following way:

(1) *Analysis of Variance Model.* When the matrix $\boldsymbol{X}$ consists only of zeros and ones, then the linear model will be called analysis of variance model. In this case the variables $x_1, \ldots, x_p$ are *qualitative.*
If y_i denotes the i-th measurement and x_j denotes the j-th treatment, then $x_{i,j}$ is either 1 or 0, depending on whether the i-th unit has received the j-th treatment or not.

(2) *Regression Model.* If all variables $x_1, \ldots, x_p$ are *quantitative*, then the model will be called regression model. The term 'regression' has originally been introduced by Galton [42] in a paper on laws of heredity.

In econometrics, one will usually be confronted with such variables (e.g. when x_1 stands for the total income from trading and x_2 is the total income from wages). If there exists one variable which is a constant equal to 1 and all other variables are quantitative, then the model is called regression with *intercept*.

(3) *Analysis of Covariance Model.* If some of the variables are $(0, 1)$ qualitative and others are quantitative, then the model is sometimes called analysis of covariance model.

In this book we assume that the underlying model is a linear regression model (with or without intercept). Nonetheless, every statement will be true under any type of model which satisfies the required assumptions.

1.1.3 Proceeding with Linear Models

For investigating the relationship between variables via linear models, an appropriate procedure is outlined by the following three stages:

Stage 1: Model Building. In a first step, the relationship one wishes to investigate must be specified. Having formulated a linear model equation

$$y_i = \beta_1 x_{i,1} + \cdots + \beta_p x_{i,p} + \varepsilon_i, \quad i = 1, \ldots, n ,$$

motivated from formal reasons, in a second step it should be clarified whether there exists some evidence in favor or against the model assumptions (*model diagnostics*). For example residual plots from computer outputs may be used to find some indication in favor or against the assumptions concerning the errors ε_i. If it is not sure whether a specific variable x_j, $j \in \{1, \ldots, p\}$, should be incorporated into the model, then a test of the hypothesis $H_0 : \beta_j = 0$ can be performed. In addition, statistical software packages (as for example S-PLUS or R, cf. [27, 40, 122]) can be used used to carry out a forward and/or backward variable elimination process, based e.g. on Akaikes 'An Information Criterion' (AIC). Other plots produced by such software will aid to find extreme values (outliers and/or high leverage points), which could be worth eliminating from the analysis. The stage of model building and diagnostics is rather delicate and complicated, since there are many aspects to consider. One should be aware, that most of the model assumptions cannot be proved, but only be more or less supported by diagnostic methods. Chapter 6 gives a review on regression diagnostics.

Stage 2: Inference. If a certain model has been fixed, then the relationship between the dependent variable y and the independent variables $x_1, \ldots, x_p$ can be examined. That is, statements about the unknown parameters (like point estimates) can be concluded from the observations. In practice, this stage cannot always be clearly distinguished from Stage 1, since for model building/diagnostic purposes it is already required to derive statements about the unknowns. If Stage 1 is performed, then usually some statistical knowledge about the relevant parameters will automatically be given. (For example testing $H_0 : \beta_j = 0$ usually requires a point estimate of β_j.) In addition, inference under a given model might lead to conclusions which question the choice of model, thus making it necessary to return to Stage 1.

Stage 3: Prediction. Frequently, the ultimate goal of an analysis via linear models is the prediction of y given some values $(x_{m,1}, \ldots, x_{m,p})$. Such a prediction will be based on the results from Stages 1 and 2. The more adequate the chosen model and the more precise the obtained estimates, the better one can hope to predict y.

In the following Chapters 2 to 5 we assume that the stage of model building has been completed. Our interest focuses on inference about the unknown regression parameters under a given model (Stage 2). Here we are mainly interested in comparisons of the performance of different point estimators. Chapter 6 gives a short review on regression diagnostic methods and Appendix C demonstrates the analysis of a linear regression model with the statistical-computing environment R.

1.1.4 A Preliminary Example

The following numerical example will give an impression about the main topic of this text, being the consideration of different estimators for regression parameters and the appropriate evaluation of their performance.

The Model

Consider a linear regression model with $p = 3$ independent (explanatory) variables and $n = 12$ observations, described by

$$y_i = \beta_1 x_{i,1} + \beta_2 x_{i,2} + \beta_3 x_{i,3} + \varepsilon_i, \qquad i = 1, \ldots, 12\,,$$

where $\mathrm{E}(\varepsilon_i) = 0$, $\mathrm{Var}(\varepsilon_i) = \sigma^2$ and $\mathrm{Cov}(\varepsilon_i, \varepsilon_k) = 0$ for $i \neq k = 1, \ldots, 12$. The 12 respective observations of the dependent and the three independent variables are given in Table 1.2.

Table 1.2. Observed values of variables y and x_1, x_2, x_3

i	1	2	3	4	5	6	7	8	9	10	11	12
y_i	2.275	3.658	.9561	3.329	2.462	3.843	.8117	.9014	2.715	3.882	3.196	.7634
$x_{i,1}$	1	1	1	1	1	1	1	1	1	1	1	1
$x_{i,2}$	2.965	2.839	3.466	2.538	1.993	3.670	1.011	2.972	3.504	1.872	2.160	3.218
$x_{i,3}$	1.776	2.365	.1971	1.442	1.419	3.215	1.140	.3691	.3799	.9427	.7657	.3075

Obviously, since $x_{i,1} = 1$ for each i, x_1 is the intercept variable. In matrix notation we can write

$$\boldsymbol{y} = \boldsymbol{X}\beta + \varepsilon, \qquad \varepsilon \sim (\mathbf{0}, \sigma^2 \boldsymbol{I}_{12}) ,$$

where the elements y_i of $\boldsymbol{y}$ are random random variables for which observations are available from Table 1.2 and the elements $x_{i,j}$ ($i = 1, \ldots, 12$, $j = 1, 2, 3$) of $\boldsymbol{X}$ are non-stochastic and determined by Table 1.2.

Different Estimators

Since the matrix $\boldsymbol{X}$ is of full column rank, the *ordinary least squares estimator* for $\boldsymbol{\beta}$ exists and is given by

$$\widehat{\beta} = (\boldsymbol{X}'\boldsymbol{X})^{-1}\boldsymbol{X}'\boldsymbol{y} .$$

By using the values from Table 1.2 we compute

$$\widehat{\beta} = (1.4570, -0.0233, 0.8423)'$$

as the least squares estimate for the parameter vector $\boldsymbol{\beta}$.

An alternative estimator for $\boldsymbol{\beta}$ is the so-called *ridge estimator*, given by

$$\widehat{\boldsymbol{\beta}}_k = (\boldsymbol{X}'\boldsymbol{X} + k\boldsymbol{I}_p)^{-1}\boldsymbol{X}'\boldsymbol{y}, \quad k \geq 0 .$$

We compute the scalar k from the formula

$$k = \frac{p\,\widehat{\sigma}^2}{\widehat{\boldsymbol{\beta}}'\widehat{\boldsymbol{\beta}}}, \qquad \widehat{\sigma}^2 = \frac{1}{n-p}(\boldsymbol{y} - \boldsymbol{X}\widehat{\beta})'(\boldsymbol{y} - \boldsymbol{X}\widehat{\beta}) ,$$

and obtain $k = 1.2288$. Then, the ridge estimate for $\boldsymbol{\beta}$ is

$$\widehat{\boldsymbol{\beta}}_k = (0.6715, 0.2518, 0.8242)' .$$

As can be seen, the ridge estimator $\widehat{\boldsymbol{\beta}}_k$ yields a fairly different estimate for the first element β_1 of the vector $\boldsymbol{\beta}$ compared to the ordinary least squares estimator $\widehat{\boldsymbol{\beta}}$. Similarly, the estimates for β_2 are different, while the estimates for β_3 are almost the same.

As a second alternative one may consider the so-called *shrinkage estimator*

$$\widehat{\beta}(\varrho) = \frac{1}{1+\varrho}\widehat{\beta}, \quad \varrho \geq 0 .$$

We compute ϱ from the formula

$$\varrho = \frac{\widehat{\sigma}^2 \operatorname{tr}[(\boldsymbol{X}'\boldsymbol{X})^{-1}]}{\widehat{\boldsymbol{\beta}}'\widehat{\boldsymbol{\beta}}}$$

and obtain $\varrho = 0.6057$. Then the shrinkage estimate for $\boldsymbol{\beta}$ is

$$\widehat{\boldsymbol{\beta}}(\varrho) = (0.9074, -0.0145, 0.5246)' .$$

As can be seen, the shrinkage estimator $\widehat{\boldsymbol{\beta}}(\varrho)$ is designed such that it simply multiplies the estimates of β_1, β_2 and β_3 obtained from $\widehat{\boldsymbol{\beta}}$ by the same factor $(1+0.6057)^{-1} = 0.6228$ and thus shrinks the ordinary least squares estimates. This is similar to the behavior of the *Stein estimator* (also called *James-Stein estimator*, see [61]), given as

$$\widehat{\boldsymbol{\beta}}_{\mathrm{S}} = \gamma\widehat{\boldsymbol{\beta}}, \quad \gamma = 1 - \frac{(p-2)(n-p)}{n-p+2}\,\frac{\widehat{\sigma}^2}{\widehat{\boldsymbol{\beta}}'\boldsymbol{X}'\boldsymbol{X}\widehat{\boldsymbol{\beta}}} .$$

We compute $\gamma = 0.9875$ and obtain

$$\widehat{\boldsymbol{\beta}}_{\mathrm{S}} = (1.4387, -0.0230, 0.8317)'$$

as an estimate for $\boldsymbol{\beta}$. Since the computed factor γ is close to 1, the estimates from the Stein estimator $\widehat{\boldsymbol{\beta}}_{\mathrm{S}}$ hardly differ from the estimates from the ordinary least squares estimator $\widehat{\boldsymbol{\beta}}$.

Observed Loss

The data in Table 1.2 is obtained as a result from a small simulation process, where in a first step the values of $x_{i,1}$, $x_{i,2}$ and $x_{i,3}$ had been fixed. Then 12 realizations of independent $N(0,1)$ distributed random variables ε_i have been obtained by a random generator. The corresponding value of y_i has been computed from the model equation by using the vector

$$\boldsymbol{\beta} = (0.2, 0.4, 0.7)' .$$

Hence, in distinction to a real-world application, the true parameter vector is known, and we are able to compare the true values with the estimated ones. Fig. 1.1 shows the vector $\boldsymbol{\beta}$ as a point in a 3-dimensional coordinate system together with the estimates $\widehat{\boldsymbol{\beta}}$, $\widehat{\boldsymbol{\beta}}_k$ and $\widehat{\boldsymbol{\beta}}(\varrho)$. The value of $\widehat{\boldsymbol{\beta}}_{\mathrm{S}}$ is not shown, since it cannot visually be distinguished from the least squares estimate $\widehat{\boldsymbol{\beta}}$.

Although, due to the graphic presentation, the differences in direction of β_1 are not easy to make out, it can be seen that the ridge estimate and the shrinkage estimate are nearer to the true parameter(vector) than the least squares estimate. We compute the squared distances from the respective estimates to $\boldsymbol{\beta} = (0.2, 0.4, 0.7)'$, and obtain

$$\begin{aligned}
l_1 &= \|\widehat{\boldsymbol{\beta}} - \boldsymbol{\beta}\|^2 = (\widehat{\boldsymbol{\beta}} - \boldsymbol{\beta})'(\widehat{\boldsymbol{\beta}} - \boldsymbol{\beta}) = 1.7794 , \\
l_2 &= \|\widehat{\boldsymbol{\beta}}_k - \boldsymbol{\beta}\|^2 = (\widehat{\boldsymbol{\beta}}_k - \boldsymbol{\beta})'(\widehat{\boldsymbol{\beta}}_k - \boldsymbol{\beta}) = 0.2597 , \\
l_3 &= \|\widehat{\boldsymbol{\beta}}(\varrho) - \boldsymbol{\beta}\|^2 = (\widehat{\boldsymbol{\beta}}(\varrho) - \boldsymbol{\beta})'(\widehat{\boldsymbol{\beta}}(\varrho) - \boldsymbol{\beta}) = 0.7030 , \\
l_4 &= \|\widehat{\boldsymbol{\beta}}_{\mathrm{S}} - \boldsymbol{\beta}\|^2 = (\widehat{\boldsymbol{\beta}}_{\mathrm{S}} - \boldsymbol{\beta})'(\widehat{\boldsymbol{\beta}}_{\mathrm{S}} - \boldsymbol{\beta}) = 1.7306 .
\end{aligned}$$

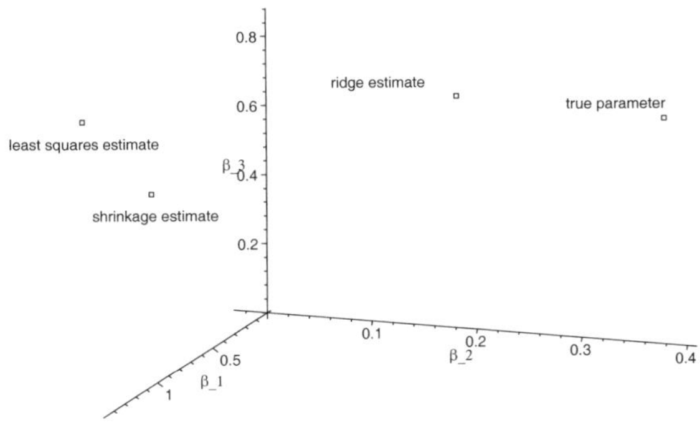

Fig. 1.1. True parameter vector $\boldsymbol{\beta} = (0.2, 0.4, 0.7)'$ and least squares, ridge and shrinkage estimates

We may interpret these squared distances as *observed losses*, which have to be admitted when the respective estimator is chosen and $\boldsymbol{\beta} = (0.2, 0.4, 0.7)'$ is the true parameter.

If we compare the values l_1, l_2, l_3 and l_4, then we will conclude that the observed loss of the ridge estimator $\widehat{\boldsymbol{\beta}}_k$ is distinctly smaller than the observed loss of the least squares estimator $\widehat{\boldsymbol{\beta}}$. The shrinkage estimator $\widehat{\boldsymbol{\beta}}(\varrho)$ yields a smaller loss compared to the loss of least squares, but a greater loss compared to the loss of ridge. Even the Stein estimator $\widehat{\boldsymbol{\beta}}_{\mathrm{S}}$ produces a smaller observed loss than $\widehat{\boldsymbol{\beta}}$, although the difference is only marginal.

Risk

Of course, the observed loss of an estimator depends on the observed realization of $\boldsymbol{y}$ (the sample value). If we wish to assess the loss of an estimator $\widetilde{\boldsymbol{\beta}}(\boldsymbol{y})$ independent of a given realization, then we can consider the *average loss*

$$\mathrm{E}\left[(\widetilde{\boldsymbol{\beta}}(\boldsymbol{y}) - \boldsymbol{\beta})'(\widetilde{\boldsymbol{\beta}}(\boldsymbol{y}) - \boldsymbol{\beta})\right] .$$

Here we do not consider a specific estimate $\widetilde{\boldsymbol{\beta}}(\boldsymbol{y})$ depending on the observed value $\boldsymbol{y}$, but the random vector $\widetilde{\boldsymbol{\beta}}(\boldsymbol{y})$ depending on the random vector $\boldsymbol{y}$. This expected loss is usually called the *risk* of the estimator $\widetilde{\boldsymbol{\beta}}(\boldsymbol{y})$ with respect to $\boldsymbol{\beta}$.

In the above situation an estimator can be called *better* than another, if it has a smaller risk with respect to $\boldsymbol{\beta} = (0.2, 0.4, 0.7)'$. Of course, in practice we do not know the true parameter vector, so we are interested in risk comparisons which do not only hold for some fixed value of $\boldsymbol{\beta}$, but for at least a certain set of possible values $\boldsymbol{\beta} \in \mathbb{R}^p$ and a certain set of possible values $\sigma^2 > 0$. Clearly it would be most helpful to have risk inequalities between estimators for $\boldsymbol{\beta}$ being valid for all possible parameter values $\boldsymbol{\beta}$, but this will rarely turn out to be the case if we consider reasonable estimators.

Some Questions

As we have demonstrated above, there are situations in which certain alternatives to the least squares estimator deliver smaller observed losses than $\widehat{\boldsymbol{\beta}}$. This, however, does not imply that expected losses must behave similarly, meaning that we cannot draw any reasonable conclusions about the behavior of the respective estimators from the above example. Nonetheless, the above results raise some questions:

- Is it possible that a better estimator than $\widehat{\boldsymbol{\beta}}$ exists for all possible $\boldsymbol{\beta} \in \mathbb{R}^p$ and all $\sigma^2 > 0$? In other words, does there exist an estimator which makes $\widehat{\boldsymbol{\beta}}$ *inadmissible* for estimating $\boldsymbol{\beta}$?
- Do there exist different possibilities to define 'better'? Are there different reasonable losses and risks?
- Do there exist estimators of $\boldsymbol{\beta}$ which are better than $\widehat{\boldsymbol{\beta}}$ for certain sets of $\boldsymbol{\beta} \in \mathbb{R}^p$ and $\sigma^2 > 0$? In other words, do there exist estimators which turn out to be *admissible* compared to $\widehat{\boldsymbol{\beta}}$?
- If there existed such admissible estimators, under what conditions should they be used?
- Can we find estimators which are admissible compared to any other estimator? In such a case, we can never find a *uniformly* better estimator, i.e. an estimator which is better for all $\boldsymbol{\beta} \in \mathbb{R}^p$ and all $\sigma^2 > 0$

Terms like 'loss', 'risk' or 'admissibility' are widely used in decision theory. Therefore, before trying to give some answers to the above questions, we present a short introduction into this theory.

1.2 Decision Theory and Point Estimation

Statistical decision theory has been established by Abraham Wald with a series of papers in the 1940s being combined 1950 in his famous book *Statistical Decision Functions* [124]. Further contributions on this topic are for example provided by Ferguson [38] and Berger [13].

1.2.1 Decision Rule

Suppose we have to reach a decision d depending on p unknown quantities $\theta_1, \ldots, \theta_p$. These quantities are combined in the parameter vector $\boldsymbol{\theta} = (\theta_1, \ldots, \theta_p)'$. Then:

- the set of all possible parameter vectors $\boldsymbol{\theta}$ is called *parameter space* and is denoted by Θ;
- the set of all possible decisions d is called *decision space* and is denoted by D.

Example 1.3. Suppose that we want to decide whether to use a certain coin for a game of chance or not. The set of decisions D consists of the two elements 'use coin' and 'reject coin'. The decision depends on the unknown probability p for the appearance of the event 'face side'. Hence the parameter space is $\Theta = [0,1]$. □

To make a specific decision, claims about the unknown parameters $\theta_1, \ldots, \theta_p$ are required. To obtain such claims, a statistical experiment is conducted, being designed such that some conclusions about the vector $\boldsymbol{\theta}$ are possible. More precisely, the experiment will yield an observation $\boldsymbol{y} = (y_1, \ldots, y_n)'$ of a random vector $\boldsymbol{Y} = (Y_1, \ldots, Y_n)'$ whose distribution depends on $\boldsymbol{\theta}$. Given a certain *decision rule* δ and given the observation $\boldsymbol{y}$, a decision d is chosen out of the decision space D.

Definition 1.1. *A mapping*

$$\begin{aligned} \delta : \mathbb{R}^n &\to \quad D \\ \boldsymbol{y} &\mapsto \delta(\boldsymbol{y}) = d \end{aligned}$$

which assigns each observed value $\boldsymbol{y}$ *exactly one decision* d *is called* decision rule.

Since $\boldsymbol{Y}$ is a random vector, it is quite reasonable to demand the same property for the function δ, i.e. $\delta(\boldsymbol{Y})$ should be a random vector/variable, too. Hence, the decision space D must come along with an appropriate σ-algebra $\mathcal{D}$, so that $(D, \mathcal{D})$ is a measurable space and δ is $(\mathcal{B}^n, \mathcal{D})$ measurable, where $\mathcal{B}^n$ stands for the n-dimensional Borel algebra.

Assumption 1.1. *In the following we will only consider decision rules* δ *such that* $\delta(\boldsymbol{Y})$ *is a random vector/variable.*

Remark 1.1. In point estimation theory, the decision simply consists in accepting one of the possible values from Θ. Thus $D = \Theta$ and decision rules δ are point estimators of $\boldsymbol{\theta}$.

Example 1.4. We can transform the decision problem from Example 1.3 to a point estimation problem simply by requiring a decision for a specific value p in $[0,1]$. Then, D becomes $D = \Theta = [0,1]$.

To obtain some information about the unknown p we can flip the coin 1000 times. Each time we note a 1 for the event 'face side' and a 0 otherwise. By this, we get 1000 realizations $\boldsymbol{y} = (y_1, \ldots, y_{1000})'$ of $n = 1000$ independent random variables $\boldsymbol{Y} = (Y_1, \ldots, Y_n)'$, each one being Bernoulli distributed with parameter p.

Let the set of decision rules Δ be the set of all point estimators for p. We choose the rule

$$\delta(\boldsymbol{Y}) = \frac{1}{n} \sum_{i=1}^{n} Y_i$$

and come to the decision $\delta(\boldsymbol{y}) = \frac{1}{n} \sum_{i=1}^{n} y_i = \frac{516}{1000} = 0.516$ (say). □

1.2.2 Non-operational Decision Rule

Decision rules $\delta(\boldsymbol{Y})$ are random variables/vectors which yield a decision from the decision space D as soon as the sample value $\boldsymbol{y}$ has been observed. The value $\boldsymbol{y}$ is the realization of the random vector $\boldsymbol{Y}$ whose distribution depends on some unknown vector of parameters $\boldsymbol{\theta} \in \Theta$. Sometimes, the distribution of $\boldsymbol{Y}$ does not only depend on the vector $\boldsymbol{\theta}$ of interest, but also on some additional unknown vector $\boldsymbol{\xi} \in \Xi$.

If we are solely interested in point estimation of the vector $\boldsymbol{\theta}$, then we can consider functions

$$\delta(\boldsymbol{Y}, \boldsymbol{\xi}) ,$$

which should have been reasonable point estimators for $\boldsymbol{\theta}$, if $\boldsymbol{\xi}$ had been known. But since this is not the case, these functions do not yield a specific value from the parameter space Θ and hence are *non-operational.*

If we replace the vector $\boldsymbol{\xi}$ in $\delta(\boldsymbol{Y}, \boldsymbol{\xi})$ by an appropriate point estimator $\delta_0(\boldsymbol{Y})$ of $\boldsymbol{\xi}$, then we obtain an *operational variant*

$$\delta(\boldsymbol{Y}, \delta_0(\boldsymbol{Y})) ,$$

which (given an appropriate choice of δ_0) yields a value from Θ when $\boldsymbol{Y}$ is replaced by the sample value $\boldsymbol{y}$.

Example 1.5. We wish to estimate the unknown variance σ^2 of a normally distributed random variable. Suppose that an experiment delivers the realizations of n independent $N(\mu, \sigma^2)$ distributed random variables $Y_1, \ldots, Y_n$, where neither $\mu \in \mathbb{R}$ nor $\sigma^2 \in (0, \infty)$ are known. For some reasons it appears that

$$\delta(\boldsymbol{Y}, \mu) = \frac{1}{n} \sum_{i=1}^{n} (Y_i - \mu)^2$$

could be an appropriate 'point estimator' of σ^2. But since μ is not known, this 'decision rule' does not yield a specific value from $(0, \infty)$ when we replace $\boldsymbol{Y}$ by its observation. Hence, by this procedure we cannot come to the desired decision, so that $\delta(\boldsymbol{Y}, \mu)$ is not operational as a decision rule. When we replace the unknown parameter μ in δ by the point estimator $\delta_0(\boldsymbol{Y}) = \overline{Y} = \frac{1}{n} \sum_{i=1}^{n} Y_i$, then we obtain

$$\delta(\boldsymbol{Y}) = \frac{1}{n} \sum_{i=1}^{n} (Y_i - \overline{Y})^2$$

as an operational variant of $\delta(\boldsymbol{Y}, \mu)$. Note that $\delta(\boldsymbol{Y})$ is a random variable which yields a specific value from the parameter space $\Theta = (0, \infty)$ of σ^2 when the observed $y_1, \ldots, y_n$ are given. This shows that in contrast to $\delta(\boldsymbol{Y}, \mu)$ the function $\delta(\boldsymbol{Y})$ can be called a point estimator. The two functions $\delta(\boldsymbol{Y}, \mu)$ for known μ and $\delta(\boldsymbol{Y})$ do not necessarily have similar properties. For example, if μ is known, then $\delta(\boldsymbol{Y}, \mu)$ is unbiased for σ^2, while the operational variant $\delta(\boldsymbol{Y})$ is always biased. □

Remark 1.2. The above example shows that there might exist situations when it can be useful to consider a non-operational decision rule as the basis for some operational rule.

In the following we will not use the terms 'non-operational decision rule' or 'non-operational estimator', since they comprise some kind of contradiction. The term 'non-operational' indicates that no decision can be made, while the terms 'decision rule' or 'estimator' imply by definition that a specific decision or estimate is assigned to a given sample value $\boldsymbol{y}$. Moreover, if we extend non-operational decision rules to functions not only depending on some unknown $\boldsymbol{\xi} \in \Xi$, but also on the parameter of interest $\boldsymbol{\theta} \in \Theta$, then the most appropriate decision rule for $\boldsymbol{\theta}$ would be $\boldsymbol{\theta}$ itself, which obviously is not very helpful. (Note that the parameter $\boldsymbol{\theta}$ is in fact an element from the parameter space, but not a specific value unless known and thus not a decision.)

1.2.3 Loss and Risk

A concomitant of any decision rule is a *loss function*, assigning a value to the parameter $\boldsymbol{\theta}$ and the decision $d = \delta(\boldsymbol{y})$. This value determines the extent of loss we have to admit when $\boldsymbol{\theta}$ is the true parameter and the decision $d = \delta(\boldsymbol{y})$ is taken. A loss function is usually designed such that the loss is zero when the decision is correct, while it is growing with an increasing level of incorrectness. The choice of a specific loss function depends on the regarded decision problem.

Definition 1.2. *A mapping*

$$L : \Theta \times D \to \mathrm{I\!R} ,$$

where $L(\boldsymbol{\theta}, d)$ gives the loss when $\boldsymbol{\theta}$ is the true parameter and the decision d is taken, is called loss function.

Assumption 1.2. *In the following we only consider loss functions L such that $L(\boldsymbol{\theta}, \delta(\boldsymbol{Y}))$ is a random variable (in the second component).*

If an observation $\boldsymbol{y}$ is given and a specific value of $\boldsymbol{\theta}$ is considered, then the loss $L(\boldsymbol{\theta}, \delta(\boldsymbol{y}))$, arising when this specific value of $\boldsymbol{\theta}$ is the true parameter and the decision $\delta(\boldsymbol{y})$ is taken, can be computed (*observed loss*).

If we are interested in assessing a decision rule $\delta(\boldsymbol{Y})$ independent of a given realization of $\boldsymbol{Y}$, then it is nearby to consider the average loss with respect to all possible realizations of $\boldsymbol{Y}$.

Definition 1.3. *The expected loss*

$$\rho(\boldsymbol{\theta}, \delta) = \mathrm{E}\left[L(\boldsymbol{\theta}, \delta(\boldsymbol{Y}))\right] ,$$

of a decision rule $\delta(\boldsymbol{Y})$ is called the risk *of $\delta(\boldsymbol{Y})$ when $\boldsymbol{\theta}$ is the true parameter.*

Example 1.6. In Example 1.3 we can compute the loss of an estimator $\delta(\boldsymbol{Y})$ by using the function

$$L(p, \delta(\boldsymbol{Y})) = (\delta(\boldsymbol{Y}) - p)^2 .$$

Clearly this loss function is a random variable in the second component. If $p = 0.5$ (say) had been the true parameter and $\delta(\boldsymbol{Y})$ had delivered the decision $\delta(\boldsymbol{y}) = 0.5$, then the observed loss would have been 0. The farther a decision had been away from 0.5, the greater would have been the corresponding loss.

If we make the decision $\delta(\boldsymbol{y}) = 0.516$, as before, then the loss will be $L(0.5, 0.516) = (0.516 - 0.5)^2 = 0.000256$ when $p = 0.5$ is the true parameter.

What about the loss independent of a given realization $\boldsymbol{y}$? For our decision rule $\delta(\boldsymbol{Y}) = \frac{1}{n}\sum_{i=1}^{n} Y_i$ with $n = 1000$ we have

$$\mathrm{E}(\delta(\boldsymbol{Y})) = p$$

and therefore

$$\rho(p, \delta) = \mathrm{E}\left[L(p, \delta(\boldsymbol{Y})\right] = \mathrm{E}\left[(\delta(\boldsymbol{Y}) - p)^2\right] = \mathrm{Var}(\delta(\boldsymbol{Y})) = \frac{p(1-p)}{1000} .$$

Hence $\rho(0.5, \delta) = 0.00025$ is the risk of δ when $p = 0.5$ is the true parameter. Of course we can determine the risk for every possible value $p \in [0, 1]$, see also Fig. 1.2. □

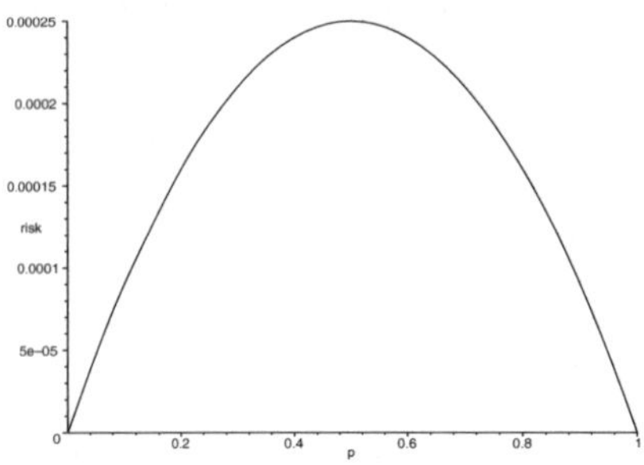

Fig. 1.2. Risk of the decision rule $\delta(\boldsymbol{Y}) = \frac{1}{1000}\sum_{i=1}^{1000} Y_i$ in Example 1.6

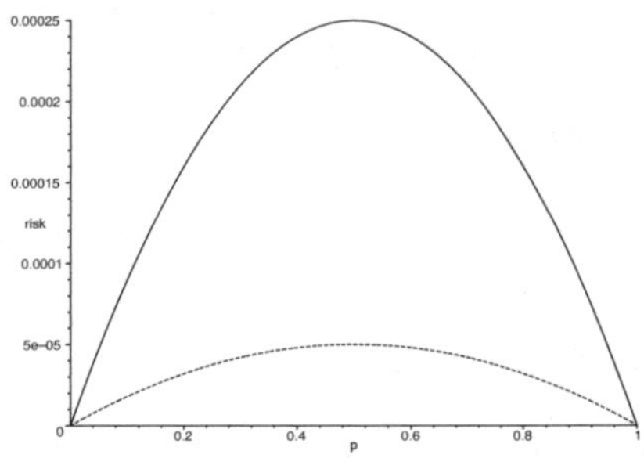

Fig. 1.3. Risk of the decision rule $\delta(\boldsymbol{Y}) = \frac{1}{1000}\sum_{i=1}^{1000} Y_i$ (solid line) and risk of $\delta_*(\boldsymbol{Y})$ (dotted line) in Example 1.7

1.2.4 Choosing a Decision Rule

In Example 1.6 we have considered only *one* decision rule to come to a decision (point estimation of p). In many cases, however, we can choose between many different decision rules to solve a specific decision problem. Then the problem is to find the most appropriate one. A possible way is to compare the risks of the respective rules for all possible values from the parameter space.

Definition 1.4. *Let* $(\Theta, D, L, \boldsymbol{Y})$ *denote a decision problem and let* $\delta_1(\boldsymbol{Y})$ *and* $\delta_2(\boldsymbol{Y})$ *be two decision rules. Then:*

(i) *The rule* $\delta_1(\boldsymbol{Y})$ *is called* uniformly not worse *than the rule* $\delta_2(\boldsymbol{Y})$ *if*

$$\rho(\boldsymbol{\theta}, \delta_1) \leq \rho(\boldsymbol{\theta}, \delta_2)$$

is satisfied for all $\boldsymbol{\theta} \in \Theta$.

(ii) *The rule* $\delta_1(\boldsymbol{Y})$ *is called* uniformly better *than the rule* $\delta_2(\boldsymbol{Y})$ *if* $\delta_1(\boldsymbol{Y})$ *is uniformly not worse than* $\delta_2(\boldsymbol{Y})$ *and in addition*

$$\rho(\boldsymbol{\theta}, \delta_1) < \rho(\boldsymbol{\theta}, \delta_2)$$

for at least one $\boldsymbol{\theta} \in \Theta$.

In general, the term 'uniformly better' could also be interpreted in the sense that the risk of one rule is strictly smaller than the risk of another rule for *all* possible values from the parameter space. From the above definition, however, 'uniformly better' essentially means 'uniformly not worse', but 'better' for at least one parameter. One may call $\delta_1(\boldsymbol{Y})$ *uniformly strictly better* than $\delta_2(\boldsymbol{Y})$ if

$$\rho(\boldsymbol{\theta}, \delta_1) < \rho(\boldsymbol{\theta}, \delta_2)$$

for all $\boldsymbol{\theta} \in \Theta$.

Example 1.7. Suppose that in Example 1.3 we have a second decision rule $\delta_*(\boldsymbol{Y})$ at hand with a risk function $\rho(p, \delta_*)$ as shown in Fig. 1.3 by the dotted line. In this case the rule δ_* is a uniformly better estimator of p than δ. (The risk of δ_* is strictly smaller than the risk of δ for all possible values of p, except for $p = 0$ and $p = 1$, where both risks are equal to zero.) □

Note that in Fig. 1.3 the dotted risk line $\rho(p, \delta_*)$ is given by $\rho(p, \delta_*) = p(1-p)/5000$. Therefore, in fact, Fig. 1.3 actually does not show a comparison of the risks of two decision functions given a single decision problem, but one decision rule $\delta(\boldsymbol{Y}) = \frac{1}{n}\sum_{i=1}^{n} Y_i$ given two decision problems (flipping a coin $n = 1000$ times and flipping a coin $n = 5000$ times). The motivation for Fig. 1.3 is simply to show two visually distinct risk functions, one dominating the other.

Remark 1.3. In the following we only consider the comparison of different decision rules given *a specific* decision problem and not the comparison of different decision problems. This also implies that we compare different decision rules for a fixed sample size n.

Example 1.8. Suppose that in Example 1.3 we wish to compare the two decision rules

$$\delta_1(\boldsymbol{Y}) = \overline{Y} = \frac{1}{n}\sum_{i=1}^{n} Y_i \quad \text{and} \quad \delta_2(\boldsymbol{Y}) = \frac{n\overline{Y} + \sqrt{n/4}}{n + \sqrt{n}}$$

for $n = 1000$. As noted before, the risk of δ_1 is given by

$$\rho(p, \delta_1) = \mathrm{E}\left[(\delta_1(\boldsymbol{Y}) - p)^2\right] = \frac{p(1-p)}{n} ,$$

while the risk of δ_2 comes up to

$$\rho(p, \delta_2) = \mathrm{E}\left[(\delta_2(\boldsymbol{Y}) - p)^2\right] = \frac{n}{4(n + \sqrt{n})^2}$$

(Problem 1.3). Here we neither have $\rho(p, \delta_1) \leq \rho(p, \delta_2)$ nor $\rho(p, \delta_2) \leq \rho(p, \delta_1)$ for *all* values $p \in [0, 1]$, so that none of the two estimators is uniformly not worse than the other, see also Fig. 1.4. □

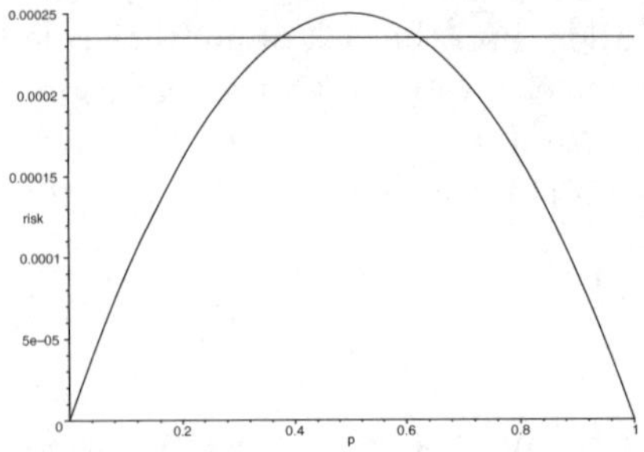

Fig. 1.4. Risk of δ_1 (parabola) and risk of δ_2 (horizontal line) in Example 1.8 for $n = 1000$

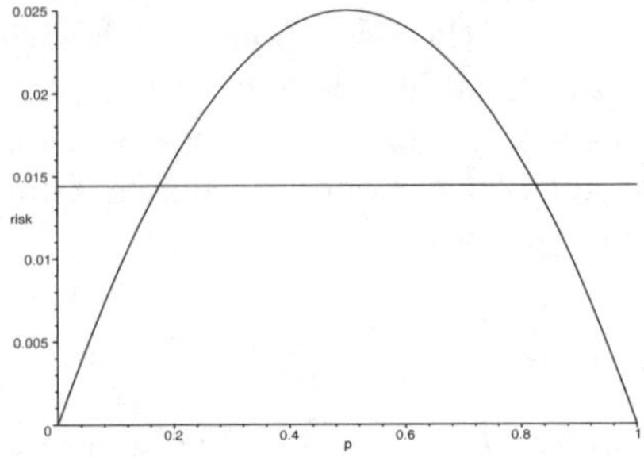

Fig. 1.5. Risk of δ_1 (parabola) and risk of δ_2 (horizontal line) in Example 1.8 for $n = 10$

In the above example none of the two decision rules δ_1 and δ_2 is preferable to the other on the basis of the squared error risk. However, we can see that δ_1 has a smaller risk for all those values of p which are not close to 0.5. Thus, if we have no prior knowledge that p is somewhere near to 0.5, we are inclined to use δ_1 rather than δ_2. The situation had been quite different if the number n would have been much smaller than $n = 1000$. In this case the risk of δ_2 can be seen to be smaller than the risk of δ_1 for a broad range of values of p, see Fig. 1.5, and one is then inclined to use δ_2 as a decision function. This shows that even in those cases when two estimators are compared with respect to their risks and none of them turns out to be uniformly not worse (or better) than the other, the risk comparison can give valuable information about the decision rules in question, see also [19, p. 332].

1.2.5 Admissibility

The choice of a decision rule depends on the given loss and corresponding risk function as well as on the set of given decision rules we can choose from.

This set will be denoted by Δ in the following. If we have a set Δ_1 at hand, then the most appropriate rule from this set can be different from the most appropriate rule from another set Δ_2.

Example 1.9. If in a given decision problem $(\Theta, D, L, \boldsymbol{Y})$ the set Δ contains the three decision rules $\delta_1(\boldsymbol{Y})$, $\delta_2(\boldsymbol{Y})$ and $\delta_3(\boldsymbol{Y})$ with corresponding risk functions as shown in Fig. 1.6, then none of the rules is uniformly better than another. Hence we can say that in $\Delta = \{\delta_1(\boldsymbol{Y}), \delta_2(\boldsymbol{Y}), \delta_3(\boldsymbol{Y})\}$ all elements are *admissible* with respect to the risk.

If, however, Δ contains the decision rules $\delta_1(\boldsymbol{Y})$, $\delta_2(\boldsymbol{Y})$, $\delta_3(\boldsymbol{Y})$ as well as a forth rule $\delta_4(\boldsymbol{Y})$ with risk functions as shown in Fig. 1.7, then $\delta_4(\boldsymbol{Y})$ is uniformly better than each of the three others. If we can choose from this set of decision rules, it is not reasonable to choose one of the rules $\delta_1(\boldsymbol{Y})$, $\delta_2(\boldsymbol{Y})$ and $\delta_3(\boldsymbol{Y})$ when the risk is a criterion. Hence we can say that these three rules are *inadmissible* in $\Delta = \{\delta_1(\boldsymbol{Y}), \delta_2(\boldsymbol{Y}), \delta_3(\boldsymbol{Y}), \delta_4(\boldsymbol{Y})\}$ with respect to the risk. □

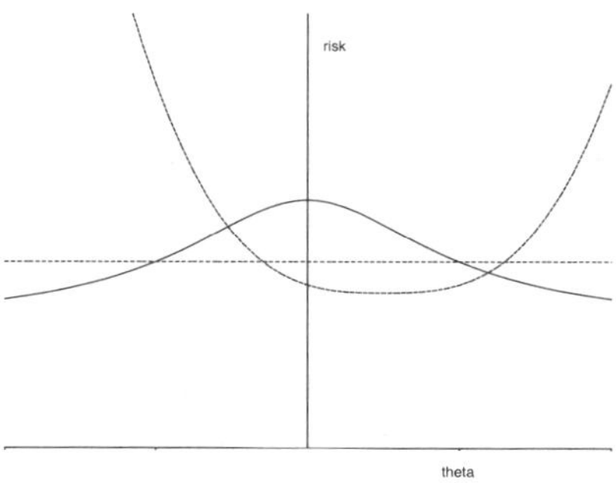

Fig. 1.6. Risks of the rules $\delta_1(\boldsymbol{Y})$, $\delta_2(\boldsymbol{Y})$ and $\delta_3(\boldsymbol{Y})$

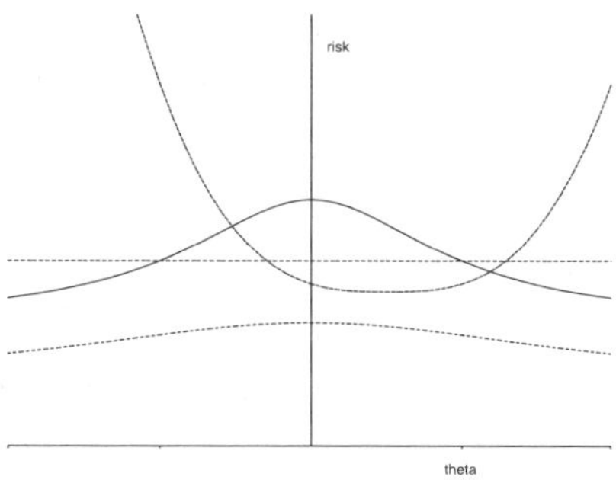

Fig. 1.7. Risks of the rules $\delta_1(\boldsymbol{Y})$, $\delta_2(\boldsymbol{Y})$, $\delta_3(\boldsymbol{Y})$ and $\delta_4(\boldsymbol{Y})$

Definition 1.5. *Let us be given a decision problem $(\Theta, D, L, \boldsymbol{Y})$ and a set of decision functions Δ. A decision rule $\boldsymbol{\delta}_0(\boldsymbol{Y}) \in \Delta$ is called* admissible *in Δ, if there does not exist a rule in Δ which is uniformly better than $\delta_0(\boldsymbol{Y})$. Otherwise $\delta_0(\boldsymbol{Y})$ is called* inadmissible *in Δ.*

Remark 1.4. Admissibility is not a criterion for choosing a decision rule. An admissible rule is not necessarily a reasonable rule.

Problem 1.3 gives an example for a hardly reasonable but nonetheless admissible estimator in a set Δ. On the other hand, if we want to use a specific decision rule which appears to be practical for some reason, then this rule should be admissible within some set Δ, since otherwise there exists a more practical (i.e. uniformly better) rule in this set.

1.2.6 Squared Error Loss

Remark 1.1 implies that in point estimation problems the set of all possible decisions D coincides with the set of of all possible parameters Θ. In this case, decision rules $\delta(\boldsymbol{Y})$ are point estimators of the unknown parameter vector $\boldsymbol{\theta}$, i.e. functions of the random vector $\boldsymbol{Y}$ which assign an observation $\boldsymbol{y}$ of $\boldsymbol{Y}$ to a specific value $\delta(\boldsymbol{y})$ from the parameter space Θ. In point estimation problems we will assume that the $p \times 1$ parameter vector $\boldsymbol{\theta}$ is real-valued.

A frequently used criterion to evaluate the performance of an estimator $\delta(\boldsymbol{Y})$ is the *weighted squared error loss function*

$$L(\boldsymbol{\theta}, \delta(\boldsymbol{Y})) = (\delta(\boldsymbol{Y}) - \boldsymbol{\theta})'\boldsymbol{W}(\delta(\boldsymbol{Y}) - \boldsymbol{\theta})$$

with corresponding weighted squared error risk

$$\rho(\boldsymbol{\theta}, \delta) = \mathrm{E}\left[(\delta(\boldsymbol{Y}) - \boldsymbol{\theta})'\boldsymbol{W}(\delta(\boldsymbol{Y}) - \boldsymbol{\theta})\right] .$$

The $p \times p$ weight matrix $\boldsymbol{W}$ is assumed to be symmetric nonnegative definite.

Squared Error Loss as Distance

If $\boldsymbol{W}$ is a $p \times p$ symmetric nonnegative definite matrix, then the mapping

$$\begin{aligned} \|\cdot\|_{\boldsymbol{W}} : \mathbb{R}^p \rightarrow & \quad \mathbb{R} \\ \boldsymbol{x} \mapsto & \|\boldsymbol{x}\|_{\boldsymbol{W}} = \sqrt{\boldsymbol{x}'\boldsymbol{W}\boldsymbol{x}} \end{aligned}$$

satisfies the conditions of a *vector seminorm*, see Sect. A.3.4. If in addition $\boldsymbol{W}$ is nonsingular, then $\|\cdot\|_{\boldsymbol{W}}$ satisfies the conditions of a *vector norm*. Hence, the weighted squared error loss is the squared distance between $\delta(\boldsymbol{Y})$ and $\boldsymbol{\theta}$ with respect to a vector (semi)norm $\|\cdot\|_{\boldsymbol{W}}$.

Different Weight Matrices

The application of weight matrices $\boldsymbol{W} \neq \boldsymbol{I}_p$ can sometimes be useful under specific estimation problems. For example, $\boldsymbol{W}$ could be a diagonal matrix with different weights on its main diagonal such that individual squared distances $(\delta(\boldsymbol{Y})_i - \boldsymbol{\theta}_i)^2$ enter the total loss with higher or lower magnitude.

Different weight matrices can cause seriously different losses. For example the observed unweighted squared error loss ($\boldsymbol{W} = \boldsymbol{I}_p$) of an estimator δ_1 can be smaller than the observed unweighted squared error loss of an estimator δ_2, while this relation can be reversed when an alternative weight matrix $\boldsymbol{W} \neq \boldsymbol{I}_p$ is used, see Problem 1.5.

Weighted Squared Error Loss and Admissibility

Although the choice of the weight matrix $\boldsymbol{W}$ can be of great importance when the performance of two estimators is compared, it turns out to be reasonable to consider the choice $\boldsymbol{W} = \boldsymbol{I}_p$ when admissibility of an estimator is investigated, see the following theorem. For this, it is assumed that the set of decision rules Δ containing the estimators of a $p \times 1$ parameter vector $\boldsymbol{\theta}$ satisfies

$$\delta_0, \delta_1 \in \Delta \quad \Rightarrow \quad \delta_0 + \boldsymbol{A}(\delta_1 - \delta_0) \in \Delta$$

for any $p \times p$ matrix $\boldsymbol{A}$. Compare also Problems 1.7 and 1.8.

Theorem 1.1. *Let us be given an estimation problem $(\Theta, L, \boldsymbol{Y})$ with weighted squared error loss function L. If $\delta_0(\boldsymbol{Y})$ is admissible for $\boldsymbol{\theta}$ in the case $\boldsymbol{W} = \boldsymbol{I}_p$, then $\delta_0(\boldsymbol{Y})$ is admissible for $\boldsymbol{\theta}$ in the case $\boldsymbol{W} = \boldsymbol{W}_*$, where $\boldsymbol{W}_*$ is an arbitrary $p \times p$ symmetric nonnegative definite matrix.*

Proof. For simplicity we will write δ to denote an estimator $\delta(\boldsymbol{Y})$ of $\boldsymbol{\theta}$.

Let δ_0 be admissible for $\boldsymbol{\theta}$ with $\boldsymbol{W} = \boldsymbol{I}_p$. We assume that δ_0 is not admissible for $\boldsymbol{\theta}$ with $\boldsymbol{W} = \boldsymbol{W}_*$ and show that this assumption leads to a contradiction.

If δ_0 is not admissible for $\boldsymbol{\theta}$ in the case $\boldsymbol{W} = \boldsymbol{W}_*$, then there exists an estimator δ_1, which is uniformly better than δ_0 in the case $\boldsymbol{W} = \boldsymbol{W}_*$. This means that

$$c(\boldsymbol{\theta}, \boldsymbol{W}_*) := \mathrm{E}\left[(\delta_1 - \boldsymbol{\theta})'\boldsymbol{W}_*(\delta_1 - \boldsymbol{\theta})\right] - \mathrm{E}\left[(\delta_0 - \boldsymbol{\theta})'\boldsymbol{W}_*(\delta_0 - \boldsymbol{\theta})\right]$$

is smaller than or equal to 0 for all $\boldsymbol{\theta} \in \Theta$ and strictly smaller than 0 for at least one $\boldsymbol{\theta} \in \Theta$.

Let $\boldsymbol{F} := (1/\lambda_{\max})\boldsymbol{W}_*$, where $\lambda_{\max}$ denotes the largest eigenvalue of $\boldsymbol{W}_*$. It follows that

$$c(\boldsymbol{\theta}, \boldsymbol{F}) \begin{cases} \leq 0 & \text{for all } \boldsymbol{\theta} \in \Theta \\ < 0 & \text{for at least one } \boldsymbol{\theta} \in \Theta \end{cases} .$$

Now, define

$$\delta := \delta_0 + \boldsymbol{F}(\delta_1 - \delta_0) .$$

Then

$$\begin{aligned} &\mathrm{E}\left[(\delta - \boldsymbol{\theta})'(\delta - \boldsymbol{\theta})\right] \\ &= \mathrm{E}\left[(\delta_0 + \boldsymbol{F}(\delta_1 - \delta_0) - \boldsymbol{\theta})'(\delta_0 + \boldsymbol{F}(\delta_1 - \delta_0) - \boldsymbol{\theta})\right] \\ &= \mathrm{E}\left[((\delta_0 - \boldsymbol{\theta})' + (\delta_1 - \delta_0)'\boldsymbol{F})((\delta_0 - \boldsymbol{\theta}) + \boldsymbol{F}(\delta_1 - \delta_0))\right] \\ &= \mathrm{E}\left[(\delta_0 - \boldsymbol{\theta})'(\delta_0 - \boldsymbol{\theta})\right] + q , \end{aligned}$$

where $q := \mathrm{E}\left[(\delta_1 - \delta_0)'\boldsymbol{F}^2(\delta_1 - \delta_0) + 2(\delta_1 - \delta_0)'\boldsymbol{F}(\delta_0 - \boldsymbol{\theta})\right]$. Since

$$\boldsymbol{x}'\boldsymbol{F}^2\boldsymbol{x} \leq \boldsymbol{x}'\boldsymbol{F}\boldsymbol{x}$$

is satisfied for all $p \times 1$ vectors $\boldsymbol{x}$ (see Problem 1.6), it follows

$$\mathrm{E}\left[(\delta_1 - \delta_0)'\boldsymbol{F}^2(\delta_1 - \delta_0)\right] \leq \mathrm{E}\left[(\delta_1 - \delta_0)'\boldsymbol{F}(\delta_1 - \delta_0)\right],$$

see e.g. [79, p. 70]. Thus,

$$q \leq \mathrm{E}\left[(\delta_1 - \delta_0)'\boldsymbol{F}(\delta_1 - \delta_0) + 2(\delta_1 - \delta_0)'\boldsymbol{F}(\delta_0 - \boldsymbol{\theta})\right].$$

By some simple transformations it can be shown that the right-hand side of this inequality equals $c(\boldsymbol{\theta}, \boldsymbol{F})$. Therefore

$$\mathrm{E}\left[(\delta - \boldsymbol{\theta})'(\delta - \boldsymbol{\theta})\right] \leq \mathrm{E}\left[(\delta_0 - \boldsymbol{\theta})'(\delta_0 - \boldsymbol{\theta})\right] + c(\boldsymbol{\theta}, \boldsymbol{F}),$$

and

$$\mathrm{E}\left[(\delta - \boldsymbol{\theta})'(\delta - \boldsymbol{\theta})\right] \begin{cases} \leq \mathrm{E}[(\delta_0 - \boldsymbol{\theta})'(\delta_0 - \boldsymbol{\theta})] & \text{for all } \boldsymbol{\theta} \in \Theta \\ < \mathrm{E}[(\delta_0 - \boldsymbol{\theta})'(\delta_0 - \boldsymbol{\theta})] & \text{for at least one } \boldsymbol{\theta} \in \Theta \end{cases}.$$

This shows that δ is uniformly better than δ_0 in case $\boldsymbol{W} = \boldsymbol{I}_p$, which is a contradiction to the assumed admissibility of δ_0 in the case $\boldsymbol{W} = \boldsymbol{I}_p$. Hence δ_0 must also be admissible for the case $\boldsymbol{W} = \boldsymbol{W}_*$, which concludes the proof. □

Remark 1.5. If $\boldsymbol{W}_*$ is symmetric positive definite, then the assertion in the above theorem can also be reversed: If $\delta_0(\boldsymbol{Y})$ is admissible for $\boldsymbol{\theta}$ in the case $\boldsymbol{W} = \boldsymbol{W}_*$, then $\delta_0(\boldsymbol{Y})$ is admissible for $\boldsymbol{\theta}$ in the case $\boldsymbol{W} = \boldsymbol{I}_p$, see Problem 1.9(b).

Remark 1.6. If we consider admissibility with respect to some weighted squared error loss, then the relevant claims can be deduced from the unweighted case, irrespective of the choice of weight matrix.

1.2.7 Matrix Valued Squared Error Loss

According to Definition 1.2, a loss function takes its values in the set of real numbers $\mathbb{R}$. In connection with estimating a $p \times 1$ parameter vector $\boldsymbol{\theta}$, the *matrix-valued squared error loss function*

$$L(\boldsymbol{\theta}, \delta(\boldsymbol{Y})) = (\delta(\boldsymbol{Y}) - \boldsymbol{\theta})(\delta(\boldsymbol{Y}) - \boldsymbol{\theta})'$$

is often considered, taking its values in $\mathbb{R}^{p \times p}$. The corresponding risk is given by

$$\mathrm{MSE}(\boldsymbol{\theta}, \delta) = \mathrm{E}\left[(\delta(\boldsymbol{Y}) - \boldsymbol{\theta})(\delta(\boldsymbol{Y}) - \boldsymbol{\theta})'\right].$$

The i-th diagonal element of this matrix is the *mean squared error*

$$\mathrm{mse}(\theta_i, \delta(\boldsymbol{Y})_i) = \mathrm{E}\left[(\delta(\boldsymbol{Y})_i - \theta_i)^2\right]$$

of the estimator $\delta(\boldsymbol{Y})_i$ for the i-th element θ_i of the vector $\boldsymbol{\theta}$, $i = 1, \ldots, p$. Furthermore, for any $\boldsymbol{\theta} \in \Theta$, the trace

$$\mathrm{tr}\left[\boldsymbol{W}\,\mathrm{MSE}(\boldsymbol{\theta}, \delta)\right] = \rho(\boldsymbol{\theta}, \delta) = \mathrm{E}\left[(\delta(\boldsymbol{Y}) - \boldsymbol{\theta})'\boldsymbol{W}(\delta(\boldsymbol{Y}) - \boldsymbol{\theta})\right]$$

is the risk of $\delta(\boldsymbol{Y})$ with respect to the weighted squared error loss with weight matrix $\boldsymbol{W}$.

Comparison of Estimators

If we use $\mathrm{MSE}(\boldsymbol{\theta}, \delta)$ for a comparison of estimators, then the corresponding previous definitions must be altered accordingly.

Definition 1.6. *Let us be given an estimation problem $(\Theta, L, \boldsymbol{Y})$ with matrix-valued squared error loss function L. Let $\delta_1(\boldsymbol{Y})$ and $\delta_2(\boldsymbol{Y})$ be two estimators for $\boldsymbol{\theta}$. Then:*

(i) *The estimator $\delta_1(\boldsymbol{Y})$ is called* uniformly not worse *than $\delta_2(\boldsymbol{Y})$, if the symmetric matrix*
$$\mathrm{MSE}(\boldsymbol{\theta}, \delta_2) - \mathrm{MSE}(\boldsymbol{\theta}, \delta_1)$$
is nonnegative definite for all $\boldsymbol{\theta} \in \Theta$.

(ii) *The estimator $\delta_1(\boldsymbol{Y})$ is called* uniformly better *than $\delta_2(\boldsymbol{Y})$, if $\delta_1(\boldsymbol{Y})$ is uniformly not worse than $\delta_2(\boldsymbol{Y})$ and in addition*
$$\mathrm{MSE}(\boldsymbol{\theta}, \delta_1) \neq \mathrm{MSE}(\boldsymbol{\theta}, \delta_2)$$
for at least one $\boldsymbol{\theta} \in \Theta$.

Note that some authors define $\delta_1(\boldsymbol{Y})$ to be uniformly better than $\delta_2(\boldsymbol{Y})$ if the difference $\mathrm{MSE}(\boldsymbol{\theta}, \delta_2) - \mathrm{MSE}(\boldsymbol{\theta}, \delta_1)$ is positive definite for all $\boldsymbol{\theta} \in \Theta$. Correspondingly, we would then call $\delta_1(\boldsymbol{Y})$ *uniformly strictly better* than $\delta_2(\boldsymbol{Y})$, similarly to the real-valued case. It is essential not to mix up both definitions since they are not equivalent. Clearly, if the difference $\mathrm{MSE}(\boldsymbol{\theta}, \delta_2) - \mathrm{MSE}(\boldsymbol{\theta}, \delta_1)$ is positive definite for some $\boldsymbol{\theta} \in \Theta$, then this difference is also nonnegative definite and in addition nonzero, but the converse is not true, i.e. from 'uniformly better' in our sense does not necessarily follow 'uniformly strictly better'.

Matrix Valued and Real Valued Risks

To denote the nonnegative definiteness of the symmetric matrix $\mathrm{MSE}(\boldsymbol{\theta}, \delta_2) - \mathrm{MSE}(\boldsymbol{\theta}, \delta_1)$ we will also write
$$\mathrm{MSE}(\boldsymbol{\theta}, \delta_1) \leq_{\mathrm{L}} \mathrm{MSE}(\boldsymbol{\theta}, \delta_2) \;,$$
where $\leq_{\mathrm{L}}$ is the Löwner partial ordering in the set of square matrices, see also Sect. A.5. The following theorem, see [115], shows the connection between a relationship of this type and a corresponding relationship between weighted real-valued risks.

Theorem 1.2. *For an arbitrary $\boldsymbol{\theta} \in \Theta$, the inequality*
$$\mathrm{MSE}(\boldsymbol{\theta}, \delta_1) \leq_{\mathrm{L}} \mathrm{MSE}(\boldsymbol{\theta}, \delta_2)$$
is satisfied if and only if
$$\mathrm{E}\left[(\delta_1(\boldsymbol{Y}) - \boldsymbol{\theta})'\boldsymbol{W}(\delta_1(\boldsymbol{Y}) - \boldsymbol{\theta})\right] \leq \mathrm{E}\left[(\delta_2(\boldsymbol{Y}) - \boldsymbol{\theta})'\boldsymbol{W}(\delta_2(\boldsymbol{Y}) - \boldsymbol{\theta})\right]$$
is valid for every $p \times p$ symmetric nonnegative definite matrix $\boldsymbol{W}$.

Proof. For an arbitrary estimator $\delta(\boldsymbol{Y})$ of $\boldsymbol{\theta}$ we have

$$\begin{aligned}&\mathrm{E}\,[(\delta(\boldsymbol{Y})-\boldsymbol{\theta})'\boldsymbol{W}(\delta(\boldsymbol{Y})-\boldsymbol{\theta})]\\&=\mathrm{E}\,[\mathrm{tr}\{\boldsymbol{W}(\delta(\boldsymbol{Y})-\boldsymbol{\theta})(\delta(\boldsymbol{Y})-\boldsymbol{\theta})'\}]=\mathrm{tr}\,[\boldsymbol{W}\,\mathrm{MSE}(\boldsymbol{\theta},\delta(\boldsymbol{Y}))]\ .\end{aligned}$$

This shows

$$\begin{aligned}&\mathrm{tr}\,[\boldsymbol{W}\{\mathrm{MSE}(\boldsymbol{\theta},\delta_2)-\mathrm{MSE}(\boldsymbol{\theta},\delta_1)\}]\\&=\mathrm{E}\,[(\delta_2(\boldsymbol{Y})-\boldsymbol{\theta})'\boldsymbol{W}(\delta_2(\boldsymbol{Y})-\boldsymbol{\theta})]-\mathrm{E}\,[(\delta_1(\boldsymbol{Y})-\boldsymbol{\theta})'\boldsymbol{W}(\delta_1(\boldsymbol{Y})-\boldsymbol{\theta})]\ ,\end{aligned}$$

and the assertion follows from Theorem A.54. □

Note that in the above theorem, given $\boldsymbol{\theta} \in \Theta$, it is not enough that

$$\mathrm{E}\,[(\delta_1(\boldsymbol{Y})-\boldsymbol{\theta})'\boldsymbol{W}(\delta_1(\boldsymbol{Y})-\boldsymbol{\theta})]\leq\mathrm{E}\,[(\delta_2(\boldsymbol{Y})-\boldsymbol{\theta})'\boldsymbol{W}(\delta_2(\boldsymbol{Y})-\boldsymbol{\theta})]$$

for *some* symmetric nonnegative definite $\boldsymbol{W}$ to guarantee

$$\mathrm{MSE}(\boldsymbol{\theta},\delta_1)\leq_{\mathrm{L}}\mathrm{MSE}(\boldsymbol{\theta},\delta_2)\ ,$$

but the real-valued inequality must hold for *all* weight matrices $\boldsymbol{W}$ to imply the matrix-valued inequality.

Moreover, note that if for some $\boldsymbol{\theta} \in \Theta$ we have

$$\mathrm{MSE}(\boldsymbol{\theta},\delta_1)\leq_{\mathrm{L}}\mathrm{MSE}(\boldsymbol{\theta},\delta_2),\quad \mathrm{MSE}(\boldsymbol{\theta},\delta_1)\neq\mathrm{MSE}(\boldsymbol{\theta},\delta_2)\ ,$$

then we do not necessarily have

$$\mathrm{E}\,[(\delta_1(\boldsymbol{Y})-\boldsymbol{\theta})'\boldsymbol{W}(\delta_1(\boldsymbol{Y})-\boldsymbol{\theta})]<\mathrm{E}\,[(\delta_2(\boldsymbol{Y})-\boldsymbol{\theta})'\boldsymbol{W}(\delta_2(\boldsymbol{Y})-\boldsymbol{\theta})]$$

for an arbitrary $p \times p$ symmetric nonnegative definite matrix $\boldsymbol{W}$.

Remark 1.7. If an estimator δ_1 is uniformly not worse than δ_2 with respect to $\mathrm{MSE}(\boldsymbol{\theta}, \delta)$, then it is always uniformly not worse than δ_2 with respect to an arbitrary real-valued weighted squared error risk. If an estimator δ_1 is uniformly better than δ_2 with respect to $\mathrm{MSE}(\boldsymbol{\theta}, \delta)$, then it is always uniformly not worse *but not necessarily uniformly better* than δ_2 with respect to an arbitrary real-valued weighted squared error risk.

Nonetheless, from the relation $\mathrm{MSE}(\boldsymbol{\theta}, \delta_1) \leq_{\mathrm{L}} \mathrm{MSE}(\boldsymbol{\theta}, \delta_2)$ together with $\mathrm{MSE}(\boldsymbol{\theta}, \delta_1) \neq \mathrm{MSE}(\boldsymbol{\theta}, \delta_2)$ we can always conclude that $\mathrm{tr}[\mathrm{MSE}(\boldsymbol{\theta}, \delta_1)] < \mathrm{tr}[\mathrm{MSE}(\boldsymbol{\theta}, \delta_2)]$.

Remark 1.8. If an estimator δ_1 is uniformly not worse/better than δ_2 with respect to $\mathrm{MSE}(\boldsymbol{\theta}, \delta)$, then it is uniformly not worse/better with respect to the real-valued *unweighted* squared error risk.

Comparison of Elements of Estimators

If for some $\boldsymbol{\theta} \in \Theta$ the inequality

$$\mathrm{MSE}(\boldsymbol{\theta}, \delta_1) \leq_{\mathrm{L}} \mathrm{MSE}(\boldsymbol{\theta}, \delta_2)$$

holds true, then for every $i \in \{1, \ldots, p\}$ it follows

$$\mathrm{E}\left[(\delta_1(\boldsymbol{Y})_i - \theta_i)^2\right] \leq \mathrm{E}\left[(\delta_2(\boldsymbol{Y})_i - \theta_i)^2\right] .$$

On the other hand, if for some $\boldsymbol{\theta} \in \Theta$ the relations

$$\mathrm{MSE}(\boldsymbol{\theta}, \delta_1) \leq_{\mathrm{L}} \mathrm{MSE}(\boldsymbol{\theta}, \delta_2), \quad \mathrm{MSE}(\boldsymbol{\theta}, \delta_1) \neq \mathrm{MSE}(\boldsymbol{\theta}, \delta_2)$$

hold true, then the inequality

$$\mathrm{E}\left[(\delta_1(\boldsymbol{Y})_i - \theta_i)^2\right] < \mathrm{E}\left[(\delta_2(\boldsymbol{Y})_i - \theta_i)^2\right]$$

does not necessarily hold for every $i \in \{1, \ldots, p\}$. The latter would have been only the case if the difference $\mathrm{MSE}(\boldsymbol{\theta}, \delta_2) - \mathrm{MSE}(\boldsymbol{\theta}, \delta_1)$ had been *positive definite.*

Remark 1.9. If an estimator δ_1 is uniformly not worse than δ_2 with respect to $\mathrm{MSE}(\boldsymbol{\theta}, \delta)$, then each element of δ_1 is uniformly not worse than the corresponding element of δ_2 with respect to the usual mean squared error. If an estimator δ_1 is uniformly better than δ_2 with respect to $\mathrm{MSE}(\boldsymbol{\theta}, \delta)$, then each element of δ_1 is uniformly not worse *but not necessarily uniformly better* than the corresponding element of δ_2 with respect to the usual mean squared error.

Admissibility of Estimators

According to Remark 1.8, for the comparison of two estimators, the matrix-valued squared risk $\mathrm{MSE}(\boldsymbol{\theta}, \delta)$ is a stronger criterion than the real-valued unweighted squared error risk $\rho(\boldsymbol{\theta}, \delta) = \mathrm{tr}[\mathrm{MSE}(\boldsymbol{\theta}, \delta)]$. Nonetheless, with regard to admissibility of estimators it is reasonable to consider the latter rather than the former, as shown by the following theorem.

Theorem 1.3. *If an estimator $\delta_0(\boldsymbol{Y})$ is admissible for $\boldsymbol{\theta}$ with respect to $\rho(\boldsymbol{\theta}, \delta) = \mathrm{E}[(\delta(\boldsymbol{Y}) - \boldsymbol{\theta})'(\delta(\boldsymbol{Y}) - \boldsymbol{\theta})]$, then $\delta_0(\boldsymbol{Y})$ is admissible for $\boldsymbol{\theta}$ with respect to* $\mathrm{MSE}(\boldsymbol{\theta}, \delta)$.

Proof. We assume that δ_0 is admissible with respect to $\rho(\boldsymbol{\theta}, \delta)$ but not with respect to $\mathrm{MSE}(\boldsymbol{\theta}, \delta)$ in a set Δ, and show that this assumptions leads to a contradiction.

If $\delta_0 \in \Delta$ is not admissible with respect to $\mathrm{MSE}(\boldsymbol{\theta}, \delta)$, then there exists a better estimator $\delta_1 \in \Delta$ with respect to this criterion. By Definition 1.6, this means that the difference

$$\boldsymbol{D} = \mathrm{E}\,[(\delta_0 - \boldsymbol{\theta})(\delta_0 - \boldsymbol{\theta})'] - \mathrm{E}\,[(\delta_1 - \boldsymbol{\theta})(\delta_1 - \boldsymbol{\theta})']$$

is nonnegative definite for all $\boldsymbol{\theta} \in \Theta$ and nonzero for at least one $\boldsymbol{\theta} \in \Theta$. Now, the trace of a nonnegative definite matrix is nonnegative and equals zero if and only if the matrix is the zero matrix. Hence, $0 \leq \mathrm{tr}(\boldsymbol{D})$ for all $\boldsymbol{\theta} \in \Theta$ and $0 \neq \mathrm{tr}(\boldsymbol{D})$ for at least one $\boldsymbol{\theta} \in \Theta$. This implies

$$\mathrm{tr}\,[\mathrm{E}[(\delta_1 - \boldsymbol{\theta})(\delta_1 - \boldsymbol{\theta})']] \leq \mathrm{tr}\,[\mathrm{E}[(\delta_0 - \boldsymbol{\theta})(\delta_0 - \boldsymbol{\theta})']]$$

for all $\boldsymbol{\theta} \in \Theta$ and

$$\mathrm{tr}\,[\mathrm{E}[(\delta_1 - \boldsymbol{\theta})(\delta_1 - \boldsymbol{\theta})']] < \mathrm{tr}\,[\mathrm{E}[(\delta_0 - \boldsymbol{\theta})(\delta_0 - \boldsymbol{\theta})']]$$

for at least one $\boldsymbol{\theta} \in \Theta$. Since $\mathrm{tr}[\mathrm{MSE}(\boldsymbol{\theta}, \delta)] = \rho(\boldsymbol{\theta}, \delta)$, the two inequalities show that δ_1 is uniformly better than δ_0 with respect to $\rho(\boldsymbol{\theta}, \delta)$. This contradicts the assumption and shows that $\delta_0 \in \Delta$ must also be admissible with respect to $\mathrm{MSE}(\boldsymbol{\theta}, \delta)$. □

We have shown that if an estimator is admissible with respect to the real-valued unweighted squared error loss, then it is also admissible with respect to the matrix-valued squared error loss. The converse is not true in general, i.e. there might exist an estimator which is admissible in some set Δ with respect to the matrix-valued squared error risk $\mathrm{MSE}(\boldsymbol{\theta}, \delta)$, but which is not admissible in the same set Δ with respect to the real-valued unweighted squared error risk $\rho(\boldsymbol{\theta}, \delta) = \mathrm{tr}[\mathrm{MSE}(\boldsymbol{\theta}, \delta)]$.

1.2.8 Alternative Loss Functions

Traditionally, the previously described real-valued (weighted) squared error loss and the matrix-valued squared error loss are considered for investigation of the performance of an estimator $\delta(\boldsymbol{Y})$ for a parameter vector $\boldsymbol{\theta} \in \Theta$. Nonetheless it is sometimes reasonable to take alternative loss functions into account. To give an impression of the possibilities, we introduce some of these functions in the following.

Vector Valued Squared Error Loss

The matrix-valued loss function $L(\boldsymbol{\theta}, \delta(\boldsymbol{Y})) = (\delta(\boldsymbol{Y}) - \boldsymbol{\theta})(\delta(\boldsymbol{Y}) - \boldsymbol{\theta})'$ might be replaced by the vector-valued loss function

$$L(\boldsymbol{\theta}, \delta(\boldsymbol{Y})) = \mathrm{dg}\,[(\delta(\boldsymbol{Y}) - \boldsymbol{\theta})(\delta(\boldsymbol{Y}) - \boldsymbol{\theta})']\ ,$$

where $\mathrm{dg}(\boldsymbol{A})$ denotes the vector whose i-th element is the i-th element of the main diagonal of the matrix $\boldsymbol{A}$. Then the expected loss is given by $\mathrm{dg}[\mathrm{MSE}(\boldsymbol{\theta}, \delta)]$, and a decision rule $\delta_1(\boldsymbol{Y})$ is called uniformly not worse than a rule $\delta_2(\boldsymbol{Y})$ if the difference

$$\mathrm{dg}[\mathrm{MSE}(\boldsymbol{\theta}, \delta_2)] - \mathrm{dg}[\mathrm{MSE}(\boldsymbol{\theta}, \delta_1)]$$

is a vector whose every element is nonnegative for all $\boldsymbol{\theta} \in \Theta$. We may write

$$\mathrm{dg}[\mathrm{MSE}(\boldsymbol{\theta}, \delta_1)] \leq \mathrm{dg}[\mathrm{MSE}(\boldsymbol{\theta}, \delta_2)] \quad \forall \boldsymbol{\theta} \in \Theta\,.$$

If in addition $\mathrm{dg}[\mathrm{MSE}(\boldsymbol{\theta}, \delta_2(\boldsymbol{Y}))] \neq \mathrm{dg}[\mathrm{MSE}(\boldsymbol{\theta}, \delta_1(\boldsymbol{Y})]$ for at least one $\boldsymbol{\theta} \in \Theta$, then $\delta_1(\boldsymbol{Y})$ is called uniformly better than $\delta_2(\boldsymbol{Y})$.

This criterion is stronger than $\rho(\boldsymbol{\theta}, \delta) = \mathrm{tr}[\mathrm{MSE}(\boldsymbol{\theta}, \delta)]$ but weaker than $\mathrm{MSE}(\boldsymbol{\theta}, \delta)$. For an arbitrary given $\boldsymbol{\theta} \in \Theta$ it follows

$$\begin{aligned}
&\mathrm{MSE}(\boldsymbol{\theta}, \delta_1) \leq_{\mathrm{L}} \mathrm{MSE}(\boldsymbol{\theta}, \delta_2) \\
&\Rightarrow \mathrm{dg}[\mathrm{MSE}(\boldsymbol{\theta}, \delta_1)] \leq \mathrm{dg}[\mathrm{MSE}(\boldsymbol{\theta}, \delta_2)] \\
&\Rightarrow \mathrm{tr}[\mathrm{MSE}(\boldsymbol{\theta}, \delta_1)] \leq \mathrm{tr}[\mathrm{MSE}(\boldsymbol{\theta}, \delta_2)]\,.
\end{aligned}$$

Similarly we can conclude

$$\begin{aligned}
&\mathrm{MSE}(\boldsymbol{\theta}, \delta_1) \neq \mathrm{MSE}(\boldsymbol{\theta}, \delta_2) \\
&\Rightarrow \mathrm{dg}[\mathrm{MSE}(\boldsymbol{\theta}, \delta_1)] \neq \mathrm{dg}[\mathrm{MSE}(\boldsymbol{\theta}, \delta_2)] \\
&\Rightarrow \mathrm{tr}[\mathrm{MSE}(\boldsymbol{\theta}, \delta_1)] \neq \mathrm{tr}[\mathrm{MSE}(\boldsymbol{\theta}, \delta_2)]\,,
\end{aligned}$$

showing that the vector-valued square loss is in between the matrix-valued and the real-valued squared error loss.

Alternative Distances as Loss Functions

As noted before, a real-valued weighted squared error loss gives the squared distance between $\delta(\boldsymbol{Y})$ and $\boldsymbol{\theta}$ with respect to a vector (semi)norm $\|\cdot\|_{\boldsymbol{W}}$. Of course it is also possible to consider alternative measures of distance as loss functions. For example we may consider the *absolute distance*

$$L(\boldsymbol{\theta}, \delta(\boldsymbol{Y})) = \sum_{i=1}^{p} |\delta(\boldsymbol{Y})_i - \boldsymbol{\theta}_i|\,.$$

Another example is the *maximal distance*

$$L(\boldsymbol{\theta}, \delta(\boldsymbol{Y})) = \max\{|\delta(\boldsymbol{Y})_i - \boldsymbol{\theta}_i|,\ i = 1, \ldots, p\}\,.$$

Although such measures could be reasonable loss functions, they will, on the whole, cause more mathematical difficulties than the usual squared error distance and thus are rarely applied.

Balanced Loss Function

Suppose that the distribution of the random vector $\boldsymbol{Y}$ depends on $\boldsymbol{\theta}$ only via $\mathrm{E}(\boldsymbol{Y}) = \boldsymbol{\theta}$. For this case Zellner [132] proposes a *balanced loss function*

$$L(\boldsymbol{\theta}, \delta(\boldsymbol{Y})) = \alpha||\boldsymbol{Y} - \delta(\boldsymbol{Y})||^2 + (1-\alpha)||\boldsymbol{\theta} - \delta(\boldsymbol{Y})||^2, \quad 0 \le \alpha < 1\,.$$

Here $||\boldsymbol{Y} - \delta(\boldsymbol{Y})||^2$ measures how good $\delta(\boldsymbol{Y})$ fits $\boldsymbol{Y}$. If we have an observation $\boldsymbol{y}$ of $\boldsymbol{Y}$, then $||\boldsymbol{y} - \delta(\boldsymbol{y})||^2$ gives the squared distance between the realization $\boldsymbol{y}$ and the decision $\delta(\boldsymbol{y})$. Given $\boldsymbol{y}$, this function yields the same value irrespective of $\boldsymbol{\theta} \in \Theta$ Thus, standing alone, $||\boldsymbol{Y} - \delta(\boldsymbol{Y})||^2$ is not a loss function in our sense. Therefore, the case $\alpha = 1$ is excluded in the above definition.

The balanced loss function consists of two components. First, a measure of *goodness of fit* $||\boldsymbol{Y} - \delta(\boldsymbol{Y})||^2$ with a relative weight α, and, second, a measure of *goodness of estimation* $||\boldsymbol{\theta} - \delta(\boldsymbol{Y})||^2$ with a relative weight $1-\alpha$. The latter is already known as the unweighted squared error loss of the estimator $\delta(\boldsymbol{Y})$ for $\boldsymbol{\theta}$. Dey, Ghosh and Strawderman [29] study the performance of point estimators with respect to the balanced loss function and compare it with their performance under the unweighted squared error loss.

Asymmetric Loss Function

If we want to estimate a real-valued parameter θ, then

$$L(\theta, \delta(\boldsymbol{Y})) = b(\delta(\boldsymbol{Y}) - \theta)^2, \quad b > 0\,,$$

is a squared loss of the point estimator $\delta(\boldsymbol{Y})$. Using this loss function, it doesn't matter whether the error

$$\Phi = \delta(\boldsymbol{Y}) - \theta$$

is positive or negative, i.e. over- and underestimation of the parameter θ are equally treated. Now it is easy to fancy a situation in which overestimation (say) could do more harm than underestimation, e.g. when overestimation is more expensive. To take this into account, one may consider asymmetric loss functions, which assign more loss to overestimation than to the same amount of underestimation (or vice versa). For example the so-called LINEX loss function may be used, given as

$$L(\theta, \delta(\boldsymbol{Y})) = b[\mathrm{e}^{a\Phi} - a\Phi - 1], \quad a \neq 0, b > 0, \quad \Phi = \delta(\boldsymbol{Y}) - \theta,$$

cf. [131]. The two parameters a and b can be chosen according to a specific problem. With the parameter b we can set an appropriate scale, while we can use the parameter a to determine the shape of the function. Fig. 1.8 shows the LINEX loss function for $b = 0.5$ and $a = 1$ depending on $\Phi = \delta(\boldsymbol{Y}) - \theta$.

In this case overestimation of θ yields a greater loss than underestimation. If for example $\Phi = 3$, then the loss comes up to $L(\theta, \delta(\boldsymbol{Y})) = 8.0428$, while for $\Phi = -3$ the loss is only $L(\theta, \delta(\boldsymbol{Y})) = 1.0249$, compare also Fig.1.8.

For small values of $|a|$ the LINEX loss function is almost symmetric and yields similar results as a corresponding squared loss function. For estimating a *vector* of parameters $\boldsymbol{\theta}$, one may use the LINEX loss to evaluate the loss of the single elements θ_i, or one might consider the estimation of a linear combination $\boldsymbol{\lambda}'\boldsymbol{\theta}$, where $\boldsymbol{\lambda}$ is a known vector.

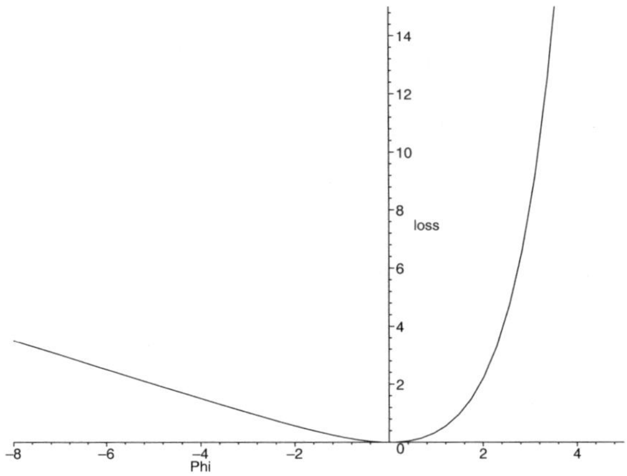

Fig. 1.8. LINEX loss function with $b = 0.5$ and $a = 1$

1.3 Problems

1.1. Let us be given $p \geq 3$ independent normally distributed random variables $Y_i \sim N(\mu_i, 1)$, $i = 1, \ldots, p$. Let $\boldsymbol{Y} = (Y_1, \ldots, Y_p)'$ and $\boldsymbol{\mu} = (\mu_1, \ldots, \mu_p)'$. Let the loss of a decision rule $\delta(\boldsymbol{Y})$ for estimating the parameter vector $\boldsymbol{\mu}$ be $L(\boldsymbol{\mu}, \delta(\boldsymbol{Y})) = (\delta(\boldsymbol{Y}) - \boldsymbol{\mu})'(\delta(\boldsymbol{Y}) - \boldsymbol{\mu}) = \sum_{i=1}^{p} (\delta(Y)_i - \mu_i)^2$.

(a) Determine the risk of the estimator $\delta_1(\boldsymbol{Y}) = \boldsymbol{Y}$.
(b) Show that the risk of the estimator

$$\delta_2(\boldsymbol{Y}) = (1 - \frac{(p-2)}{Z})\boldsymbol{Y}, \quad Z = \boldsymbol{Y}'\boldsymbol{Y} ,$$

is given as

$$\rho(\boldsymbol{\mu}, \delta_2) = p - (p-2)^2 \, \mathrm{E}(1/Z) .$$

[Hint: Use the identity $\mathrm{E}[(\boldsymbol{Y} - \boldsymbol{\mu})'\boldsymbol{Y}/Z] = (p-2)\,\mathrm{E}(1/Z)$.]
(c) Compare both risks. What can be said?

1.2. Consider the situation from Problem 1.1 for the case $p = 10$. Let

$$\boldsymbol{y} = (1.9665, .5456, 2.2983, .6172, 1.9213, -2.1017, -.2727, 1.3510, 2.6029, 1.4598)'$$

be a given observation of $\boldsymbol{Y}$.

(a) Compute the two possible decisions $\delta_1(\boldsymbol{y})$ and $\delta_2(\boldsymbol{y})$ (estimates of $\boldsymbol{\mu}$) corresponding to the decision rules $\delta_1(\boldsymbol{Y})$ and $\delta_2(\boldsymbol{Y})$ from Problem 1.1.
(b) Compute the observed losses of $\delta_1(\boldsymbol{y})$ and $\delta_2(\boldsymbol{y})$ when $\mu_i = 1$, $i = 1, \ldots, 10$, are the true parameters.
(c) The i-th element of $\delta_1(\boldsymbol{y})$ and $\delta_2(\boldsymbol{y})$ can respectively be seen as an estimate of μ_i. Check whether each element of $\delta_2(\boldsymbol{y})$ has a smaller observed loss than the corresponding element of $\delta_1(\boldsymbol{y})$ when $\mu_i = 1$ is the true parameter.

1.3. Consider the situation from Example 1.3 for an arbitrary sample size n. Assume that the set Δ contains the three rules

$$\delta_1(\boldsymbol{Y}) = \overline{Y} = \frac{1}{n}\sum_{i=1}^{n} Y_i \quad \delta_2(\boldsymbol{Y}) = \frac{n\overline{Y} + \sqrt{n/4}}{n + \sqrt{n}}$$

and $\delta_{0.5}(\boldsymbol{Y}) = 0.5$.

(a) Determine the risks of δ_2 and $\delta_{0.5}$.
(b) Is $\delta_{0.5}(\boldsymbol{Y})$ admissible in $\Delta = \{\delta_1(\boldsymbol{Y}), \delta_2(\boldsymbol{Y}), \delta_{0.5}(\boldsymbol{Y})\}$? (Plot the three risk functions for $n = 10$.) Is $\delta_{0.5}(\boldsymbol{Y})$ a reasonable decision rule?

1.4. Let $Y_i \sim N(\mu, 1)$, $i = 1, \ldots, n$, be n independent distributed random variables. Show that $\overline{Y} = \frac{1}{n}\sum_{i=1}^{n} Y_i$ is admissible for μ in the class

$$\Delta = \{\delta(\boldsymbol{Y}) : \delta(\boldsymbol{Y}) = c\overline{Y}, c \in \mathbb{R}\}$$

with respect to the loss function $L(\mu, \delta(\boldsymbol{Y})) = (\delta(\boldsymbol{Y}) - \mu)^2$.

1.5. Consider the situation from Problem 1.2 and compute the observed losses of the estimates $\delta_1(\boldsymbol{y})$ and $\delta_2(\boldsymbol{y})$ when $\mu_i = 1$, $i = 1, \ldots, 10$ are the true parameters. Use the weighted squared error loss function

$$L(\boldsymbol{\theta}, \delta(\boldsymbol{Y})) = (\delta(\boldsymbol{Y}) - \boldsymbol{\theta})'\boldsymbol{W}(\delta(\boldsymbol{Y}) - \boldsymbol{\theta})\ ,$$

where

$$\boldsymbol{W} = \text{diag}(0.01, 0.46, 0.01, 0.46, 0.01, 0.01, 0.01, 0.01, 0.01, 0.01)\ .$$

1.6. (See the proof of Theorem 1.1) Let $\boldsymbol{W}$ be a $p \times p$ symmetric nonnegative definite matrix. Let $\lambda_{\max}$ denote the largest eigenvalue of $\boldsymbol{W}$.

(a) Show that all eigenvalues of $\boldsymbol{F} = (1/\lambda_{\max})\boldsymbol{W}$ lie in the closed interval $[0, 1]$.
(b) Show that $\boldsymbol{F} - \boldsymbol{F}^2$ is symmetric nonnegative definite.

[Hint: Use the spectral decomposition of $\boldsymbol{W}$.]

1.7. Explain why the condition

$$\delta_0, \delta_1 \in \Delta \quad \Rightarrow \quad \delta_0 + \boldsymbol{A}(\delta_1 - \delta_0) \in \Delta$$

is necessary for the proof of Theorem 1.1. Explain in addition, why it is actually enough to demand the condition to be true only for symmetric matrices $\boldsymbol{A}$ with all eigenvalues in $[0, 1]$.

1.8. Is the condition

$$\delta_0, \delta_1 \in \Delta \quad \Rightarrow \quad \delta_0 + \boldsymbol{A}(\delta_1 - \delta_0) \in \Delta\ ,$$

where $\boldsymbol{A}$ is an arbitrary $p \times p$ matrix, satisfied for the following sets of decision rules?

(i) $\Delta = \{\delta(\boldsymbol{Y}) : \delta(\boldsymbol{Y}) = \boldsymbol{F}\boldsymbol{Y} + \boldsymbol{f}\}$, where $\boldsymbol{F}$ is a $p \times n$ matrix and $\boldsymbol{f}$ is a $p \times 1$ vector;
(ii) $\Delta = \{\delta(\boldsymbol{Y}) : \mathrm{E}[\delta(\boldsymbol{Y})] = \boldsymbol{\theta}\}$.

1.9. Let $(\Theta, L, \boldsymbol{Y})$ be an estimation problem with a weighted squared error loss function L and symmetric nonnegative definite weight matrix $\boldsymbol{W}$. Let $\boldsymbol{W}_*$ denote an arbitrary but fixed symmetric positive definite matrix.

(a) Show that if $\delta_0(\boldsymbol{Y})$ is admissible for $\boldsymbol{\theta}$ for the case $\boldsymbol{W} = \boldsymbol{W}_*$, then $\boldsymbol{W}_*^{1/2}\delta_0(\boldsymbol{Y})$ is admissible for $\boldsymbol{W}_*^{1/2}\boldsymbol{\theta}$ for the case $\boldsymbol{W} = \boldsymbol{I}_p$. Note that $\boldsymbol{W}_*^{1/2}$ is the (uniquely determined) symmetric positive matrix satisfying $\boldsymbol{W}_*^{1/2}\boldsymbol{W}_*^{1/2} = \boldsymbol{W}_*$.
(b) Use part (a) and Theorem 1.1 to derive the following claim: If $\delta_0(\boldsymbol{Y})$ is admissible for $\boldsymbol{\theta}$ for the case $\boldsymbol{W} = \boldsymbol{W}_*$, then $\delta_0(\boldsymbol{Y})$ is admissible for $\boldsymbol{\theta}$ for the case $\boldsymbol{W} = \boldsymbol{I}_p$.

1.10. Show that

$$\mathrm{MSE}(\boldsymbol{\theta}, \delta) = \mathrm{Cov}(\delta(\boldsymbol{Y})) + \mathrm{bias}(\delta(\boldsymbol{Y}))\,\mathrm{bias}(\delta(\boldsymbol{Y}))' ,$$

where $\mathrm{bias}(\delta(\boldsymbol{Y})) = \mathrm{E}(\delta(\boldsymbol{Y})) - \boldsymbol{\theta}$.

1.11. Let $\delta_1(\boldsymbol{Y})$ and $\delta_2(\boldsymbol{Y})$ be two decision rules such that the matrix-valued inequality $\mathrm{MSE}(\boldsymbol{\theta}, \delta_1) \leq_{\mathrm{L}} \mathrm{MSE}(\boldsymbol{\theta}, \delta_2)$ is satisfied for all $\boldsymbol{\theta} \in \Theta$. Show that if for some given $\boldsymbol{\theta} \in \Theta$ the difference $\mathrm{MSE}(\boldsymbol{\theta}, \delta_2) - \mathrm{MSE}(\boldsymbol{\theta}, \delta_1)$ is positive definite, then for this $\boldsymbol{\theta}$ the following two statements hold true:

(i) $\mathrm{E}\left[(\delta_1(\boldsymbol{Y})_i - \theta_i)^2\right] < \mathrm{E}\left[(\delta_2(\boldsymbol{Y})_i - \theta_i)^2\right]$ for every $i \in \{1, \ldots, p\}$.
(ii) $\mathrm{E}\left[(\delta_1(\boldsymbol{Y}) - \boldsymbol{\theta})'\boldsymbol{W}(\delta_1(\boldsymbol{Y}) - \boldsymbol{\theta})\right] < \mathrm{E}\left[(\delta_2(\boldsymbol{Y}) - \boldsymbol{\theta})'\boldsymbol{W}(\delta_2(\boldsymbol{Y}) - \boldsymbol{\theta})\right]$ for every $p \times p$ symmetric positive definite matrix $\boldsymbol{W}$.

1.12. Consider the situation from Problem 1.1. Check whether $\delta(\boldsymbol{Y}) = \boldsymbol{0}$ is admissible for $\boldsymbol{\mu}$ in

$$\Delta = \{\delta(\boldsymbol{Y}) : \delta(\boldsymbol{Y}) = \boldsymbol{F}\boldsymbol{Y} + \boldsymbol{f}\}$$

with respect to the unweighted squared error loss. Here $\boldsymbol{F}$ denotes a $p \times p$ matrix and $\boldsymbol{f}$ denotes a $p \times 1$ vector.

2 The Linear Regression Model

In this chapter, we consider point estimation of the parameters $\beta \in \mathbb{R}^p$ and $\sigma^2 \in (0, \infty)$ in the linear regression model

$$y = X\beta + \varepsilon, \quad \varepsilon \sim (0, \sigma^2 I_n) \, .$$

We will focus our attention to the *ordinary least squares estimator*

$$\widehat{\beta} = (X'X)^{-1}X'y$$

and the *least squares variance estimator*

$$\widehat{\sigma}^2 = \frac{1}{n-p}(y - X\widehat{\beta})'(y - X\widehat{\beta}) \, ,$$

both estimators being unbiased for β and σ^2, respectively.

We will demonstrate that the ordinary least squares estimator $\widehat{\beta}$ is *optimal* within the class of linear unbiased estimators but nonetheless can be *unreliable* when collinearity is present. In addition, under certain extra conditions $\widehat{\beta}$ can uniformly be outperformed, and, thus, this estimator turns out to be *inadmissible.* Both shortcomings (unreliability and inadmissibility) do not necessarily mean that the ordinary least squares estimator is inappropriate for estimating β, but they make clear that in certain situations it could be beneficial to consider alternative biased estimators.

2.1 Assumptions

Inference under a linear regression model is carried out with respect to certain model assumptions, implying that all statements are true only when these assumptions are satisfied. In practice, an analysis via linear models should make these assumptions as safe as possible, but clearly a complete certainty about each and every assumption cannot be achieved. Chapter 6 gives a review on regression diagnostics, being useful for finding evidence in favor or against the usual model assumptions.

Fundamental Assumptions

The ordinary linear regression model is described by the equation

$$\boldsymbol{y} = \boldsymbol{X\beta} + \boldsymbol{\varepsilon} ,$$

where

(i) $\boldsymbol{X}$ is a non-stochastic $n \times p$ matrix with $p < n$;
(ii) the matrix $\boldsymbol{X}$ has rank p, i.e. $\boldsymbol{X}$ is of full column rank;
(iii) the elements of the $n \times 1$ vector $\boldsymbol{y}$ are observable random vectors;
(iv) the elements of the $n \times 1$ vector $\boldsymbol{\varepsilon}$ are non-observable random variables such that $\mathrm{E}(\boldsymbol{\varepsilon}) = \boldsymbol{0}$ and $\mathrm{Cov}(\boldsymbol{\varepsilon}) = \sigma^2 \boldsymbol{I}_n$ with $\sigma^2 > 0$. We will write $\boldsymbol{\varepsilon} \sim (\boldsymbol{0}, \sigma^2 \boldsymbol{I}_n)$ for short.

As mentioned before, due to common practice we do not use different symbols for the random vector $\boldsymbol{y}$ and an observed outcome of $\boldsymbol{y}$. We will give a specific hint when the respective meaning is not clear from the context.

Notation 2.1. *The matrix $\boldsymbol{X}$ is called* model matrix *or* design matrix. *The term 'design matrix' is common under variance analysis models. When $\boldsymbol{y}$ denotes an observed outcome, the matrix $(\boldsymbol{y}, \boldsymbol{X})$ is called* observation matrix.

Condition (i) means that the number p of independent variables is strictly smaller than the number n of observations. There are good reasons supporting this assumption. If we had $n < p$, then there would exist fewer observations $(y_i, x_{i,1}, \dots, x_{i,p})$, $i = 1, \dots, n$, than unknowns $\beta_1, \dots, \beta_p$, making it hard to think of any reasonable estimation process. The special case $n = p$ is also excluded, mainly for mathematical reasons (e.g. to avoid division by $0 = n - p$). In practice it is desirable to consider only those independent variables which tell a lot about the dependent variable. Usually, the model is therefore be built due to the *principle of parsimony*, implying that $p < n$ will be satisfied.

If we consider the elements of $\boldsymbol{X}$, then in applications they often turn out to be outcomes of random variables. This collides with our assumption that the elements of $\boldsymbol{X}$ are non-stochastic. However, we proceed by neglecting the possible stochastic character of independent variables. Nonetheless, we keep in mind that our investigations are exactly true only for the given set of values of the independent variables. In the literature, this is often emphasized by writing $\mathrm{E}(\boldsymbol{y}|\boldsymbol{X}) = \boldsymbol{X\beta}$ instead of $\mathrm{E}(\boldsymbol{y}) = \boldsymbol{X\beta}$.

In econometrics the assumption of a non-stochastic nature of the elements in $\boldsymbol{X}$ is often weakened to the assumption that the source of data in $\boldsymbol{X}$ is unrelated to the source of $\boldsymbol{\varepsilon}$, see e.g. [46, Sect. 2.3.5]. This is seen to be crucial for a reasonable application of the classical linear regression methods (assuming non-randomness of the elements of $\boldsymbol{X}$). Whenever there is a relationship between $\boldsymbol{X}$ and $\boldsymbol{\varepsilon}$, then classical regression methods are supposed to yield unsatisfactory results and alternative approaches are to be considered.

Condition (ii) is usually satisfied in any linear regression model with or without intercept when the the values of the independent variables are qualitative measurements. In this case one cannot expect exact linear dependencies between columns of $\boldsymbol{X}$. There may also be a time trend and/or certain dummy variables included, but this will in general not disturb the full rank assumption. In variance analysis models, assumption (ii) is often violated.

Condition (iii) implies that the dependent variable y is stochastic and a set of n outcomes is available. The stochastic character of this variable follows from the fact that the dependence on the independent variables is not exact, but additively disturbed by a non-observable random error.

Condition (iv) postulates certain rules, the non-observable random vector ε has to obey. These assumptions should not be trusted without carrying out some diagnostics in the first step of model building. Using residual plots, it may be tried, for example, to find evidence for correlatedness and/or varying variances of errors. If such evidence cannot be found, then it is convenient to rely on (iv), although knowledge about the true random behavior of the errors can only be ascertained to a limited extent. See Appendix C for a review on regression diagnostics. In Sect. 5.2, additional assumptions concerning the dispersion of ε are discussed.

Additional Assumptions

We will assume in the following that conditions (i) to (iv) are satisfied. Then the model is also called *classical linear regression model.* Additional assumptions will be used when necessary:

(v) the vector ε is n-variate normally distributed (see Definition B.5);
(vi) the inequality $p \geq 3$ is true;
(vii) the identity $\boldsymbol{X}'\boldsymbol{X} = \boldsymbol{I}_p$ is true;
(viii) the parameter vector $\boldsymbol{\beta}$ satisfies $\boldsymbol{R}\boldsymbol{\beta} = \boldsymbol{r}$ for a given $m \times p$ matrix $\boldsymbol{R}$ of full row rank and a given $m \times 1$ vector $\boldsymbol{r}$.

Condition (v) is usually introduced when the test of a linear hypothesis with respect to $\boldsymbol{\beta}$ is required. In practice, it should be tried to find evidence in favor or against the normality assumption, e.g. by inspecting a normal quantile plot, see Sect. 6.4.1. There are applications, where normality of the random error cannot be assumed. Conditions (iv) and (v) are both satisfied if and only if the elements ε_i of ε are independently $N(0, \sigma^2)$ distributed.

Condition (vi) is not very restrictive and will often be true.

Condition (vii) is usually not satisfied, but can always be achieved by an appropriate reparameterization of the model as shown is Sect. 2.5.1.

Condition (viii) means that further information concerning the parameter vector $\boldsymbol{\beta}$ is given in form of a linear system. We will consider this situation in Sect. 3.1.

Regression with Intercept

Quite often regression analysis is based on a model with intercept, implying that one of the independent variables is equal to 1 for $i = 1, \ldots, n$ (constant). Then one of the columns (usually the first) of the matrix $\boldsymbol{X}$ is the vector of n ones, denoted by $\mathbf{1}_n$. In that case the linear regression model is sometimes written as

$$\boldsymbol{y} = \mathbf{1}_n \mu + \boldsymbol{X}\boldsymbol{\beta} + \boldsymbol{\varepsilon} \,,$$

in order to obtain a model matrix $\boldsymbol{X}$ which is only related to the 'real' independent variables, and to distinguish between the analysis of the parameter vector $\boldsymbol{\beta}$ and the intercept parameter μ. Although this may sometimes be helpful, in many respects the constant can be (and should be) treated as any other independent variable, see also the discussion in [11, Sect. 6.8]. We will follow this line, and, unless otherwise stated, the model matrix $\boldsymbol{X}$ may or may not have $\mathbf{1}_n$ as a column.

For most applications the inclusion of an intercept parameter is advisable, although there may also be situations when this is not the case. Omission of the intercept term may be viewed as additional information that the intercept parameter is equal to zero and will enhance the analysis when this prior information is (almost) true. Hence, some attention should be given to this question at the stage of model building.

Linear Models and Decision Theory

Under the linear regression model, we are interested in inference about the unknown parameters $\boldsymbol{\beta} \in \mathbb{R}^p$ and $\sigma^2 > 0$. Thus, the parameter space is given by $\Theta = \mathbb{R}^p \times (0, \infty)$. It is clear that the distribution of the stochastic vector $\boldsymbol{y}$ depends on the unknown parameters, since

$$\mathrm{E}(\boldsymbol{y}) = \boldsymbol{X}\boldsymbol{\beta} \quad \text{and} \quad \mathrm{Cov}(\boldsymbol{y}) = \sigma^2 \boldsymbol{I}_n \,.$$

The observable vector $\boldsymbol{y}$ will be used to estimate the unknowns, so that the decision rules in question are estimators $\widetilde{\boldsymbol{\beta}}(\boldsymbol{y})$ and $\widetilde{\sigma}^2(\boldsymbol{y})$ of $\boldsymbol{\beta}$ and σ^2, respectively. For the purpose of comparing different estimators, we will use either a (weighted) squared error loss, or a matrix-valued squared error loss, the latter being only appropriate for estimating $\boldsymbol{\beta}$.

2.2 Ordinary Least Squares Estimation

The principle of ordinary least squares is presumably the best known and most applied method for estimating the regression parameters.

2.2.1 The Principle of Least Squares

Under the linear regression model described by

$$y = X\beta + \varepsilon, \quad \varepsilon \sim (0, \sigma^2 I_n) ,$$

we can take hold of an observed outcome of the random vector y. If for this outcome there had existed a vector β^0 such that $y = X\beta^0$, then this vector would have been a natural replacement for the unknown parameter β. However, usually the system

$$y = X\beta_*$$

is not solvable with respect to $\beta_* \in \mathbb{R}^p$ for a given y. In other words, the observable random vector y will not realize within the column space $\mathcal{C}(X)$. Based on these two equivalent descriptions of the same fact, we can develop two slightly different approaches which both yield the same replacement (= point estimator) of β.

The Principle of Least Squares. (How to solve a non-solvable linear system?) If we assume that the system of equations $y = X\beta_*$ is solvable with respect to β_*, then a solution β^0 clearly satisfies $\|y - X\beta^0\|^2 = 0$. On the other hand, when we assume that $y = X\beta_*$ is not solvable, then we can nonetheless determine a vector $\widehat{\beta}$ such that

$$\|y - X\widehat{\beta}\|^2 \leq \|y - X\beta_*\|^2$$

for every vector $\beta_* \in \mathbb{R}^p$. Such a vector $\widehat{\beta}$ is called *least squares solution* of $y = X\beta_*$, since the above condition can equivalently be reexpressed as

$$\sum_{i=1}^{n} (y - X\widehat{\beta})_i^2 \leq \sum_{i=1}^{n} (y - X\beta_*)_i^2 .$$

If we consider $\varepsilon_{*i} := (y - X\beta_*)_i$ as the i-th residual of the solution β_*, then the sum of squared residuals is minimized for $\beta_* = \widehat{\beta}$, so that $\widehat{\beta}$ has the smallest sum of squared residuals.

How can we obtain the solution $\widehat{\beta}$? A direct approach is by differentiating the function $\|y - X\beta_*\|^2$ with respect to β_*. By using the rules

$$\frac{\partial}{\partial a} a'Aa = (A + A')a \quad \text{and} \quad \frac{\partial}{\partial a} a'b = b ,$$

it follows that

$$\frac{\partial}{\partial \beta_*} \|y - X\beta_*\|^2 = 2X'X\beta_* - 2X'y .$$

If we put the right-hand side equal to $\mathbf{0}$ and solve for β_*, then the solution is given by $\widehat{\beta} = (X'X)^{-1}X'y$. See also Problem 2.1 for a more algebraic proof of the following theorem.

Theorem 2.1. *Under the linear regression model with assumptions (i) to (iv), the function*

$$f(\boldsymbol{\beta}_*) = \|\boldsymbol{y} - \boldsymbol{X}\boldsymbol{\beta}_*\|^2 = (\boldsymbol{y} - \boldsymbol{X}\boldsymbol{\beta}_*)'(\boldsymbol{y} - \boldsymbol{X}\boldsymbol{\beta}_*)$$

is minimized for $\boldsymbol{\beta}_* = \widehat{\boldsymbol{\beta}}$, *where* $\widehat{\boldsymbol{\beta}} = (\boldsymbol{X}'\boldsymbol{X})^{-1}\boldsymbol{X}'\boldsymbol{y}$.

It is clear that the solution $\widehat{\boldsymbol{\beta}} = (\boldsymbol{X}'\boldsymbol{X})^{-1}\boldsymbol{X}'\boldsymbol{y}$ obtained for a specific observed outcome $\boldsymbol{y}$ can also represent a general decision rule (point estimator for $\boldsymbol{\beta}$) when $\boldsymbol{y}$ represents a random vector.

Notation 2.2. *Under the linear regression model with assumptions (i) to (iv), the vector* $\widehat{\boldsymbol{\beta}} = (\boldsymbol{X}'\boldsymbol{X})^{-1}\boldsymbol{X}'\boldsymbol{y}$ *is called* ordinary least squares estimator *of* $\boldsymbol{\beta}$.

The Closest Point Principle. If the system $\boldsymbol{y} = \boldsymbol{X}\boldsymbol{\beta}_*$ had been solvable with respect to $\boldsymbol{\beta}_*$, then $\boldsymbol{y} \in \mathcal{C}(\boldsymbol{X})$ would hold true. Since we assume non-solvability, we can try to find the vector $\widehat{\boldsymbol{y}} \in \mathcal{C}(\boldsymbol{X})$ which has the closest squared distance to $\boldsymbol{y} \notin \mathcal{C}(\boldsymbol{X})$. In other words, we may seek for $\widehat{\boldsymbol{y}} \in \mathcal{C}(\boldsymbol{X})$ such that

$$\|\boldsymbol{y} - \widehat{\boldsymbol{y}}\|^2 \leq \|\boldsymbol{y} - \boldsymbol{y}_*\|^2$$

holds true for every vector $\boldsymbol{y}_* \in \mathcal{C}(\boldsymbol{X})$.

Theorem 2.2. *Under the linear regression model with assumptions (i) to (iv), the function*

$$g(\boldsymbol{y}_*) = \|\boldsymbol{y} - \boldsymbol{y}_*\|^2$$

is minimized for $\boldsymbol{y}_* = \widehat{\boldsymbol{y}}$, *where* $\widehat{\boldsymbol{y}} = \boldsymbol{X}(\boldsymbol{X}'\boldsymbol{X})^{-1}\boldsymbol{X}'\boldsymbol{y}$.

Proof. The assertion follows from Theorem A.30. □

Now, $\widehat{\boldsymbol{y}}$ delivers the vector in $\mathcal{C}(\boldsymbol{X})$ which has the smallest squared distance from the observed $\boldsymbol{y}$. In a second step, we can determine a vector $\widehat{\boldsymbol{\beta}}$ with $\widehat{\boldsymbol{y}} = \boldsymbol{X}\widehat{\boldsymbol{\beta}}$. It is easy to see that there is exactly one such vector, given as $\widehat{\boldsymbol{\beta}} = (\boldsymbol{X}'\boldsymbol{X})^{-1}\boldsymbol{X}'\boldsymbol{y}$. Hence, again, we obtain the ordinary least squares estimator for $\boldsymbol{\beta}$.

Example 2.1. Consider a linear regression model

$$y_i = \beta_1 + \beta_2 x_i + \varepsilon_i, \quad i = 1, \ldots, n$$

with $n = 9$ observations from Table 2.1.

A specific line described by $y = \widehat{\beta}_1 + \widehat{\beta}_2 x$ can be drawn through the cloud of points (x_i, y_i) from Table 2.1 such that

$$\sum_{i=1}^{n}(y_i - \widehat{\beta}_1 - \widehat{\beta}_2 x_i)^2 = \min_{\beta_{*1}, \beta_{*2}} \sum_{i=1}^{n}(y_i - \beta_{*1} - \beta_{*2} x_i)^2$$

Table 2.1. Observations in Example 2.1

i	1	2	3	4	5	6	7	8	9
y_i	3.545	3.307	12.48	3.848	9.796	2.619	4.284	22.20	15.02
x_i	3.575	7.153	13.58	4.148	3.141	6.830	5.430	13.48	10.50

holds true, see Fig. 2.1. This means that the line is chosen in such a way that the sum of squared vertical deviations of y_i from the respective values on the line is smallest, compared to the sum of squared deviations arising from other possible lines. For the values from the table we obtain $\widehat{\boldsymbol{\beta}} = (\boldsymbol{X}'\boldsymbol{X})^{-1}\boldsymbol{X}'\boldsymbol{y} = (\widehat{\beta}_1, \widehat{\beta}_2)' = (-1.0981, 1.2822)'$. □

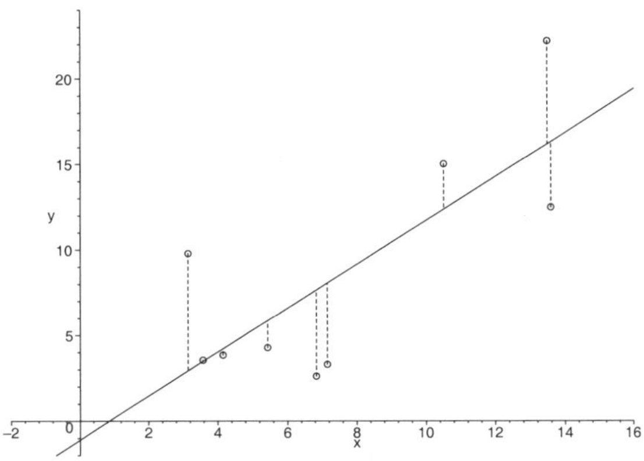

Fig. 2.1. Deviations of the points (x_i, y_i) from the line of regression $y = \widehat{\beta}_1 + \widehat{\beta}_2 x$

2.2.2 Coefficient of Determination R^2

Consider the vector $\widehat{\varepsilon} = \boldsymbol{y} - \widehat{\boldsymbol{y}}$, which is also called *vector of least squares residuals*. Then the *Pythagorean Theorem*

$$||\boldsymbol{y}||^2 = ||\widehat{\boldsymbol{y}}||^2 + ||\widehat{\varepsilon}||^2,$$

holds true, see Fig. 2.2, from which we can conclude the following:

If for an observed outcome $\boldsymbol{y}$ the value $||\widehat{\boldsymbol{y}}||^2$ is large, then the squared length of the corresponding vector of residuals $||\widehat{\varepsilon}||^2$ is small and $\widehat{\boldsymbol{y}}$ is an appropriate replacement for $\boldsymbol{y}$.

If we consider $||\widehat{\boldsymbol{y}}||^2$ in proportion to $||\boldsymbol{y}||^2$, then we obtain a measure of goodness for the fit of $\widehat{\boldsymbol{y}}$ on $\boldsymbol{y}$.

Definition 2.1. *The ratio*

$$R^2 = \frac{||\hat{y}||^2}{||y||^2} = \frac{y'Py}{y'y}, \qquad P = X(X'X)^{-1}X',$$

is called (classical) coefficient of determination.

The squared length $||\hat{\varepsilon}||^2$ of $\hat{\varepsilon}$ is also called the *residual sum of squares* (RSS) and is given by

$$\text{RSS} = y'My, \quad M = I_n - P.$$

Hence the coefficient of determination R^2 may also be written as

$$R^2 = 1 - \frac{\text{RSS}}{y'y}.$$

Since the coefficient of determination R^2 can be shown to be restricted to the interval $[0, 1]$, its value may give some clue to the intensity of the fit.

Theorem 2.3. *Under the linear regression model with assumptions (i) to (iv) the following statements hold true:*

(i) $0 \leq R^2 \leq 1$.
(ii) *The identity* $R^2 = 0$ *hold true if and only if* $\hat{y} = \mathbf{0}$*, in which case the vector of residuals* $\hat{\varepsilon}$ *has maximal length.*
(iii) *The identity* $R^2 = 1$ *holds true if and only if* $\hat{y} = y$*, in which case the vector of residuals* $\hat{\varepsilon}$ *has minimal length.*

Proof. See Problem 2.6 and Sect. 6.2. □

If the model building process is completed and the model has been chosen, then the value of R^2 is fixed for the given observation y. Since we assume this situation here, the value of R^2 is nothing but an interpretable number. Nonetheless, for most practical applications the value of R^2 should lie within an 'acceptable' region, meaning somewhere between 0.5 and 1. The extremal cases $R^2 = 0$ and $R^2 = 1$ will not occur for real data.

Assumption 2.1. *In the following we assume that* R^2 *is strictly greater than* 0 *and strictly smaller than* 1.

By the above assumption we exclude the cases $\hat{y} = y$ and $\hat{y} = \mathbf{0}$. The condition $\hat{y} \neq \mathbf{0}$ is also seen to be equivalent to $\hat{\beta} \neq \mathbf{0}$. A more detailed analysis of the coefficient of determination is given in Sec 6.2.

Example 2.2. For the data from Example 2.1 we obtain $R^2 = 0.8492$. Let us consider the estimator

$$\tilde{\beta} = c\hat{\beta}, \quad c = \frac{(n-p)R^2}{p + (n-2p)R^2},$$

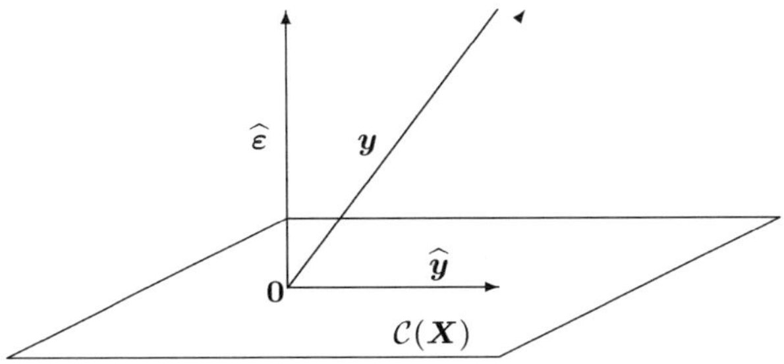

Fig. 2.2. Orthogonal projection of $\boldsymbol{y}$ onto $\mathcal{C}(\boldsymbol{X})$

for $\boldsymbol{\beta}$. We compute $c = 0.9517$ and thus obtain $\widetilde{\boldsymbol{\beta}} = (-1.0451, 1.2203)'$ as an estimate of $\boldsymbol{\beta}$. Corresponding to our remarks from the preceding section, the sum of squared deviations $f(\widetilde{\boldsymbol{\beta}}) = (\boldsymbol{y} - \boldsymbol{X}\widetilde{\boldsymbol{\beta}})'(\boldsymbol{y} - \boldsymbol{X}\widetilde{\boldsymbol{\beta}})$ is necessarily greater than $f(\widehat{\boldsymbol{\beta}}) = (\boldsymbol{y} - \boldsymbol{X}\widehat{\boldsymbol{\beta}})'(\boldsymbol{y} - \boldsymbol{X}\widehat{\boldsymbol{\beta}})$. As a matter of fact, we obtain $f(\widetilde{\boldsymbol{\beta}}) = 157.8795$ and $f(\widehat{\boldsymbol{\beta}}) = 155.8340$.

Since we have simulated the data, we know the true parameter vector $\boldsymbol{\beta}$, given as

$$\boldsymbol{\beta} = \begin{pmatrix} 0.5 \\ 1 \end{pmatrix} .$$

Thus, we can compute the observed losses of $\widehat{\boldsymbol{\beta}}$ and $\widetilde{\boldsymbol{\beta}}$, defined as the Euclidean distances of the respective estimates from the true parameter vector. We obtain

$$L(\boldsymbol{\beta}, \widehat{\boldsymbol{\beta}}) = (\widehat{\boldsymbol{\beta}} - \boldsymbol{\beta})'(\widehat{\boldsymbol{\beta}} - \boldsymbol{\beta}) = 2.6337$$

and

$$L(\boldsymbol{\beta}, \widetilde{\boldsymbol{\beta}}) = (\widetilde{\boldsymbol{\beta}} - \boldsymbol{\beta})'(\widetilde{\boldsymbol{\beta}} - \boldsymbol{\beta}) = 2.4360 .$$

We can see that although $\widetilde{\boldsymbol{\beta}}$ is not chosen due to the principle of least squares, its actual loss is smaller than the actual loss of $\widehat{\boldsymbol{\beta}}$ when $\boldsymbol{\beta} = (0.5, 1)'$ is the true parameter. □

The estimator $\widetilde{\boldsymbol{\beta}}$ from Example 2.2 does not always yield a smaller observed loss than $\widehat{\boldsymbol{\beta}}$. In addition, the estimator does not yield a smaller expected loss (= risk) for all possible values of $\boldsymbol{\beta}$ and σ^2. However, the above result shows the *possibility* of improving upon $\widehat{\boldsymbol{\beta}}$. In Sect. 2.5 we will demonstrate that there do in fact exist estimators which, under certain conditions, have a smaller risk than $\widehat{\boldsymbol{\beta}}$ for all possible values $(\boldsymbol{\beta}, \sigma^2) \in \mathbb{R}^p \times (0, \infty)$.

2.2.3 Predictive Loss

Under the linear regression model the estimate of $\mathrm{E}(\boldsymbol{y}) = \boldsymbol{X}\boldsymbol{\beta}$ is sometimes called *prediction* of $\boldsymbol{y}$. If we use an estimator of the form $\boldsymbol{X}\widetilde{\boldsymbol{\beta}}$, where $\widetilde{\boldsymbol{\beta}}$ is an estimator of $\boldsymbol{\beta}$, then the unweighted squared error loss of $\boldsymbol{X}\widetilde{\boldsymbol{\beta}}$ is

$$L(\boldsymbol{X}\boldsymbol{\beta}, \boldsymbol{X}\widetilde{\boldsymbol{\beta}}) = (\boldsymbol{X}\widetilde{\boldsymbol{\beta}} - \boldsymbol{X}\boldsymbol{\beta})'(\boldsymbol{X}\widetilde{\boldsymbol{\beta}} - \boldsymbol{X}\boldsymbol{\beta}) ,$$

which can be rewritten as the weighted loss

$$L(\boldsymbol{X}\boldsymbol{\beta}, \boldsymbol{X}\widetilde{\boldsymbol{\beta}}) = L(\boldsymbol{\beta}, \widetilde{\boldsymbol{\beta}}) = (\widetilde{\boldsymbol{\beta}} - \boldsymbol{\beta})'\boldsymbol{W}(\widetilde{\boldsymbol{\beta}} - \boldsymbol{\beta}), \quad \boldsymbol{W} = \boldsymbol{X}'\boldsymbol{X} ,$$

of $\widetilde{\boldsymbol{\beta}}$. This weighted loss function is also called *predictive squared error loss* of $\widetilde{\boldsymbol{\beta}}$ and provides an alternative to the unweighted squared error loss $L(\boldsymbol{\beta}, \widetilde{\boldsymbol{\beta}}) = (\widetilde{\boldsymbol{\beta}} - \boldsymbol{\beta})'(\widetilde{\boldsymbol{\beta}} - \boldsymbol{\beta})$ of $\widetilde{\boldsymbol{\beta}}$.

Predictive Versus Unweighted Loss

Using either $\boldsymbol{W} = \boldsymbol{I}_p$ or $\boldsymbol{W} = \boldsymbol{X}'\boldsymbol{X}$ can yield different conclusions when regarding the performance of estimators. If for two estimates $\widetilde{\boldsymbol{\beta}}_1$ and $\widetilde{\boldsymbol{\beta}}_2$ the relation

$$(\widetilde{\boldsymbol{\beta}}_1 - \boldsymbol{\beta})'(\widetilde{\boldsymbol{\beta}}_1 - \boldsymbol{\beta}) \le (\widetilde{\boldsymbol{\beta}}_2 - \boldsymbol{\beta})'(\widetilde{\boldsymbol{\beta}}_2 - \boldsymbol{\beta})$$

holds true, we do *not* necessarily have

$$(\widetilde{\boldsymbol{\beta}}_1 - \boldsymbol{\beta})'\boldsymbol{X}'\boldsymbol{X}(\widetilde{\boldsymbol{\beta}}_1 - \boldsymbol{\beta}) \le (\widetilde{\boldsymbol{\beta}}_2 - \boldsymbol{\beta})'\boldsymbol{X}'\boldsymbol{X}(\widetilde{\boldsymbol{\beta}}_2 - \boldsymbol{\beta}) .$$

Similarly, the reverse conclusion is not valid in general. Therefore, the actual choice $\boldsymbol{W} = \boldsymbol{I}_p$ or $\boldsymbol{W} = \boldsymbol{X}'\boldsymbol{X}$ depends on whether we are more interested in estimation of $\boldsymbol{\beta}$ or estimation of $\boldsymbol{X}\boldsymbol{\beta}$.

Example 2.3. In Example 2.2 we have estimated $\boldsymbol{\beta}$ by the ordinary least squares estimator $\widehat{\boldsymbol{\beta}}$ as well as by

$$\widetilde{\boldsymbol{\beta}} = c\widehat{\boldsymbol{\beta}}, \quad c = \frac{(n-p)R^2}{p + (n-2p)R^2} .$$

The observed unweighted squared error loss of $\widetilde{\boldsymbol{\beta}}$ turns out to be smaller than the observed loss of $\widehat{\boldsymbol{\beta}}$ when $\boldsymbol{\beta} = (0.5, 1)'$ is the true parameter vector. If we compute the weighted losses with $\boldsymbol{W} = \boldsymbol{X}'\boldsymbol{X}$, then

$$L(\boldsymbol{\beta}, \widehat{\boldsymbol{\beta}}) = (\widehat{\boldsymbol{\beta}} - \boldsymbol{\beta})'\boldsymbol{X}'\boldsymbol{X}(\widehat{\boldsymbol{\beta}} - \boldsymbol{\beta}) = 13.0503$$

and

$$L(\boldsymbol{\beta}, \widetilde{\boldsymbol{\beta}}) = (\widetilde{\boldsymbol{\beta}} - \boldsymbol{\beta})'\boldsymbol{X}'\boldsymbol{X}(\widetilde{\boldsymbol{\beta}} - \boldsymbol{\beta}) = 6.5384 ,$$

showing that also the observed predictive squared error loss of $\widetilde{\boldsymbol{\beta}}$ is smaller than the observed predictive squared error loss of $\widehat{\boldsymbol{\beta}}$ when $\boldsymbol{\beta} = (0.5, 1)'$ is the true parameter vector.

On the other hand, for another observed value of $\boldsymbol{y}$ the situation can be different. Suppose that our observed value of the vector $\boldsymbol{y}$ is given in the following Table 2.2.

Then we obtain $R^2 = 0.8942$ and the two estimators in question $\widehat{\boldsymbol{\beta}}$ and $\widetilde{\boldsymbol{\beta}}$ yield

Table 2.2. A new observation $\boldsymbol{y}$ in Example 2.3

i	1	2	3	4	5	6	7	8	9
y_i	6.895	1.987	12.31	5.524	5.931	3.989	2.554	14.32	8.486

$$\widehat{\boldsymbol{\beta}} = \begin{pmatrix} 0.9860 \\ 0.7831 \end{pmatrix} \quad \text{and} \quad \widetilde{\boldsymbol{\beta}} = \begin{pmatrix} 0.9537 \\ 0.7575 \end{pmatrix} .$$

The observed losses for estimating $\boldsymbol{\beta} = (0.5, 1)'$ are given in Table 2.3.

Here, the observed unweighted squared error loss of $\widetilde{\boldsymbol{\beta}}$ is smaller than the observed loss of $\widehat{\boldsymbol{\beta}}$ under $\boldsymbol{W} = \boldsymbol{I}_p$, while the reverse relation holds for $\boldsymbol{W} = \boldsymbol{X}'\boldsymbol{X}$. □

Table 2.3. Observed losses with respect to the new observation in Example 2.3

	$\widehat{\boldsymbol{\beta}}$	$\widetilde{\boldsymbol{\beta}}$
$\boldsymbol{W} = \boldsymbol{I}_p$	0.2832	0.2646
$\boldsymbol{W} = \boldsymbol{X}'\boldsymbol{X}$	18.1021	24.7695

Predictive Loss and Balanced Loss Function

Zellner's balanced loss function

$$L(\boldsymbol{\beta}, \widetilde{\boldsymbol{\beta}}) = \alpha \|\boldsymbol{y} - \boldsymbol{X}\widetilde{\boldsymbol{\beta}}\|^2 + (1-\alpha)(\widetilde{\boldsymbol{\beta}} - \boldsymbol{\beta})'\boldsymbol{X}'\boldsymbol{X}(\widetilde{\boldsymbol{\beta}} - \boldsymbol{\beta}), \quad 0 \le \alpha < 1 ,$$

see also Sect. 1.2.8, is a convex combination of the predictive squared error loss of $\widetilde{\boldsymbol{\beta}}$ and the value of the least squares function $\|\boldsymbol{y} - \boldsymbol{X}\widetilde{\boldsymbol{\beta}}\|^2$. The number $1-\alpha$ represents the relative weight which is laid on the precision of estimation $\boldsymbol{X}\widetilde{\boldsymbol{\beta}}$ of $\mathrm{E}(\boldsymbol{y}) = \boldsymbol{X}\boldsymbol{\beta}$, compared to the relative weight α which is laid on the adjustment (fit) of $\boldsymbol{X}\widetilde{\boldsymbol{\beta}}$ to $\boldsymbol{y}$.

Note that the adjustment of $\boldsymbol{X}\widetilde{\boldsymbol{\beta}}$ to an observed value $\boldsymbol{y}$ and the estimation of $\boldsymbol{X}\boldsymbol{\beta}$ via $\boldsymbol{X}\widetilde{\boldsymbol{\beta}}$ are two different concepts. The ordinary least squares estimator has specific properties with respect to both. On the one hand, $\widehat{\boldsymbol{\beta}}$ delivers the best adjustment $\widehat{\boldsymbol{y}} = \boldsymbol{X}\widehat{\boldsymbol{\beta}}$ to $\boldsymbol{y}$ (see Theorem 2.2), and, on the other hand, $\widehat{\boldsymbol{\beta}}$ is an estimator for $\boldsymbol{\beta}$ (and thus $\boldsymbol{X}\widehat{\boldsymbol{\beta}}$ is an estimator for $\boldsymbol{X}\boldsymbol{\beta}$) with specific optimality properties, see Sect. 2.3. Any alternative estimator $\widetilde{\boldsymbol{\beta}}$ of $\boldsymbol{\beta}$ can only yield a worse adjustment $\boldsymbol{X}\widetilde{\boldsymbol{\beta}}$ to $\boldsymbol{y}$, but the observed loss can be smaller. Thus the precision of estimation can be better, see Example 2.1. By using a balanced loss function one can try to account for both, the goodness of adjustment and the precision of estimation.

2.2.4 Least Squares Variance Estimator

The most common estimator for the unknown σ^2 in a linear regression model is

$$\widehat{\sigma}^2 = \frac{1}{n-p}(\boldsymbol{y} - \boldsymbol{X}\widehat{\boldsymbol{\beta}})'(\boldsymbol{y} - \boldsymbol{X}\widehat{\boldsymbol{\beta}}) .$$

It is based on the vector of least squares residuals $\widehat{\boldsymbol{\varepsilon}} = \boldsymbol{y} - \boldsymbol{X}\widehat{\boldsymbol{\beta}}$ and can be shown to be unbiased for σ^2.

Theorem 2.4. *Under the linear regression model with assumptions (i) to (iv) the identity* $\mathrm{E}(\widehat{\sigma}^2) = \sigma^2$ *holds true for every* $\sigma^2 \in (0, \infty)$.

Proof. It is known that under the linear regression model with assumptions (i) to (iv) the expectation of a quadratic form $\boldsymbol{y}'\boldsymbol{A}\boldsymbol{y}$ is given by

$$\mathrm{E}(\boldsymbol{y}'\boldsymbol{A}\boldsymbol{y}) = \sigma^2 \operatorname{tr}(\boldsymbol{A}) + \boldsymbol{\beta}'\boldsymbol{X}'\boldsymbol{A}\boldsymbol{X}\boldsymbol{\beta} .$$

We have $\widehat{\sigma}^2 = \boldsymbol{y}'\boldsymbol{A}\boldsymbol{y}$ for $(n-p)\boldsymbol{A} = \boldsymbol{M}$, where $\boldsymbol{M} = \boldsymbol{I}_n - \boldsymbol{X}(\boldsymbol{X}'\boldsymbol{X})^{-1}\boldsymbol{X}'$ (see Problem 2.7), and

$$\operatorname{tr}(\boldsymbol{M}) = \operatorname{tr}(\boldsymbol{I}_n) - \operatorname{tr}[\boldsymbol{X}(\boldsymbol{X}'\boldsymbol{X})^{-1}\boldsymbol{X}'] = n - \operatorname{tr}[\underbrace{(\boldsymbol{X}'\boldsymbol{X})^{-1}\boldsymbol{X}'\boldsymbol{X}}_{\boldsymbol{I}_p}] = n - p .$$

Therefore $\operatorname{tr}(\boldsymbol{A}) = 1$, and the assertion follows in view of $\boldsymbol{A}\boldsymbol{X} = \boldsymbol{0}$. □

Remark 2.1. The above estimator $\widehat{\sigma}^2 = \frac{1}{n-p}(\boldsymbol{y} - \boldsymbol{X}\widehat{\boldsymbol{\beta}})'(\boldsymbol{y} - \boldsymbol{X}\widehat{\boldsymbol{\beta}})$ is also called *least squares variance estimator*, since it is an unbiased estimator for σ^2, which is based on the residual vector $\widehat{\boldsymbol{\varepsilon}} = \boldsymbol{y} - \boldsymbol{X}\widehat{\boldsymbol{\beta}}$ of the least squares estimator $\widehat{\boldsymbol{\beta}}$.

Apart from the least squares variance estimator for σ^2, there exist further alternatives, e.g. Theil's [113] estimator

$$\widehat{\sigma}^2_{\mathrm{T}} = \frac{1}{n-p+2}(\boldsymbol{y} - \boldsymbol{X}\widehat{\boldsymbol{\beta}})'(\boldsymbol{y} - \boldsymbol{X}\widehat{\boldsymbol{\beta}}) ,$$

or the maximum likelihood estimator

$$\widehat{\sigma}^2_{\mathrm{ML}} = \frac{1}{n}(\boldsymbol{y} - \boldsymbol{X}\widehat{\boldsymbol{\beta}})'(\boldsymbol{y} - \boldsymbol{X}\widehat{\boldsymbol{\beta}}) .$$

These two estimators differ from the least squares variance estimator with respect to a factor multiplied by $(\boldsymbol{y} - \boldsymbol{X}\widehat{\boldsymbol{\beta}})'(\boldsymbol{y} - \boldsymbol{X}\widehat{\boldsymbol{\beta}}) = \boldsymbol{y}'\boldsymbol{M}\boldsymbol{y}$ and thus are not unbiased for σ^2. Estimation of σ^2 is considered in more detail in Sect. 5.1.

2.2.5 Properties of the Ordinary Least Squares Estimator

The ordinary least squares estimator $\widehat{\boldsymbol{\beta}} = (\boldsymbol{X}'\boldsymbol{X})^{-1}\boldsymbol{X}'\boldsymbol{y}$ of $\boldsymbol{\beta}$ is a *linear transformation* of $\boldsymbol{y}$, that is $\widehat{\boldsymbol{\beta}} = \boldsymbol{A}\boldsymbol{y}$ for some $p \times n$ non-stochastic matrix $\boldsymbol{A}$. For such a linear function, expectation vector and covariance matrix are easy to determine.

Theorem 2.5. *Under the linear regression model with assumptions (i) to (iv),*

$$\mathrm{E}(\widehat{\boldsymbol{\beta}}) = \boldsymbol{\beta} \quad \textit{and} \quad \mathrm{Cov}(\widehat{\boldsymbol{\beta}}) = \sigma^2(\boldsymbol{X}'\boldsymbol{X})^{-1}$$

hold true for $\widehat{\boldsymbol{\beta}} = (\boldsymbol{X}'\boldsymbol{X})^{-1}\boldsymbol{X}'\boldsymbol{y}$.

Proof. The assertion follows immediately from the identities $\mathrm{E}(\boldsymbol{A}\boldsymbol{y}) = \boldsymbol{A}\,\mathrm{E}(\boldsymbol{y})$ and $\mathrm{Cov}(\boldsymbol{A}\boldsymbol{y}) = \boldsymbol{A}\,\mathrm{Cov}(\boldsymbol{y})\boldsymbol{A}'$, where $\boldsymbol{A} = (\boldsymbol{X}'\boldsymbol{X})^{-1}\boldsymbol{X}'$. □

From the above theorem we can also conclude that the least squares estimator is *unbiased* for $\boldsymbol{\beta}$, i.e. we have $\mathrm{E}(\widehat{\boldsymbol{\beta}}) = \boldsymbol{\beta}$ for every $\boldsymbol{\beta} \in \mathbb{R}^p$. In addition to its expectation, we can derive the expected squared length of $\widehat{\boldsymbol{\beta}}$.

Theorem 2.6. *Under the linear regression model with assumptions (i) to (iv),*

$$\mathrm{E}(\widehat{\boldsymbol{\beta}}'\widehat{\boldsymbol{\beta}}) = \sigma^2\,\mathrm{tr}[(\boldsymbol{X}'\boldsymbol{X})^{-1}] + \boldsymbol{\beta}'\boldsymbol{\beta}\,,$$

where $\widehat{\boldsymbol{\beta}} = (\boldsymbol{X}'\boldsymbol{X})^{-1}\boldsymbol{X}'\boldsymbol{y}$.

Proof. The assertion follows from the formula

$$\mathrm{E}(\boldsymbol{y}'\boldsymbol{A}\boldsymbol{y}) = \sigma^2\,\mathrm{tr}(\boldsymbol{A}) + \boldsymbol{\beta}'\boldsymbol{X}'\boldsymbol{A}\boldsymbol{X}\boldsymbol{\beta}\,,$$

where $\boldsymbol{A} = \boldsymbol{X}(\boldsymbol{X}'\boldsymbol{X})^{-1}(\boldsymbol{X}'\boldsymbol{X})^{-1}\boldsymbol{X}'$. □

We note that the number $\sigma^2\,\mathrm{tr}[(\boldsymbol{X}'\boldsymbol{X})^{-1}]$ is always positive. Hence we can state the following:

Remark 2.2. Under the linear regression model with assumptions (i) to (iv), the least squares estimator $\widehat{\boldsymbol{\beta}}$ is unbiased for $\boldsymbol{\beta}$, but its expected squared length is greater than the squared length of $\boldsymbol{\beta}$.

Now, we have stated that although the least squares estimator is unbiased for $\boldsymbol{\beta}$, it is at the same time 'too long' for estimating $\boldsymbol{\beta}$. A similar observation led Charles Stein [111] to derive an alternative to the maximum likelihood estimator for the expectation of a normally distributed random vector. This was presumably the first time that a maximum likelihood estimator has been shown to be uniformly worse than a different estimator (*Stein Paradox*). Also for estimation of $\boldsymbol{\beta}$ in linear regression, an appropriate Stein estimator can be derived, turning out to be uniformly better than $\widehat{\boldsymbol{\beta}}$ under certain conditions, see Sect. 2.5. Other alternative estimators $\widetilde{\boldsymbol{\beta}}$ presented in Chap. 3 prove to be not *uniformly* better than the least squares estimator but nonetheless shorten $\widehat{\beta}$ in the sense that $||\widetilde{\beta}||^2 \leq ||\widehat{\beta}||^2$ or even $||\widetilde{\beta}||^2 < ||\widehat{\beta}||^2$ for any observed value y.

The relations

$$||\beta||^2 \leq \mathrm{E}\left(||\widehat{\beta}||^2\right) \quad \text{and} \quad ||\beta|| \leq \mathrm{E}\left(||\widehat{\beta}||\right)$$

can also be concluded from Jensen's inequality

$$g\left(\mathrm{E}[z]\right) \leq \mathrm{E}\left[g(z)\right] ,$$

where z is a $p \times 1$ stochastic vector and $g : \mathbb{R}^p \to \mathbb{R}$ is a *convex* function. Equality holds if and only if there exists a number $a \in \mathbb{R}$ and a $p \times 1$ vector b, such that $g(z) = a + b'z$ with probability one, see for example [15, p. 518]. Since the functions $||\cdot||^2$ and $||\cdot||$ are convex, the above mentioned inequalities follow for the choice $z = \widehat{\beta}$.

To illustrate these inequalities, we assume for simplicity that under a model with $p = 2$ the least squares estimator is symmetrically distributed around the true parameter vector and takes values within some prescribed ellipsoid with probability one. This assumption is is accordance with the unbiasedness of the least squares estimator. If, in addition, we consider a circle with center $\mathbf{0}$ and radius equal to the length of the true parameter vector, then all vectors within the circle are shorter than the true parameter vector. It can be seen that the surface of intersection between ellipsoid and circle is less than the remaining surface of the ellipsoid. Since the latter has more probability mass than the former, the expected length of the least squares estimator will turn out to be greater than the length of the parameter vector itself, compare Fig. 2.3.

2.2.6 Properties Under Normality

If under the linear regression model the assumptions (i) to (iv) are satisfied, and in addition (v) holds true, i.e. ε is an n-variate normally distributed vector, then the following statements concerning the distributions of $\widehat{\beta}$ and $\widehat{\sigma}^2$ are true.

Theorem 2.7. *Under the linear regression model with assumptions (i) to (iv) and (v), the following two statements hold:*

(i) $\widehat{\beta} \sim N(\beta, \sigma^2(X'X)^{-1})$ *and* $\dfrac{n-p}{\sigma^2}\widehat{\sigma}^2 \sim \chi^2_{(n-p)}$*;*

(ii) $\widehat{\beta}$ *and* $\widehat{\sigma}^2$ *are stochastically independent.*

Proof. From Theorem B.9 we know that every linear transformation of $\varepsilon \sim N_n(\mathbf{0}, I_n)$ is normally distributed, so that $y \sim N_n(X\beta, \sigma^2 I_n)$ and $\widehat{\beta} = (X'X)^{-1}X'y \sim N_p(\beta, \sigma^2(X'X)^{-1})$. For $z = \sigma^{-1}y$ we have $z \sim N_n(\sigma^{-1}X\beta, I_n)$ and $(n - p/\sigma^2)\widehat{\sigma}^2$ can be written as

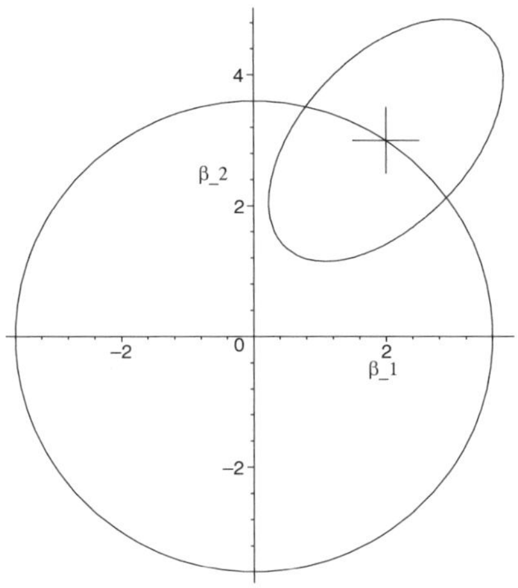

Fig. 2.3. Illustration of the expected length of least squares estimation

$$\frac{n-p}{\sigma^2}\widehat{\sigma}^2 = \boldsymbol{z}'\boldsymbol{M}\boldsymbol{z}, \quad \boldsymbol{M} = \boldsymbol{I}_n - \boldsymbol{X}(\boldsymbol{X}'\boldsymbol{X})^{-1}\boldsymbol{X}' .$$

Since $\boldsymbol{M}$ is symmetric and idempotent, from Theorem B.18 $\boldsymbol{z}'\boldsymbol{M}\boldsymbol{z} \sim \chi^2_{(a,\lambda)}$, where $a = \mathrm{rk}(\boldsymbol{M}) = n - p$ and $\lambda = \boldsymbol{\beta}'\boldsymbol{X}'\boldsymbol{M}\boldsymbol{X}\boldsymbol{\beta}/2\sigma^2 = 0$. Moreover, from Theorem B.13 $\boldsymbol{z}'\boldsymbol{M}\boldsymbol{z}$ and $(\boldsymbol{X}'\boldsymbol{X})^{-1}\boldsymbol{X}'\boldsymbol{z}$ are stochastically independent, which also establishes the independence of $\widehat{\sigma}^2$ and $\widehat{\boldsymbol{\beta}}$. □

Since, as shown above, under the additional assumption (v) the distribution of $\widehat{\sigma}^2$ is known, it is also possible to give the variance of $\widehat{\sigma}^2$.

Corollary 2.1. *Under the linear regression model with assumptions (i) to (iv) and (v),*

$$\mathrm{Var}(\widehat{\sigma}^2) = \frac{2\sigma^4}{n-p} .$$

Proof. From Theorem 2.7 (i) it follows that $\mathrm{Var}\left(\frac{n-p}{\sigma^2}\widehat{\sigma}^2\right) = 2(n-p)$, thus showing the assertion. □

2.3 Optimality of Least Squares Estimation

In this section the ordinary least squares estimator is shown to be *optimal* within the class of linear unbiased estimators with respect to the real-valued or matrix-valued squared error loss. If, in addition, the error vector is normally distributed, then the considered optimality is even valid in the class of unbiased not necessarily linear estimators. This optimality implies, of course, that the ordinary least squares estimator $\widehat{\boldsymbol{\beta}}$ is the only admissible estimator when the set of considered decision functions is the set of (linear) unbiased estimators for $\boldsymbol{\beta}$.

2.3.1 Linear Unbiased Estimation

Linear estimators for the $p \times 1$ vector $\boldsymbol{\beta}$ are of the form

$$\widetilde{\boldsymbol{\beta}} = \boldsymbol{A}\boldsymbol{y} + \boldsymbol{a} \, ,$$

where $\boldsymbol{A}$ is a $p \times n$ non-stochastic matrix and $\boldsymbol{a}$ is a $p \times 1$ non-stochastic vector. If $\boldsymbol{a}$ equals the $p \times 1$ zero vector, then the linear estimator is called *homogeneous*, while in case $\boldsymbol{a} \neq \boldsymbol{0}$ the linear estimator is called *non-homogeneous*.

The ordinary least squares estimator $\widehat{\boldsymbol{\beta}}$ is a homogeneous linear estimator with $\boldsymbol{A} = (\boldsymbol{X}'\boldsymbol{X})^{-1}\boldsymbol{X}'$. As shown before, the identity $\mathrm{E}(\widehat{\boldsymbol{\beta}}) = \mathrm{E}(\boldsymbol{A}\boldsymbol{y}) = \boldsymbol{\beta}$ holds true for all $\boldsymbol{\beta} \in \mathbb{R}^p$.

Definition 2.2. *Under the linear regression model with assumptions (i) to (iv), a linear estimator $\boldsymbol{A}\boldsymbol{y} + \boldsymbol{a}$ is called* unbiased *for $\boldsymbol{\beta}$ if*

$$\mathrm{E}(\boldsymbol{A}\boldsymbol{y} + \boldsymbol{a}) = \boldsymbol{\beta}$$

is satisfied for all $\boldsymbol{\beta} \in \mathbb{R}^p$.

Theorem 2.8. *Under the linear regression model with assumptions (i) to (iv), a linear estimator $\boldsymbol{A}\boldsymbol{y} + \boldsymbol{a}$ is unbiased for $\boldsymbol{\beta}$ if and only if $\boldsymbol{A}\boldsymbol{X} = \boldsymbol{I}_p$ and $\boldsymbol{a} = \boldsymbol{0}$.*

Proof. Problem 2.10. □

The characterization of unbiasedness given in the above theorem depends on the choice of the parameter space $\Theta_{\boldsymbol{\beta}}$ in which we assume $\boldsymbol{\beta}$ to lie in. In a linear regression model with assumptions (i) to (iv) we naturally consider $\Theta_{\boldsymbol{\beta}} = \mathbb{R}^p$. If additional assumptions are accepted, e.g. model assumption (viii), then the parameter space can change to some 'smaller' space $\Theta_{\boldsymbol{\beta}} \subset \mathbb{R}^p$. In this case a linear estimator $\boldsymbol{A}\boldsymbol{y} + \boldsymbol{a}$ satisfying $\boldsymbol{A}\boldsymbol{X} = \boldsymbol{I}_p$ and $\boldsymbol{a} = \boldsymbol{0}$ will also be unbiased for $\boldsymbol{\beta} \in \Theta_{\boldsymbol{\beta}}$, but there might also exist estimators which are unbiased for all $\boldsymbol{\beta} \in \Theta_{\boldsymbol{\beta}} \subset \mathbb{R}^p$ but not unbiased for all $\boldsymbol{\beta} \in \mathbb{R}^p$. We consider this situation in Sect. 3.1.

Notation 2.3. *The set of linear unbiased estimators for $\boldsymbol{\beta}$ is denoted by $\mathcal{U}(\boldsymbol{\beta})$, that is*

$$\mathcal{U}(\boldsymbol{\beta}) = \{\boldsymbol{A}\boldsymbol{y} + \boldsymbol{a} : \, \boldsymbol{A}\boldsymbol{X} = \boldsymbol{I}_p, \, \boldsymbol{a} = \boldsymbol{0}\} \, ,$$

where $\boldsymbol{A}$ is a $p \times n$ non-stochastic matrix.

Since $\boldsymbol{X}$ is of full column rank, the system $\boldsymbol{AX} = \boldsymbol{I}_p$ is always solvable with respect to $\boldsymbol{A}$. The possible solutions can be described by the identity

$$\boldsymbol{A} = (\boldsymbol{X}'\boldsymbol{X})^{-1}\boldsymbol{X}' + \boldsymbol{Z}(\boldsymbol{I}_n - \boldsymbol{X}(\boldsymbol{X}'\boldsymbol{X})^{-1}\boldsymbol{X}') ,$$

where $\boldsymbol{Z}$ is an arbitrary $p \times n$ matrix. Hence, we have infinitely many linear unbiased estimators for $\boldsymbol{\beta}$ at our disposal. The least squares estimator $\widehat{\boldsymbol{\beta}} = (\boldsymbol{X}'\boldsymbol{X})^{-1}\boldsymbol{X}'\boldsymbol{y}$ is only one among those, but it is the one which has been derived due to the principle of least squares. What if we consider the performance of $\widehat{\boldsymbol{\beta}}$ in the class $\mathcal{U}(\boldsymbol{\beta})$ with respect to loss and corresponding risk? Before addressing this problem in the following section, we reconsider the situation from Example 2.1.

Example 2.4. For the situation from Example 2.1 we construct an unbiased estimator for $\boldsymbol{\beta}$ by choosing $\boldsymbol{Z} = (\boldsymbol{I}_p, \boldsymbol{0}_{p\times(n-p)})$ in the above equation. Then the unbiased estimator

$$\widehat{\boldsymbol{\beta}}_e = \widehat{\boldsymbol{\beta}} + \boldsymbol{Z}(\boldsymbol{I}_n - \boldsymbol{X}(\boldsymbol{X}'\boldsymbol{X})^{-1}\boldsymbol{X}')\boldsymbol{y}$$

yields $\widehat{\boldsymbol{\beta}}_e = (-1.0390, -3.4844)'$. If we compute the observed losses of $\widehat{\boldsymbol{\beta}}$ and $\widehat{\boldsymbol{\beta}}_e$ with respect to the loss function

$$L(\boldsymbol{\beta}, \widetilde{\boldsymbol{\beta}}) = (\boldsymbol{\beta} - \widetilde{\boldsymbol{\beta}})'(\boldsymbol{\beta} - \widetilde{\boldsymbol{\beta}})$$

when $\boldsymbol{\beta} = (0.5, 1)'$ is the true parameter vector, then we obtain $L(\boldsymbol{\beta}, \widehat{\boldsymbol{\beta}}) = 2.6337$ and $L(\boldsymbol{\beta}, \widehat{\boldsymbol{\beta}}_e) = 22.4781$. Thus, the estimator $\widehat{\boldsymbol{\beta}}_e$ turns out to be distinctly worse than $\widehat{\boldsymbol{\beta}}$ with respect to the *observed* unweighted squared error loss. □

Of course the numerical result from the above example cannot be used to draw any conclusions about the *expected* loss of $\widehat{\boldsymbol{\beta}}_e$, since individual observed losses can differ substantially from the expected losses. Nonetheless, in the next section we show that the least squares estimator is the most appropriate one in $\mathcal{U}(\boldsymbol{\beta})$ when the expected loss is a measure of goodness.

2.3.2 Gauss-Markov Theorem

In the previous section we have shown that $\widehat{\boldsymbol{\beta}}$ belongs to the set $\mathcal{U}(\boldsymbol{\beta})$ of linear estimators being unbiased for $\boldsymbol{\beta}$. In this set we have infinitely many other estimators for $\boldsymbol{\beta}$ available. Let $g(\boldsymbol{y})$ denote one such member, i.e. $g(\boldsymbol{y}) \in \mathcal{U}(\boldsymbol{\beta})$. Then we can write

$$g(\boldsymbol{y}) = \widehat{\boldsymbol{\beta}} + z(\boldsymbol{y}) ,$$

where $z(\boldsymbol{y}) = g(\boldsymbol{y}) - \widehat{\boldsymbol{\beta}}$. Since $g(\boldsymbol{y})$ and $\widehat{\boldsymbol{\beta}}$ are both linear, also $z(\boldsymbol{y})$ is linear. Moreover, $\mathrm{E}[z(\boldsymbol{y})] = \boldsymbol{0}$ for every $\boldsymbol{\beta} \in \mathbb{R}^p$, meaning that $z(\boldsymbol{y})$ is unbiased for $\boldsymbol{0}$. Let us denote the class of all linear estimators which are unbiased for $\boldsymbol{0}$ by

$\mathcal{U}(\mathbf{0})$. It is easy to see that $\boldsymbol{A}\boldsymbol{y}+\boldsymbol{a} \in \mathcal{U}(\mathbf{0})$ if and only if $\boldsymbol{A}\boldsymbol{X}=\mathbf{0}$ and $\boldsymbol{a}=\mathbf{0}$. Hence, we have

$$\mathcal{U}(\mathbf{0}) = \{\boldsymbol{A}\boldsymbol{y}+\boldsymbol{a} : \boldsymbol{A}\boldsymbol{X}=\mathbf{0},\, \boldsymbol{a}=\mathbf{0}\}\,.$$

An element from $\mathcal{U}(\mathbf{0})$ is also called *linear zero function* . If $z(\boldsymbol{y})$ is an arbitrary element from $\mathcal{U}(\mathbf{0})$, then $\mathrm{E}[\widehat{\boldsymbol{\beta}}+z(\boldsymbol{y})]=\boldsymbol{\beta}$ for all $\boldsymbol{\beta} \in \mathrm{IR}^p$. Thus, we can conclude

$$\mathcal{U}(\boldsymbol{\beta}) = \widehat{\boldsymbol{\beta}} + \mathcal{U}(\mathbf{0})\,.$$

Now let us consider the covariance matrix of an arbitrary element $g(\boldsymbol{y})$ from $\mathcal{U}(\boldsymbol{\beta})$, so that $g(\boldsymbol{y}) = \widehat{\boldsymbol{\beta}} + z(\boldsymbol{y})$ for some $z(\boldsymbol{y}) \in \mathcal{U}(\mathbf{0})$. Then

$$\mathrm{Cov}[g(\boldsymbol{y})] = \mathrm{Cov}(\widehat{\boldsymbol{\beta}}) + \mathrm{Cov}[z(\boldsymbol{y})] + \mathrm{Cov}[\widehat{\boldsymbol{\beta}}, z(\boldsymbol{y})] + (\mathrm{Cov}[\widehat{\boldsymbol{\beta}}, z(\boldsymbol{y})])'\,.$$

Since $z(\boldsymbol{y}) = \boldsymbol{A}\boldsymbol{y}$ for some $\boldsymbol{A}$ such that $\boldsymbol{A}\boldsymbol{X}=\mathbf{0}$, it follows

$$\mathrm{Cov}[\widehat{\boldsymbol{\beta}}, z(\boldsymbol{y})] = \mathrm{Cov}[(\boldsymbol{X}'\boldsymbol{X})^{-1}\boldsymbol{X}'y, \boldsymbol{A}\boldsymbol{y}] = \sigma^2(\boldsymbol{X}'\boldsymbol{X})^{-1}\boldsymbol{X}'\boldsymbol{A}' = \mathbf{0}\,.$$

This shows that the ordinary least squares estimator $\widehat{\boldsymbol{\beta}}$ is uncorrelated with every linear zero function. Hence, for any $g(\boldsymbol{y}) \in \mathcal{U}(\boldsymbol{\beta})$ there exists $z(\boldsymbol{y}) \in U(\mathbf{0})$ such that the identity

$$\mathrm{Cov}[g(\boldsymbol{y})] = \mathrm{Cov}(\widehat{\boldsymbol{\beta}}) + \mathrm{Cov}[z(\boldsymbol{y})]$$

holds true. This is the crucial fact which can easily be used to demonstrate that $\widehat{\boldsymbol{\beta}} = (\boldsymbol{X}'\boldsymbol{X})^{-1}\boldsymbol{X}'\boldsymbol{y}$ is uniformly not worse than any element from $\mathcal{U}(\boldsymbol{\beta})$.

Lemma 2.1. *Under the linear regression model with assumptions (i) to (iv), let $\widetilde{\boldsymbol{\beta}} \in \mathcal{U}(\boldsymbol{\beta})$. Then the identity*

$$\mathrm{MSE}(\boldsymbol{\beta}, \widetilde{\boldsymbol{\beta}}) = \mathrm{Cov}(\widetilde{\boldsymbol{\beta}}) = \sigma^2\boldsymbol{B} + \sigma^2(\boldsymbol{X}'\boldsymbol{X})^{-1}$$

holds true for some $p \times p$ symmetric nonnegative definite matrix $\boldsymbol{B}$.

Proof. In view of $\widetilde{\boldsymbol{\beta}} \in \mathcal{U}(\boldsymbol{\beta})$, the identity

$$\mathrm{MSE}(\boldsymbol{\beta}, \widetilde{\boldsymbol{\beta}}) = \mathrm{E}\left[(\widetilde{\boldsymbol{\beta}}-\boldsymbol{\beta})(\widetilde{\boldsymbol{\beta}}-\boldsymbol{\beta})'\right] = \mathrm{E}\left[(\widetilde{\boldsymbol{\beta}}-\mathrm{E}(\widetilde{\boldsymbol{\beta}}))(\widetilde{\boldsymbol{\beta}}-\mathrm{E}(\widetilde{\boldsymbol{\beta}}))'\right] = \mathrm{Cov}(\widetilde{\boldsymbol{\beta}})$$

is true, see also Problem 1.10. Hence, the assertion is simply a reformulation of the already derived identity $\mathrm{Cov}[g(\boldsymbol{y})] = \mathrm{Cov}(\widehat{\boldsymbol{\beta}}) + \mathrm{Cov}[z(\boldsymbol{y})]$ for any $g(\boldsymbol{y}) \in \mathcal{U}(\boldsymbol{\beta})$. □

Since we have $\mathrm{MSE}(\boldsymbol{\beta}, \widehat{\boldsymbol{\beta}}) = \mathrm{Cov}(\widehat{\boldsymbol{\beta}}) = \sigma^2(\boldsymbol{X}'\boldsymbol{X})^{-1}$, we can easily compare the risks $\mathrm{MSE}(\boldsymbol{\beta}, \widetilde{\boldsymbol{\beta}})$ and $\mathrm{MSE}(\boldsymbol{\beta}, \widehat{\boldsymbol{\beta}})$ for some $\widetilde{\boldsymbol{\beta}} \in \mathcal{U}(\boldsymbol{\beta})$.

Theorem 2.9. *(Gauss-Markov Theorem) Under the linear regression model with assumptions (i) to (iv), the least squares estimator*

$$\widehat{\beta} = (X'X)^{-1}X'y$$

is uniformly not worse than any estimator $\widetilde{\beta}$ *from* $\mathcal{U}(\beta)$ *with respect to* $\mathrm{MSE}(\beta, \widetilde{\beta})$*, as well as with respect to*

$$\rho(\beta, \widetilde{\beta}) = \mathrm{E}\left[(\widetilde{\beta} - \beta)'W(\widetilde{\beta} - \beta)\right],$$

where W *is an arbitrary* $p \times p$ *symmetric nonnegative definite weight matrix.*

Proof. For every estimator $\widetilde{\beta} \in \mathcal{U}(\beta)$ the identity

$$\begin{aligned}\mathrm{MSE}(\beta, \widetilde{\beta}) - \mathrm{MSE}(\beta, \widehat{\beta}) &= \mathrm{Cov}(\widetilde{\beta}) - \mathrm{Cov}(\widehat{\beta}) \\ &= \sigma^2 B + \sigma^2(X'X)^{-1} - \sigma^2(X'X)^{-1}\end{aligned}$$

holds true for some symmetric nonnegative definite matrix B, see Lemma 2.1. Hence the assertion follows from Definition 1.6 and Theorem 1.2. □

2.3.3 Normality Assumption

So far, we have considered $\widehat{\beta}$ as an element from the set $\mathcal{U}(\beta)$ being a subset of $\mathcal{U}^*(\beta)$, where

$$\mathcal{U}^*(\beta) = \{\widetilde{\beta} : \mathrm{E}(\widetilde{\beta}) = \beta \quad \text{for all } \beta \in \mathbb{R}^p \text{ and all } \sigma^2 > 0\}$$

is the set of all unbiased but not necessarily linear estimators for β. If $g(y)$ denotes an arbitrary element from $\mathcal{U}^*(\beta)$, then $g(y)$ can be written in the form $g(y) = \widehat{\beta} + z(y)$ for some $z(y) \in \mathcal{U}^*(\mathbf{0})$, where

$$\mathcal{U}^*(\mathbf{0}) = \{\widetilde{\beta} : \mathrm{E}(\widetilde{\beta}) = \mathbf{0} \quad \text{for all } \beta \in \mathbb{R}^p \text{ and all } \sigma^2 > 0\}.$$

Hence, for any $g(y) \in \mathcal{U}^*(\beta)$ we can write

$$\mathrm{Cov}[g(y)] = \mathrm{Cov}(\widehat{\beta}) + \mathrm{Cov}[z(y)] + \mathrm{Cov}[\widehat{\beta}, z(y)] + (\mathrm{Cov}[\widehat{\beta}, z(y)])'$$

for some $z(y) \in \mathcal{U}^*(\mathbf{0})$. Now, let us consider the matrix $\mathrm{Cov}[\widehat{\beta}, z(y)]$ in detail. Clearly,

$$\mathrm{Cov}[\widehat{\beta}, z(y)] = \mathrm{E}[\widehat{\beta}(z(y))'] - \mathrm{E}(\widehat{\beta})\,\mathrm{E}[(z(y))'] = \mathrm{E}[\widehat{\beta}(z(y))']$$

holds true, in view of $\mathrm{E}[z(y)] = \mathbf{0}$ for all $(\beta, \sigma^2) \in \mathbb{R}^p \times (0, \infty)$. Thus,

$$\mathrm{Cov}[\widehat{\beta}, z(y)] = (X'X)^{-1}\,\mathrm{E}[X'y(z(y))'].$$

The (i, j)th element of the matrix $\mathrm{E}[X'y(z(y))']$ is given by

$$\mathrm{E}[\boldsymbol{x}_i'\boldsymbol{y}\, z_j(\boldsymbol{y})]\,,$$

where $\boldsymbol{x}_i$ is the i-th column of the matrix $\boldsymbol{X}$, $i = 1, \ldots, p$, and $z_j(\boldsymbol{y})$ is the j-th element of the vector $z(\boldsymbol{y})$, $j = 1, \ldots, p$. Now, whenever

$$\mathrm{E}[\boldsymbol{x}_i'\boldsymbol{y}\, z_j(\boldsymbol{y})] = 0 \quad \forall i, j = 1, \ldots, p$$

for any $z(\boldsymbol{y}) \in \mathcal{U}^*(\mathbf{0})$ and all $(\boldsymbol{\beta}, \sigma^2) \in \mathrm{I\!R}^p \times (0, \infty)$, then

$$\mathrm{Cov}[g(\boldsymbol{y})] = \mathrm{Cov}(\widehat{\boldsymbol{\beta}}) + \mathrm{Cov}[z(\boldsymbol{y})]\,.$$

As noted in [89, p. 318], the former is satisfied when $\boldsymbol{y}$ is normally distributed. In view of the identity $\mathrm{Cov}[g(\boldsymbol{y})] = \mathrm{MSE}[\boldsymbol{\beta}, g(\boldsymbol{y})]$ for $g(\boldsymbol{y}) \in \mathcal{U}^*(\boldsymbol{\beta})$, the following is true:

Theorem 2.10. *Under the linear regression model with assumptions (i) to (iv) and (v), the Gauss-Markov Theorem remains valid when $\mathcal{U}(\boldsymbol{\beta})$ is replaced by $\mathcal{U}^*(\boldsymbol{\beta})$.*

2.3.4 Admissibility

The Gauss-Markov Theorem states that under the linear regression model with assumptions (i) to (iv), there cannot exist some estimator from $\mathcal{U}(\boldsymbol{\beta})$ which is uniformly better than $\widehat{\boldsymbol{\beta}}$. In other words, $\widehat{\boldsymbol{\beta}}$ is seen to be admissible for $\boldsymbol{\beta}$ in the set of decisions $\Delta = \mathcal{U}(\boldsymbol{\beta})$. Moreover, if there existed other estimators $\boldsymbol{A}\boldsymbol{y} \neq \widehat{\boldsymbol{\beta}}$ in $\mathcal{U}(\boldsymbol{\beta})$ being admissible, too, then they would necessarily have the same risk as $\widehat{\boldsymbol{\beta}}$ for all values $(\boldsymbol{\beta}, \sigma^2) \in \mathrm{I\!R}^p \times (0, \infty)$. As shown in the following, we can neglect the existence of such estimators when using $\mathrm{MSE}(\boldsymbol{\beta}, \widetilde{\boldsymbol{\beta}})$ or $\rho(\boldsymbol{\beta}, \widetilde{\boldsymbol{\beta}}) = \mathrm{tr}[\mathrm{MSE}(\boldsymbol{\beta}, \widetilde{\boldsymbol{\beta}})]$ as risk functions.

Theorem 2.11. *Under the linear regression model with assumptions (i) to (iv), let $\boldsymbol{A}\boldsymbol{y} \in \mathcal{U}(\boldsymbol{\beta})$. Then*

$$\mathrm{MSE}(\boldsymbol{\beta}, \boldsymbol{A}\boldsymbol{y}) = \mathrm{MSE}(\boldsymbol{\beta}, \widehat{\boldsymbol{\beta}}) \qquad \forall (\boldsymbol{\beta}, \sigma^2) \in \mathrm{I\!R}^p \times (0, \infty)$$

is satisfied if and only if $\boldsymbol{A} = (\boldsymbol{X}'\boldsymbol{X})^{-1}\boldsymbol{X}'$ holds true.

Proof. From Lemma 2.1 we have $\mathrm{MSE}(\boldsymbol{\beta}, \boldsymbol{A}\boldsymbol{y}) = \mathrm{MSE}(\boldsymbol{\beta}, \widehat{\boldsymbol{\beta}})$ for all $(\boldsymbol{\beta}, \sigma^2) \in \mathrm{I\!R}^p \times (0, \infty)$ if and only if $\mathrm{Cov}(\boldsymbol{A}\boldsymbol{y}) = \mathrm{Cov}(\widehat{\boldsymbol{\beta}})$ for all $\sigma^2 > 0$. Since $\boldsymbol{A}\boldsymbol{y} \in \mathcal{U}(\boldsymbol{\beta})$, there exists some matrix $\boldsymbol{B}$ with $\boldsymbol{B}\boldsymbol{X} = \mathbf{0}$ such that $\boldsymbol{A}\boldsymbol{y} = \widehat{\boldsymbol{\beta}} + \boldsymbol{B}\boldsymbol{y}$ and $\mathrm{Cov}(\boldsymbol{A}\boldsymbol{y}) = \mathrm{Cov}(\widehat{\boldsymbol{\beta}}) + \mathrm{Cov}(\boldsymbol{B}\boldsymbol{y})$. Hence, the identity $\mathrm{Cov}(\boldsymbol{A}\boldsymbol{y}) = \mathrm{Cov}(\widehat{\boldsymbol{\beta}})$ can hold only if $\mathrm{Cov}(\boldsymbol{B}\boldsymbol{y}) = \mathbf{0}$, implying that $\boldsymbol{B} = \mathbf{0}$. This shows the assertion. □

Similarly to the above theorem we can conclude that for $\boldsymbol{A}\boldsymbol{y} \in \mathcal{U}(\boldsymbol{\beta})$ and $\rho(\boldsymbol{\beta}, \boldsymbol{A}\boldsymbol{y}) = \mathrm{tr}[\mathrm{MSE}(\boldsymbol{\beta}, \boldsymbol{A}\boldsymbol{y})]$, the identity

$$\rho(\boldsymbol{\beta}, \boldsymbol{A}\boldsymbol{y}) = \rho(\boldsymbol{\beta}, \widehat{\boldsymbol{\beta}}) \qquad \forall (\boldsymbol{\beta}, \sigma^2) \in \mathrm{I\!R}^p \times (0, \infty)$$

is satisfied if and only if $\boldsymbol{A} = (\boldsymbol{X}'\boldsymbol{X})^{-1}\boldsymbol{X}'$ holds true, see Problem 2.14.

Theorem 2.12. *Under the linear regression model with assumptions (i) to (iv), the least squares estimator $\widehat{\boldsymbol{\beta}}$ is the only admissible estimator in $\mathcal{U}(\boldsymbol{\beta})$ with respect to the real-valued unweighted squared error loss or with respect to the matrix-valued squared error loss.*

2.4 Unreliability of Least Squares Estimation

In the previous section we have shown that $\widehat{\boldsymbol{\beta}}$ is uniformly better than any other linear unbiased estimator for $\boldsymbol{\beta}$. Thus $\widehat{\boldsymbol{\beta}}$ is *optimal* within this set of estimators. This optimality, however, does not mean that the ordinary least squares estimator, as a whole, is necessarily a good estimator. In fact, there exist situations in which the ordinary least squares estimator can be *unreliable* in the sense that its expected loss is quite large. Of course, the expected loss of any other linear unbiased estimator is even greater than the loss of the least squares estimator, but nonetheless estimation with least squares in such situations can take the risk of an individual estimate $\widehat{\boldsymbol{\beta}}$ being far away from the true parameter $\boldsymbol{\beta}$.

2.4.1 Estimation of the Covariance Matrix

As can be seen from Theorem 2.5, the covariance matrix of the ordinary least squares estimator is given by

$$\mathrm{Cov}(\widehat{\boldsymbol{\beta}}) = \sigma^2(\boldsymbol{X}'\boldsymbol{X})^{-1} .$$

In order to obtain some impression of the preciseness of $\widehat{\boldsymbol{\beta}}$, the individual estimates $\widehat{\beta}_i$ are usually computed along with their estimated standard deviations $\sqrt{\widehat{\mathrm{Var}}(\widehat{\beta}_i)}$. The latter are simply obtained as the square roots of the main diagonal elements of $\widehat{\mathrm{Cov}}(\widehat{\boldsymbol{\beta}})$, where $\widehat{\mathrm{Cov}}(\widehat{\boldsymbol{\beta}})$ is obtained from $\mathrm{Cov}(\widehat{\boldsymbol{\beta}})$ by replacing σ^2 by the least squares variance estimator $\widehat{\sigma}^2$. Hence,

$$\sqrt{\widehat{\mathrm{Var}}(\widehat{\beta}_i)} = \widehat{\sigma}\, \boldsymbol{e}_i'(\boldsymbol{X}'\boldsymbol{X})^{-1}\boldsymbol{e}_i ,$$

where $\widehat{\sigma} = \sqrt{\widehat{\sigma}^2}$ and $\boldsymbol{e}_i$ is the i-th $p \times 1$ unit vector. The order of magnitude of estimated standard deviations is comparable to the order of magnitude of the estimated parameters.

Quite often, the estimated standard deviation $\sqrt{\widehat{\mathrm{Var}}(\widehat{\beta}_i)}$ is simply called the *standard error* of $\widehat{\beta}_i$. Since the standard error is an estimator, it is also exposed to possible impreciseness.

When the actual estimate of σ^2 is much too small/great compared to the true σ^2, then also the standard error may give a wrong impression about the preciseness of ordinary least squares.

Example 2.5. Consider the linear regression model $\boldsymbol{y} = \boldsymbol{X}\boldsymbol{\beta} + \boldsymbol{\varepsilon}$ with

$$\boldsymbol{y} = \begin{pmatrix} 5.4621 \\ 6.4557 \\ 6.2638 \\ 6.0826 \\ 5.0569 \end{pmatrix}, \quad \boldsymbol{X} = \begin{pmatrix} 1 & 1.9 \\ 1 & 2.1 \\ 1 & 2 \\ 1 & 2 \\ 1 & 1.8 \end{pmatrix}, \quad \boldsymbol{\beta} = \begin{pmatrix} \beta_1 \\ \beta_2 \end{pmatrix}.$$

The ordinary least squares estimates for β_1 and β_2 together with the estimated standard errors are given in Table 2.4.

Table 2.4. Ordinary least squares estimates and standard errors in Example 2.5

parameter	estimate	standard error
β_1	-3.9674	1.2415
β_2	5.0161	0.6326

Since the data is simulated, the true values of $\boldsymbol{\beta}$ and σ^2 are known, being $\boldsymbol{\beta} = (2, 2)'$ and $\sigma^2 = 1$. Hence, as can also be seen from a comparison with Table 2.5, the ordinary least squares estimates look much more reliable than they actually are, here. The reason for this is, that the actual estimate $\widehat{\sigma}^2 = 0.0208$ is considerably smaller than the true $\sigma^2 = 1$. □

Table 2.5. True parameter values and true standard deviations of the corresponding individual ordinary least squares estimators in Example 2.5

parameter	value	standard deviation ($\sigma^2 = 1$)
β_1	2	8.6068
β_2	2	4.3853

Although, especially when n is not small, the actual estimate $\widehat{\sigma}^2$ is of course not always as different from the true σ^2 as in the above example, the standard errors do not in any case give a proper impression about the preciseness of the individual ordinary least squares estimators. In addition, the estimator $\widehat{\sigma} = \sqrt{\widehat{\sigma}^2}$ is not an unbiased estimator for σ. As a matter of fact, since the function $-\sqrt{\cdot}$ is strictly convex on $(0, \infty)$, it follows from the well-known Jensen inequality that

$$\mathrm{E}[\sqrt{\widehat{\sigma}^2}] < \sqrt{\mathrm{E}(\widehat{\sigma}^2)},$$

i.e. $\mathrm{E}(\widehat{\sigma}) < \sigma$. See also Problem 2.21 for an explicit formula for $\mathrm{E}(\widehat{\sigma})$ under normality. Thus, $\widehat{\sigma}$ *underestimates* the true σ on average and hence the standard error of $\widehat{\beta}_i$ underestimates the true standard deviation of $\widehat{\beta}_i$ on average.

(In general, however, the order of magnitude of underestimation will only be relevant for rather small n.)

2.4.2 Unbiased Versus Biased Estimation

The ordinary least squares estimator can be called to imprecise or unreliable, when it has a large expected loss. Then, individual ordinary least squares estimates might be far away from the true parameter vector, while this is not necessarily the case for an estimate obtained from a biased estimator.

Example 2.6. Consider the linear regression model $\boldsymbol{y} = \boldsymbol{X}\boldsymbol{\beta} + \boldsymbol{\varepsilon}$ with

$$\boldsymbol{y} = \begin{pmatrix} 6.0521 \\ 7.0280 \\ 7.1230 \\ 4.4441 \\ 5.0813 \end{pmatrix}, \quad \boldsymbol{X} = \begin{pmatrix} 1 & 1.9 \\ 1 & 2.1 \\ 1 & 2 \\ 1 & 2 \\ 1 & 1.8 \end{pmatrix}, \quad \boldsymbol{\beta} = \begin{pmatrix} \beta_1 \\ \beta_2 \end{pmatrix} .$$

This is the same model as in Example 2.5 but with a different observation vector $\boldsymbol{y}$. The least squares estimate $\boldsymbol{\beta}$ computes to

$$\widehat{\boldsymbol{\beta}} = (\boldsymbol{X}'\boldsymbol{X})^{-1}\boldsymbol{X}'\boldsymbol{y} = \begin{pmatrix} -4.2489 \\ 5.2013 \end{pmatrix} ,$$

while the ridge estimate $\widehat{\boldsymbol{\beta}}_k = (\boldsymbol{X}'\boldsymbol{X} + k\boldsymbol{I}_2)^{-1}\boldsymbol{X}'\boldsymbol{y}$ for the choice $k = 1/5$ is

$$\widehat{\boldsymbol{\beta}}_{1/5} = \begin{pmatrix} 0.9396 \\ 2.5349 \end{pmatrix} .$$

Since the data is simulated, we know the true parameter vector $\boldsymbol{\beta}$ given as $\boldsymbol{\beta} = (2, 2)'$. Then the estimate obtained from the least squares estimator is much farther away from the true parameter than the estimate obtained from the ridge estimator, see also Fig. 2.4. This is similar to a result in Sect. 1.1.4 for a different (stochastic!) choice of k. The appropriate choice of k is the critical factor in ridge estimation. We will discuss this point in Sect. 3.4.3. Note that the ridge estimator with some fixed $k > 0$ is biased for $\boldsymbol{\beta}$, see Problem 2.15. □

As demonstrated by the above example, unbiasedness of an estimator does not automatically imply an individual estimate being close to the true parameter. We can even have a biased estimator which produces a nearer estimate. This can also carry over to the case of more than one computed estimate (when more than one observation $\boldsymbol{y}$ is obtainable). Thus, an unbiased estimator can be quite imprecise compared to a biased estimator $\widetilde{\boldsymbol{\beta}}$, provided the bias $\mathrm{E}(\widetilde{\boldsymbol{\beta}}) - \boldsymbol{\beta}$ is small enough. The following example illustrates such a situation.

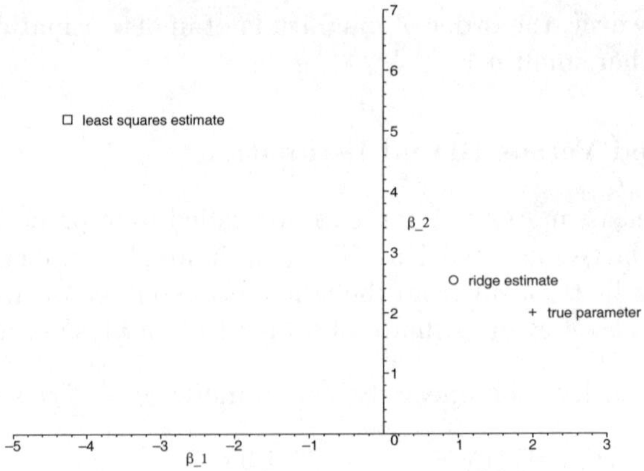

Fig. 2.4. Two estimates and the true parameter vector in Example 2.6

Example 2.7. Suppose we want to estimate the unknown expectation vector $\boldsymbol{\mu} = (\mu_1, \mu_2)' \in \mathbb{R}^2$ of a bivariate normally distributed random vector $\boldsymbol{z} \sim N(\boldsymbol{\mu}, \sigma^2 \boldsymbol{I}_2)$. We can either use the unbiased estimator $\widehat{\boldsymbol{\mu}} = \boldsymbol{z}$ for $\boldsymbol{\mu}$ (i.e. the observed realization of $\boldsymbol{z}$ is itself the considered estimate), or we can use the shrinkage estimator $\widetilde{\boldsymbol{\mu}} = \frac{1}{3}\widehat{\boldsymbol{\mu}}$ (i.e. each element of $\widehat{\boldsymbol{\mu}}$ is decreased by multiplication with 1/3). Now suppose that we are able to obtain not only one but 30 realizations of $\boldsymbol{z}$. Then we can compute 30 estimates $\widehat{\boldsymbol{\mu}}$ of $\boldsymbol{\mu}$ as well as 30 estimates $\widetilde{\boldsymbol{\mu}}$. Fig. 2.5 and 2.6 show the respective estimates.

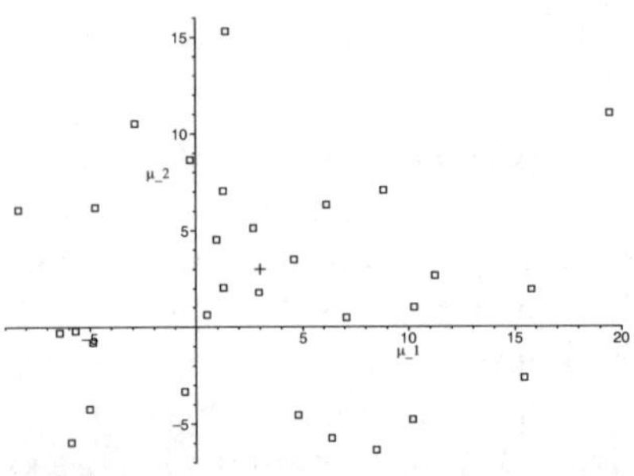

Fig. 2.5. True parameter (cross) and 30 estimates $\widehat{\boldsymbol{\mu}}$ (boxes) in Example 2.7

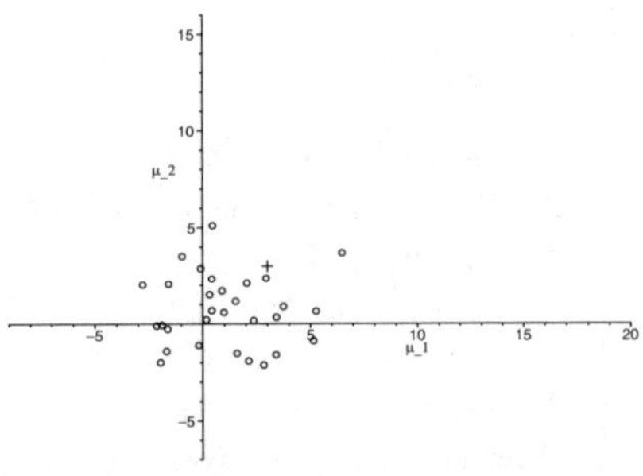

Fig. 2.6. True parameter (cross) and 30 estimates $\widetilde{\boldsymbol{\mu}}$ (circles) in Example 2.7

It can be seen that the estimates $\widehat{\boldsymbol{\mu}}$ scatter around the true parameter vector

$$\boldsymbol{\mu} = \begin{pmatrix} 3 \\ 3 \end{pmatrix}$$

but several individual estimates $\widehat{\boldsymbol{\mu}}$ are quite far away from the true $\boldsymbol{\mu}$. (The true σ^2 is $\sigma^2 = 36$.) On the other hand, the estimates $\widetilde{\boldsymbol{\mu}}$ do not scatter around the true $\boldsymbol{\mu}$ but somewhere nearby, and the area generated by all estimates is distinctly smaller than the corresponding area produced by the estimates $\widehat{\boldsymbol{\mu}}$. In this situation the biased estimator appears to be the more appropriate choice, since a large distance between estimate and parameter is less likely to occur from the biased estimator than from the unbiased estimator. □

The factor 1/3 is used in Example 2.7 to shrink $\widehat{\boldsymbol{\mu}}$. This appropriate choice has been motivated by knowledge of the true values $\boldsymbol{\mu}$ and σ^2. Clearly, in practice these parameters are not known, but to the contrary the unknown objectives of interest. Hence it is desired to have a method for obtaining an appropriate shrinkage factor without knowledge of the parameters. Similarly, in Example 2.6 it is of interest to have a method for obtaining an appropriate ridge parameter k. Here we recognize the critical point in connection with biased estimation of parameters.

Remark 2.3. Often it can be expected that there exist biased estimators which yield smaller expected distances from the true parameters than unbiased estimators. Difficulty in application, however, arises from the correct identification of these estimators.

The difficulty in identifying an appropriate biased estimator should not cause complete refusal of biased estimators. In certain situations, for example when collinearity is present, there exist applicable methods which can yield better estimates.

2.4.3 Collinearity

It is known that the least squares estimator $\widehat{\boldsymbol{\beta}} = (\boldsymbol{X}'\boldsymbol{X})^{-1}\boldsymbol{X}'\boldsymbol{y}$ will not be a precise estimator when *near* linear dependencies between some columns of the matrix $\boldsymbol{X}$ exist. This is the case e.g. in Example 2.6, where the second column of $\boldsymbol{X}$ equals nearly twice its first column.

Usually one refers to such a case as an incidence of *collinearity* or *multicollinearity*, but often without accurately specifying these terms (what exactly does 'near linear dependency' mean?).

What makes collinearity a particular problem is the behavior of the inverse of the matrix $\boldsymbol{X}'\boldsymbol{X}$. If collinearity is present, then small relative changes in the matrix $\boldsymbol{X}'\boldsymbol{X}$ will produce large relative changes in the matrix $(\boldsymbol{X}'\boldsymbol{X})^{-1}$. Moreover, at least some main diagonal elements will be quite large, which in view of $\mathrm{Cov}(\widehat{\boldsymbol{\beta}}) = \sigma^2(\boldsymbol{X}'\boldsymbol{X})^{-1}$ means that some elements of the least squares estimator will have a large variance.

Condition

Based on the above considerations it is natural to identify collinearity with the *condition* of a matrix. A nonsingular matrix $\boldsymbol{A}$ is called *ill-conditioned* if small relative changes in $\boldsymbol{A}$ can produce large relative changes in $\boldsymbol{A}^{-1}$. As a measure for the degree of ill-condition one can use a *condition number* $\kappa(\boldsymbol{A}) = ||\boldsymbol{A}||||\boldsymbol{A}^{-1}||$, where $||\cdot||$ is a given matrix norm. The greater $\kappa(\boldsymbol{A})$, the more the ill-condition of $\boldsymbol{A}$. If we use the *spectral norm*, then the corresponding measure of conditioning of $\boldsymbol{A}$ is the number

$$\kappa(\boldsymbol{A}) = \sqrt{\frac{\lambda_{\max}(\boldsymbol{A}'\boldsymbol{A})}{\lambda_{\min}(\boldsymbol{A}'\boldsymbol{A})}} ,$$

where $\kappa(\boldsymbol{A}) \in [1, \infty)$. If $\kappa(\boldsymbol{A}) = 1$, then the matrix $\boldsymbol{A}$ is called *perfectly conditioned.* We now consider this condition number in the light of three examples, see also [10, p. 103] and [78, p. 466].

- Let $\boldsymbol{A} = \begin{pmatrix} 1 & \alpha \\ \alpha & 1 \end{pmatrix}$ for $0 \leq \alpha < 1$. In case $\alpha \to 1$ the matrix $\boldsymbol{A}$ converges to the singular matrix $\begin{pmatrix} 1 & 1 \\ 1 & 1 \end{pmatrix}$, while in case $\alpha = 0$, the matrix $\boldsymbol{A}$ equals the perfectly conditioned identity matrix. If we consider the condition number

$$\kappa(\boldsymbol{A}) = \frac{1+\alpha}{1-\alpha}$$

  of this matrix, then we can see that the condition number equals 1 in case $\alpha = 0$, while in case $\alpha \to 1$ the condition number becomes very large. As expected, the matrix $\boldsymbol{A}$ is badly conditioned when α is close to 1.
- Let $\boldsymbol{A} = \begin{pmatrix} \alpha & 0 \\ 0 & \alpha \end{pmatrix}$ for $\alpha > 0$. For arbitrary small α, a relative small change in $\boldsymbol{A}$ produces a corresponding relative small change in $\boldsymbol{A}^{-1}$. Hence the matrix $\boldsymbol{A}$ is perfectly conditioned, which is also reflected in the condition number

$$\kappa(\boldsymbol{A}) = \frac{\alpha}{\alpha} = 1 .$$

- Let $\boldsymbol{A} = \begin{pmatrix} \alpha & 0 \\ 0 & 1/\alpha \end{pmatrix}$ for $\alpha \geq 1$. The larger α, the more becomes $\boldsymbol{A}$ a singular matrix. If we consider the condition number

$$\kappa(\boldsymbol{A}) = \alpha^2 ,$$

  then the matrix $\boldsymbol{A}$ is perfectly conditioned for $\alpha = 1$, while the conditioning becomes worse for higher values α.

The concept of condition number for a square matrix naturally carries over to rectangular matrices. Referring to the condition number $\kappa(\boldsymbol{X})$ of the matrix $\boldsymbol{X}$ in the sequel, we mean

$$\kappa(\boldsymbol{X}) = \sqrt{\frac{\lambda_{\max}(\boldsymbol{X}'\boldsymbol{X})}{\lambda_{\min}(\boldsymbol{X}'\boldsymbol{X})}} ,$$

taking its values in $[1, \infty)$.

Consequences of Collinearity

As noted before, collinearity can cause unreliability of the least squares estimator with respect to different aspects.

Sensitivity in Computation. Small relative changes in $\boldsymbol{X}'\boldsymbol{X}$ can produce large relative changes in $(\boldsymbol{X}'\boldsymbol{X})^{-1}$.

Example 2.8. In Example 2.6 the matrix $\boldsymbol{X}'\boldsymbol{X}$ and its inverse are given as

$$\boldsymbol{X}'\boldsymbol{X} = \begin{pmatrix} 5 & 9.8 \\ 9.8 & 19.26 \end{pmatrix} \quad \text{and} \quad (\boldsymbol{X}'\boldsymbol{X})^{-1} = \begin{pmatrix} 74.0769 & -37.6923 \\ -37.6923 & 19.2308 \end{pmatrix} .$$

Suppose now that instead of $\boldsymbol{X}$ from Example 2.6, the matrix $\boldsymbol{X}$ is given as

$$\boldsymbol{X}' = \begin{pmatrix} 1 & 1 & 1 & 1 & 1 \\ 1.9 & 2.05 & 2 & 2 & 1.85 \end{pmatrix} .$$

Then

$$\boldsymbol{X}'\boldsymbol{X} = \begin{pmatrix} 5 & 9.8 \\ 9.8 & 19.235 \end{pmatrix} \quad \text{and} \quad (\boldsymbol{X}'\boldsymbol{X})^{-1} = \begin{pmatrix} 142.4815 & -72.5926 \\ -72.5926 & 37.0370 \end{pmatrix} ,$$

showing that a small relative change in $\boldsymbol{X}'\boldsymbol{X}$ produces a large relative change in $(\boldsymbol{X}'\boldsymbol{X})^{-1}$. This carries over to ordinary least squares estimation here. The new estimate $\widehat{\boldsymbol{\beta}} = (-6.6226, 6.4124)'$ is quite different from the former estimate $\widehat{\boldsymbol{\beta}} = (-4.2489, 5.2013)'$ in Example 2.6. □

Note that the difference in the inverses of $\boldsymbol{X}'\boldsymbol{X}$ is *not* caused by some inaccuracy in computing the inverse. On the contrary, we are able to precisely compute the inverse of $\boldsymbol{X}'\boldsymbol{X}$ in both cases. However, the small difference between both matrices yields a large relative difference between the corresponding inverses.

As we have seen, when strong collinearity is present, the least squares estimator may not be robust with respect to small changes in values of independent variables. Since usually in regression analysis these values are measurements exposed to measurement errors, this non-robustness is an undesired property.

Statistical Unreliability. If collinearity is present, then one or more main diagonal elements of the matrix $(\boldsymbol{X}'\boldsymbol{X})^{-1}$ are quite large. In this case the corresponding element of the least squares estimator has a large variance and thus the estimate is unreliable.

Example 2.9. In Example 2.6, the matrix $(\boldsymbol{X}'\boldsymbol{X})^{-1}$ is given as

$$(\boldsymbol{X}'\boldsymbol{X})^{-1} = \begin{pmatrix} 74.0769 & -37.6923 \\ -37.6923 & 19.2308 \end{pmatrix} .$$

Since we know that $\sigma^2 = 1$, we have $\sqrt{\mathrm{Var}(\widehat{\beta}_1)} = 8.6068$, showing that large deviations of the estimates $\widehat{\beta}_1$ from the true parameter $\beta_1 = 2$ are likely to occur. □

A further characteristic feature of collinearity is that usually some of the elements of the least squares estimator are highly correlated. Then the corresponding elements of the parameter vector may be considered as weakly identifiable.

Example 2.10. In Example 2.6, from the elements of $\mathrm{Cov}(\widehat{\boldsymbol{\beta}}) = \sigma^2(\boldsymbol{X}'\boldsymbol{X})^{-1}$ we can compute the correlation coefficient

$$\mathrm{Corr}(\widehat{\beta}_1, \widehat{\beta}_2) = \frac{\mathrm{Cov}(\widehat{\beta}_1, \widehat{\beta}_2)}{\sqrt{\mathrm{Var}(\widehat{\beta}_1)}\sqrt{\mathrm{Var}(\widehat{\beta}_2)}} = \frac{-37.6923}{\sqrt{74.0769}\sqrt{19.2308}} = -0.9986\,,$$

i.e. the estimators $\widehat{\beta}_1$ and $\widehat{\beta}_2$ are highly negative correlated. □

The statistical implications of collinearity may also be graphically illustrated in Example 2.6. To this end, we generate 30 realizations of $\boldsymbol{y}$ and compute the corresponding estimates of $\boldsymbol{\beta}$.

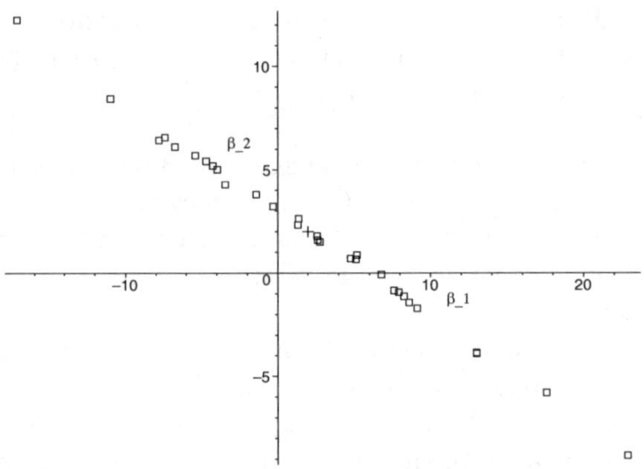

Fig. 2.7. True parameter (cross) and 30 ordinary least squares estimates (boxes) in Example 2.6

Fig. 2.7 shows the 30 computed estimates. Obviously a good amount of the estimates is far away from the true parameter $\boldsymbol{\beta} = (2, 2)'$, underlining the statistical unreliability of $\widehat{\boldsymbol{\beta}}$ in this situation. Moreover, the high negative correlation between $\widehat{\beta}_1$ and $\widehat{\beta}_2$ becomes evident from the fact that the estimates almost lie on a straight line.

Length of Estimation. If in Fig. 2.7 we examine the distances between the estimates and the origin of the coordinate system, we can conclude that almost all least squares estimates have greater length than the true parameter vector. From Theorem 2.6 the expected squared length of the least squares estimator is

$$\mathrm{E}\left(\|\widehat{\boldsymbol{\beta}}\|^2\right) = \sigma^2 \operatorname{tr}[(\boldsymbol{X}'\boldsymbol{X})^{-1}] + \|\boldsymbol{\beta}\|^2 .$$

In presence of collinearity some eigenvalues of $\boldsymbol{X}'\boldsymbol{X}$ are quite small, and hence some eigenvalues of $(\boldsymbol{X}'\boldsymbol{X})^{-1}$ are quite large. Therefore $\operatorname{tr}[(\boldsymbol{X}'\boldsymbol{X})^{-1}]$ will also be large, showing that an individual estimate $\widehat{\boldsymbol{\beta}}$ will often have a distinctly greater length than the true parameter vector $\boldsymbol{\beta}$.

Better Estimates. Negative consequences of collinearity are to be expected when estimation of the whole parameter vector $\boldsymbol{\beta}$ is considered. However, when specific linear combinations $\boldsymbol{\lambda}'\boldsymbol{\beta}$ for some given $p \times 1$ vector $\boldsymbol{\lambda}$ are of interest, then collinearity can even have a positive effect, as shown by the following example, see also [44, Sect. 23.4].

Example 2.11. Consider the linear regression model $\boldsymbol{y} = \boldsymbol{X}\boldsymbol{\beta} + \boldsymbol{\varepsilon}$ with assumptions (i) to (iv) and $p = 2$ such that

$$\boldsymbol{X}'\boldsymbol{X} = \begin{pmatrix} 1 & r \\ r & 1 \end{pmatrix}, \quad -1 \leq r \leq 1 .$$

For $r = 0$ no collinearity is present, while this is not the case for $r = 0.9$. From the representation

$$\operatorname{Cov}(\widehat{\boldsymbol{\beta}}) = \frac{\sigma^2}{1 - r^2} \begin{pmatrix} 1 & -r \\ -r & 1 \end{pmatrix}$$

it may easily be concluded that $\operatorname{Var}(\widehat{\beta}_1) = \operatorname{Var}(\widehat{\beta}_2) = \sigma^2$ in case $r = 0$, while $\operatorname{Var}(\widehat{\beta}_1) = \operatorname{Var}(\widehat{\beta}_2) = \sigma^2 = \sigma^2/(1 - 0.9^2) = 5.2632\sigma^2$ in case $r = 0.9$. Thus the estimators $\widehat{\beta}_1$ and $\widehat{\beta}_2$ are less precise in presence of collinearity.

Suppose now that we are not interested in estimation of the individual parameters β_1 and β_2 but in the linear combination $\beta_1 + \beta_2$. The estimator $\widehat{\beta}_1 + \widehat{\beta}_2$ has variance

$$\operatorname{Var}(\widehat{\beta}_1 + \widehat{\beta}_2) = \operatorname{Var}(\widehat{\beta}_1) + \operatorname{Var}(\widehat{\beta}_2) + 2\operatorname{Cov}(\widehat{\beta}_1, \widehat{\beta}_2) ,$$

which amounts to $\operatorname{Var}(\widehat{\beta}_1 + \widehat{\beta}_2) = 2\sigma^2$ in case $r = 0$, and to $\operatorname{Var}(\widehat{\beta}_1 + \widehat{\beta}_2) = 1.0527\sigma^2$ in case $r = 0.9$. Hence, the estimator $\widehat{\beta}_1 + \widehat{\beta}_2$ is more precise in presence of collinearity than in complete absence. □

Micronumerosity. Goldberger [44, Sect. 23] criticizes that sometimes the effect of collinearity is dramatized by authors who write about this topic. He compares the situation of collinearity with a situation when one is faced with estimation of a univariate population mean and the sample size n is only small. In order to find an 'exotic polysyllabic name' for 'small sample size' he introduces the term *micronumerosity.* The presence of micronumerosity has similar implications on the preciseness of usual estimators as collinearity has on the preciseness on the ordinary least squares estimator, but usually one does not complain tediously about the former.

Although this is true, one may acknowledge that micronumerosity can of course be a motive for considering alternative estimators. In Sect. 1.2.4 we have already encountered such a situation, see the discussion following Example 1.8. If $Y_1, \dots, Y_n$ are independently Bernoulli distributed with parameter $p \in [0,1]$ (tossing a coin), then in case of a small n one might want to consider the estimator

$$\frac{n\overline{Y} + \sqrt{n/4}}{n + \sqrt{n}}$$

rather than the usual estimator $\overline{Y}$ for the population mean p, which can be justified by the risk performance of the two estimators. For small n the former estimator has a smaller unweighted squared error risk than the latter for a wide range of parameter values p. Nonetheless it is also true that

- collinearity can be, but must not be the reason for large variances of elements of $\widehat{\boldsymbol{\beta}}$;
- collinearity can be, but must not be a problem. When the true σ^2 is quite small, then ordinary least squares estimates will turn out to be precise enough. Moreover, when we are interested in estimation of parametric functions $\boldsymbol{\lambda}'\boldsymbol{\beta}$ rather than in estimation of the whole parameter vector $\boldsymbol{\beta}$, then there can occur situations in which $\boldsymbol{\lambda}'\widehat{\boldsymbol{\beta}}$ has a rather small variance, while some or even all individual $\widehat{\beta}_i$ have rather large variances.

Critical Values

The condition number $\kappa(\boldsymbol{X})$ takes its values in the interval $[1, \infty)$. The greater $\kappa(\boldsymbol{X})$, the worse the condition of $\boldsymbol{X}$. It is, however, not possible to give values of general validity which characterize bad condition of a matrix.

As some critical values, Belsley, Kuh and Welsch [10, Chap. 3] consider condition numbers between 5 and 10 as an indication for weak collinearity, while condition numbers greater than 30 are seen to indicate strong collinearity.

For this, it is suggested that not the original condition number $\kappa(\boldsymbol{X})$ is applied but the so-called *scaled condition number*

$$\widetilde{\kappa}(\boldsymbol{X}) = \kappa(\boldsymbol{X}\boldsymbol{D}) ,$$

where $\boldsymbol{D}$ is a $p \times p$ diagonal matrix whose j-th main diagonal element d_j is given by

$$d_j = \frac{1}{\sqrt{\boldsymbol{e}_j' \boldsymbol{X}' \boldsymbol{X} \boldsymbol{e}_j}} ,$$

where $\boldsymbol{e}_j$ is the j-th $p \times 1$ unit vector. See also [11, Sect. 3.3] for a discussion on scaled condition numbers.

We can use the scaled condition number to judge the degree of collinearity in our previous examples, see Table 2.6.

Table 2.6. Scaled condition numbers in previous examples

Example	$\widetilde{\kappa}(\boldsymbol{X})$	collinearity
Sect. 1.1.4	8.5144	weak (to medium)
2.1	4.1728	weak
2.6	38.4648	strong

Determinant

A square matrix $\boldsymbol{A}$ is singular if and only if $\det(\boldsymbol{A}) = 0$ is satisfied. Based on this fact it is conceivable that the determinant could be an appropriate number for assessing how far $\boldsymbol{A}$ is close to singularity. This is, however, not the case in general. Consider for example the matrix

$$\boldsymbol{A} = \begin{pmatrix} \alpha & 0 \\ 0 & 1/\alpha \end{pmatrix}$$

for some $\alpha > 1$. Then clearly $\det(\boldsymbol{A}) = 1$ for every possible α, but for large values of α the matrix $\boldsymbol{A}$ is close to singularity. Moreover, it is seen that small values of $\det(\boldsymbol{A})$ do not imply that $\boldsymbol{A}$ is close to singularity, consider e.g.

$$\boldsymbol{A} = \begin{pmatrix} \alpha & 0 \\ 0 & \alpha \end{pmatrix}$$

for $\alpha > 0$. Even for very small values of α the matrix $\boldsymbol{A}$ is perfectly conditioned while $\det(\boldsymbol{A}) = \alpha^2$ becomes a very small number for small α. Compare also [78, p. 466] for a discussion on determinants.

In the literature, sometimes the formula

$$(\boldsymbol{X}'\boldsymbol{X})^{-1} = \frac{1}{\det(\boldsymbol{X}'\boldsymbol{X})} \operatorname{adj}(\boldsymbol{X}'\boldsymbol{X}) ,$$

where $\operatorname{adj}(\boldsymbol{X}'\boldsymbol{X})$ denotes the *adjugate* (also called *adjoint*) of $\boldsymbol{X}'\boldsymbol{X}$, is used as a justification for the reasoning that if $\det(\boldsymbol{X}'\boldsymbol{X})$ is small, then the main diagonal elements of

$$\operatorname{Cov}(\widehat{\boldsymbol{\beta}}) = \sigma^2(\boldsymbol{X}'\boldsymbol{X})^{-1}$$

can be large and, as a consequence, the least squares estimator can be unreliable. It should be noted, however, that emphasis in this reasoning is on 'can be large', since the main diagonal elements of the adjugate can of course annul this effect. As a matter of fact, in presence of collinearity we often observe that only some but not all main diagonal elements of $\sigma^2(\boldsymbol{X}'\boldsymbol{X})^{-1}$ are relatively large.

In addition, it should be noted that the above formula cannot be recruited as a cause for numerical problems in case that $\det(\boldsymbol{X}'\boldsymbol{X})$ is small (in contrast to what is sometimes stated in the literature), since simply the formula is just inappropriate for numerical computation of the inverse of $\boldsymbol{X}'\boldsymbol{X}$, see [78, p. 480].

Complete Collinearity

If $\det(\boldsymbol{X}'\boldsymbol{X}) = 0$, then we have *complete (exact) collinearity*. In this case the matrix $\boldsymbol{X}'\boldsymbol{X}$ is singular and thus $\boldsymbol{X}$ is not of full column rank. Complete collinearity is therefore equivalent to violation of model assumption (ii). In this case we cannot find a linear unbiased estimator for the vector $\boldsymbol{\beta}$, so that the set $\mathcal{U}(\boldsymbol{\beta})$ is empty, see Problem 2.20.

In regression analysis complete collinearity is quite rare, mainly because the values of the matrix $\boldsymbol{X}$ are results of measurements exposed to measurement error and thus an *exact* linear dependency between the variables cannot be expected.

Diagnose and Remedy of Collinearity

The scaled condition number $\widetilde{\kappa}(\boldsymbol{X})$ is a helpful tool for indicating collinearity. In the literature a number of further tools and diagnostic methods are known, see also Sect. 6.3.

To remedy collinearity, different procedures are conceivable. One possibility is to build a new model and consider a different set of variables (if reasonable) which will not produce the same magnitude of collinearity as the actual ones (but see also Remark 6.5, p. 308). Another way of remedy is to obtain additional sets of observations (if possible) or to incorporate further information about the parameters (if available) into the model. One may also consider the estimation of specific linear combinations $\boldsymbol{\lambda}'\boldsymbol{\beta}$ rather than estimation of the complete parameter vector $\boldsymbol{\beta}$.

In the following we will not pursue such methods, but we will tackle the problem of collinearity via the point estimation approach, i.e. we will consider alternative estimators such as ridge or principal components estimator, which can produce better results than the ordinary least squares estimator in presence of collinearity.

2.4.4 Consistency

As we have already seen, unbiasedness of an estimator for $\boldsymbol{\beta}$ does not guarantee that an actual estimate must be close to the true parameter vector. Even more, there can be situations under which it is *likely* that an actual estimate of an unbiased estimator realizes quite far away from the true $\boldsymbol{\beta}$. The idea behind the principle of unbiasedness is, that given a great number of vectors $(y_1, \ldots, y_n)'_1, \ldots, (y_1, \ldots, y_n)'_r$, the mean of the corresponding estimates will almost coincide with the true parameter vector. In practice we usually cannot take advantage of this fact, since we will only have *one* observed vector $(y_1, \ldots, y_n)'$. In many cases it would simply not be possible to obtain a number of realizations $\boldsymbol{y}$ for the same matrix $\boldsymbol{X}$.

On the other hand, it is sometimes possible to increase the sample size n, which, however, will also require an additional effort (time and cost). For this procedure it is necessary to obtain additional values of the independent variables (increase the number of rows of $\boldsymbol{X}$) and to obtain corresponding additional values of the dependent variable (increase the number of elements of the vector $\boldsymbol{y}$). It is then hoped that from additional data some benefit in precision of estimators follows.

In view of this, we are interested in the question whether our estimates will come nearer to the true parameter vector when the sample size n is increased. This is the question whether the estimator is *consistent* for the true parameter vector.

Definition 2.3. *An estimator $\widetilde{\boldsymbol{\beta}}$ is called* consistent *for $\boldsymbol{\beta}$, if* $\operatorname{plim}_{n\to\infty} \widetilde{\boldsymbol{\beta}} = \boldsymbol{\beta}$ *holds true. The symbol* plim *denotes convergence in probability.*

Remark 2.4. A series $Y_1, \ldots, Y_n, \ldots$ of random variables converges in probability to a fixed c, if

$$\lim_{n\to\infty} \mathrm{P}[|Y_n - c| > \varepsilon] = 0$$

for every $\varepsilon > 0$. This concept carries over to the case that Y_n denotes a random vector and c denotes a fixed vector. Then Y_n converges in probability to c, if each element of Y_n converges in probability to the corresponding element of c.

Remark 2.5. Let $\widetilde{\mu}_1, \ldots, \widetilde{\mu}_n, \ldots$ denote a series of estimators for some parameter $\mu \in \mathrm{I\!R}$. If $\mathrm{E}(\widetilde{\mu}_n) = \mu$ for every n and $\lim_{n\to\infty} \mathrm{Var}(\widetilde{\mu}_n) = 0$, then $\widetilde{\mu}_n$ is consistent for μ. The converse of this statement is not true in general.

Theorem 2.13. *Under the linear regression model with assumptions (i) to (iv), the following statements hold true:*

(i) *If* $\lim_{n\to\infty} \frac{1}{n}\boldsymbol{X}'\boldsymbol{X} = \boldsymbol{Q}$, *where $\boldsymbol{Q}$ is symmetric positive definite, then $\widehat{\boldsymbol{\beta}}$ is consistent for $\boldsymbol{\beta}$.*
(ii) *If assumption (v) is satisfied, then $\widehat{\sigma}^2$ is consistent for σ^2.*

Proof. Under the assumptions (i) to (iv) we have $\mathrm{E}(\widehat{\boldsymbol{\beta}}) = \boldsymbol{\beta}$ for every $\boldsymbol{\beta} \in \mathbb{R}^p$ and every positive integer n. Moreover,

$$\lim_{n\to\infty} \mathrm{Cov}(\widehat{\boldsymbol{\beta}}) = \lim_{n\to\infty} \sigma^2(\boldsymbol{X}'\boldsymbol{X})^{-1} = \lim_{n\to\infty} \frac{\sigma^2}{n}(\frac{1}{n}\boldsymbol{X}'\boldsymbol{X})^{-1} = 0\boldsymbol{Q}^{-1} = \boldsymbol{0}.$$

Consequently each element of $\widehat{\boldsymbol{\beta}}$ is unbiased for the corresponding element of $\boldsymbol{\beta}$ and has a variance which converges to 0 for $n \to \infty$. This shows the consistency of $\widehat{\boldsymbol{\beta}}$. Furthermore, $\widehat{\sigma}^2$ is unbiased for σ^2 and has variance $2\sigma^4/(n-p)$, provided assumption (v) is satisfied, see Corollary 2.1. Since $2\sigma^4/(n-p)$ converges to 0 for $n \to \infty$, consistency of $\widehat{\sigma}^2$ for σ^2 follows. □

If $\widehat{\boldsymbol{\beta}}$ is consistent for $\boldsymbol{\beta}$, then we can conclude that the least squares estimator $\widehat{\boldsymbol{\beta}}$ becomes a more precise estimator for the parameter vector $\boldsymbol{\beta}$ when the sample size is increased. In Sect. 1.2.4 we have already noted a similar behavior. In Example 1.3 (tossing a coin) it becomes evident that increasing the number n of tosses yields a uniformly smaller risk of the decision rule $\delta(\boldsymbol{Y}) = \frac{1}{n}\sum_{i=1}^{n} Y_i$, see p. 16.

Note that the condition $\lim_{n\to\infty} \frac{1}{n}\boldsymbol{X}'\boldsymbol{X} = \boldsymbol{Q}$ is sufficient but not necessary for consistency of $\widehat{\boldsymbol{\beta}}$. A simple example where the condition is violated is provided by a linear regression model with $\boldsymbol{X}' = (1, 2, \ldots, n)$. Nonetheless, under this model the least squares estimator $\widehat{\boldsymbol{\beta}}$ is consistent for $\boldsymbol{\beta}$, see also [113, p. 362ff].

Consistency and Unbiasedness

The following two diagrams illustrate the two concepts of unbiasedness and consistency.

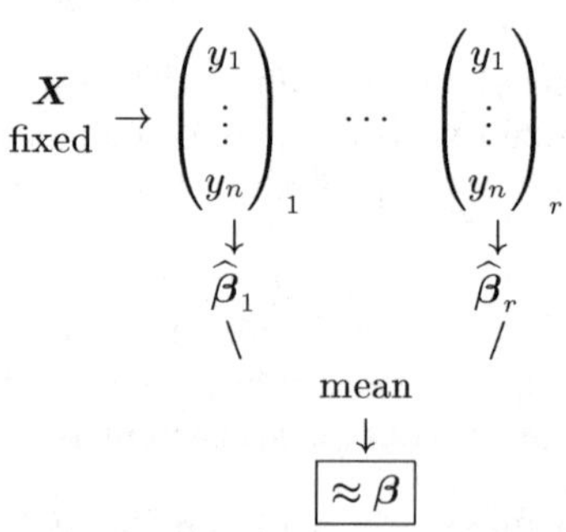

Fig. 2.8. Unbiasedness of the least squares estimator

If we compute the mean of r least squares estimates based on r sample vectors $\boldsymbol{y}$ given a fixed matrix $\boldsymbol{X}$, then we will approximately obtain the true parameter vector (the larger r, the more precise the mean), see Fig. 2.8. This corresponds to computation of the expectation of $\widehat{\boldsymbol{\beta}}$, yielding exactly the true parameter vector $\boldsymbol{\beta}$.

If we increase the number of observations of independent and dependent variables, then the least squares estimates will more and more approach the true parameter vector (the more the increase, the nearer the estimate), see Fig. 2.9. This corresponds to computation of the limit (in probability) of $\widehat{\boldsymbol{\beta}}$ for $n \to \infty$, which is exactly the true parameter vector $\boldsymbol{\beta}$.

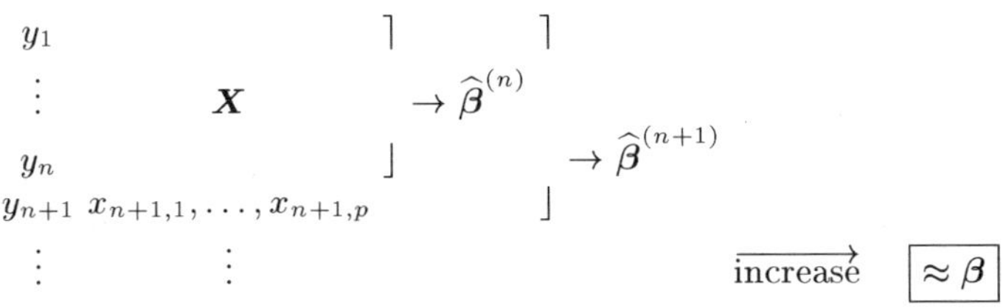

Fig. 2.9. Consistency of the least squares estimator

Note that the least squares estimator $\widehat{\boldsymbol{\beta}}$ is unbiased as well as consistent for $\boldsymbol{\beta}$ (at least under a certain condition). In general, these two properties must not come along together. Consistent estimators are not necessarily unbiased and unbiased estimators are not necessarily consistent, see Problem 2.22.

In the following we focus on the comparison of *different* estimators of $\boldsymbol{\beta}$ for a fixed sample size n. Therefore we will not consider consistency of estimators, which, however, should not imply that consistency is not a meaningful property.

2.4.5 Biased Estimation

If one or more main diagonal elements of $\text{Cov}(\widehat{\boldsymbol{\beta}}) = \sigma^2(\boldsymbol{X}'\boldsymbol{X})^{-1}$ are considerably large, i.e. $\widehat{\boldsymbol{\beta}}$ is imprecise, then by Lemma 2.1 the main diagonal elements of $\text{Cov}(\boldsymbol{Ay})$ are even larger (or at least not smaller) for any estimator $\boldsymbol{Ay} \in \mathcal{U}(\boldsymbol{\beta})$. Thus, linear unbiased estimators for $\boldsymbol{\beta}$ can by no means be more precise than the least squares estimator $\widehat{\boldsymbol{\beta}}$.

The ridge estimator

$$\widehat{\boldsymbol{\beta}}_k = (\boldsymbol{X}'\boldsymbol{X} + k\boldsymbol{I}_p)^{-1}\boldsymbol{X}'\boldsymbol{y} ,$$

which is biased for $k > 0$, can be seen as an example for an estimator which has been developed to ease the effect of collinearity. Here, however, occurs the problem of the appropriate choice of k, which is discussed in Sect. 3.4.3

If the ridge parameter k is chosen non-stochastically, then $\widehat{\boldsymbol{\beta}}_k$ is a homogeneous linear biased estimator of $\boldsymbol{\beta}$. In this case the ridge estimator belongs to the set

$$\mathcal{L}(\boldsymbol{\beta}) = \{\boldsymbol{A}\boldsymbol{y} : \boldsymbol{A} \in \mathbb{R}^{p \times n}\}\,.$$

A further example for an estimator from this set is the so-called *shrinkage estimator*

$$\widehat{\boldsymbol{\beta}}(\varrho) = \frac{1}{1+\varrho}\widehat{\boldsymbol{\beta}}, \qquad \varrho > 0\,,$$

which *shrinks* the elements of the least squares estimator. Here we encounter the problem of the appropriate choice of the shrinkage parameter ϱ, which is discussed in Sect. 3.5.3.

These two examples show that it is reasonable to consider not only estimators from $\mathcal{U}(\boldsymbol{\beta})$ but also from the superset $\mathcal{L}(\boldsymbol{\beta})$. Chapters 3 and 4 deal with estimators and admissibility of estimators from the set $\Delta = \mathcal{L}(\boldsymbol{\beta})$.

Linear estimators have the advantage that their properties (like expectation vector and covariance matrix) are more or less easy to determine. In practice, however, linear estimators are only the initials for more advanced estimation procedures. (The ordinary least squares estimator seems to be the only reasonable linear estimator which does not require additional effort. This is presumably one of the reasons for its popularity.) As we will see for example, the required ridge and shrinkage parameters are usually determined on the basis of the observation $\boldsymbol{y}$, thus making ridge and shrinkage estimator non-linear. Hence the set of *all* possible estimators $\{\widetilde{\boldsymbol{\beta}}(\boldsymbol{y})\}$ cannot be dismissed. The estimators

$$\widehat{\boldsymbol{\beta}}_{\widehat{k}} = (\boldsymbol{X}'\boldsymbol{X} + \widehat{k}\boldsymbol{I}_p)^{-1}\boldsymbol{X}'\boldsymbol{y}, \quad \widehat{k} = \frac{p\widehat{\sigma}^2}{\widehat{\boldsymbol{\beta}}'\widehat{\boldsymbol{\beta}}}\,,$$

$$\widehat{\boldsymbol{\beta}}(\widehat{\varrho}) = \frac{1}{1+\widehat{\varrho}}\widehat{\boldsymbol{\beta}}, \quad \widehat{\varrho} = \frac{\widehat{\sigma}^2\,\mathrm{tr}[(\boldsymbol{X}'\boldsymbol{X})^{-1}]}{\widehat{\boldsymbol{\beta}}'\widehat{\boldsymbol{\beta}}}\,,$$

and

$$\widehat{\boldsymbol{\beta}}_{\mathrm{S}} = \left[1 - \frac{(p-2)(n-p)}{n-p+2}\,\frac{\widehat{\sigma}^2}{\widehat{\boldsymbol{\beta}}'\boldsymbol{X}'\boldsymbol{X}\widehat{\boldsymbol{\beta}}}\right]\widehat{\boldsymbol{\beta}}$$

from Sect. 1.1.4 are examples for non-linear biased estimators of $\boldsymbol{\beta}$, not belonging to $\mathcal{L}(\boldsymbol{\beta})$ but to $\{\widetilde{\boldsymbol{\beta}}(\boldsymbol{y})\}$.

In contrast to the ridge estimator, the Stein estimator $\widehat{\boldsymbol{\beta}}_{\mathrm{S}}$ has *not* been created to solve the problem of unreliability of the ordinary least squares estimator, but to demonstrate that the ordinary least squares estimator can uniformly be outperformed. If in addition to model assumptions (i) to (iv) also condition (v) is imposed, as it is the case in Sect. 2.5, then the possibility for a uniformly better estimator is only in the set of biased estimators.

Remark 2.6. The use of alternatives to the ordinary least squares estimator can be beneficial when the least squares estimator is unreliable or can uniformly be outperformed. Such alternatives will be biased estimators for $\boldsymbol{\beta}$.

2.5 Inadmissibility of the Ordinary Least Squares Estimator

No matter if the least squares estimator is unreliable or not, it can be shown that under certain conditions $\widehat{\boldsymbol{\beta}}$ is inadmissible with respect to the risk function

$$\rho(\boldsymbol{\beta}, \widetilde{\boldsymbol{\beta}}) = \mathrm{E}\left[(\widetilde{\boldsymbol{\beta}} - \boldsymbol{\beta})'(\widetilde{\boldsymbol{\beta}} - \boldsymbol{\beta})\right] .$$

This is achieved by demonstrating that the Stein estimator $\widehat{\boldsymbol{\beta}}_{\mathrm{S}}$ is a uniformly better estimator. The aforementioned 'certain conditions' are the additional model assumptions (v) to (vii). As it turns out, inadmissibility remains true even if condition (vii), i.e. $\boldsymbol{X}'\boldsymbol{X} = \boldsymbol{I}_p$, is violated, but in that case the Stein estimator is uniformly better than the least squares estimator only with respect to the weighted risk function

$$\rho(\boldsymbol{\beta}, \widetilde{\boldsymbol{\beta}}) = \mathrm{E}\left[(\widetilde{\boldsymbol{\beta}} - \boldsymbol{\beta})'\boldsymbol{X}'\boldsymbol{X}(\widetilde{\boldsymbol{\beta}} - \boldsymbol{\beta})\right] .$$

On the other hand, condition (vii) can always be achieved by some reparameterization of the original regression model.

2.5.1 The Reparameterized Regression Model

Let $(\boldsymbol{X}'\boldsymbol{X})^{1/2}$ be the uniquely determined symmetric positive definite matrix such that

$$(\boldsymbol{X}'\boldsymbol{X})^{1/2}(\boldsymbol{X}'\boldsymbol{X})^{1/2} = \boldsymbol{X}'\boldsymbol{X} ,$$

and let $(\boldsymbol{X}'\boldsymbol{X})^{-1/2} = [(\boldsymbol{X}'\boldsymbol{X})^{1/2}]^{-1}$. If we define

$$\boldsymbol{\vartheta} := (\boldsymbol{X}'\boldsymbol{X})^{1/2}\boldsymbol{\beta} \quad \text{and} \quad \boldsymbol{Z} := \boldsymbol{X}(\boldsymbol{X}'\boldsymbol{X})^{-1/2} ,$$

then the linear regression model can be written in the form

$$\boldsymbol{y} = \boldsymbol{X}\boldsymbol{\beta} + \boldsymbol{\varepsilon} = \boldsymbol{Z}\boldsymbol{\vartheta} + \boldsymbol{\varepsilon} .$$

Notation 2.4. *The model described by the equation $\boldsymbol{y} = \boldsymbol{Z}\boldsymbol{\vartheta} + \boldsymbol{\varepsilon}$ is called the* reparameterized model *corresponding to the original regression model $\boldsymbol{y} = \boldsymbol{X}\boldsymbol{\beta} + \boldsymbol{\varepsilon}$.*

Remark 2.7. The following statements hold true:

- If in the original model $\boldsymbol{y} = \boldsymbol{X}\boldsymbol{\beta} + \boldsymbol{\varepsilon}$ the model assumptions (i) to (iv) are satisfied, then in the reparameterized model the conditions (i) to (iv) and in addition (vii) hold true for $\boldsymbol{\vartheta}$ instead of $\boldsymbol{\beta}$ and $\boldsymbol{Z}$ instead of $\boldsymbol{X}$.
- The least squares estimator in the reparameterized model is given as

$$\widehat{\boldsymbol{\vartheta}} = (\boldsymbol{Z}'\boldsymbol{Z})^{-1}\boldsymbol{Z}'\boldsymbol{y} = \boldsymbol{Z}'\boldsymbol{y} .$$

- From the identity

$$\boldsymbol{P} = \boldsymbol{X}(\boldsymbol{X}'\boldsymbol{X})^{-1}\boldsymbol{X}' = \boldsymbol{Z}\boldsymbol{Z}' = \boldsymbol{Z}(\boldsymbol{Z}'\boldsymbol{Z})^{-1}\boldsymbol{Z}'$$

  it follows that

$$\boldsymbol{X}\widehat{\boldsymbol{\beta}} = \widehat{\boldsymbol{y}} = \boldsymbol{Z}\widehat{\boldsymbol{\vartheta}} \quad \text{as well as} \quad \boldsymbol{y} - \boldsymbol{X}\widehat{\boldsymbol{\beta}} = \widehat{\boldsymbol{\varepsilon}} = \boldsymbol{y} - \boldsymbol{Z}\widehat{\boldsymbol{\vartheta}}\,.$$

- The least squares variance estimator $\widehat{\sigma}^2$ in the original regression model is identical to the least squares variance estimator in the reparameterized model, since in both models we have

$$\widehat{\sigma}^2 = \frac{1}{n-p}\boldsymbol{y}'(\boldsymbol{I}_n - \boldsymbol{P})\boldsymbol{y} = \frac{1}{n-p}\widehat{\boldsymbol{\varepsilon}}'\widehat{\boldsymbol{\varepsilon}}\,.$$

- The coefficient of determination R^2 in the original model is identical to the coefficient of determination in the reparameterized model, since in both models $R^2 = ||\boldsymbol{P}\boldsymbol{y}||/||\boldsymbol{y}||$.

The Stein estimator for $\boldsymbol{\beta}$ in the original regression model is

$$\widehat{\boldsymbol{\beta}}_{\mathrm{S}} = \gamma\widehat{\boldsymbol{\beta}}, \quad \gamma = 1 - \frac{(p-2)(n-p)}{n-p+2}\frac{\widehat{\sigma}^2}{\widehat{\boldsymbol{\beta}}'\boldsymbol{X}'\boldsymbol{X}\widehat{\boldsymbol{\beta}}}\,.$$

Thus, the Stein estimator for $\boldsymbol{\vartheta}$ in the reparameterized model is given as

$$\widehat{\boldsymbol{\vartheta}}_{\mathrm{S}} = \gamma\widehat{\boldsymbol{\vartheta}}\,,$$

where the factor γ is identical in both models due to Remark 2.7.

Theorem 2.14. *Under the linear regression model with assumptions (i) to (iv), we have* $\widehat{\boldsymbol{\beta}} = (\boldsymbol{X}'\boldsymbol{X})^{-1/2}\widehat{\boldsymbol{\vartheta}}$ *and* $\widehat{\boldsymbol{\beta}}_{\mathrm{S}} = (\boldsymbol{X}'\boldsymbol{X})^{-1/2}\widehat{\boldsymbol{\vartheta}}_{\mathrm{S}}$.

If, in the reparameterized model, we estimate the vector $\boldsymbol{\vartheta}$ by the estimator $\widetilde{\boldsymbol{\vartheta}}$, then we can estimate the parameter vector $\boldsymbol{\beta}$ by $\widetilde{\boldsymbol{\beta}} = (\boldsymbol{X}'\boldsymbol{X})^{-1/2}\widetilde{\boldsymbol{\vartheta}}$, where the use of the unweighted squared error loss function $L(\boldsymbol{\vartheta},\widetilde{\boldsymbol{\vartheta}}) = (\widetilde{\boldsymbol{\vartheta}} - \boldsymbol{\vartheta})'(\widetilde{\boldsymbol{\vartheta}} - \boldsymbol{\vartheta})$ in the reparameterized model corresponds to the use of the weighted squared error loss function $L(\boldsymbol{\beta},\widetilde{\boldsymbol{\beta}}) = (\widetilde{\boldsymbol{\beta}} - \boldsymbol{\beta})'\boldsymbol{W}(\widetilde{\boldsymbol{\beta}} - \boldsymbol{\beta})$ with weight matrix $\boldsymbol{W} = \boldsymbol{X}'\boldsymbol{X}$ in the corresponding original regression model. This fact is expressed in the following theorem.

Theorem 2.15. *Under the linear regression model with assumptions (i) to (iv), the two identities*

$$L(\boldsymbol{\vartheta},\widetilde{\boldsymbol{\vartheta}}) = (\widetilde{\boldsymbol{\vartheta}} - \boldsymbol{\vartheta})'(\widetilde{\boldsymbol{\vartheta}} - \boldsymbol{\vartheta}) = (\widetilde{\boldsymbol{\beta}} - \boldsymbol{\beta})'\boldsymbol{X}'\boldsymbol{X}(\widetilde{\boldsymbol{\beta}} - \boldsymbol{\beta})$$

and

$$L(\boldsymbol{\beta},\widetilde{\boldsymbol{\beta}}) = (\widetilde{\boldsymbol{\beta}} - \boldsymbol{\beta})'(\widetilde{\boldsymbol{\beta}} - \boldsymbol{\beta}) = (\widetilde{\boldsymbol{\vartheta}} - \boldsymbol{\vartheta})'(\boldsymbol{X}'\boldsymbol{X})^{-1}(\widetilde{\boldsymbol{\vartheta}} - \boldsymbol{\vartheta}).$$

hold true for $\widetilde{\boldsymbol{\beta}} = (\boldsymbol{X}'\boldsymbol{X})^{-1/2}\widetilde{\boldsymbol{\vartheta}}$.

Remark 2.8. By the above described reparameterization of the linear regression model it can always be achieved that model assumption (vii) is satisfied in the reparameterized model. The applied loss function of an estimator for the reparameterized parameter vector, however, does not coincide with the same kind of loss function of the *corresponding* estimator for the original parameter in the regression model.

A reparameterization of the regression model is especially useful when we wish to use a weighted squared error loss function

$$L(\boldsymbol{\beta}, \widetilde{\boldsymbol{\beta}}) = (\widetilde{\boldsymbol{\beta}} - \boldsymbol{\beta})' \boldsymbol{X}' \boldsymbol{X} (\widetilde{\boldsymbol{\beta}} - \boldsymbol{\beta})$$

in the original model. In this case, the necessary derivations and computations are easier to obtain under the reparameterized model with an unweighted squared error loss.

2.5.2 Risk Comparison of Least Squares and Stein Estimator

We can show that the Stein estimator $\widehat{\boldsymbol{\vartheta}}_{\mathrm{S}}$ for $\boldsymbol{\vartheta}$ is uniformly better than the ordinary least squares estimator $\widehat{\boldsymbol{\vartheta}}$ with respect to

$$\rho(\boldsymbol{\vartheta}, \widetilde{\boldsymbol{\vartheta}}) = \mathrm{E}\left[(\widetilde{\boldsymbol{\vartheta}} - \boldsymbol{\vartheta})'(\widetilde{\boldsymbol{\vartheta}} - \boldsymbol{\vartheta})\right]$$

when considering a linear regression model with model assumptions (i) to (vi), so that the corresponding reparameterized models satisfies assumptions (i) to (vii).

The Risks of Both Estimators

To start with, we determine the risk of the ordinary least squares estimator $\widehat{\boldsymbol{\vartheta}}$ for $\boldsymbol{\vartheta}$ under the reparameterized model.

Theorem 2.16. *Under the linear regression model with assumptions (i) to (iv), for $\widehat{\boldsymbol{\vartheta}} = \boldsymbol{Z}'\boldsymbol{y}$ the following identity holds true:*

$$\rho(\boldsymbol{\vartheta}, \widehat{\boldsymbol{\vartheta}}) = \mathrm{E}\left[(\widehat{\boldsymbol{\vartheta}} - \boldsymbol{\vartheta})'(\widehat{\boldsymbol{\vartheta}} - \boldsymbol{\vartheta})\right] = \sigma^2 p\,.$$

Proof. We have $\rho(\boldsymbol{\vartheta}, \widehat{\boldsymbol{\vartheta}}) = \mathrm{tr}[\mathrm{Cov}(\widehat{\boldsymbol{\vartheta}})] = \mathrm{tr}(\sigma^2 \boldsymbol{Z}'\boldsymbol{Z}) = \sigma^2\, \mathrm{tr}(\boldsymbol{I}_p) = \sigma^2 p.$ □

The derivation of the risk of $\widehat{\boldsymbol{\vartheta}}_{\mathrm{S}}$ is considerably more complicated. Before we can state the desired identity, we give the following lemma and corollary. Note that we will also use the symbol $\chi^2_{(m,\lambda)}$ to denote a $\chi^2_{(m,\lambda)}$ distributed random variable.

Lemma 2.2. *Under the linear regression model with assumptions (i) to (iv) and (v), let $s(\boldsymbol{y}) = \boldsymbol{y}'\boldsymbol{M}\boldsymbol{y}$, where $\boldsymbol{M} = \boldsymbol{I}_n - \boldsymbol{Z}\boldsymbol{Z}'$, and let $\widehat{\boldsymbol{\vartheta}} = \boldsymbol{Z}'\boldsymbol{y}$. Then the following statements hold true:*

(i) $\sigma^{-2}s(\boldsymbol{y}) \sim \chi^2_{(n-p)}$;
(ii) $\sigma^{-2}s(\boldsymbol{y})$ *and* $\widehat{\boldsymbol{\vartheta}}$ *are stochastically independent;*
(iii) $\sigma^{-2}s(\boldsymbol{y})$ *and* $(\widehat{\boldsymbol{\vartheta}}'\widehat{\boldsymbol{\vartheta}})^{-1}\widehat{\boldsymbol{\vartheta}}$ *are stochastically independent;*
(iv) $\sigma^{-4}s(\boldsymbol{y})^2$ *and* $(\widehat{\boldsymbol{\vartheta}}'\widehat{\boldsymbol{\vartheta}})^{-1}\widehat{\boldsymbol{\vartheta}}$ *are stochastically independent;*
(v) $\mathrm{E}\left[(\widehat{\boldsymbol{\vartheta}}'\widehat{\boldsymbol{\vartheta}})^{-1}\widehat{\boldsymbol{\vartheta}}\right] = \sigma^{-2}\boldsymbol{\vartheta}\,\mathrm{E}\left[1/\chi^2_{(p+2,\lambda)}\right]$, *where* $\lambda = \boldsymbol{\vartheta}'\boldsymbol{\vartheta}/2\sigma^2$;
(vi) $\mathrm{E}\left[(\widehat{\boldsymbol{\vartheta}}'\widehat{\boldsymbol{\vartheta}})^{-1}\right] = \sigma^{-2}\,\mathrm{E}\left[1/\chi^2_{(p,\lambda)}\right]$, *where* $\lambda = \boldsymbol{\vartheta}'\boldsymbol{\vartheta}/2\sigma^2$.

Proof. Statements (i) and (ii) immediately follow from Theorem 2.7. Statements (iii) and (iv) follow from (ii), since also functions $f(\sigma^{-2}s(\boldsymbol{y}))$ and $g(\widehat{\boldsymbol{\vartheta}})$ of random variables/vectors are stochastically independent when these functions $f(\cdot)$ and $g(\cdot)$ produce again random variables/vectors, see [79, p. 151]. Statement (v) follows by applying Theorem B.20. Statement (vi) can be deduced from $\sigma^{-2}\widehat{\boldsymbol{\vartheta}}'\widehat{\boldsymbol{\vartheta}} \sim \chi^2_{(p,\lambda)}$, in view of $\sigma^{-1}\widehat{\boldsymbol{\vartheta}} = \sigma^{-1}\boldsymbol{Z}'\boldsymbol{y} \sim N(\sigma^{-1}\boldsymbol{\vartheta}, \boldsymbol{I}_p)$. □

Corollary 2.2. *Under the linear regression model with assumptions (i) to (iv) and (v), let* $s(\boldsymbol{y}) = \boldsymbol{y}'\boldsymbol{M}\boldsymbol{y}$, *where* $\boldsymbol{M} = \boldsymbol{I}_n - \boldsymbol{Z}\boldsymbol{Z}'$, *and let* $\widehat{\boldsymbol{\vartheta}} = \boldsymbol{Z}'\boldsymbol{y}$. *Then the following statements hold true:*

(i) $\mathrm{E}\left[s(\boldsymbol{y})(\widehat{\boldsymbol{\vartheta}}'\widehat{\boldsymbol{\vartheta}})^{-1}\widehat{\boldsymbol{\vartheta}}\right] = (n-p)\boldsymbol{\vartheta}\,\mathrm{E}\left[1/\chi^2_{(p+2,\lambda)}\right]$, *where* $\lambda = \boldsymbol{\vartheta}'\boldsymbol{\vartheta}/2\sigma^2$;
(ii) $\mathrm{E}\left[s(\boldsymbol{y})^2(\widehat{\boldsymbol{\vartheta}}'\widehat{\boldsymbol{\vartheta}})^{-1}\right] = \sigma^2(n-p+2)(n-p)\,\mathrm{E}\left[1/\chi^2_{(p,\lambda)}\right]$, *where* $\lambda = \boldsymbol{\vartheta}'\boldsymbol{\vartheta}/2\sigma^2$.

Proof. Problem 2.28. □

The above statements allow to give the risk of $\widehat{\boldsymbol{\vartheta}}_{\mathrm{S}}$ under the reparameterized model.

Theorem 2.17. *Under the linear regression model with assumptions (i) to (iv) and (v),(vi), for* $\widehat{\boldsymbol{\vartheta}}_{\mathrm{S}}$ *the following identity holds true:*

$$\rho(\boldsymbol{\vartheta}, \widehat{\boldsymbol{\vartheta}}_{\mathrm{S}}) = \mathrm{E}\left[(\widehat{\boldsymbol{\vartheta}}_{\mathrm{S}} - \boldsymbol{\vartheta})'(\widehat{\boldsymbol{\vartheta}}_{\mathrm{S}} - \boldsymbol{\vartheta})\right] = \sigma^2 p - \sigma^2 \frac{(p-2)^2(n-p)}{n-p+2}\,\mathrm{E}\left[\frac{1}{\chi^2_{(p,\lambda)}}\right],$$

where $\lambda = \boldsymbol{\vartheta}'\boldsymbol{\vartheta}/2\sigma^2$.

Proof. Let

$$a := \frac{p-2}{n-p+2}, \quad s(\boldsymbol{y}) := \boldsymbol{y}'\boldsymbol{M}\boldsymbol{y}, \quad \widetilde{\boldsymbol{\vartheta}} = a\, s(\boldsymbol{y})(\widehat{\boldsymbol{\vartheta}}'\widehat{\boldsymbol{\vartheta}})^{-1}\widehat{\boldsymbol{\vartheta}}\,,$$

where $\boldsymbol{M} = \boldsymbol{I}_n - \boldsymbol{Z}\boldsymbol{Z}'$. We can write $\widehat{\boldsymbol{\vartheta}}_{\mathrm{S}} = \widehat{\boldsymbol{\vartheta}} - \widetilde{\boldsymbol{\vartheta}}$, and the risk of $\widehat{\boldsymbol{\vartheta}}_{\mathrm{S}}$ is given by

$$\begin{aligned}
\rho(\boldsymbol{\vartheta},\widehat{\boldsymbol{\vartheta}}_{\mathrm{S}}) &= \mathrm{E}\left[(\widehat{\boldsymbol{\vartheta}}_{\mathrm{S}}-\boldsymbol{\vartheta})'(\widehat{\boldsymbol{\vartheta}}_{\mathrm{S}}-\boldsymbol{\vartheta})\right] \\
&= \mathrm{E}\left[(\widehat{\boldsymbol{\vartheta}}-\widetilde{\boldsymbol{\vartheta}}-\boldsymbol{\vartheta})'(\widehat{\boldsymbol{\vartheta}}-\widetilde{\boldsymbol{\vartheta}}-\boldsymbol{\vartheta})\right] \\
&= \mathrm{E}\left[\left((\widehat{\boldsymbol{\vartheta}}-\boldsymbol{\vartheta})'-\widetilde{\boldsymbol{\vartheta}}'\right)\left((\widehat{\boldsymbol{\vartheta}}-\boldsymbol{\vartheta})-\widetilde{\boldsymbol{\vartheta}}\right)\right] \\
&= \rho\left(\boldsymbol{\vartheta},\widehat{\boldsymbol{\vartheta}}\right)-2\,\mathrm{E}\left[(\widehat{\boldsymbol{\vartheta}}-\boldsymbol{\vartheta})'\widetilde{\boldsymbol{\vartheta}}\right]+\mathrm{E}\left(\widetilde{\boldsymbol{\vartheta}}'\widetilde{\boldsymbol{\vartheta}}\right) \\
&= \underbrace{\rho\left(\boldsymbol{\vartheta},\widehat{\boldsymbol{\vartheta}}\right)}_{=\sigma^2 p}-2\underbrace{\mathrm{E}\left(\widehat{\boldsymbol{\vartheta}}'\widetilde{\boldsymbol{\vartheta}}\right)}_{=:A}+2\underbrace{\mathrm{E}\left(\boldsymbol{\vartheta}'\widetilde{\boldsymbol{\vartheta}}\right)}_{=:B}+\underbrace{\mathrm{E}\left(\widetilde{\boldsymbol{\vartheta}}'\widetilde{\boldsymbol{\vartheta}}\right)}_{=:C}.
\end{aligned}$$

It remains to determine A, B and C. We have

$$A=\mathrm{E}\left(\widehat{\boldsymbol{\vartheta}}'\widetilde{\boldsymbol{\vartheta}}\right)=\mathrm{E}\left[a\,s(\boldsymbol{y})(\widehat{\boldsymbol{\vartheta}}'\widehat{\boldsymbol{\vartheta}})^{-1}\widehat{\boldsymbol{\vartheta}}'\widehat{\boldsymbol{\vartheta}}\right]=\mathrm{E}[as(\boldsymbol{y})]=a\sigma^2\,\mathrm{E}\left[\sigma^{-2}s(\boldsymbol{y})\right]\;,$$

and by Lemma 2.2 (i) it follows

$$A=\sigma^2 a(n-p)\;.$$

For B we can conclude

$$B=\mathrm{E}\left(\boldsymbol{\vartheta}'\widetilde{\boldsymbol{\vartheta}}\right)=\boldsymbol{\vartheta}'\,\mathrm{E}\left[a\,s(\boldsymbol{y})(\widehat{\boldsymbol{\vartheta}}'\widehat{\boldsymbol{\vartheta}})^{-1}\widehat{\boldsymbol{\vartheta}}\right]=a\,\boldsymbol{\vartheta}'\,\mathrm{E}\left[s(\boldsymbol{y})(\widehat{\boldsymbol{\vartheta}}'\widehat{\boldsymbol{\vartheta}})^{-1}\widehat{\boldsymbol{\vartheta}}\right]\;.$$

Using Corollary 2.2 (i), it follows

$$B=a(n-p)\boldsymbol{\vartheta}'\boldsymbol{\vartheta}\,\mathrm{E}\left[\frac{1}{\chi^2_{(p+2,\lambda)}}\right]\;,$$

where $\lambda=\boldsymbol{\vartheta}'\boldsymbol{\vartheta}/2\sigma^2$. Eventually for C we obtain

$$C=\mathrm{E}\left(\widetilde{\boldsymbol{\vartheta}}'\widetilde{\boldsymbol{\vartheta}}\right)=\mathrm{E}\left[(a\,s(\boldsymbol{y})(\widehat{\boldsymbol{\vartheta}}'\widehat{\boldsymbol{\vartheta}})^{-1})^2\,\widehat{\boldsymbol{\vartheta}}'\widehat{\boldsymbol{\vartheta}}\right]=a^2\,\mathrm{E}\left[s(\boldsymbol{y})^2(\widehat{\boldsymbol{\vartheta}}'\widehat{\boldsymbol{\vartheta}})^{-1}\right]\;.$$

Hence, by Corollary 2.2 (ii), it follows

$$C=\sigma^2a^2(n-p+2)(n-p)\,\mathrm{E}[\frac{1}{\chi^2_{(p,\lambda)}}]=\sigma^2\frac{(p-2)^2(n-p)}{n-p+2}\,\mathrm{E}\left[\frac{1}{\chi^2_{(p,\lambda)}}\right]\;.$$

It can be shown that $A=B+C$. For this we apply Corollary B.2 for the case $m=p-2$ and $\lambda=\boldsymbol{\vartheta}'\boldsymbol{\vartheta}/2\sigma^2$, yielding

$$1=\mathrm{E}\left[\frac{1}{\chi^2_{(p,\lambda)}}\right](p-2)+\mathrm{E}\left[\frac{1}{\chi^2_{(p+2,\lambda)}}\right]\sigma^{-2}\boldsymbol{\vartheta}'\boldsymbol{\vartheta}\;.$$

Therefore we can conclude

$$\begin{aligned} A &= \sigma^2 a(n-p) \left[\mathrm{E}\left[\frac{1}{\chi^2_{(p,\lambda)}}\right](p-2) + \mathrm{E}\left[\frac{1}{\chi^2_{(p+2,\lambda)}}\right] \sigma^{-2}\boldsymbol{\vartheta}'\boldsymbol{\vartheta} \right] \\ &= \sigma^2 a(n-p)(p-2)\,\mathrm{E}\left[\frac{1}{\chi^2_{(p,\lambda)}}\right] + B \\ &= \sigma^2 \frac{(p-2)^2(n-p)}{n-p+2}\,\mathrm{E}\left[\frac{1}{\chi^2_{(p,\lambda)}}\right] + B = C + B\,. \end{aligned}$$

By this identity we get

$$\rho(\boldsymbol{\vartheta}, \widehat{\boldsymbol{\vartheta}}_{\mathrm{S}}) = \sigma^2 p - 2A + 2B + C = \sigma^2 p - C\,,$$

which shows the assertion. □

Comparison of Both Risks

Based on the above results we can now easily compare the risks of $\widehat{\boldsymbol{\vartheta}}_{\mathrm{S}}$ and $\widehat{\boldsymbol{\vartheta}}$.

Corollary 2.3. *Consider the linear regression model with assumptions (i) to (iv) and (v),(vi). Then $\widehat{\boldsymbol{\vartheta}}_{\mathrm{S}}$ is uniformly better than $\widehat{\boldsymbol{\vartheta}}$ with respect to $\rho(\boldsymbol{\vartheta}, \widetilde{\boldsymbol{\vartheta}}) = \mathrm{E}\left[(\widetilde{\boldsymbol{\vartheta}} - \boldsymbol{\vartheta})'(\widetilde{\boldsymbol{\vartheta}} - \boldsymbol{\vartheta})\right]$.*

Proof. From Theorems 2.16 and 2.17, the difference between the risks of $\widehat{\boldsymbol{\vartheta}}$ and $\widehat{\boldsymbol{\vartheta}}_{\mathrm{S}}$ can be written as

$$\rho(\boldsymbol{\vartheta}, \widehat{\boldsymbol{\vartheta}}) - \rho(\boldsymbol{\vartheta}, \widehat{\boldsymbol{\vartheta}}_{\mathrm{S}}) = C, \quad C = \sigma^2 \frac{(p-2)^2(n-p)}{n-p+2}\,\mathrm{E}\left[\frac{1}{\chi^2_{(p,\lambda)}}\right].$$

Clearly we have $C \geq 0$ for all $(\boldsymbol{\vartheta}, \sigma^2) \in \mathbb{R}^p \times (0,\infty)$, implying that $\widehat{\boldsymbol{\vartheta}}_{\mathrm{S}}$ is uniformly not worse than $\widehat{\boldsymbol{\vartheta}}$. Hence, the assertion follows if C is strictly greater than zero for at least one $(\boldsymbol{\vartheta}, \sigma^2) \in \mathbb{R}^p \times (0,\infty)$. Setting $\boldsymbol{\vartheta} = \mathbf{0}$ and noting $p \geq 3$, from Theorem B.19 we obtain

$$C = \sigma^2 \frac{(p-2)(n-p)}{n-p+2}\,.$$

This shows that in view of $p \geq 3$ and $n > p$, the number C is strictly greater than zero in case $\boldsymbol{\vartheta} = \mathbf{0}$ (for any $\sigma^2 > 0$), and the assertion follows. □

Figure 2.10 exemplarily shows the risks of $\widehat{\boldsymbol{\vartheta}}$ and $\widehat{\boldsymbol{\vartheta}}_{\mathrm{S}}$ for the case

$$p = 3, \quad n = 12 \quad \text{and} \quad \sigma^2 = 1$$

as functions of $\lambda = \boldsymbol{\vartheta}'\boldsymbol{\vartheta}/2\sigma^2$.

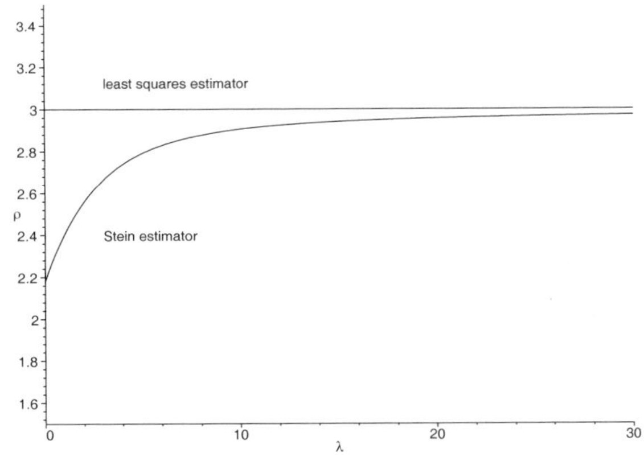

Fig. 2.10. Risks of least squares and Stein estimator

Note that, as indicated by Fig. 2.10, the risk of the Stein estimator $\widehat{\boldsymbol{\vartheta}}_{\mathrm{S}}$ is strictly smaller than the risk of the least squares estimator $\widehat{\boldsymbol{\vartheta}}$ for any finite λ, and thus for any $\boldsymbol{\vartheta}$ such that $||\boldsymbol{\vartheta}||^2 < \infty$. In fact, the difference of risks

$$C = \sigma^2 \frac{(p-2)^2(n-p)}{n-p+2} \mathrm{E}\left[\frac{1}{\chi^2_{(p,\lambda)}}\right]$$

is strictly greater than zero for any $(\boldsymbol{\vartheta}, \sigma^2) \in \mathbb{R}^p \times (0, \infty)$ such that $||\boldsymbol{\vartheta}||^2 < \infty$ and not only for any $(\boldsymbol{\vartheta}, \sigma^2) \in \{0\} \times (0, \infty)$ as noted in the proof of Corollary 2.3.

Remark 2.9. Given the assumptions of Corollary 2.3 and $||\boldsymbol{\vartheta}||^2 < \infty$, the Stein estimator $\widehat{\boldsymbol{\vartheta}}_{\mathrm{S}}$ is uniformly *strictly* better than the least squares estimator $\widehat{\boldsymbol{\vartheta}}$.

Remark 2.10. Under the linear regression model with assumptions (i) to (iv) and (v),(vi), the inequality

$$\rho(\vartheta_i, (\widehat{\vartheta}_{\mathrm{S}})_i) \le \rho(\vartheta_i, \widehat{\vartheta}_i) \quad \forall (\vartheta_i, \sigma^2) \in \mathbb{R} \times (0, \infty)$$

does not necessarily hold for the elements $(\widehat{\vartheta}_{\mathrm{S}})_i$ and $\widehat{\vartheta}_i$ of $\widehat{\boldsymbol{\vartheta}}_{\mathrm{S}}$ and $\widehat{\boldsymbol{\vartheta}}$, the former being estimators for the individual parameters ϑ_i, $i = 1, \ldots, p$. Consequently, it might happen that the i-th element of the Stein estimator as an estimator for ϑ_i can be worse than the corresponding element of the least squares estimator for specific values from the parameter space with respect to $\rho(\vartheta_i, \widetilde{\vartheta}_i) = \mathrm{E}[(\widetilde{\vartheta}_i - \vartheta_i)^2]$, while this cannot be the case for $\widehat{\boldsymbol{\vartheta}}_{\mathrm{S}}$ as an estimator for the whole vector $\boldsymbol{\vartheta}$ with respect to $\rho(\boldsymbol{\vartheta}, \widetilde{\boldsymbol{\vartheta}}) = \mathrm{E}\left[(\widetilde{\boldsymbol{\vartheta}} - \boldsymbol{\vartheta})'(\widetilde{\boldsymbol{\vartheta}} - \boldsymbol{\vartheta})\right]$.

Convergence Behavior

While the risk $\rho(\boldsymbol{\vartheta}, \widehat{\boldsymbol{\vartheta}})$ only depends on the number of regressors p and the parameter σ^2, the risk $\rho(\boldsymbol{\vartheta}, \widehat{\boldsymbol{\vartheta}}_{\mathrm{S}})$ additionally depends on the number of observations n and on the unknown parameter vector $\boldsymbol{\vartheta}$, or rather on the squared length $||\boldsymbol{\vartheta}||^2 = \boldsymbol{\vartheta}'\boldsymbol{\vartheta}$ of the parameter vector.

If we consider the behavior of the risk of the Stein estimator for the two cases $n \to \infty$ and $||\boldsymbol{\vartheta}||^2 \to \infty$, we obtain the following statements.

Theorem 2.18. *Under the linear regression model with assumptions (i) to (vi), the following statements hold true.*

(i) *For any fixed* $\sigma^2 \in (0, \infty)$, *the identity*

$$\lim_{\boldsymbol{\vartheta}'\boldsymbol{\vartheta} \to \infty} \rho(\boldsymbol{\vartheta}, \widehat{\boldsymbol{\vartheta}}_{\mathrm{S}}) = \rho(\boldsymbol{\vartheta}, \widehat{\boldsymbol{\vartheta}})$$

is valid.

(ii) *For any fixed* $(\boldsymbol{\vartheta}, \sigma^2) \in \mathbb{R}^p \times (0, \infty)$, *the identity*

$$\lim_{n \to \infty} \rho(\boldsymbol{\vartheta}, \widehat{\boldsymbol{\vartheta}}_{\mathrm{S}}) = \sigma^2 p - \sigma^2 (p-2)^2 \, \mathrm{E}\left[\frac{1}{\chi^2_{(p,\lambda)}}\right]$$

is valid.

From Theorem 2.18 (i) we can conclude that the improvement of $\widehat{\boldsymbol{\vartheta}}_{\mathrm{S}}$ upon $\widehat{\boldsymbol{\vartheta}}$ is substantial when $||\boldsymbol{\vartheta}||^2 = \boldsymbol{\vartheta}'\boldsymbol{\vartheta}$ is small. In addition, Theorem 2.18 (ii) shows that the risk of $\widehat{\boldsymbol{\vartheta}}_{\mathrm{S}}$ decreases for increasing n, see also Fig 2.11.

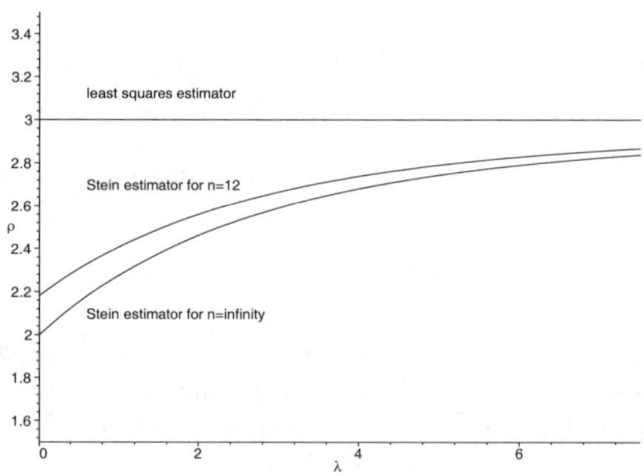

Fig. 2.11. Risks of least squares and Stein estimator

Remark 2.11. The application of the Stein estimator $\widehat{\boldsymbol{\vartheta}}_{\mathrm{S}}$ turns out to be particularly useful when $||\boldsymbol{\vartheta}||^2$ is small and n is large.

2.5.3 An Example for Stein Estimation

Consider the introductory regression example from Sect. 1.1.4 with $n = 12$, $p = 3$, and the observations given in Table 1.2 on page 8. In this model the true parameters are given by

$$(\boldsymbol{\beta}, \sigma^2) = (\begin{pmatrix} 0.2 \\ 0.4 \\ 0.7 \end{pmatrix}, 1) \,.$$

If we reparameterize the model, then the parameter vector $\boldsymbol{\vartheta}$ is given as

$$\boldsymbol{\vartheta} = (\boldsymbol{X}'\boldsymbol{X})^{1/2}\boldsymbol{\beta} = \begin{pmatrix} 2.2496 \\ 6.0110 \\ 4.2567 \end{pmatrix} \,.$$

Then we compute $\lambda = 29.6561$. If we look at Fig. 2.10, then we see that the difference between the risks of $\widehat{\boldsymbol{\vartheta}}$ and $\widehat{\boldsymbol{\vartheta}}_{\mathrm{S}}$ at this value of λ is only small. As a matter of fact we compute $\rho(\boldsymbol{\vartheta}, \widehat{\boldsymbol{\vartheta}}_{\mathrm{S}}) = 2.9860$ and $\rho(\boldsymbol{\vartheta}, \widehat{\boldsymbol{\vartheta}}) = 3$. In this situation, use of $\widehat{\boldsymbol{\vartheta}}_{\mathrm{S}}$ will yield only a slightly smaller average observed loss than use of $\widehat{\boldsymbol{\vartheta}}$ for estimating $\boldsymbol{\vartheta}$. The same will be true for the use of $\widehat{\boldsymbol{\beta}}_{\mathrm{S}}$ compared to the use of $\widehat{\boldsymbol{\beta}}$ for estimating $\boldsymbol{\beta}$ with respect to a weighted squared error loss $L(\boldsymbol{\beta}, \widetilde{\boldsymbol{\beta}}) = (\widetilde{\boldsymbol{\beta}} - \boldsymbol{\beta})'\boldsymbol{X}'\boldsymbol{X}(\widetilde{\boldsymbol{\beta}} - \boldsymbol{\beta})$. In fact, for the true parameters we can compute the observed losses

$$L(\boldsymbol{\beta}, \widehat{\boldsymbol{\beta}}) = (\widehat{\boldsymbol{\beta}} - \boldsymbol{\beta})'\boldsymbol{X}'\boldsymbol{X}(\widehat{\boldsymbol{\beta}} - \boldsymbol{\beta}) = 2.3865$$

and

$$L(\boldsymbol{\beta}, \widehat{\boldsymbol{\beta}}_{\mathrm{S}}) = (\widehat{\boldsymbol{\beta}}_{\mathrm{S}} - \boldsymbol{\beta})'\boldsymbol{X}'\boldsymbol{X}(\widehat{\boldsymbol{\beta}}_{\mathrm{S}} - \boldsymbol{\beta}) = 2.1635 \,.$$

Here, the observed losses behave somehow similar to the expected error losses, which, however, is not always the case. For some different observed $\boldsymbol{y}$ it may even happen that the observed weighted squared error loss of $\widehat{\boldsymbol{\beta}}_{\mathrm{S}}$ is greater than the observed weighted squared error loss of $\widehat{\boldsymbol{\beta}}$.

In Sect. 1.1.4 we have compared estimators for $\boldsymbol{\beta}$ not with respect to the above weighted squared error loss but with respect to the unweighted squared error loss $L(\boldsymbol{\beta}, \widetilde{\boldsymbol{\beta}}) = (\widetilde{\boldsymbol{\beta}} - \boldsymbol{\beta})'(\widetilde{\boldsymbol{\beta}} - \boldsymbol{\beta})$. We have already noted that also the observed unweighted squared error loss of $\widehat{\boldsymbol{\beta}}_{\mathrm{S}}$ is slightly smaller than the observed unweighted squared error loss of $\widehat{\boldsymbol{\beta}}$. We have

$$L(\boldsymbol{\beta}, \widehat{\boldsymbol{\beta}}) = (\widehat{\boldsymbol{\beta}} - \boldsymbol{\beta})'(\widehat{\boldsymbol{\beta}} - \boldsymbol{\beta}) = 1.7794$$

and

$$L(\boldsymbol{\beta}, \widehat{\boldsymbol{\beta}}_{\mathrm{S}}) = (\widehat{\boldsymbol{\beta}}_{\mathrm{S}} - \boldsymbol{\beta})'(\widehat{\boldsymbol{\beta}}_{\mathrm{S}} - \boldsymbol{\beta}) = 1.7306 \,.$$

As noted before, we cannot draw any conclusions about expected losses from individual observed losses. We have no statements about $\rho(\boldsymbol{\beta}, \widehat{\boldsymbol{\beta}}_{\mathrm{S}}) =$

$\mathrm{E}[(\widehat{\boldsymbol{\beta}}_{\mathrm{S}}-\boldsymbol{\beta})'(\widehat{\boldsymbol{\beta}}_{\mathrm{S}}-\boldsymbol{\beta})]$ available. To obtain an impression of the behavior of $\widehat{\boldsymbol{\beta}}_{\mathrm{S}}$ with respect to this risk function, we can generate not only one but 1000 realizations of $\boldsymbol{y}$ and compute for the corresponding 1000 estimates $\widehat{\boldsymbol{\beta}}_{\mathrm{S}}$ the average observed loss.

Table 2.8 gives a summary about the respective unweighted and weighted average observed losses for the true parameters $\boldsymbol{\beta}=(0.2,0.4,0.7)'$ and $\sigma^2=1$.

Table 2.7. Average observed losses

	$\widehat{\boldsymbol{\beta}}$	$\widehat{\boldsymbol{\beta}}_{\mathrm{S}}$
$\boldsymbol{W}=\boldsymbol{X}'\boldsymbol{X}$	3.0158	3.0004
$\boldsymbol{W}=\boldsymbol{I}_p$	1.4662	1.4303

If we compare the average observed losses with the known expected losses, see Table 2.8, then the average observed unweighted loss 1.4303 of $\widehat{\boldsymbol{\beta}}_{\mathrm{S}}$ can obviously be taken as a good 'estimate' for the unknown expected unweighted loss.

As can be seen, the average observed unweighted loss of $\widehat{\boldsymbol{\beta}}_{\mathrm{S}}$ is smaller than the average observed unweighted loss of $\widehat{\boldsymbol{\beta}}$. Of course, the differences between the average losses of $\widehat{\boldsymbol{\beta}}$ and $\widehat{\boldsymbol{\beta}}_{\mathrm{S}}$ are extremely small in this example so that we do not obtain a considerable improvement by using the Stein estimator here.

Table 2.8. Expected losses

	$\widehat{\boldsymbol{\beta}}$	$\widehat{\boldsymbol{\beta}}_{\mathrm{S}}$
$\boldsymbol{W}=\boldsymbol{X}'\boldsymbol{X}$	$\sigma^2 p=3$	$\sigma^2 p-C=2.9860$
$\boldsymbol{W}=\boldsymbol{I}_p$	$\sigma^2\,\mathrm{tr}[(\boldsymbol{X}'\boldsymbol{X})^{-1}]=1.4787$	??

It should also be noted that this example cannot be taken as an evidence for a claim that $\widehat{\boldsymbol{\beta}}_{\mathrm{S}}$ has always a smaller unweighted squared error risk than $\widehat{\boldsymbol{\beta}}$. For this statement (if it is correct), the computation of the exact unweighted squared error risk of $\widehat{\boldsymbol{\beta}}_{\mathrm{S}}$ is necessary.

2.5.4 Admissibility

By comparing the risks of $\widehat{\boldsymbol{\vartheta}}$ and $\widehat{\boldsymbol{\vartheta}}_{\mathrm{S}}$, we have stated in Corollary 2.3 that $\widehat{\boldsymbol{\vartheta}}_{\mathrm{S}}$ is a uniformly better estimator. This shows that $\widehat{\boldsymbol{\vartheta}}$ is inadmissible in the set of all possible point estimators for $\boldsymbol{\vartheta}$.

Remark 2.12. Consider the linear regression model with assumptions (i) to (iv) and (v),(vi). Then $\widehat{\boldsymbol{\vartheta}}$ is inadmissible in the set $\{\widetilde{\boldsymbol{\vartheta}}(\boldsymbol{y})\}$ of all point estimators for $\boldsymbol{\vartheta}$ with respect to the unweighted squared error loss $L(\boldsymbol{\vartheta}, \widetilde{\boldsymbol{\vartheta}}) = (\widetilde{\boldsymbol{\vartheta}} - \boldsymbol{\vartheta})'(\widetilde{\boldsymbol{\vartheta}} - \boldsymbol{\vartheta})$.

Similarly the estimator $\widehat{\boldsymbol{\beta}}$ is inadmissible in the set $\{\widetilde{\boldsymbol{\beta}}(\boldsymbol{y})\}$ of all point estimators for $\boldsymbol{\beta}$ with respect to the weighted squared error loss $L(\boldsymbol{\beta}, \widetilde{\boldsymbol{\beta}}) = (\widetilde{\boldsymbol{\beta}} - \boldsymbol{\beta})'\boldsymbol{X}'\boldsymbol{X}(\widetilde{\boldsymbol{\beta}} - \boldsymbol{\beta})$, if model assumptions (i) to (iv) and (v), (vi) hold true.

Positive Part Stein Estimator

The Stein estimator $\widehat{\boldsymbol{\vartheta}}_{\mathrm{S}}$ uniformly improves upon the ordinary least squares estimator $\widehat{\boldsymbol{\vartheta}}$, implying that $\widehat{\boldsymbol{\vartheta}}$ is inadmissible. But what about admissibility of $\widehat{\boldsymbol{\vartheta}}_{\mathrm{S}}$ itself? As a matter of fact, it can be shown that $\widehat{\boldsymbol{\vartheta}}_{\mathrm{S}}$ is inadmissible, too, in the set $\{\widetilde{\boldsymbol{\vartheta}}(\boldsymbol{y})\}$. Let us write

$$\widehat{\boldsymbol{\vartheta}}_{\mathrm{S}} = \gamma\,\widehat{\boldsymbol{\vartheta}}, \quad \gamma = 1 - \frac{p-2}{n-p+2}\frac{\boldsymbol{y}'\boldsymbol{M}\boldsymbol{y}}{\widehat{\boldsymbol{\vartheta}}'\widehat{\boldsymbol{\vartheta}}}, \qquad \boldsymbol{M} = \boldsymbol{I}_n - \boldsymbol{Z}\boldsymbol{Z}' \,.$$

Then the so-called *positive part Stein estimator* of $\boldsymbol{\vartheta}$ is given by

$$\widehat{\boldsymbol{\vartheta}}_{\mathrm{S}}^{+} = \delta\,\widehat{\boldsymbol{\vartheta}}, \quad \delta = \max\{0, \gamma\} \,.$$

Since $\gamma \le 1$, it follows $0 \le \delta \le 1$. When making use of the coefficient of determination R^2, then we have

$$\widehat{\boldsymbol{\vartheta}}_{\mathrm{S}}^{+} = \begin{cases} \boldsymbol{0} & \text{for } R^2 \le \dfrac{p-2}{n} \\ \widehat{\boldsymbol{\vartheta}}_{\mathrm{S}} & \text{otherwise} \end{cases} ,$$

compare also Problem 2.29. The positive part Stein estimator $\widehat{\boldsymbol{\vartheta}}_{\mathrm{S}}^{+}$ is uniformly better than the Stein estimator with respect to

$$\rho(\boldsymbol{\vartheta}, \widetilde{\boldsymbol{\vartheta}}) = \mathrm{E}\left[(\widetilde{\boldsymbol{\vartheta}} - \boldsymbol{\vartheta})'(\widetilde{\boldsymbol{\vartheta}} - \boldsymbol{\vartheta})\right] ,$$

see [63, Chap. 8.5]. Nonetheless, this estimator is also inadmissible, see [101].

Obviously, similar statements are true for the positive part Stein estimator $\widehat{\boldsymbol{\beta}}_{\mathrm{S}}^{+}$ for $\boldsymbol{\beta}$ with respect to $\rho(\boldsymbol{\beta}, \widetilde{\boldsymbol{\beta}}) = \mathrm{E}\left[(\widetilde{\boldsymbol{\beta}} - \boldsymbol{\beta})'\boldsymbol{X}'\boldsymbol{X}(\widetilde{\boldsymbol{\beta}} - \boldsymbol{\beta})\right]$.

Berger Estimator

From Remark 2.12 the ordinary least squares estimator $\widehat{\boldsymbol{\beta}} = (\boldsymbol{X}'\boldsymbol{X})^{-1}\boldsymbol{X}'\boldsymbol{y}$ is inadmissible in $\{\widetilde{\boldsymbol{\beta}}(\boldsymbol{y})\}$ with respect to the weighted squared error loss $L(\boldsymbol{\beta}, \widetilde{\boldsymbol{\beta}}) = (\widetilde{\boldsymbol{\beta}} - \boldsymbol{\beta})'\boldsymbol{X}'\boldsymbol{X}(\widetilde{\boldsymbol{\beta}} - \boldsymbol{\beta})$, provided model assumptions (i) to (vi) hold true.

A natural question in this context is, whether $\widehat{\beta}$ is also inadmissible with respect to the unweighted squared error loss $L(\beta, \widetilde{\beta}) = (\widetilde{\beta} - \beta)'(\widetilde{\beta} - \beta)$. This is in fact the case. Consider the estimator from Berger [12], given as

$$\widehat{\beta}_{\mathrm{B}} = \left(I_p - \frac{p-2}{n-p+2} \frac{y'My}{\widehat{\beta}'(X'X)^2\widehat{\beta}} X'X \right) \widehat{\beta} ,$$

with $M = I_n - X(X'X)^{-1}X'$, where we assume that model assumptions (i) to (vi) hold true. From the exposition in [63, Chap. 10.3.3], the unweighted squared error risk of $\widehat{\beta}_{\mathrm{B}}$ is given by

$$\rho(\beta, \widehat{\beta}_{\mathrm{B}}) = \sigma^2 \operatorname{tr}[(X'X)^{-1}] - \sigma^4 \frac{(p-2)^2(n-p)}{n-p+2} \mathrm{E}\left[\frac{1}{\widehat{\beta}'(X'X)^2\widehat{\beta}} \right] ,$$

and the Berger estimator $\widehat{\beta}_{\mathrm{B}}$ is therefore uniformly better than the ordinary least squares estimator $\widehat{\beta}$ with respect to the unweighted squared error risk.

Further Estimators

Besides the estimators $\widehat{\beta}_{\mathrm{S}}$, $\widehat{\beta}_{\mathrm{S}}^{+}$, and $\widehat{\beta}_{\mathrm{B}}$ presented in this section, there are further estimators of similar type, some of them described in Sect. 3.5. Refer to [47, 55, 63, 123] for additional explanations, as well as [57] for a survey on Stein estimation.

2.6 Problems

2.1. (See also Theorem 2.1) Show that under the linear regression model with assumptions (i) to (iv), the identity

$$f(\beta_*) = f(\widehat{\beta}) + (\widehat{\beta} - \beta_*)'X'X(\widehat{\beta} - \beta_*)$$

is satisfied for any vector $\beta_* \in \mathbb{R}^p$.

2.2. Consider the linear regression model

$$y_i = \beta_1 + \beta_2 x_i + \varepsilon_i, \qquad i = 1, \ldots, n ,$$

with $n = 8$ observations

i	1	2	3	4	5	6	7	8
y_i	1.764	1.672	1.796	1.548	-0.719	-0.635	0.484	3.133
x_i	15.00	15.89	23.39	6.07	7.63	8.57	14.63	18.79

(a) Plot the pairs (x_i, y_i), $i = 1, \ldots, 8$.

(b) Determine the estimates $\widehat{\beta}_1$ and $\widehat{\beta}_2$ for β_1 and β_2, using the least squares principle, and insert the regression line $\widehat{y} = \widehat{\beta}_1 + \widehat{\beta}_2 x$ into the picture.
(c) The true parameters are $\beta_1 = 0$ and $\beta_2 = 0.1$. Insert the line $y = \beta_1 + \beta_2 x$ into the picture.

2.3. Consider the linear regression model with assumption (i) to (iv) and $\mathbf{1}_n \in \mathcal{C}(\boldsymbol{X})$, where $\mathbf{1}_n$ denotes the $n \times 1$ vector whose elements are all equal to one. (The condition $\mathbf{1}_n \in \mathcal{C}(\boldsymbol{X})$ is always satisfied in a linear regression model with intercept.)

(a) Show that the condition $\mathbf{1}_n \in \mathcal{C}(\boldsymbol{X})$ is equivalent to $\boldsymbol{P}\mathbf{1}_n = \mathbf{1}_n$, where $\boldsymbol{P} = \boldsymbol{X}(\boldsymbol{X}'\boldsymbol{X})^{-1}\boldsymbol{X}'$.
(b) Show that the identities $\sum_{i=1}^n \widehat{y}_i = \sum_{i=1}^n y_i$ and $\sum_{i=1}^n \widehat{\varepsilon}_i = 0$ are satisfied. (Here $\widehat{y}_i$, y_i and $\widehat{\varepsilon}_i$ are the elements of the vectors $\widehat{\boldsymbol{y}} = \boldsymbol{X}\widehat{\boldsymbol{\beta}}$, $\boldsymbol{y}$ and $\widehat{\boldsymbol{\varepsilon}} = \boldsymbol{y} - \widehat{\boldsymbol{y}}$ respectively.)

2.4. Consider the linear regression model with assumptions (i) to (iv) and $\mathbf{1}_n \in \mathcal{C}(\boldsymbol{X})$. Under such a model often the *corrected coefficient of determination*

$$R_{\mathrm{c}}^2 = \frac{\boldsymbol{y}'\boldsymbol{P}\boldsymbol{y} - \frac{1}{n}(\mathbf{1}_n'\boldsymbol{y})^2}{\boldsymbol{y}'\boldsymbol{y} - \frac{1}{n}(\mathbf{1}_n'\boldsymbol{y})^2}$$

is applied instead of the usual coefficient of determination R^2. Show that R_{c}^2 can be written as

$$R_{\mathrm{c}}^2 = \frac{\widehat{\boldsymbol{y}}'\boldsymbol{C}\widehat{\boldsymbol{y}}}{\boldsymbol{y}'\boldsymbol{C}\boldsymbol{y}} ,$$

where $\boldsymbol{C} = \boldsymbol{I}_n - \frac{1}{n}\mathbf{1}_n\mathbf{1}_n'$ (*centering matrix*).

2.5. Consider the situation from Problem 2.2.

(a) Compute the coefficient of determination R^2.
(b) Compute the unweighted squared error loss of the ordinary least squares estimator $\widehat{\boldsymbol{\beta}}$, when $\boldsymbol{\beta} = (0, 0.1)'$ is the true parameter vector.
(c) Compute the unweighted squared error loss of the estimator $\widetilde{\boldsymbol{\beta}} = c\widehat{\boldsymbol{\beta}}$, where

$$c = \frac{(n-p)R^2}{1 + (n-p-1)R^2} ,$$

when $\boldsymbol{\beta} = (0, 0.1)'$ is the true parameter. Compare the losses from (b) and (c).
(d) Insert the line obtained from the estimates $\widetilde{\boldsymbol{\beta}}$ into the picture from Problem 2.2. What can be said about the estimation?

2.6. Prove statements (i) to (iii) in Theorem 2.3.

2.7. Show that under the linear regression model with assumptions (i) to (iv), the identity

$$\widehat{\sigma}^2 = \frac{1}{n-p}\boldsymbol{y}'\boldsymbol{M}\boldsymbol{y}, \quad \boldsymbol{M} = \boldsymbol{I}_n - \boldsymbol{X}(\boldsymbol{X}'\boldsymbol{X})^{-1}\boldsymbol{X}' ,$$

is true.

2.8. Consider the linear regression model with

$$\boldsymbol{y} = \begin{pmatrix} y_1 \\ y_2 \\ y_3 \\ y_4 \end{pmatrix}, \quad \boldsymbol{X} = \begin{pmatrix} 1 & 0 \\ 1 & 0 \\ 0 & 1 \\ 0 & 1 \end{pmatrix}, \quad \boldsymbol{\beta} = \begin{pmatrix} \beta_1 \\ \beta_2 \end{pmatrix}.$$

(a) Determine the least squares estimator $\widehat{\boldsymbol{\beta}}$ of $\boldsymbol{\beta}$ as a function of y_i, $i = 1, \ldots, 4$.
(b) Find a further unbiased estimator for $\boldsymbol{\beta}$.
(c) Consider the estimator

$$\widetilde{\boldsymbol{\beta}} = \frac{1}{2}\begin{pmatrix} y_1 + y_2 + y_1y_3 + y_1y_4 - y_2y_3 - y_2y_4 \\ y_3 + y_4 + y_1y_3 + y_1y_4 - y_2y_3 - y_2y_4 \end{pmatrix}$$

for $\boldsymbol{\beta}$. Explain why $\widetilde{\boldsymbol{\beta}}$ is a non-linear estimator and check if $\widetilde{\boldsymbol{\beta}}$ is unbiased for $\boldsymbol{\beta}$.

2.9. Consider the estimator $c\widehat{\boldsymbol{\beta}}$ from Problem 2.5 (c). Is this a linear estimator for $\boldsymbol{\beta}$?

2.10. Show that under the linear regression model with assumptions (i) to (iv), an estimator $\boldsymbol{A}\boldsymbol{y} + \boldsymbol{a}$ is unbiased for $\boldsymbol{\beta}$ if and only if $\boldsymbol{A}\boldsymbol{X} = \boldsymbol{I}_p$ and $\boldsymbol{a} = \boldsymbol{0}$.

2.11. Consider the linear regression model from Problem 2.8 and the estimator

$$\widetilde{\boldsymbol{\beta}} = \frac{1}{4}\begin{pmatrix} y_1 + y_2 - y_3 - y_4 \\ y_3 + y_4 - y_1 - y_2 \end{pmatrix}$$

for $\boldsymbol{\beta}$. Show that $\widetilde{\boldsymbol{\beta}}$ has a strictly smaller risk than $\widehat{\boldsymbol{\beta}}$ with respect to the real-valued unweighted squared error loss, when $\boldsymbol{\beta} = \boldsymbol{0}$ is the true parameter value. Is this a contradiction to the Gauss-Markov Theorem?

2.12. Consider the linear regression model from Problem 2.8 and the two estimators

$$\widetilde{\boldsymbol{\beta}}_1 = \frac{1}{2}\begin{pmatrix} 2y_1 - y_3 + y_4 \\ -y_1 + y_2 + y_3 + y_4 \end{pmatrix}, \quad \widetilde{\boldsymbol{\beta}}_2 = \frac{1}{2}\begin{pmatrix} 2y_1 \\ -y_1 + y_2 - y_3 + 3y_4 \end{pmatrix}$$

for $\boldsymbol{\beta}$.

(a) Show that $\widetilde{\boldsymbol{\beta}}_1$ is uniformly better than $\widetilde{\boldsymbol{\beta}}_2$ with respect to the real-valued unweighted squared error loss.
(b) Show that a similar statement does not hold with respect to the matrix-valued squared error loss.

2.13. Let $\boldsymbol{T}$ be a symmetric positive definite matrix and let

$$f_{\boldsymbol{T}}(\boldsymbol{\beta}_*) = (\boldsymbol{y} - \boldsymbol{X}\boldsymbol{\beta}_*)'\boldsymbol{T}(\boldsymbol{y} - \boldsymbol{X}\boldsymbol{\beta}_*) \,.$$

(a) Show that

$$\min_{\boldsymbol{\beta}_*} f_{\boldsymbol{T}}(\boldsymbol{\beta}_*) = f_{\boldsymbol{T}}(\widehat{\boldsymbol{\beta}}_{\boldsymbol{T}})$$

holds true for $\widehat{\boldsymbol{\beta}}_{\boldsymbol{T}} = (\boldsymbol{X}'\boldsymbol{T}\boldsymbol{X})^{-1}\boldsymbol{X}'\boldsymbol{T}\boldsymbol{y}$ (weighted least squares estimator). [Hint: $\boldsymbol{T} = \boldsymbol{T}^{1/2}\boldsymbol{T}^{1/2}$.]
(b) Explain why under a linear regression model with assumptions (i) to(iv), the least squares estimator $\widehat{\boldsymbol{\beta}}$ is uniformly not worse than a weighted least squares estimator $\widehat{\boldsymbol{\beta}}_{\boldsymbol{T}}$ (with respect to the matrix-valued squared error loss or the real-valued unweighted squared error loss).

2.14. Consider the linear regression model with assumptions (i) to (iv). Show that for $\boldsymbol{A}\boldsymbol{y} \in \mathcal{U}(\boldsymbol{\beta})$ and $\rho(\boldsymbol{\beta}, \widetilde{\boldsymbol{\beta}}) = \mathrm{E}[(\widetilde{\boldsymbol{\beta}} - \boldsymbol{\beta})'(\widetilde{\boldsymbol{\beta}} - \boldsymbol{\beta})]$ the identity

$$\rho(\boldsymbol{\beta}, \boldsymbol{A}\boldsymbol{y}) = \rho(\boldsymbol{\beta}, \widehat{\boldsymbol{\beta}}) \qquad \forall(\boldsymbol{\beta}, \sigma^2) \in \mathrm{I\!R}^p \times (0, \infty)$$

is satisfied if and only if $\boldsymbol{A} = (\boldsymbol{X}'\boldsymbol{X})^{-1}\boldsymbol{X}'$.

2.15. Show that under the linear regression model the estimator $\widehat{\boldsymbol{\beta}}_k = (\boldsymbol{X}'\boldsymbol{X} + k\boldsymbol{I}_p)^{-1}\boldsymbol{X}'\boldsymbol{y}$, where k is fixed, satisfies the two conditions

(i) $\mathrm{E}(\widehat{\boldsymbol{\beta}}_k) = \boldsymbol{\beta} - k(\boldsymbol{X}'\boldsymbol{X} + k\boldsymbol{I}_p)^{-1}\boldsymbol{\beta}$;
(ii) $\mathrm{Cov}(\widehat{\boldsymbol{\beta}}_k) = \sigma^2\boldsymbol{X}'\boldsymbol{X}(\boldsymbol{X}'\boldsymbol{X} + k\boldsymbol{I}_p)^{-2}$.

2.16. Show that for a fixed k the estimator $\widehat{\boldsymbol{\beta}}_k = (\boldsymbol{X}'\boldsymbol{X} + k\boldsymbol{I}_p)^{-1}\boldsymbol{X}'\boldsymbol{y}$ is unbiased for $\boldsymbol{\beta}$ if and only if $k = 0$.

2.17. Recall the situation from Example 2.6 and consider the individual elements $(\widehat{\boldsymbol{\beta}})_i$ and $(\widehat{\boldsymbol{\beta}}_{1/5})_i$, $i = 1, 2$, of $\widehat{\boldsymbol{\beta}}$ and $\widehat{\boldsymbol{\beta}}_{1/5}$ as estimators for β_1 and β_2. Compute

(i) $\mathrm{E}[(\widehat{\boldsymbol{\beta}}_{1/5})_i]$, $i = 1, 2$;
(ii) $\mathrm{Var}[(\widehat{\boldsymbol{\beta}})_i]$ and $\mathrm{Var}[(\widehat{\boldsymbol{\beta}}_{1/5})_i]$, $i = 1, 2$.

2.18. Consider the linear regression model with assumptions (i) to (iv).

(a) Determine $\rho(\boldsymbol{\beta}, \widehat{\boldsymbol{\beta}}) = \sum_{i=1}^p \mathrm{Var}(\widehat{\beta}_i)$ as a function of the eigenvalues of $\boldsymbol{X}'\boldsymbol{X}$. What effect does a small eigenvalue of $\boldsymbol{X}'\boldsymbol{X}$ on the sum $\sum_{i=1}^p \mathrm{Var}(\widehat{\beta}_i)$ has?

(b) Determine $\sum_{i=1}^{p} \mathrm{Var}((\widehat{\boldsymbol{\beta}}_k)_i)$ as a function of the eigenvalues of $\boldsymbol{X}'\boldsymbol{X}$.
(c) Show that $\sum_{i=1}^{p} \mathrm{Var}((\widehat{\boldsymbol{\beta}}_k)_i) \leq \sum_{i=1}^{p} \mathrm{Var}(\widehat{\boldsymbol{\beta}}_i)$ is always satisfied.

[Hint: spectral decomposition of $\boldsymbol{X}'\boldsymbol{X}$.]

2.19. Consider the linear regression model with $n = 10$ observations and $p = 4$ independent variables

i	1	2	3	4	5	6	7	8	9	10
$x_{i,1}$	1.9	1.8	1.8	1.8	2.0	2.1	2.1	2.2	2.3	2.3
$x_{i,2}$	2.2	2.2	2.4	2.4	2.5	2.6	2.6	2.6	2.8	2.7
$x_{i,3}$	1.9	2.0	2.1	2.2	2.3	2.4	2.6	2.6	2.8	2.8
$x_{i,4}$	3.7	3.8	3.6	3.8	3.8	3.7	3.8	4.0	3.7	3.8

(percentage of research and development expenses on the gross national product of four countries in ten different years).

(a) Compute the eigenvalues of $\boldsymbol{X}'\boldsymbol{X}$.
(b) Compute $\sum_{i=1}^{p} \mathrm{Var}(\widehat{\boldsymbol{\beta}}_i)$.
(c) Compute $v := \sum_{i=1}^{p} \mathrm{Var}((\widehat{\boldsymbol{\beta}}_{0.02})_i)$. What is the amount of reduction of v compared to $\sum_{i=1}^{p} \mathrm{Var}(\widehat{\boldsymbol{\beta}}_i)$?

2.20. Show that if in the linear regression model the matrix $\boldsymbol{X}$ is not of full column rank, i.e. assumption (ii) is violated, then $\mathcal{U}(\boldsymbol{\beta}) = \emptyset$.

2.21. Bias of the positive square root of the least squares variance estimator:

(a) Let $X \sim \chi^2_{(k)}$. Show that

$$\mathrm{E}(\sqrt{X}) = \sqrt{2}\frac{\Gamma(\frac{k+1}{2})}{\Gamma(\frac{k}{2})},$$

where $\Gamma(\cdot)$ is the Gamma function.

(b) Use Theorem 2.7 to show that under the linear regression model with assumptions (i) to (iv) and (v)

$$\mathrm{E}(\widehat{\sigma}) = \left(\frac{n-p}{2}\right)^{-1/2} \frac{\Gamma(\frac{n-p+1}{2})}{\Gamma(\frac{n-p}{2})}\,\sigma ,$$

where $\widehat{\sigma}$ is the positive square root of the least squares variance estimator $\widehat{\sigma}^2$.

(c) Show that for

$$c(\tau) = \frac{\Gamma(\tau + \frac{1}{2})}{\sqrt{\tau}\,\Gamma(\tau)}, \quad \tau > 0 ,$$

it follows that $c(\tau) < 1$ for $\tau < \infty$ and $\lim_{\tau\to\infty} c(\tau) = 1$.

(d) Conclude that under the assumptions of (b) it follows that $\mathrm{E}(\widehat{\sigma}) < \sigma$ and $\lim_{n\to\infty} \mathrm{E}(\widehat{\sigma}) = \sigma$.

2.22. To see that unbiasedness and consistency must not come along together, consider the following:

(a) Let $Y_1, \ldots, Y_n, \ldots$ be a series of random variables such that $P(Y_n = \theta) = 1 - \frac{1}{\sqrt{n}}$ and $P(Y_n = n) = \frac{1}{\sqrt{n}}$ for any n, where θ is not an integer. Show that Y_n converges in probability to θ but $\mathrm{E}(Y_n) \neq \theta$.
(b) Let $Y_1, \ldots, Y_n, \ldots$ be a series of random variables such that $P(Y_n = 2\theta) = \frac{1}{2} = P(Y_n = 0)$ for any n, where $\theta > 1$. Show that $\mathrm{E}(Y_n) = \theta$, but Y_n does not converge in probability to θ.

2.23. Consider a linear regression model with $n = 9$, $p = 3$, and normally distributed errors, i.e. model assumptions (i) to (vi) are satisfied. The matrix $\boldsymbol{X}$ is given as

$$\boldsymbol{X}' = \begin{pmatrix} 1\,1\,1\,0\,0\,0\,0\,0\,0 \\ 0\,0\,0\,1\,1\,1\,0\,0\,0 \\ 0\,0\,0\,0\,0\,0\,1\,1\,1 \end{pmatrix}.$$

(a) Show that the ordinary least squares estimator $\widehat{\boldsymbol{\vartheta}}$ for $\boldsymbol{\vartheta}$ is given by

$$\widehat{\boldsymbol{\vartheta}} = \frac{1}{\sqrt{3}} \begin{pmatrix} y_1 + y_2 + y_3 \\ y_4 + y_5 + y_6 \\ y_7 + y_8 + y_9 \end{pmatrix}$$

in the reparameterized model.
(b) Show that the Stein estimator $\widehat{\boldsymbol{\vartheta}}_{\mathrm{S}}$ for $\boldsymbol{\vartheta}$ is given by

$$\widehat{\boldsymbol{\vartheta}}_{\mathrm{S}} = \frac{1}{8}\left(9 - \frac{3\sum_{i=1}^{9} y_i^2}{\sum_{j=1}^{3}(\sum_{i=1}^{3} y_{i+3(j-1)})^2}\right)\widehat{\boldsymbol{\vartheta}}$$

in the reparameterized model.

2.24. Consider the linear regression model from Problem 2.23. Show that if in the reparameterized model the observed loss of $\widehat{\boldsymbol{\vartheta}}_{\mathrm{S}}$ is smaller than the observed loss of $\widehat{\boldsymbol{\vartheta}}$, then the same relation holds true for the observed losses of $\widehat{\boldsymbol{\beta}}_{\mathrm{S}}$ and $\widehat{\boldsymbol{\beta}}$ in the original model, when in both models the unweighted squared error loss is considered.

2.25. Consider the linear regression model from Problem 2.23 and a realization

$$\boldsymbol{y} = (-.8946, -.8882, -.0525, -.5289, .6568, .0842, 1.072, -.0459, .4267)'.$$

(a) Compute the estimates $\widehat{\boldsymbol{\vartheta}}$ and $\widehat{\boldsymbol{\vartheta}}_{\mathrm{S}}$ for $\boldsymbol{\vartheta}$ in the reparameterized model.
(b) Compute the estimates $\widehat{\boldsymbol{\beta}}$ and $\widehat{\boldsymbol{\beta}}_{\mathrm{S}}$ for $\boldsymbol{\beta}$ in the original model.
(c) Compute the losses of the estimates $\widehat{\boldsymbol{\vartheta}}$, $\widehat{\boldsymbol{\vartheta}}_{\mathrm{S}}$ for $\boldsymbol{\vartheta}$ and the losses of the estimates $\widehat{\boldsymbol{\beta}}$, $\widehat{\boldsymbol{\beta}}_{\mathrm{S}}$ for $\boldsymbol{\beta}$, each with respect to the unweighted squared error loss. The true parameter in the original model is $\boldsymbol{\beta} = (0, 0.1, 0.2)'$.

2.26. Consider the reparameterized regression model. If σ^2 is known, then

$$\widehat{\boldsymbol{\vartheta}}_{\mathrm{S}} = \left(1 - \frac{(p-2)\sigma^2}{\widehat{\boldsymbol{\vartheta}}'\widehat{\boldsymbol{\vartheta}}}\right)\widehat{\boldsymbol{\vartheta}}$$

can be used as an estimator for $\boldsymbol{\vartheta}$ (Stein estimation for known σ^2). Show that

$$\rho(\boldsymbol{\vartheta}, \widehat{\boldsymbol{\vartheta}}_{\mathrm{S}}) = \sigma^2 p - \sigma^2 (p-2)^2 \, \mathrm{E}\left[\frac{1}{\chi^2_{(p,\lambda)}}\right],$$

where $\lambda = \boldsymbol{\vartheta}'\boldsymbol{\vartheta}/2\sigma^2$.

2.27. Consider the situation from Problem 1.1. Determine $\mathrm{E}\left(\|\boldsymbol{Y}\|^2\right)$ and explain why $\delta_1(\boldsymbol{Y}) = \boldsymbol{Y}$ is 'too long' as an estimator for $\boldsymbol{\mu}$. In addition, explain why $\delta_2(\boldsymbol{y})$ can be seen as the Stein estimator of $\boldsymbol{\mu}$.

2.28. Prove Corollary 2.2 considering Lemma 2.2.

2.29. Consider the linear regression model with assumptions (i) to (iv) and (vi), where $R^2 > 0$. Let

$$\gamma = 1 - \frac{p-2}{n-p+2} \frac{\boldsymbol{y}'\boldsymbol{M}\boldsymbol{y}}{\widehat{\boldsymbol{\beta}}'\boldsymbol{X}'\boldsymbol{X}\widehat{\boldsymbol{\beta}}},$$

i.e. $\widehat{\boldsymbol{\beta}}_{\mathrm{S}} = \gamma\widehat{\boldsymbol{\beta}}$. Show that the equivalence

$$0 < \gamma \leq 1 \quad \Leftrightarrow \quad \frac{p-2}{n} < R^2$$

is satisfied.

2.30. Consider the model from Problem 2.23 with a sample value

$$\boldsymbol{y} = (.4900, .3102, -.5425, -.2422, -.6419, .1904, .9029, .2579, -.9080)'.$$

(a) Compute the estimates $\widehat{\boldsymbol{\vartheta}}$, $\widehat{\boldsymbol{\vartheta}}_{\mathrm{S}}$, and $\widehat{\boldsymbol{\vartheta}}^{+}_{\mathrm{S}}$ for $\boldsymbol{\vartheta}$ in the reparameterized model.
(b) Compute the losses of the estimates $\widehat{\boldsymbol{\vartheta}}$, $\widehat{\boldsymbol{\vartheta}}_{\mathrm{S}}$ and $\widehat{\boldsymbol{\vartheta}}^{+}_{\mathrm{S}}$ with respect to the unweighted squared error loss. The true parameter in the original model is $\boldsymbol{\beta} = (0, 0.1, 0.2)'$.

2.31. Show that in the regression model from Problem 2.23 the identity $\widehat{\boldsymbol{\beta}}_{\mathrm{B}} = \widehat{\boldsymbol{\beta}}_{\mathrm{S}}$ holds true.

Part II

Alternatives to Least Squares Estimation

3 Alternative Estimators

In the previous chapter we have considered the non-linear Stein estimator as an example for an alternative to the ordinary least squares estimator which can have specific advantages. Now, we present further alternative estimators whose use can be shown to benefit certain situations. In general, we proceed by presenting an alternative as a *linear* estimator and comparing its properties with those of the ordinary least squares estimator. Then we demonstrate the use of one or more *non-linear* variants of the estimator in question in practice.

3.1 Restricted Least Squares Estimation

Under the ordinary linear regression model, sometimes additional information about the unknown parameter vector $\boldsymbol{\beta}$ is available, which can be expressed in terms of a linear equation

$$\boldsymbol{R\beta} = \boldsymbol{r} \ ,$$

where $\boldsymbol{R}$ is a known $m \times p$ matrix of full row rank and $\boldsymbol{r}$ is a known $m \times 1$ vector. Examples for such additional information are:

- knowledge about an element of $\boldsymbol{\beta}$, e.g. $\beta_1 = 1$, so that $\boldsymbol{R} = (1, 0, \ldots, 0)$ and $\boldsymbol{r} = (1)$;
- knowledge about the sum of elements of $\boldsymbol{\beta}$, e.g. $\beta_1 + \beta_2 = 1$ so that $\boldsymbol{R} = (1, 1)$ and $\boldsymbol{r} = (1)$;
- knowledge about the proportion of elements, e.g. $\beta_1 : \beta_2 : \beta_3 = 6 : 2 : 1$ so that $\beta_2 = 2\beta_3$ and $\beta_1 = 3\beta_2$, and therefore $\boldsymbol{R} = \begin{pmatrix} 1 & -3 & 0 \\ 0 & 1 & -2 \end{pmatrix}$ and $\boldsymbol{r} = \begin{pmatrix} 0 \\ 0 \end{pmatrix}$.

Such information can for example stem from 'knowledge' of an expert or from earlier, similar investigations. In both cases the additional information is more or less trustworthy, but should not be neglected from the outset, since in general additional information may result in more precise estimators. Of course, it may also be known that parameters *must* satisfy certain linear equations, in which case the restrictions are exactly true.

3.1.1 The Principle of Restricted Least Squares

An obvious method to construct an appropriate estimator $\widehat{\boldsymbol{\beta}}_{\mathrm{R}}$ for $\boldsymbol{\beta}$ under additional restrictions is to find the vector $\widehat{\boldsymbol{\beta}}_{\mathrm{R}}$, such that the length of $\boldsymbol{y} - \boldsymbol{X}\widehat{\boldsymbol{\beta}}_{\mathrm{R}}$ is smallest compared to the length of any other vector $\boldsymbol{y} - \boldsymbol{X}\boldsymbol{\beta}_*$, where only those vectors $\boldsymbol{\beta}_*$ are considered which satisfy $\boldsymbol{R}\boldsymbol{\beta}_* = \boldsymbol{r}$. The vector $\widehat{\boldsymbol{\beta}}_{\mathrm{R}}$ can then be said to be chosen according to the principle of restricted least squares. In other words, the vector $\widehat{\boldsymbol{\beta}}_{\mathrm{R}}$ is assumed to satisfy $\boldsymbol{R}\widehat{\boldsymbol{\beta}}_{\mathrm{R}} = \boldsymbol{r}$ and in addition

$$(\boldsymbol{y} - \boldsymbol{X}\widehat{\boldsymbol{\beta}}_{\mathrm{R}})'(\boldsymbol{y} - \boldsymbol{X}\widehat{\boldsymbol{\beta}}_{\mathrm{R}}) = \min_{\boldsymbol{\beta}_*:\, \boldsymbol{R}\boldsymbol{\beta}_* = \boldsymbol{r}} (\boldsymbol{y} - \boldsymbol{X}\boldsymbol{\beta}_*)'(\boldsymbol{y} - \boldsymbol{X}\boldsymbol{\beta}_*) .$$

How can we actually obtain $\widehat{\boldsymbol{\beta}}_{\mathrm{R}}$? By using the method of Lagrange multipliers, an explicit solution to this problem can be derived. For this we consider the function

$$L = (\boldsymbol{y} - \boldsymbol{X}\boldsymbol{\beta}_*)'(\boldsymbol{y} - \boldsymbol{X}\boldsymbol{\beta}_*) - 2\boldsymbol{\lambda}'(\boldsymbol{R}\boldsymbol{\beta}_* - \boldsymbol{r}) ,$$

where $\boldsymbol{\lambda}$ is the $p \times 1$ vector of Lagrange multipliers. By using the rules

$$\frac{\partial}{\partial \boldsymbol{a}} \boldsymbol{a}'\boldsymbol{A}\boldsymbol{a} = (\boldsymbol{A} + \boldsymbol{A}')\boldsymbol{a} \quad \text{and} \quad \frac{\partial}{\partial \boldsymbol{a}} \boldsymbol{a}'\boldsymbol{b} = \boldsymbol{b} ,$$

it follows that

$$\frac{\partial}{\partial \boldsymbol{\beta}_*} L = 2\boldsymbol{X}'\boldsymbol{X}\boldsymbol{\beta}_* - 2\boldsymbol{X}'\boldsymbol{y} + \boldsymbol{R}'\boldsymbol{\lambda} \quad \text{and} \quad \frac{\partial}{\partial \boldsymbol{\lambda}} L = \boldsymbol{R}\boldsymbol{\beta}_* - \boldsymbol{r}.$$

Now let $\widehat{\boldsymbol{\beta}}_{\mathrm{R}}$ and $\widehat{\boldsymbol{\lambda}}_{\mathrm{R}}$ denote the solution to

$$\frac{\partial}{\partial \boldsymbol{\beta}_*} L = \boldsymbol{0} \quad \text{and} \quad \frac{\partial}{\partial \boldsymbol{\lambda}} L = \boldsymbol{0}$$

with respect to $\boldsymbol{\beta}_*$ and $\boldsymbol{\lambda}$. Then

$$\widehat{\boldsymbol{\beta}}_{\mathrm{R}} = \widehat{\boldsymbol{\beta}} - \frac{1}{2}(\boldsymbol{X}'\boldsymbol{X})^{-1}\boldsymbol{R}'\widehat{\boldsymbol{\lambda}}_{\mathrm{R}} ,$$

where $\widehat{\boldsymbol{\beta}} = (\boldsymbol{X}'\boldsymbol{X})^{-1}\boldsymbol{X}'\boldsymbol{y}$. If we multiply this identity by $\boldsymbol{R}$ from the left and set $\boldsymbol{R}\widehat{\boldsymbol{\beta}}_{\mathrm{R}}$ equal to $\boldsymbol{r}$, then we can write $\widehat{\boldsymbol{\lambda}}_{\mathrm{R}}$ in the form

$$\widehat{\boldsymbol{\lambda}}_{\mathrm{R}} = -2[\boldsymbol{R}(\boldsymbol{X}'\boldsymbol{X})^{-1}\boldsymbol{R}']^{-1}(\boldsymbol{r} - \boldsymbol{R}\widehat{\boldsymbol{\beta}}) .$$

Thus, we obtain

$$\widehat{\boldsymbol{\beta}}_{\mathrm{R}} = \widehat{\boldsymbol{\beta}} - (\boldsymbol{X}'\boldsymbol{X})^{-1}\boldsymbol{R}'[\boldsymbol{R}(\boldsymbol{X}'\boldsymbol{X})^{-1}\boldsymbol{R}']^{-1}(\boldsymbol{R}\widehat{\boldsymbol{\beta}} - \boldsymbol{r}) .$$

Compare also Problem 3.2 and Problem 3.1 for a more algebraic oriented proof of the following theorem.

Theorem 3.1. *Under the linear regression model with assumptions (i) to (iv), the function*

$$f(\boldsymbol{\beta}_*) = (\boldsymbol{y} - \boldsymbol{X}\boldsymbol{\beta}_*)'(\boldsymbol{y} - \boldsymbol{X}\boldsymbol{\beta}_*)$$

is minimized with respect to $\boldsymbol{\beta}_*|_{\boldsymbol{R}\boldsymbol{\beta}_*=\boldsymbol{r}}$ *for* $\boldsymbol{\beta}_* = \widehat{\boldsymbol{\beta}}_{\mathrm{R}}$, *where*

$$\widehat{\boldsymbol{\beta}}_{\mathrm{R}} = \widehat{\boldsymbol{\beta}} - (\boldsymbol{X}'\boldsymbol{X})^{-1}\boldsymbol{R}'[\boldsymbol{R}(\boldsymbol{X}'\boldsymbol{X})^{-1}\boldsymbol{R}']^{-1}(\boldsymbol{R}\widehat{\boldsymbol{\beta}} - \boldsymbol{r}) ,$$

and $\widehat{\boldsymbol{\beta}} = (\boldsymbol{X}'\boldsymbol{X})^{-1}\boldsymbol{X}'\boldsymbol{y}$ *is the ordinary least squares estimator of* $\boldsymbol{\beta}$.

Notation 3.1. *Under the linear regression model with assumptions (i) to (iv), the vector* $\widehat{\boldsymbol{\beta}}_{\mathrm{R}}$ *is called* restricted least squares estimator *of* $\boldsymbol{\beta}$.

3.1.2 The Parameter Space

If we have sure knowledge about restrictions of the form $\boldsymbol{R}\boldsymbol{\beta} = \boldsymbol{r}$ for some known matrix $\boldsymbol{R}$ of full row rank and some known vector $\boldsymbol{r}$, then we can add the condition

$$\text{(viii)} \quad \boldsymbol{R}\boldsymbol{\beta} = \boldsymbol{r}$$

to the list of model assumptions concerning our regression model. The additional condition (viii) changes the parameter space for $\boldsymbol{\beta}$. While under the model with assumptions (i) to (iv) the parameter space Θ for $(\boldsymbol{\beta}, \sigma^2)$ is given as

$$\Theta = \mathbb{R}^p \times (0, \infty) ,$$

under the model with assumptions (i) to (iv) and (viii), Θ becomes

$$\Theta = \mathcal{B} \times (0, \infty), \quad \mathcal{B} = \{\boldsymbol{\beta} \in \mathbb{R}^p : \boldsymbol{R}\boldsymbol{\beta} = \boldsymbol{r}\} .$$

We have different classes of linear unbiased estimators for $\boldsymbol{\beta}$ for different parameter spaces:

- Under the linear regression model with assumptions (i) to (iv), a linear estimator $\boldsymbol{A}\boldsymbol{y} + \boldsymbol{a}$ is unbiased for $\boldsymbol{\beta}$ if and only if

  $$\mathrm{E}(\boldsymbol{A}\boldsymbol{y} + \boldsymbol{a}) = \boldsymbol{\beta} \quad \forall \boldsymbol{\beta} \in \mathbb{R}^p .$$

  As we have seen before, this is satisfied if and only if $\boldsymbol{A}\boldsymbol{X} = \boldsymbol{I}_p$ and $\boldsymbol{a} = \boldsymbol{0}$.
- Under the linear regression model with assumptions (i) to (iv) and (viii), a linear estimator $\boldsymbol{A}\boldsymbol{y} + \boldsymbol{a}$ is unbiased for $\boldsymbol{\beta}$ if and only if

  $$\mathrm{E}(\boldsymbol{A}\boldsymbol{y} + \boldsymbol{a}) = \boldsymbol{\beta} \quad \forall \boldsymbol{\beta} \in \mathcal{B} .$$

  The latter is equivalent to

  $$\boldsymbol{A}\boldsymbol{X}(\boldsymbol{I}_p - \boldsymbol{R}'(\boldsymbol{R}\boldsymbol{R}')^{-1}\boldsymbol{R}) + \boldsymbol{R}'(\boldsymbol{R}\boldsymbol{R}')^{-1}\boldsymbol{R} = \boldsymbol{I}_p$$

  and

  $$\boldsymbol{a} = (\boldsymbol{I}_p - \boldsymbol{A}\boldsymbol{X})\boldsymbol{R}'(\boldsymbol{R}\boldsymbol{R}')^{-1}\boldsymbol{r} ,$$

  see Problem 3.3.

It is not difficult to see that the elements $\boldsymbol{Ay}+\boldsymbol{a}$ from the former set also belong to the latter one.

Remark 3.1. If $\boldsymbol{Ay}+\boldsymbol{a}$ is unbiased for $\boldsymbol{\beta}$ under the linear regression model with assumptions (i) to (iv) and corresponding parameter space $\Theta = \mathbb{R}^p \times (0,\infty)$, then $\boldsymbol{Ay}+\boldsymbol{a}$ is also unbiased for $\boldsymbol{\beta}$ under the additional assumption (viii) and corresponding parameter space $\Theta = \mathcal{B}\times(0,\infty)$ with $\mathcal{B} = \{\boldsymbol{\beta} \in \mathbb{R}^p : \boldsymbol{R\beta} = \boldsymbol{r}\}$.

Example 3.1. Consider the Cobb-Douglas production function $P = f(L,K)$ with $f(L,K) = \alpha L^\gamma K^\delta$, where P denotes the output of a firm and L and K are two inputs, L standing for labor and K standing for capital. The letters α, γ and δ are unknown parameters. It is often desired that the production function satisfies *constant-returns-to-scale*, i.e. $f(\lambda L, \lambda K) = \lambda f(L,K)$ for any $\lambda > 1$. If we take logarithms on the identity $P = f(L,K)$ and assume that the obtained relationship is additively disturbed by some random error, then we have

$$\underbrace{\ln(P)}_{=:y} = \underbrace{\alpha}_{=:\beta_1} + \underbrace{\gamma}_{=:\beta_2} \underbrace{\ln(L)}_{=:x_2} + \underbrace{\delta}_{=:\beta_3} \underbrace{\ln(K)}_{=:x_3} + \varepsilon\,,$$

which can be seen as the description of a linear regression model with intercept. (Data can be obtained by observing the variables P, L and K in a series of years.) The constant-returns-to-scale assumption is satisfied (for nonzero P) if and only if $\gamma + \delta = 1$, which also reads

$$\beta_2 + \beta_3 = 1\,.$$

This condition is usually assumed when considering the above model and is seen to be a specific linear restriction on the parameters. It does, however, not stem from additional information, but is imposed by purpose, and thus it is assumed to be *exactly* satisfied. Here, we have a situation in which the linear restrictions assumption (viii) is a definite model assumption. □

3.1.3 Properties of Restricted Least Squares Estimator

At first glance, use of the restricted least squares estimator $\widehat{\boldsymbol{\beta}}_{\mathrm{R}}$ appears to be useful only in situations where the corresponding restrictions are satisfied. If assumption (viii) is violated, nevertheless the restricted least squares estimator can be computed with respect to an arbitrary set of linear restrictions. Hence we have an infinite number of competitors to the ordinary least squares estimator at our disposal. But does any of them make sense? The answer is affirmative when we consider situations where linear restrictions $\boldsymbol{R\beta} = \boldsymbol{r}$ are supposed to be 'almost' satisfied.

To clarify this point, we start with considering the restricted least squares estimator under the linear regression model with assumptions (i) to (iv), and without caring whether the corresponding restrictions are satisfied or not.

Clearly the restricted least squares estimator is (non-homogeneous) linear, i.e. $\widehat{\boldsymbol{\beta}}_{\mathrm{R}} = \boldsymbol{A}\boldsymbol{y} + \boldsymbol{a}$, where

$$\boldsymbol{A} = \boldsymbol{S}^{-1}\boldsymbol{X}' - \boldsymbol{C}\boldsymbol{X}' \quad \text{and} \quad \boldsymbol{a} = \boldsymbol{S}^{-1}\boldsymbol{R}'(\boldsymbol{R}\boldsymbol{S}^{-1}\boldsymbol{R}')^{-1}\boldsymbol{r}$$

with

$$\boldsymbol{S} := \boldsymbol{X}'\boldsymbol{X} \quad \text{and} \quad \boldsymbol{C} := \boldsymbol{S}^{-1}\boldsymbol{R}'(\boldsymbol{R}\boldsymbol{S}^{-1}\boldsymbol{R}')^{-1}\boldsymbol{R}\boldsymbol{S}^{-1} .$$

The symbols $\boldsymbol{S}$ and $\boldsymbol{C}$ will repeatedly be used in the following. In addition, we will write

$$\boldsymbol{\delta} := \boldsymbol{R}\boldsymbol{\beta} - \boldsymbol{r} .$$

We may now give expectation vector and covariance matrix of the restricted least squares estimator.

Theorem 3.2. *Under the linear regression model with assumptions (i) to (iv), expectation vector and covariance matrix of* $\widehat{\boldsymbol{\beta}}_{\mathrm{R}}$ *are given as*

$$\mathrm{E}(\widehat{\boldsymbol{\beta}}_{\mathrm{R}}) = \boldsymbol{\beta} - \boldsymbol{S}^{-1}\boldsymbol{R}'(\boldsymbol{R}\boldsymbol{S}^{-1}\boldsymbol{R}')^{-1}\boldsymbol{\delta} \quad \textit{and} \quad \mathrm{Cov}(\widehat{\boldsymbol{\beta}}_{\mathrm{R}}) = \sigma^2\boldsymbol{S}^{-1} - \sigma^2\boldsymbol{C} .$$

Proof. From $\mathrm{E}(\widehat{\boldsymbol{\beta}}) = \boldsymbol{\beta}$ and $\mathrm{E}(\boldsymbol{R}\widehat{\boldsymbol{\beta}} - \boldsymbol{r}) = \boldsymbol{\delta}$, we conclude $\mathrm{E}(\widehat{\boldsymbol{\beta}}_{\mathrm{R}}) = \boldsymbol{\beta} - \boldsymbol{S}^{-1}\boldsymbol{R}'(\boldsymbol{R}\boldsymbol{S}^{-1}\boldsymbol{R}')^{-1}\boldsymbol{\delta}$. To determine $\mathrm{Cov}(\widehat{\boldsymbol{\beta}}_{\mathrm{R}})$, we write $\widehat{\boldsymbol{\beta}}_{\mathrm{R}} = \boldsymbol{A}\boldsymbol{y} + \boldsymbol{a}$, where

$$\boldsymbol{A} = \boldsymbol{S}^{-1}\boldsymbol{X}' - \boldsymbol{C}\boldsymbol{X}' \quad \text{and} \quad \boldsymbol{a} = \boldsymbol{S}^{-1}\boldsymbol{R}'(\boldsymbol{R}\boldsymbol{S}^{-1}\boldsymbol{R}')^{-1}\boldsymbol{r} .$$

Then it follows $\mathrm{Cov}(\widehat{\boldsymbol{\beta}}_{\mathrm{R}}) = \mathrm{Cov}(\boldsymbol{A}\boldsymbol{y} + \boldsymbol{a}) = \mathrm{Cov}(\boldsymbol{A}\boldsymbol{y}) = \sigma^2\boldsymbol{A}\boldsymbol{A}'$, and thus

$$\begin{aligned}\mathrm{Cov}(\widehat{\boldsymbol{\beta}}_{\mathrm{R}}) &= \sigma^2(\boldsymbol{S}^{-1}\boldsymbol{X}' - \boldsymbol{C}\boldsymbol{X}')(\boldsymbol{X}\boldsymbol{S}^{-1} - \boldsymbol{X}\boldsymbol{C}) \\ &= \sigma^2(\underbrace{\boldsymbol{S}^{-1}\boldsymbol{X}'\boldsymbol{X}}_{=\boldsymbol{I}_p}\boldsymbol{S}^{-1} - \boldsymbol{C}\underbrace{\boldsymbol{X}'\boldsymbol{X}\boldsymbol{S}^{-1}}_{=\boldsymbol{I}_p} - \underbrace{\boldsymbol{S}^{-1}\boldsymbol{X}'\boldsymbol{X}}_{=\boldsymbol{I}_p}\boldsymbol{C} + \underbrace{\boldsymbol{C}\boldsymbol{X}'\boldsymbol{X}\boldsymbol{C}}_{=\boldsymbol{C}}) \\ &= \sigma^2(\boldsymbol{S}^{-1} - \boldsymbol{C}) ,\end{aligned}$$

which completes the proof. □

From the above theorem, $\mathrm{MSE}(\boldsymbol{\beta}, \widehat{\boldsymbol{\beta}}_{\mathrm{R}})$ can easily be derived.

Corollary 3.1. *Under the linear regression model with assumptions (i) to (iv), the matrix-valued mean squared error risk of* $\widehat{\boldsymbol{\beta}}_{\mathrm{R}}$ *is given as*

$$\begin{aligned}&\mathrm{MSE}(\boldsymbol{\beta}, \widehat{\boldsymbol{\beta}}_{\mathrm{R}}) \\ &= \mathrm{E}\left[(\widehat{\boldsymbol{\beta}}_{\mathrm{R}} - \boldsymbol{\beta})(\widehat{\boldsymbol{\beta}}_{\mathrm{R}} - \boldsymbol{\beta})'\right] \\ &= \sigma^2\boldsymbol{S}^{-1} - \sigma^2\boldsymbol{C} + \boldsymbol{S}^{-1}\boldsymbol{R}'(\boldsymbol{R}\boldsymbol{S}^{-1}\boldsymbol{R}')^{-1}\boldsymbol{\delta}\boldsymbol{\delta}'(\boldsymbol{R}\boldsymbol{S}^{-1}\boldsymbol{R}')^{-1}\boldsymbol{R}\boldsymbol{S}^{-1} .\end{aligned}$$

Proof. We can write

$$\mathrm{MSE}(\boldsymbol{\beta}, \widehat{\boldsymbol{\beta}}_{\mathrm{R}}) = \mathrm{Cov}(\widehat{\boldsymbol{\beta}}_{\mathrm{R}}) + \mathrm{bias}(\widehat{\boldsymbol{\beta}}_{\mathrm{R}})\,\mathrm{bias}(\widehat{\boldsymbol{\beta}}_{\mathrm{R}})',$$

where $\mathrm{bias}(\widehat{\boldsymbol{\beta}}_{\mathrm{R}}) = \mathrm{E}(\widehat{\boldsymbol{\beta}}_{\mathrm{R}}) - \boldsymbol{\beta} = -\boldsymbol{S}^{-1}\boldsymbol{R}'(\boldsymbol{R}\boldsymbol{S}^{-1}\boldsymbol{R}')^{-1}\boldsymbol{\delta}$, see Problem 1.10. Then the assertion follows from the above theorem. □

3.1.4 Risk Comparison of Restricted and Ordinary Least Squares Estimator

Next, we compare $\mathrm{MSE}(\boldsymbol{\beta}, \widehat{\boldsymbol{\beta}})$ and $\mathrm{MSE}(\boldsymbol{\beta}, \widehat{\boldsymbol{\beta}}_{\mathrm{R}})$ under the assumption that linear restrictions $\boldsymbol{R\beta} = \boldsymbol{r}$ are satisfied, so that $\beta \in \mathcal{B} = \{\beta \in \mathbb{R}^p : \boldsymbol{R\beta} = \boldsymbol{r}\}$. In this case it is quite natural to expect that $\widehat{\boldsymbol{\beta}}_{\mathrm{R}}$ is a better estimator than $\widehat{\boldsymbol{\beta}}$, since the former incorporates the additional information provided by $\boldsymbol{R\beta} = \boldsymbol{r}$.

Theorem 3.3. *Under the linear regression model with assumptions (i) to (iv) and (viii), the difference*

$$\boldsymbol{\Delta} = \mathrm{MSE}(\boldsymbol{\beta}, \widehat{\boldsymbol{\beta}}) - \mathrm{MSE}(\boldsymbol{\beta}, \widehat{\boldsymbol{\beta}}_{\mathrm{R}})$$

is nonnegative definite for all $(\boldsymbol{\beta}, \sigma^2) \in \mathcal{B} \times (0, \infty)$. *Moreover,* $\boldsymbol{\Delta} \neq \mathbf{0}$ *for all* $(\boldsymbol{\beta}, \sigma^2) \in \mathcal{B} \times (0, \infty)$.

Proof. Since under model assumption (viii) we have $\boldsymbol{\delta} = \mathbf{0}$, the matrix $\boldsymbol{\Delta} = \sigma^2 \boldsymbol{C}$ is symmetric nonnegative definite by Corollary 3.1. Moreover, $\boldsymbol{C}$ is the zero matrix if and only if $\boldsymbol{R} = \mathbf{0}$. Since (viii) entails the full row rank of $\boldsymbol{R}$, the case $\boldsymbol{R} = \mathbf{0}$ cannot occur. □

Following the above theorem, $\widehat{\boldsymbol{\beta}}_{\mathrm{R}}$ is uniformly better than $\widehat{\boldsymbol{\beta}}$ with respect to $\mathrm{MSE}(\boldsymbol{\beta}, \widetilde{\boldsymbol{\beta}})$ under the linear regression model with assumptions (i) to (iv) and (viii). This confirms our expectation that if restrictions $\boldsymbol{R\beta} = \boldsymbol{r}$ are exactly satisfied, the restricted least squares estimators is advantageous over the ordinary least squares estimator.

We have already mentioned, however, the possibility of situations where linear restrictions are not exactly satisfied but the restricted least squares estimator still performs better. To investigate these cases in detail, we compare the risks of both estimators in question without imposing model assumption (viii).

Theorem 3.4. *Under the linear regression model with assumptions (i) to (iv), let*

$$\lambda = \frac{\boldsymbol{\delta}'(\boldsymbol{R S}^{-1} \boldsymbol{R}')^{-1} \boldsymbol{\delta}}{2\sigma^2}.$$

Then for the difference

$$\boldsymbol{\Delta} = \mathrm{MSE}(\boldsymbol{\beta}, \widehat{\boldsymbol{\beta}}) - \mathrm{MSE}(\boldsymbol{\beta}, \widehat{\boldsymbol{\beta}}_{\mathrm{R}})$$

the following statements hold true:

(i) *The matrix* $\boldsymbol{\Delta}$ *is nonnegative definite if and only if* $\lambda \leq \frac{1}{2}$.
(ii) *The matrix* $\boldsymbol{\Delta}$ *equals the zero matrix* $\mathbf{0}$ *if and only if* $m = 1$ *and* $\lambda = \frac{1}{2}$.
(iii) *If* $\lambda < \frac{1}{2}$, *then* $\boldsymbol{\Delta}$ *is nonnegative definite and* $\boldsymbol{\Delta} \neq \mathbf{0}$.

Proof. From Corollary 3.1 we conclude

$$\begin{aligned}\boldsymbol{\Delta} &= \sigma^2 \boldsymbol{C} - \boldsymbol{S}^{-1}\boldsymbol{R}'(\boldsymbol{R}\boldsymbol{S}^{-1}\boldsymbol{R}')^{-1}\boldsymbol{\delta}\boldsymbol{\delta}'(\boldsymbol{R}\boldsymbol{S}^{-1}\boldsymbol{R}')^{-1}\boldsymbol{R}\boldsymbol{S}^{-1} \\ &= \boldsymbol{S}^{-1}\boldsymbol{R}'(\boldsymbol{R}\boldsymbol{S}^{-1}\boldsymbol{R}')^{-1}\left[\sigma^2\boldsymbol{R}\boldsymbol{S}^{-1}\boldsymbol{R}' - \boldsymbol{\delta}\boldsymbol{\delta}'\right](\boldsymbol{R}\boldsymbol{S}^{-1}\boldsymbol{R}')^{-1}\boldsymbol{R}\boldsymbol{S}^{-1} .\end{aligned}$$

Since the matrix $(\boldsymbol{R}\boldsymbol{S}^{-1}\boldsymbol{R}')^{-1}\boldsymbol{R}\boldsymbol{S}^{-1}$ is of full row rank, the symmetric matrix $\boldsymbol{\Delta}$ is nonnegative definite if and only if the symmetric matrix

$$\sigma^2\boldsymbol{R}\boldsymbol{S}^{-1}\boldsymbol{R}' - \boldsymbol{\delta}\boldsymbol{\delta}'$$

is nonnegative definite. Theorem A.60 states that when $\boldsymbol{A}$ is symmetric positive definite, $\boldsymbol{a}$ is an arbitrary vector (of appropriate dimension) and $\alpha > 0$, then the symmetric matrix $\alpha\boldsymbol{A} - \boldsymbol{a}\boldsymbol{a}'$ is nonnegative definite if and only if $\boldsymbol{a}'\boldsymbol{A}^{-1}\boldsymbol{a} \leq \alpha$ is satisfied. By applying this theorem we can state that $\boldsymbol{\Delta}$ is nonnegative definite if and only if $\boldsymbol{\delta}'(\boldsymbol{R}\boldsymbol{S}^{-1}\boldsymbol{R}')^{-1}\boldsymbol{\delta} \leq \sigma^2$. The latter is equivalent to $\lambda \leq \frac{1}{2}$, so that statement (i) is shown. For statement (ii) note that $\boldsymbol{\Delta} = \boldsymbol{0}$ if and only if $\sigma^2\boldsymbol{R}\boldsymbol{S}^{-1}\boldsymbol{R}' - \boldsymbol{\delta}\boldsymbol{\delta}' = \boldsymbol{0}$, i.e.

$$\sigma^2\boldsymbol{R}\boldsymbol{S}^{-1}\boldsymbol{R}' = \boldsymbol{\delta}\boldsymbol{\delta}' .$$

Since the rank of the right-hand side is either 0 or 1 and the rank of the left-hand side is equal to $m \geq 1$, this identity can only hold in case $m = 1$. If $m = 1$ is satisfied, then $\boldsymbol{R}\boldsymbol{S}^{-1}\boldsymbol{R}'$ and $\boldsymbol{\delta}$ are real-valued and $\boldsymbol{R}\boldsymbol{S}^{-1}\boldsymbol{R}' \neq 0$, showing that in case $m = 1$ the identity can equivalently be expressed as

$$\sigma^2\boldsymbol{R}\boldsymbol{S}^{-1}\boldsymbol{R}' = \boldsymbol{\delta}\boldsymbol{\delta}' \Leftrightarrow 1 = \frac{\boldsymbol{\delta}\boldsymbol{\delta}'}{\sigma^2\boldsymbol{R}\boldsymbol{S}^{-1}\boldsymbol{R}'} \Leftrightarrow 1 = \frac{\boldsymbol{\delta}'(\boldsymbol{R}\boldsymbol{S}^{-1}\boldsymbol{R}')^{-1}\boldsymbol{\delta}}{\sigma^2} .$$

This shows statement (ii). Statement (iii) follows by combining statements (i) and (ii). □

If we compare the two estimators $\widehat{\boldsymbol{\beta}}$ and $\widehat{\boldsymbol{\beta}}_{\mathrm{R}}$ with respect to the matrix-valued risk $\mathrm{MSE}(\boldsymbol{\beta}, \widetilde{\boldsymbol{\beta}})$, we cannot say that under the linear regression model with assumptions (i) to (iv) one is *uniformly* better than the other . Nonetheless we conclude from Theorem 3.4 (iii) that for all $\boldsymbol{\beta}$ satisfying $\lambda < \frac{1}{2}$, the restricted least squares estimator $\widehat{\boldsymbol{\beta}}_{\mathrm{R}}$ performs better than the ordinary least squares estimator.

Thus, the case $\lambda < \frac{1}{2}$ can be identified with a situation where linear restrictions on $\boldsymbol{\beta}$ are not necessarily exactly but 'almost' true.

Considering the performance of estimators with respect to the real-valued squared error function $\rho(\boldsymbol{\beta}, \widetilde{\boldsymbol{\beta}}) = \mathrm{tr}[\mathrm{MSE}(\boldsymbol{\beta}, \widetilde{\boldsymbol{\beta}})]$, we obtain similar statements to the above.

Corollary 3.2. *Under the linear regression model with assumptions (i) to (iv), the following two statements hold true:*

(i) *If* $\lambda \leq \frac{1}{2}$, *then* $\rho(\boldsymbol{\beta}, \widehat{\boldsymbol{\beta}}_{\mathrm{R}}) \leq \rho(\boldsymbol{\beta}, \widehat{\boldsymbol{\beta}})$.

(ii) *If* $\lambda < \frac{1}{2}$, *then* $\rho(\boldsymbol{\beta}, \widehat{\boldsymbol{\beta}}_{\mathrm{R}}) < \rho(\boldsymbol{\beta}, \widehat{\boldsymbol{\beta}})$.

Proof. Statements (i) and (ii) follow from Theorem 3.4 in view of $\rho(\boldsymbol{\beta}, \widetilde{\boldsymbol{\beta}}) = \mathrm{tr}[\mathrm{MSE}(\boldsymbol{\beta}, \widetilde{\boldsymbol{\beta}})]$. □

Clearly, the inequality $\lambda < \frac{1}{2}$ is satisfied for $\boldsymbol{\delta} = \mathbf{0}$. But as shown above, even if the restrictions are not exactly satisfied but λ is small enough, the use of the restricted least squares estimator will be beneficial.

Example 3.2. Consider the linear regression model $\boldsymbol{y} = \boldsymbol{X}\boldsymbol{\beta} + \boldsymbol{\varepsilon}$ with

$$\boldsymbol{y} = \begin{pmatrix} 2.6262 \\ 2.1640 \\ 2.3589 \\ 3.9844 \\ 4.2513 \end{pmatrix}, \quad \boldsymbol{X} = \begin{pmatrix} 1 & 1 \\ 1 & 2 \\ 1 & 3 \\ 1 & 4 \\ 1 & 5 \end{pmatrix}, \quad \boldsymbol{\beta} = \begin{pmatrix} \beta_1 \\ \beta_2 \end{pmatrix}.$$

In addition, the elements ε_i of the error vector $\boldsymbol{\varepsilon}$ are independently normally distributed with $\mathrm{E}(\varepsilon_i) = 0$ and $\mathrm{Var}(\varepsilon_i) = 1$ for $i = 1, \ldots, 5$. Thus, we have a model where assumptions (i) to (iv) and (v) are satisfied and the parameter σ^2 is known to be $\sigma^2 = 1$. Moreover, for some reason, we assume that the unknown parameter vector $\boldsymbol{\beta}$ could satisfy linear restrictions of the form

$$\boldsymbol{R}\boldsymbol{\beta} = \boldsymbol{r}, \qquad \boldsymbol{R} = (1, 1), \quad \boldsymbol{r} = (1).$$

The ordinary least squares estimator of $\boldsymbol{\beta}$ is computed as

$$\widehat{\boldsymbol{\beta}} = \begin{pmatrix} 1.5558 \\ 0.5070 \end{pmatrix},$$

while the corresponding restricted least squares estimator $\widehat{\boldsymbol{\beta}}_{\mathrm{R}}$ is

$$\widehat{\boldsymbol{\beta}}_{\mathrm{R}} = \begin{pmatrix} 0.1387 \\ 0.8613 \end{pmatrix}.$$

Since we have simulated the data in this example, we know the true parameter vector $\boldsymbol{\beta}$, which is given by $\boldsymbol{\beta} = (0.2, 0.9)'$. The true parameter vector does not exactly but *almost* satisfy our imposed restrictions. If we compare the estimates $\widehat{\boldsymbol{\beta}}$ and $\widehat{\boldsymbol{\beta}}_{\mathrm{R}}$, we see that $\widehat{\boldsymbol{\beta}}_{\mathrm{R}}$ is less far away from the true parameter vector than $\widehat{\boldsymbol{\beta}}$. In other words, the ***observed loss*** of $\widehat{\boldsymbol{\beta}}_{\mathrm{R}}$ is distinctly smaller than the observed loss of $\widehat{\boldsymbol{\beta}}$ with respect to the unweighted squared error loss $L(\boldsymbol{\beta}, \widetilde{\boldsymbol{\beta}}) = (\widetilde{\boldsymbol{\beta}} - \boldsymbol{\beta})'(\widetilde{\boldsymbol{\beta}} - \boldsymbol{\beta})$, when $\boldsymbol{\beta} = (0.2, 0.9)'$ is the true parameter vector. We compute

$$L(\boldsymbol{\beta}, \widehat{\boldsymbol{\beta}}) = 1.9927 \quad \text{and} \quad L(\boldsymbol{\beta}, \widehat{\boldsymbol{\beta}}_{\mathrm{R}}) = 0.0053,$$

compare also Fig. 3.1. □

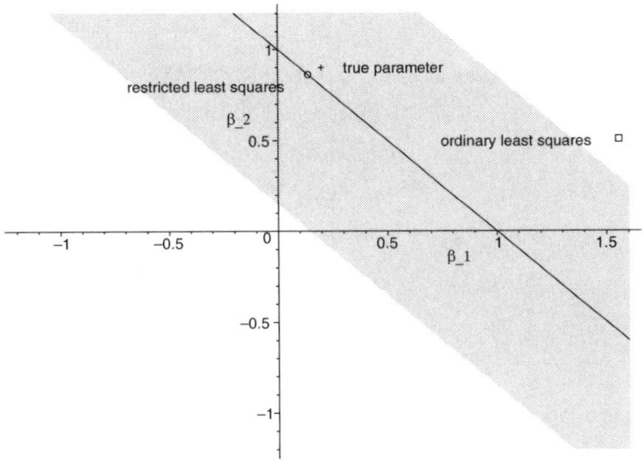

Fig. 3.1. True parameter and estimates in Example 3.2

Figure 3.1 shows the true parameter vector, the ordinary least squares and the restricted least squares estimates. The line contains all parameter values $\boldsymbol{\beta} = (\beta_1, \beta_2)'$ satisfying the restrictions. The grey zone contains all values $\boldsymbol{\beta} = (\beta_1, \beta_2)'$ for which $\lambda \leq \frac{1}{2}$ is satisfied.

Example 3.3. Consider the situation from Example 3.2, where now a different realization of $\boldsymbol{y}$ is given, namely

$$\boldsymbol{y} = (-0.4362, 1.4540, 2.9561, 2.0109, 2.9237)' .$$

In this case we compute

$$\widehat{\boldsymbol{\beta}} = \begin{pmatrix} -0.4013 \\ 0.7277 \end{pmatrix} \quad \text{and} \quad \widehat{\boldsymbol{\beta}}_{\mathrm{R}} = \begin{pmatrix} 0.4969 \\ 0.5031 \end{pmatrix}$$

as estimates for $\boldsymbol{\beta}$. The corresponding losses are given as

$$L(\boldsymbol{\beta}, \widehat{\boldsymbol{\beta}}) = 0.3913 \quad \text{and} \quad L(\boldsymbol{\beta}, \widehat{\boldsymbol{\beta}}_{\mathrm{R}}) = 0.2457 ,$$

when $\boldsymbol{\beta} = (0.2, 0.9)'$ is the true parameter. In this case the improvement of $\widehat{\boldsymbol{\beta}}_{\mathrm{R}}$ upon $\widehat{\boldsymbol{\beta}}$ with respect to the observed loss is considerably smaller than before, see also Fig 3.2. □

The above example shows again that *observed* losses can differ a lot from *expected* losses. In Example 3.2 the expected losses of $\widehat{\boldsymbol{\beta}}$ and $\widehat{\boldsymbol{\beta}}_{\mathrm{R}}$ are given as

$$\rho(\boldsymbol{\beta}, \widehat{\boldsymbol{\beta}}) = 1.2 \quad \text{and} \quad \rho(\boldsymbol{\beta}, \widehat{\boldsymbol{\beta}}_{\mathrm{R}}) = 0.0856 ,$$

when $\boldsymbol{\beta} = (0.2, 0.9)'$ is the true parameter vector. From these values, the restricted least squares estimator distinctly improves upon the ordinary least squares estimator. As demonstrated above, this does not imply that observed losses must behave in the same order of magnitude.

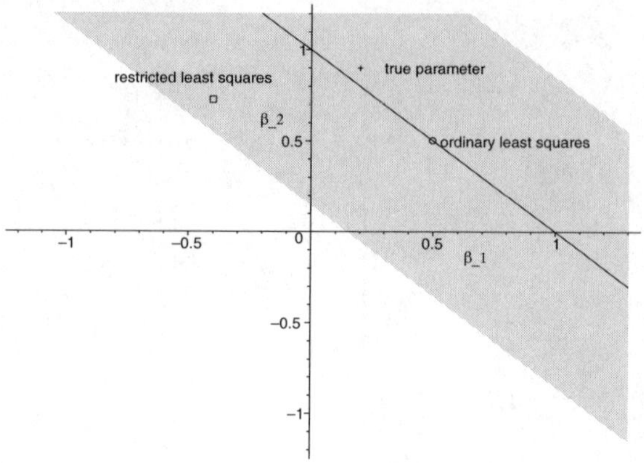

Fig. 3.2. True parameter and estimates in Example 3.3

3.1.5 Pretest Estimation

It goes without saying that the true parameter vector $\boldsymbol{\beta}$ is unknown in practice and we do not have the opportunity to compute observed and expected losses. However, to decide which of the two estimators $\widehat{\boldsymbol{\beta}}$ and $\widehat{\boldsymbol{\beta}}_{\mathrm{R}}$ is useful, it would be important to know whether the true $\boldsymbol{\beta}$ satisfies the inequality

$$\lambda = \frac{\boldsymbol{\delta}'(\boldsymbol{R}\boldsymbol{S}^{-1}\boldsymbol{R}')^{-1}\boldsymbol{\delta}}{2\sigma^2} \leq \frac{1}{2},$$

since, in this case, the expected loss of $\widehat{\boldsymbol{\beta}}_{\mathrm{R}}$ is not greater than the expected loss of $\widehat{\boldsymbol{\beta}}$.

To find some evidence for $\lambda \leq \frac{1}{2}$, we can perform a test of the hypothesis $H_0 : \lambda \leq \frac{1}{2}$ (if assumption (v) is satisfied) and choose either the ordinary or the restricted least squares estimator, depending on rejection or non-rejection of H_0 (method of Toro-Vizcarrondo and Wallace [116]).

Very often, under the linear regression model with assumptions (i) to (iv) and (v), one performs a test of the hypothesis $H_0 : \lambda = 0$, which can equivalently be written as the linear hypothesis $H_0 : \boldsymbol{R}\boldsymbol{\beta} = \boldsymbol{r}$. The corresponding test statistic is given as

$$F = \frac{(\widehat{\boldsymbol{\beta}} - \widehat{\boldsymbol{\beta}}_{\mathrm{R}})'\boldsymbol{X}'\boldsymbol{X}(\widehat{\boldsymbol{\beta}} - \widehat{\boldsymbol{\beta}}_{\mathrm{R}})}{\boldsymbol{y}'\boldsymbol{M}\boldsymbol{y}} \frac{n-p}{m}, \qquad \boldsymbol{M} = \boldsymbol{I}_n - \boldsymbol{X}(\boldsymbol{X}'\boldsymbol{X})^{-1}\boldsymbol{X}'.$$

If, for a given level α, the value of the test statistic satisfies

$$F \geq F_{m,n-p;1-\alpha},$$

then the hypothesis H_0 is rejected and the ordinary least squares estimator $\widehat{\boldsymbol{\beta}}$ is chosen for estimating $\boldsymbol{\beta}$. Otherwise, the restricted least squares estimator

$\widehat{\boldsymbol{\beta}}_{\mathrm{R}}$ is employed. The estimator obtained by proceeding in this way can be written as

$$\widehat{\boldsymbol{\beta}}_{\mathrm{P}} = I_{[0,c)}(F)\widehat{\boldsymbol{\beta}}_{\mathrm{R}} + I_{[c,\infty)}(F)\widehat{\boldsymbol{\beta}} \,,$$

where $c = F_{m,n-p;1-\alpha}$ and

$$I_{[a,b)}(u) = \begin{cases} 1 & \text{if } a \leq u < b \\ 0 & \text{otherwise} \end{cases} .$$

Notation 3.2. *Under the linear regression model with assumptions (i) to (iv) and (v), the vector* $\widehat{\boldsymbol{\beta}}_{\mathrm{P}}$ *is called* pretest estimator *of* $\boldsymbol{\beta}$.

A pretest estimator is non-linear, since it depends on the non-linear statistic F.

Example 3.4. Consider again the situation from example 3.2. We carry out the pretest estimation for $\boldsymbol{\beta}$ with level $\alpha = 0.05$ and compute

$$c = F_{m,n-p;1-\alpha} = F_{1,3;0.95} = 10.13.$$

For the observed vector $\boldsymbol{y} = (2.6262, 2.1640, 2.3589, 3.9844, 4.2513)'$ we obtain

$$F = 1.8715$$

and thus $\widehat{\boldsymbol{\beta}}_{\mathrm{P}} = \widehat{\boldsymbol{\beta}}_{\mathrm{R}} = \begin{pmatrix} 0.1387 \\ 0.8613 \end{pmatrix}$.

For the observed vector $\boldsymbol{y} = (-0.4362, 1.4540, 2.9561, 2.0109, 2.9237)'$ we compute

$$F = 0.7517$$

and thus $\widehat{\boldsymbol{\beta}}_{\mathrm{P}} = \widehat{\boldsymbol{\beta}}_{\mathrm{R}} = \begin{pmatrix} 0.4969 \\ 0.5031 \end{pmatrix}$. □

Note that although the *observed losses* of $\widehat{\boldsymbol{\beta}}_{\mathrm{R}}$ and $\widehat{\boldsymbol{\beta}}_{\mathrm{P}}$ are identical in the above example when $\boldsymbol{\beta} = (0.2, 0.9)'$ is the true parameter vector, the *expected losses* are quite different. The derivation of the risk of the pretest estimator is rather complicated, see e.g. [63, Chap. 5]. Fig. 3.3 visualizes the risks (based on the usual unweighted squared error loss) of $\widehat{\boldsymbol{\beta}}$, $\widehat{\boldsymbol{\beta}}_{\mathrm{R}}$ and $\widehat{\boldsymbol{\beta}}_{\mathrm{P}}$ (where $\alpha = 0.05$) from Example 3.2.

Now, in Example 3.2 let us consider the set $\Delta = \{\widehat{\boldsymbol{\beta}}, \widehat{\boldsymbol{\beta}}_{\mathrm{R}}, \widehat{\boldsymbol{\beta}}_{\mathrm{P}}\}$ of estimators for $\boldsymbol{\beta}$. Fig. 3.3 shows that each of the three estimators is admissible in Δ when the usual unweighted squared error loss function is used. On the other hand, $\widehat{\boldsymbol{\beta}}_{\mathrm{R}}$ and $\widehat{\boldsymbol{\beta}}_{\mathrm{P}}$ are reasonable estimators only in cases when we have some indication that the true parameters $(\boldsymbol{\beta}, \sigma^2)$ yield a small value λ. This is the case in Example 3.2 where $(\boldsymbol{\beta}, \sigma^2) = ((0.2, 0.9)', 1)$, so that $\lambda = 0.0083$. If we have no indication that λ can be small, then the use of restricted and pretest estimator is highly questionable. If in Example 3.2 the true parameters $(\boldsymbol{\beta}, \sigma^2)$ yielded $\lambda = 6$, then the risk of $\widehat{\boldsymbol{\beta}}_{\mathrm{P}}$ would be approximately five times greater

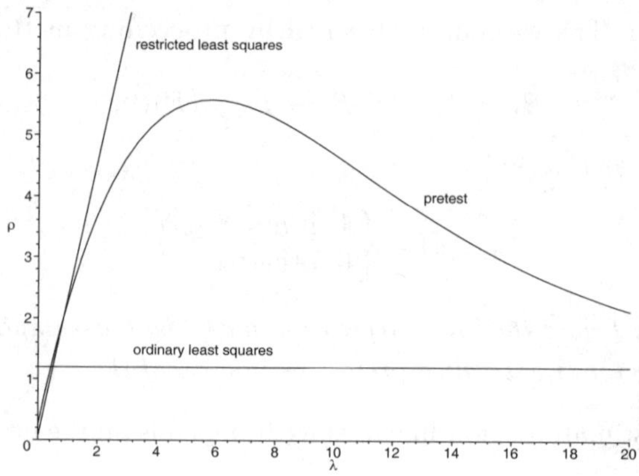

Fig. 3.3. $\rho(\boldsymbol{\beta},\widehat{\boldsymbol{\beta}})$,$\rho(\boldsymbol{\beta},\widehat{\boldsymbol{\beta}}_{\mathrm{R}})$ and $\rho(\boldsymbol{\beta},\widehat{\boldsymbol{\beta}}_{\mathrm{P}})$ in Example 3.2

than the risk of the ordinary least squares estimator. Obviously the risk of $\widehat{\boldsymbol{\beta}}_{\mathrm{R}}$ would even be greater than that of $\widehat{\boldsymbol{\beta}}_{\mathrm{P}}$.

It is clear that not only for $\lambda = 6$ but also for other values of λ, the expected losses of $\widehat{\boldsymbol{\beta}}_{\mathrm{P}}$ and $\widehat{\boldsymbol{\beta}}_{\mathrm{R}}$ are greater than the expected loss of the ordinary least squares estimator and it is not useful to apply these two estimators when the true parameters yield such values λ. This also shows once more that admissibility alone cannot be a criterion for the actual choice of an appropriate estimator, but additional information about the parameters and/or the nature of the estimators in question are necessary. On the other hand, if it would have turned out that $\widehat{\boldsymbol{\beta}}_{\mathrm{P}}$ is uniformly worse than $\widehat{\boldsymbol{\beta}}$ (which is not the case), then one would have dismissed $\widehat{\boldsymbol{\beta}}_{\mathrm{P}}$ as a reasonable estimator for $\boldsymbol{\beta}$.

Example 3.5. Consider the situation from Example 3.2 with the distinction that the true parameter vector is now given as $\boldsymbol{\beta} = (1.8, 1.9)'$. This vector does not satisfy the restrictions $\boldsymbol{R\beta} = \boldsymbol{r}$ from Example 3.2 and we compute $\lambda = 6.075$. If we generate 50 realizations of $\boldsymbol{y}$ under this new model, then we can compute 50 estimates $\widehat{\boldsymbol{\beta}}_{\mathrm{P}}$ for $\boldsymbol{\beta}$, shown in Fig. 3.4.

Here the dotted line stands for the set of all values $\boldsymbol{\beta} = (\beta_1, \beta_2)'$ satisfying the restrictions. The pretest estimator comes to the ordinary least squares estimate in 33 cases to the restricted least squares estimate in 17 cases. If we compute the mean of the observed losses of the 50 estimates, we obtain 5.4815, being not far away from the expected loss of the pretest estimator when $\lambda = 6.075$, compare also Fig. 3.3. On the other hand, the observed loss of an individual estimate will distinctly differ from the mean of the observed losses. If the pretest estimation procedure comes to the restricted least squares estimate, then the observed loss will be considerably greater than the mean, whereas in the other case the observed loss will be considerably smaller. Since in 17 out of 50 cases the pretest estimator comes

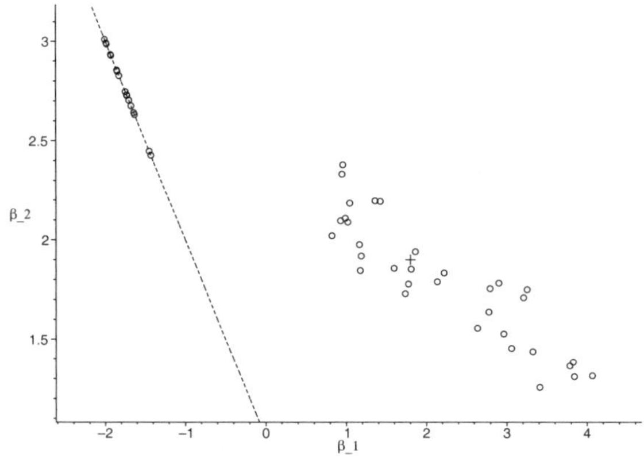

Fig. 3.4. 50 pretest estimates (circles) of the true parameter (cross) $\boldsymbol{\beta} = (1.8, 1.9)'$

to a restricted least squares estimate, we are confronted with a high probability for obtaining an estimate far away from the true parameter vector. □

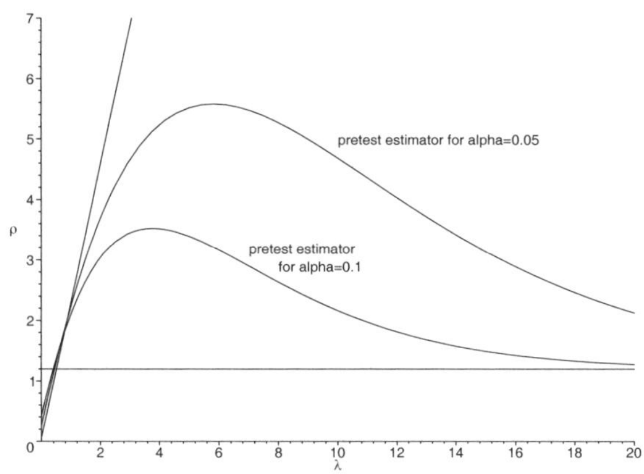

Fig. 3.5. Risks of pretest estimators with levels $\alpha = 0.05$ and $\alpha = 0.1$ in Example 3.2

Choice of the Level

The risk of the pretest estimator does, of course, also depend on the choice of the level α. Fig. 3.5 shows the risks of the pretest estimator in Example 3.2 for levels $\alpha = 0.05$ and $\alpha = 0.1$. For large values of λ the risk of the pretest estimator with level $\alpha = 0.1$ is distinctly smaller than the risk of the pretest estimator with level $\alpha = 0.05$. On the other hand, for small values of λ the situation is reversed, see Fig. 3.6. If the level α becomes greater, then the probability for rejecting the hypothesis raises. In the case $\alpha = 1$ we always reject the hypothesis and the pretest estimator coincides with the ordinary least squares estimator. Vice versa, if α becomes smaller, the probability for rejection falls. In the case $\alpha = 0$ we never reject the hypothesis, and thus the pretest estimator coincides with the restricted least squares estimator.

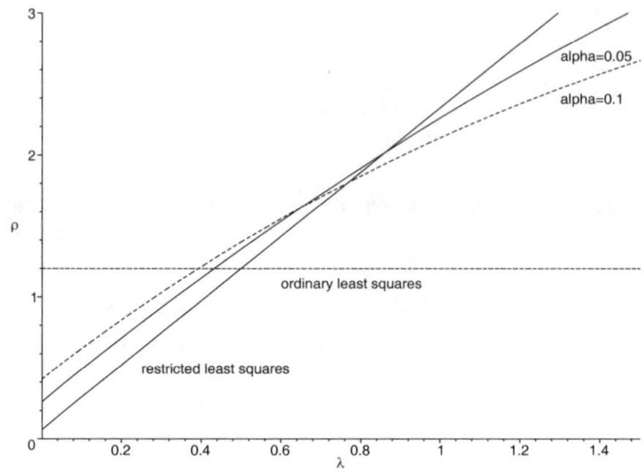

Fig. 3.6. Risks of pretest estimators with levels $\alpha = 0.05$ and $\alpha = 0.1$ in Example 3.2

3.2 Other Types of Restriction

Instead of assuming that the parameter vector $\boldsymbol{\beta}$ is the solution to a linear system, other possibilities for specifying restrictions may turn out to be appropriate to incorporate additional information.

3.2.1 Stochastic Linear Restrictions

If we consider the identity $\boldsymbol{R\beta} = \boldsymbol{r}$, we can imagine situations when this relationship does not hold exactly, but is disturbed by some random error $\boldsymbol{v}$. In other words, the given vector $\boldsymbol{r}$ could be the realization of a random vector, and the restriction would then be of the form

$$\boldsymbol{r} = \boldsymbol{R\beta} + \boldsymbol{v} ,$$

where $\mathrm{E}(\boldsymbol{v}) = \mathbf{0}$ and $\mathrm{Cov}(\boldsymbol{v}) = \boldsymbol{V}_0$, where $\boldsymbol{V}_0$ is a known $m \times m$ symmetric positive definite matrix. Then we have two linear models, both involving the unknown parameter $\boldsymbol{\beta}$. These two model can be combined into one via

$$\begin{pmatrix} \boldsymbol{y} \\ \boldsymbol{r} \end{pmatrix} = \begin{pmatrix} \boldsymbol{X} \\ \boldsymbol{R} \end{pmatrix} \boldsymbol{\beta} + \begin{pmatrix} \boldsymbol{\varepsilon} \\ \boldsymbol{v} \end{pmatrix} ,$$

where $\mathrm{E}[(\boldsymbol{\varepsilon}', \boldsymbol{v}')'] = \mathbf{0}$ and

$$\mathrm{Cov} \begin{pmatrix} \boldsymbol{\varepsilon} \\ \boldsymbol{v} \end{pmatrix} = \begin{pmatrix} \sigma^2 \boldsymbol{I}_n & \mathbf{0} \\ \mathbf{0} & \boldsymbol{V}_0 \end{pmatrix} ,$$

when we assume that $\mathrm{Cov}(\boldsymbol{\varepsilon}, \boldsymbol{v}) = \mathbf{0}$ (mixed model). In this situation, however, the vector $(\boldsymbol{\varepsilon}', \boldsymbol{v}')'$ does not satisfy model assumption (iv). Nonetheless, we can compute the *generalized least squares estimator* under this model, see Sect. 5.2, which comes up to

$$\widehat{\boldsymbol{\beta}}_* = (\sigma^{-2} \boldsymbol{X}'\boldsymbol{X} + \boldsymbol{R}'\boldsymbol{V}_0^{-1}\boldsymbol{R})^{-1}(\sigma^{-2}\boldsymbol{X}'\boldsymbol{y} + \boldsymbol{R}'\boldsymbol{V}_0^{-1}\boldsymbol{r}).$$

This estimator has been developed by Theil and Goldberger [114] and is known as the *mixed estimator*. It can easily be generalized to situations when the covariance matrix of $\boldsymbol{\varepsilon}$ is $\sigma^2 \boldsymbol{V}$ for some known $n \times n$ symmetric positive definite matrix $\boldsymbol{V}$ and/or $\mathrm{Cov}(\boldsymbol{\varepsilon}, \boldsymbol{v}) \neq \mathbf{0}$. However, in the situation described above, the mixed estimator is only practical if σ^2 is replaced by some estimator $\widehat{\sigma}^2$.

For the application of this estimator, the matrix $\boldsymbol{R}$, the vector $\boldsymbol{r}$, and in addition the matrix $\boldsymbol{V}_0$ must be known. A special case of the mixed estimator can, for example, be used when some earlier estimate $\widehat{\boldsymbol{\beta}}_{\mathrm{old}}$ of $\boldsymbol{\beta}$ and some estimate $\widehat{\boldsymbol{\Sigma}}$ of $\mathrm{Cov}(\widehat{\boldsymbol{\beta}}_{\mathrm{old}})$ are available. In that case, we may assume

$$\widehat{\boldsymbol{\beta}}_{\mathrm{old}} = \boldsymbol{\beta} + \boldsymbol{v} ,$$

where $\mathrm{E}(\boldsymbol{v}) = \mathbf{0}$ and $\mathrm{Cov}(\boldsymbol{v}) = \mathrm{Cov}(\widehat{\boldsymbol{\beta}}_{\mathrm{old}})$, and we can apply the mixed estimator for the choices $\boldsymbol{R} = \boldsymbol{I}_p$, $\boldsymbol{r} = \widehat{\boldsymbol{\beta}}_{\mathrm{old}}$, and $\boldsymbol{V}_0 = \widehat{\boldsymbol{\Sigma}}$. The resulting practical estimator is

$$\widehat{\boldsymbol{\beta}}_* = (\widehat{\sigma}^{-2} \boldsymbol{X}'\boldsymbol{X} + \widehat{\boldsymbol{\Sigma}}^{-1})^{-1}(\widehat{\sigma}^{-2}\boldsymbol{X}'\boldsymbol{y} + \widehat{\boldsymbol{\Sigma}}^{-1}\widehat{\boldsymbol{\beta}}_{\mathrm{old}}) ,$$

which can also be seen as the practical variant of the linear Bayes estimator when $\boldsymbol{\beta}$ is regarded as the realization of a random vector $\boldsymbol{b}$ with $\mathrm{E}(\boldsymbol{b}) = \widehat{\boldsymbol{\beta}}_{\mathrm{old}}$ and $\mathrm{Cov}(\boldsymbol{b}) = \widehat{\boldsymbol{\Sigma}}$, see also Sect. 3.8.

This procedure can also be applied when earlier estimates are available for some of the parameters only. Then $m < p$ and each row of the matrix $\boldsymbol{R}$ is the transpose of a unit vector, while $\widehat{\boldsymbol{\Sigma}}$ is an estimate of $\boldsymbol{R}\,\mathrm{Cov}(\widehat{\boldsymbol{\beta}}_{\mathrm{old}})\boldsymbol{R}'$. See also [117] for an extensive discussion of mixed estimation.

3.2.2 Inequality Restrictions

Another type of restrictions on $\boldsymbol{\beta}$ is given when specific linear functions of the elements of the vector $\boldsymbol{\beta}$ satisfy certain inequalities rather than equalities, so that

$$r_0 \leq R_0\beta \quad \text{and} \quad R_1\beta \leq r_1 ,$$

where the inequality sign '$\leq$' is understood to hold element-wise. These inequalities can also be written in the form

$$\begin{pmatrix} r_0 \\ -r_1 \end{pmatrix} \leq \begin{pmatrix} R_0 \\ -R_1 \end{pmatrix} \beta ,$$

so that, without loss of generality, we can always assume some given inequality restrictions of the form

$$r \leq R\beta .$$

Under such a situation we can still think of minimizing the quadratic form

$$f(\beta_*) = (y - X\beta_*)'(y - X\beta_*) ,$$

but now with respect to all vectors β_* satisfying $r \leq R\beta_*$. This optimization problem can iteratively be solved as a quadratic programming problem and the solution can be called the *inequality restricted least squares estimator* of β. For example, the statistical computer program R, see e.g. [27, 40], provides the function `solve.QP` in the package `quadprog`, which can be used to find the solution to the problem

$$\min_{x:A'x \geq x_0} (-d'x + \frac{1}{2}x'Dx)$$

for given d, D, A, and x_0. Then, this function can be used in the situation above to derive the inequality restricted least squares estimator by setting $d = 2X'y$, $D = 2X'X$, $A' = R$ and $x_0 = r$.

As shown in [34], for R of full row rank, the inequality restricted least squares estimator can be written in the form

$$\widehat{\beta}_{\mathrm{IR}} = \widehat{\beta} - (X'X)^{-1}R'_\Delta[R_\Delta(X'X)^{-1}R'_\Delta]^{-1}(R_\Delta\widehat{\beta} - r_\Delta)$$

for some set $\Delta \subseteq \{1, \ldots, m\}$. Here, A_Δ denotes the matrix which is obtained from a matrix A with m rows by deleting its i-th row for all those i for which i is not in Δ. If $\Delta = \emptyset$, then A_Δ is interpreted as absent.

Since the form of the estimator is known, it 'only' remains to find the appropriate subset Δ among 2^m possible subsets of $\{1, \ldots, m\}$. If m is large, this can be quite a lengthly task, so that in this case one would switch to numerical solutions provided by algorithms, as noted above. Compare also [34, 125] for details on how to obtain the appropriate Δ.

Note that although the above estimator $\widehat{\beta}_{\mathrm{IR}}$ is of the form of a restricted least squares estimator with respect to restrictions $R_\Delta\beta = r_\Delta$, it is nevertheless a *non-linear* estimator. This is due to the fact that the appropriate subset Δ depends on the actual value of y, and hence R_Δ and r_Δ are stochastic. This makes it difficult to derive further statistical properties (like risk performance) of the non-linear inequality restricted least squares estimator.

Example 3.6. Consider the linear regression model $\boldsymbol{y} = \boldsymbol{X}\boldsymbol{\beta} + \varepsilon$, where

$$\boldsymbol{X}' = \begin{pmatrix} 1 & 1 & 1 & 1 & 1 \\ 1 & 2 & 3 & 4 & 5 \end{pmatrix}$$

and the parameter vector $\boldsymbol{\beta} = (\beta_1, \beta_2)'$ is assumed to satisfy

$$\begin{pmatrix} 0 \\ -1 \end{pmatrix} \leq \begin{pmatrix} 1 & 0 \\ -1 & -1 \end{pmatrix} \begin{pmatrix} \beta_1 \\ \beta_2 \end{pmatrix} ,$$

i.e. $\beta_1 \geq 0$ and $\beta_1 + \beta_2 \leq 1$. The true parameters are

$$\boldsymbol{\beta} = \begin{pmatrix} 0.2 \\ 0.7 \end{pmatrix} \quad \text{and} \quad \sigma^2 = 1.$$

We generate 50 realizations of $\boldsymbol{y}$ and compute the corresponding inequality restricted least squares estimates via the above mentioned function `solve.QP`. Fig. 3.7 displays the result. As can be seen, there are several estimates corresponding to the subsets $\Delta = \{1\}$ and $\Delta = \{2\}$ (belonging to the restrictions $\beta_1 \geq 0$ and $\beta_1 + \beta_2 \leq 1$, respectively), indicated by the dotted lines. There is also one estimate which corresponds to the set $\Delta = \{1, 2\}$, and there are eight estimates which correspond to $\Delta = \emptyset$. The latter coincide with the ordinary least squares estimates. □

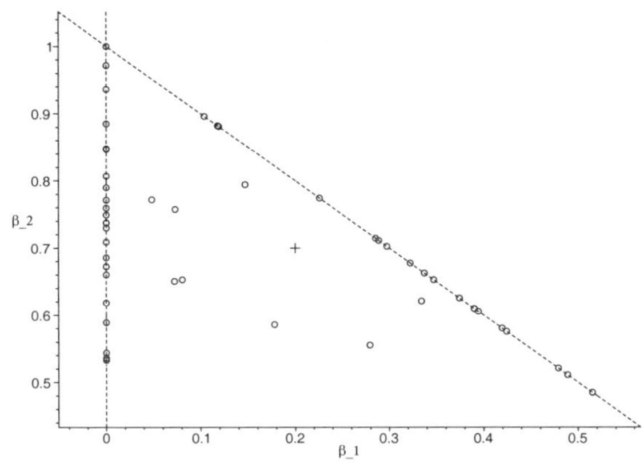

Fig. 3.7. 50 inequality restricted least squares estimates in Example 3.6

3.2.3 Elliptical Restrictions

If we know that the unknown parameter vector $\boldsymbol{\beta}$ lies in a specific region of $\mathbb{R}^p$, then we might describe this region in the form of an ellipsoid. The set of all vectors $\boldsymbol{\beta} \in \mathbb{R}^p$ satisfying

$$(\beta - \beta_0)'T(\beta - \beta_0) \leq 1$$

for a given $p \times p$ symmetric positive definite matrix T and a given $p \times 1$ vector β_0 specifies an ellipsoid in $\mathbb{R}^p$ with β_0 as its center. For example if

$$T = \begin{pmatrix} 4 & 1 \\ 1 & 2 \end{pmatrix} \quad \text{and} \quad \beta_0 = \begin{pmatrix} 1 \\ 2 \end{pmatrix},$$

then the corresponding ellipsoid is shown in Fig. 3.8.

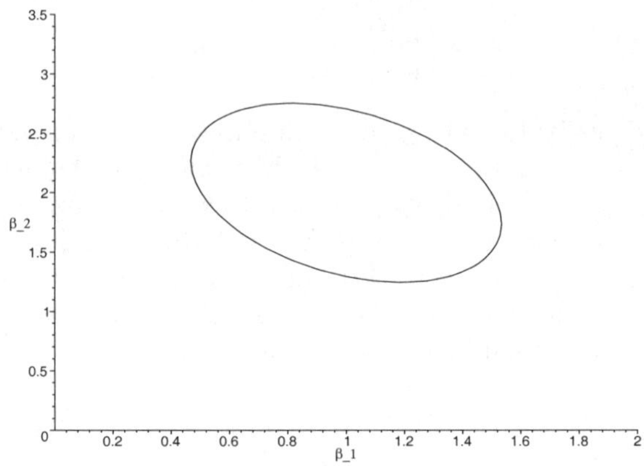

Fig. 3.8. Ellipsoid in $\mathbb{R}^2$

A possible estimator for β when elliptical restrictions are given is provided by

$$\breve{\beta} = \beta_0 + (X'X + \sigma^2 T)^{-1} X'(y - X\beta_0),$$

which, however, is only practical if σ^2 is either known or replaced by some estimator $\widehat{\sigma}^2$. In the latter case, the resulting estimator is *non-linear.*

In Sect. 3.7 we consider the estimation problem when elliptical restrictions are given in more detail.

3.3 Principal Components Estimator

The principal components estimator is a special restricted least squares estimator which can be used in presence of collinearity. By imposing appropriate restrictions on the parameter vector it is aimed to obtain an estimator which cannot take its values in those directions where the least squares estimator has large variance.

3.3.1 Preliminary Considerations

In Example 2.6 (p. 55) we consider a linear regression model $\boldsymbol{y} = \boldsymbol{X\beta} + \boldsymbol{\varepsilon}$, where

$$\boldsymbol{X}' = \begin{pmatrix} 1 & 1 & 1 & 1 & 1 \\ 1.9 & 2.1 & 2 & 2 & 1.8 \end{pmatrix} .$$

For the observed

$$\boldsymbol{y} = (6.0521, 7.0280, 7.1230, 4.4441, 5.0813)'$$

the least squares estimate

$$\widehat{\boldsymbol{\beta}} = (\boldsymbol{X}'\boldsymbol{X})^{-1}\boldsymbol{X}'\boldsymbol{y} = \begin{pmatrix} -4.2489 \\ 5.2013 \end{pmatrix}$$

lies considerably far away from the true parameter vector $\boldsymbol{\beta} = (2, 2)'$. This will also hold true for many more realizations $\boldsymbol{y}$, so that, although the least squares estimator is unbiased for $\boldsymbol{\beta}$, it is likely here that individual estimates are far away from the true parameter. The reason for the impreciseness of the least squares estimator in Example 2.6 is the bad condition of the matrix $\boldsymbol{X}'\boldsymbol{X}$. Since the second column of $\boldsymbol{X}$ equals approximately twice its first column, the matrix $\boldsymbol{X}'\boldsymbol{X}$ is almost singular.

To avoid the consequences of this bad condition one could think of building a new model. Since the values of the second independent variable (second column of $\boldsymbol{X}$) are approximately twice the values of the first independent variable (first column of $\boldsymbol{X}$) it seems to be a nearby solution to remove one of the two variables in question. It can be argued that the removed independent variable does not contribute much to the explanation of the dependent variable in presence of the other independent variable (overfitting of the model). A new model can then be obtained with only *one* variable and only *one* unknown parameter to estimate. Often, however, it is not as easy to see the origin of the bad condition as it is in Example 2.6. Then the building of a new model can become rather complicated.

Another possibility for the removal of information which is responsible for the increase of impreciseness in estimation is offered by the principal components estimator. For this, let us consider the spectral decomposition of the matrix $\boldsymbol{X}'\boldsymbol{X}$ given as

$$\boldsymbol{X}'\boldsymbol{X} = (\boldsymbol{U}_1, \boldsymbol{U}_2) \begin{pmatrix} \boldsymbol{\Lambda}_1 & \boldsymbol{0} \\ \boldsymbol{0} & \boldsymbol{\Lambda}_2 \end{pmatrix} \begin{pmatrix} \boldsymbol{U}_1' \\ \boldsymbol{U}_2' \end{pmatrix} .$$

Here $\boldsymbol{\Lambda}_1$ and $\boldsymbol{\Lambda}_2$ are diagonal matrices such that the main diagonal elements of the $r \times r$ matrix $\boldsymbol{\Lambda}_1$ are the $r \leq p$ largest eigenvalues of $\boldsymbol{X}'\boldsymbol{X}$, while the main diagonal elements of the $(p-r) \times (p-r)$ matrix $\boldsymbol{\Lambda}_2$ are the remaining $p - r$ eigenvalues. The $p \times p$ matrix $\boldsymbol{U} = (\boldsymbol{U}_1, \boldsymbol{U}_2)$ is orthogonal with $\boldsymbol{U}_1 = (\boldsymbol{u}_1, \ldots, \boldsymbol{u}_r)$ consisting of its first r columns and $\boldsymbol{U}_2 = (\boldsymbol{u}_{r+1}, \ldots, \boldsymbol{u}_p)$ consisting of the remaining $p - r$ columns of the matrix $\boldsymbol{U}$.

If $\boldsymbol{\Lambda}$ denotes the matrix

$$\boldsymbol{\Lambda} = \begin{pmatrix} \boldsymbol{\Lambda}_1 & \mathbf{0} \\ \mathbf{0} & \boldsymbol{\Lambda}_2 \end{pmatrix} = \begin{pmatrix} \lambda_1 & 0 & \dots & \dots & \dots & 0 \\ 0 & \ddots & & & & \vdots \\ \vdots & & \lambda_r & & & \vdots \\ \vdots & & & \lambda_{r+1} & & \vdots \\ \vdots & & & & \ddots & 0 \\ 0 & \dots\dots & \dots & & 0 & \lambda_p \end{pmatrix},$$

then the p columns of the matrix $\boldsymbol{U\Lambda} = (\lambda_1 \boldsymbol{u}_1, \dots, \lambda_p \boldsymbol{u}_p)$ generate the whole $\mathbb{R}^p$. Since fairly small eigenvalues of $\boldsymbol{X}'\boldsymbol{X}$ are responsible for the impreciseness of the least squares estimator, it seems to be a good strategy to prevent the estimate to move in directions $\lambda_i \boldsymbol{u}_i$ with corresponding fairly small λ_i. One such strategy is offered by the restricted least squares estimator for $\boldsymbol{\beta} \in \mathbb{R}^p$ by restricting $\boldsymbol{\beta}$ to lie in the subspace which is generated by the columns $\lambda_1 \boldsymbol{u}_1, \dots, \lambda_r \boldsymbol{u}_r$, where $\lambda_1, \dots, \lambda_r$ are the $r \leq p$ largest eigenvalues of $\boldsymbol{X}'\boldsymbol{X}$. This estimator eliminates the disadvantageous influence of the remaining columns $\lambda_{r+1}\boldsymbol{u}_{r+1}, \dots, \lambda_p \boldsymbol{u}_p$ on the estimation.

The space generated by the columns $\lambda_1 \boldsymbol{u}_1, \dots, \lambda_r \boldsymbol{u}_r$ is identical with the column space $\mathcal{C}(\boldsymbol{U}_1)$ of the matrix $\boldsymbol{U}_1$. Restricting $\boldsymbol{\beta}$ to this space is equivalent to assuming that $\boldsymbol{\beta}$ can be written as $\boldsymbol{\beta} = \boldsymbol{U}_1 \boldsymbol{\xi}$ for some vector $\boldsymbol{\xi}$. This is satisfied if and only if

$$\boldsymbol{U}_1 \boldsymbol{U}_1' \boldsymbol{\beta} = \boldsymbol{\beta} \,.$$

Since $\boldsymbol{I}_p - \boldsymbol{U}_1 \boldsymbol{U}_1' = \boldsymbol{U}_2 \boldsymbol{U}_2'$, the above identity can be written in the form $\boldsymbol{U}_2 \boldsymbol{U}_2' \boldsymbol{\beta} = \mathbf{0}$, which again is equivalent to

$$\boldsymbol{U}_2' \boldsymbol{\beta} = \mathbf{0} \,.$$

The restricted least squares estimator with respect to this identity has the form

$$\widehat{\boldsymbol{\beta}}^{(r)} = \widehat{\boldsymbol{\beta}} - (\boldsymbol{X}'\boldsymbol{X})^{-1} \boldsymbol{U}_2 [\boldsymbol{U}_2' (\boldsymbol{X}'\boldsymbol{X})^{-1} \boldsymbol{U}_2]^{-1} \boldsymbol{U}_2' \widehat{\boldsymbol{\beta}} \,,$$

where $\widehat{\boldsymbol{\beta}} = (\boldsymbol{X}'\boldsymbol{X})^{-1} \boldsymbol{X}'\boldsymbol{y}$ is the ordinary least squares estimator.

Notation 3.3. *Under the linear regression model with assumptions (i) to (iv), a restricted least squares estimator* $\widehat{\boldsymbol{\beta}}^{(r)}$ *is called* principal components estimator *of* $\boldsymbol{\beta}$.

3.3.2 Properties of the Principal Components Estimator

The principal components estimator can alternatively be written with the help of the matrix $\boldsymbol{U}_1$ instead of the matrix $\boldsymbol{U}_2$.

Theorem 3.5. *Under the linear regression model with assumptions (i) to (iv), the principal components estimator for* $\boldsymbol{\beta}$ *can be written as*

$$\widehat{\boldsymbol{\beta}}^{(r)} = \boldsymbol{U}_1(\boldsymbol{U}_1'\boldsymbol{X}'\boldsymbol{X}\boldsymbol{U}_1)^{-1}\boldsymbol{U}_1'\boldsymbol{X}'\boldsymbol{y} = \boldsymbol{U}_1\boldsymbol{\Lambda}_1^{-1}\boldsymbol{U}_1'\boldsymbol{X}'\boldsymbol{y} .$$

Proof. Because of the orthogonality of the matrix $\boldsymbol{U} = (\boldsymbol{U}_1, \boldsymbol{U}_2)$, the matrices $\boldsymbol{U}_1$ and $\boldsymbol{U}_2$ satisfy the identities $\boldsymbol{U}_1'\boldsymbol{U}_1 = \boldsymbol{I}_r$, $\boldsymbol{U}_2'\boldsymbol{U}_2 = \boldsymbol{I}_{p-r}$, and $\boldsymbol{U}_1'\boldsymbol{U}_2 = \boldsymbol{0}$, as well as $\boldsymbol{U}_1\boldsymbol{U}_1' + \boldsymbol{U}_2\boldsymbol{U}_2' = \boldsymbol{I}_p$. From

$$(\boldsymbol{X}'\boldsymbol{X})^{-1} = \boldsymbol{U}_1\boldsymbol{\Lambda}_1^{-1}\boldsymbol{U}_1' + \boldsymbol{U}_2\boldsymbol{\Lambda}_2^{-1}\boldsymbol{U}_2'$$

it follows $[\boldsymbol{U}_2'(\boldsymbol{X}'\boldsymbol{X})^{-1}\boldsymbol{U}_2]^{-1} = \boldsymbol{\Lambda}_2$,

$$(\boldsymbol{X}'\boldsymbol{X})^{-1}\boldsymbol{U}_2[\boldsymbol{U}_2'(\boldsymbol{X}'\boldsymbol{X})^{-1}\boldsymbol{U}_2]^{-1}\boldsymbol{U}_2'(\boldsymbol{X}'\boldsymbol{X})^{-1} = \boldsymbol{U}_2\boldsymbol{\Lambda}_2^{-1}\boldsymbol{U}_2' ,$$

and

$$(\boldsymbol{X}'\boldsymbol{X})^{-1} - (\boldsymbol{X}'\boldsymbol{X})^{-1}\boldsymbol{U}_2[\boldsymbol{U}_2'(\boldsymbol{X}'\boldsymbol{X})^{-1}\boldsymbol{U}_2]^{-1}\boldsymbol{U}_2'(\boldsymbol{X}'\boldsymbol{X})^{-1} = \boldsymbol{U}_1\boldsymbol{\Lambda}_1^{-1}\boldsymbol{U}_1' .$$

In view of $\boldsymbol{\Lambda}_1 = \boldsymbol{U}_1'\boldsymbol{X}'\boldsymbol{X}\boldsymbol{U}_1$ we conclude

$$\begin{aligned}\widehat{\boldsymbol{\beta}}^{(r)} &= [(\boldsymbol{X}'\boldsymbol{X})^{-1} - (\boldsymbol{X}'\boldsymbol{X})^{-1}\boldsymbol{U}_2[\boldsymbol{U}_2'(\boldsymbol{X}'\boldsymbol{X})^{-1}\boldsymbol{U}_2]^{-1}\boldsymbol{U}_2'(\boldsymbol{X}'\boldsymbol{X})^{-1}]\boldsymbol{X}'\boldsymbol{y} \\ &= \boldsymbol{U}_1(\boldsymbol{U}_1'\boldsymbol{X}'\boldsymbol{X}\boldsymbol{U}_1)^{-1}\boldsymbol{U}_1'\boldsymbol{X}'\boldsymbol{y} ,\end{aligned}$$

which shows the assertion. □

Example 3.7. Considering the situation from Example 2.6, we now compute the principal components estimator for $\boldsymbol{\beta}$. For this we consider the decomposition

$$\boldsymbol{X}'\boldsymbol{X} = \underbrace{\begin{pmatrix} -0.4537 & -0.8912 \\ -0.8912 & 0.4537 \end{pmatrix}}_{\boldsymbol{U}} \underbrace{\begin{pmatrix} 24.2493 & 0 \\ 0 & 0.0107 \end{pmatrix}}_{\boldsymbol{\Lambda}} \begin{pmatrix} -0.4537 & -0.8912 \\ -0.8912 & 0.4537 \end{pmatrix} .$$

Here the orthogonal matrix $\boldsymbol{U}$ is symmetric. The first column

$$\boldsymbol{u}_1 = (-0.4537, -0.8912)'$$

of the matrix $\boldsymbol{U}$ corresponds to the eigenvalue $\lambda_1 = 24.2493$, while the second column of $\boldsymbol{U}$ corresponds to the eigenvalue $\lambda_2 = 0.0107$. For the observation

$$\boldsymbol{y} = (6.0521, 7.0280, 7.1230, 4.4441, 5.0813)'$$

the principal components estimator $\widehat{\boldsymbol{\beta}}^{(1)}$ comes up to

$$\widehat{\boldsymbol{\beta}}^{(1)} = \boldsymbol{u}_1(\boldsymbol{u}_1'\boldsymbol{X}'\boldsymbol{X}\boldsymbol{u}_1)^{-1}\boldsymbol{u}_1'\boldsymbol{X}'\boldsymbol{y} = \begin{pmatrix} 1.2284 \\ 2.4128 \end{pmatrix} .$$

This estimate is considerably nearer to the true parameter $\boldsymbol{\beta} = (2, 2)'$ than the ordinary least squares estimate $\widehat{\boldsymbol{\beta}} = (-4.2489, 5.2013)'$. To obtain an impression of the behavior of the principal components estimator compared to the ordinary least squares estimator, we generate 30 further realizations of $\boldsymbol{y}$ and compute the respective estimates $\widehat{\boldsymbol{\beta}}$ and $\widehat{\boldsymbol{\beta}}^{(1)}$. Fig. 3.9 shows the 30 values which are obtained from the ordinary least squares estimator. The two arrows represent the two vectors $\boldsymbol{u}_1$ and $\boldsymbol{u}_2$, where for better visuality the arrows are made ten times longer than the original vectors. Note that the two vectors are perpendicular to each other, which is not seen on first sight, since both axes are differently scaled. A good amount of the estimates is quite far away from the true parameter, where, obviously, the main direction of variance is parallel to the vector $\boldsymbol{u}_2$, being the eigenvector corresponding to the small eigenvalue $\lambda_2 = 0.0107$.

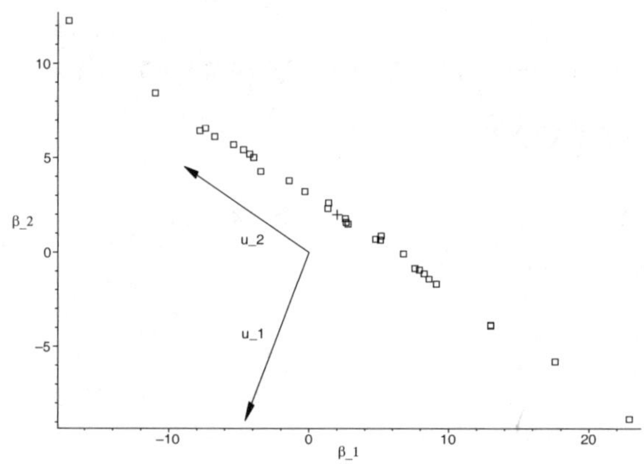

Fig. 3.9. True parameter (cross) and 30 ordinary least squares estimates (boxes) in Example 2.6

For the smaller area $0 \leq \beta_1 \leq 3$ and $0 \leq \beta_2 \leq 3$, Fig. 3.10 shows the 30 principal components estimates (circles) as well as those ordinary least squares estimates (boxes) realizing within this area. Here, again the two arrows represent the two vectors $\boldsymbol{u}_1$ and $\boldsymbol{u}_2$ but shifted from the origin $(0, 0)'$ to $(1, 1)'$. As can be seen, all principal components estimates lie on a straight line determined by the condition $\boldsymbol{u}_2'\widehat{\boldsymbol{\beta}}^{(1)} = 0$.

Here we can clearly see the effect of principal components estimation. The estimates cannot move in direction of $\boldsymbol{u}_2$ but only in direction of $\boldsymbol{u}_1$ (in contrast to the ordinary least squares estimates). This implies that large distances from the true parameter $\boldsymbol{\beta} = (2, 2)'$ are avoided. If we consider the observed unweighted squared error loss of an estimator $\widetilde{\boldsymbol{\beta}}$ for $\boldsymbol{\beta}$, then we

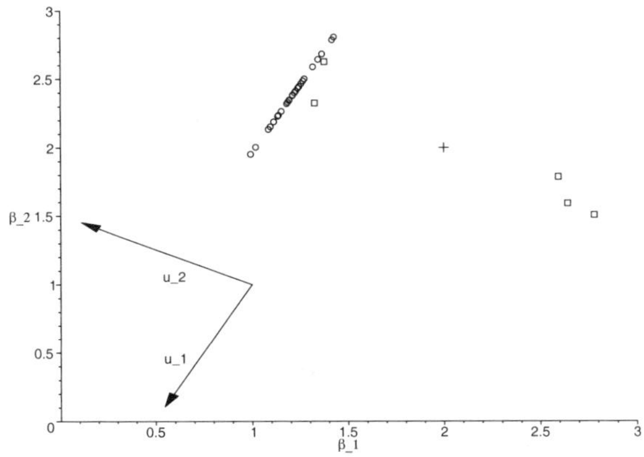

Fig. 3.10. True parameter (cross), some ordinary least squares estimates (boxes) and 30 principal components estimates (circles) in Example 2.6

compute 91.9358 as the average observed loss of $\widetilde{\boldsymbol{\beta}} = \widehat{\boldsymbol{\beta}}$, while we compute 0.8136 as the average loss for $\widetilde{\boldsymbol{\beta}} = \widehat{\boldsymbol{\beta}}^{(1)}$. □

Risk Comparison of Principal Components and Least Squares Estimator

Since the principal components estimator is a special case of the restricted least squares estimator, we can use the results from Sect. 3.1 to derive conditions under which $\widehat{\boldsymbol{\beta}}^{(r)}$ is uniformly not worse or better than the ordinary least squares estimator $\widehat{\boldsymbol{\beta}}$ with respect to $\mathrm{MSE}(\boldsymbol{\beta}, \widetilde{\boldsymbol{\beta}})$.

Theorem 3.6. *Under the linear regression model with assumptions (i) to (iv), the difference* $\boldsymbol{\Delta} = \mathrm{MSE}(\boldsymbol{\beta}, \widehat{\boldsymbol{\beta}}) - \mathrm{MSE}(\boldsymbol{\beta}, \widehat{\boldsymbol{\beta}}^{(r)})$ *is nonnegative definite if and only if the inequality*

$$\boldsymbol{\beta}'\boldsymbol{U}_2\boldsymbol{\Lambda}_2\boldsymbol{U}_2'\boldsymbol{\beta} \leq \sigma^2$$

is satisfied. If this inequality is strict, i.e. $\boldsymbol{\beta}'\boldsymbol{U}_2\boldsymbol{\Lambda}_2\boldsymbol{U}_2'\boldsymbol{\beta} < \sigma^2$, *then* $\boldsymbol{\Delta} \neq \mathbf{0}$.

Proof. The assertions follow from Theorem 3.4. □

As can be seen from the above theorem, when the main diagonal elements of $\boldsymbol{\Lambda}_2$ are rather small, then also $\boldsymbol{\beta}'\boldsymbol{U}_2\boldsymbol{\Lambda}_2\boldsymbol{U}_2'\boldsymbol{\beta}$ will be small, being a good prerequisite for the possible improvement of $\widehat{\boldsymbol{\beta}}^{(r)}$ upon $\widehat{\boldsymbol{\beta}}$.

Example 3.8. For the true parameters $(\boldsymbol{\beta}, \sigma^2) = (\begin{pmatrix} 2 \\ 2 \end{pmatrix}, 1)$ in Example 2.6, the inequality $\boldsymbol{\beta}'\boldsymbol{U}_2\boldsymbol{\Lambda}_2\boldsymbol{U}_2'\boldsymbol{\beta} \leq \sigma^2$ becomes $\boldsymbol{\beta}'\boldsymbol{u}_2\lambda_2\boldsymbol{u}_2'\boldsymbol{\beta} \leq \sigma^2$, being equivalent to

$$0.0082 \leq 1 \,.$$

Thus, the required inequality is satisfied. □

The Shrinkage Property of the Principal Components Estimator

The expected squared length of the ordinary least squares estimator is given as

$$\mathrm{E}(||\widehat{\boldsymbol{\beta}}||^2) = \sigma^2 \operatorname{tr}[(\boldsymbol{X}'\boldsymbol{X})^{-1}] + ||\boldsymbol{\beta}||^2 \,,$$

which means that the expected squared length of $\widehat{\boldsymbol{\beta}}$ is larger than the expected squared length of $\boldsymbol{\beta}$, see Theorem 2.6. In presence of collinearity, some eigenvalues of $\boldsymbol{X}'\boldsymbol{X}$ are relatively small and thus some eigenvalues of $(\boldsymbol{X}'\boldsymbol{X})^{-1}$ are relatively large. Consequently, $\operatorname{tr}[(\boldsymbol{X}'\boldsymbol{X})^{-1}]$ is also relatively large and an actual estimate $\widehat{\boldsymbol{\beta}}$ is likely to be considerably longer than the true parameter vector $\boldsymbol{\beta}$. If we look at the principal components estimator $\widehat{\boldsymbol{\beta}}^{(r)}$ and consider its expected squared length, then we can establish the fact that the expected squared length of the principal components estimator cannot become greater but only smaller with decreasing r.

Theorem 3.7. *Under the linear regression model with assumptions (i) to (iv), the inequality* $||\widehat{\boldsymbol{\beta}}^{(r-1)}||^2 \leq ||\widehat{\boldsymbol{\beta}}^{(r)}||^2$ *holds true for* $1 < r \leq p$.

Proof. For the squared length $||\widehat{\boldsymbol{\beta}}^{(r)}||^2 = (\widehat{\boldsymbol{\beta}}^{(r)})'\widehat{\boldsymbol{\beta}}^{(r)}$ of the principal components estimator $\widehat{\boldsymbol{\beta}}^{(r)} = \boldsymbol{U}_1\boldsymbol{\Lambda}_1^{-1}\boldsymbol{U}_1'\boldsymbol{X}'\boldsymbol{y}$, the identity

$$(\widehat{\boldsymbol{\beta}}^{(r)})'\widehat{\boldsymbol{\beta}}^{(r)} = \widehat{\boldsymbol{\beta}}'\boldsymbol{X}'\boldsymbol{X}\boldsymbol{U}_1\boldsymbol{\Lambda}_1^{-2}\boldsymbol{U}_1'\boldsymbol{X}'\boldsymbol{X}\widehat{\boldsymbol{\beta}} = \widehat{\boldsymbol{\beta}}'\boldsymbol{U}_1\boldsymbol{U}_1'\widehat{\boldsymbol{\beta}}$$

holds true. If $\boldsymbol{u}_i$ denotes the i-th column of the matrix $\boldsymbol{U} = (\boldsymbol{U}_1, \boldsymbol{U}_2)$, then it follows

$$(\widehat{\boldsymbol{\beta}}^{(r)})'\widehat{\boldsymbol{\beta}}^{(r)} = (\boldsymbol{u}_1'\widehat{\boldsymbol{\beta}})^2 + \cdots + (\boldsymbol{u}_r'\widehat{\boldsymbol{\beta}})^2 \,.$$

This shows the assertion. □

Since it is rather unlikely that $\boldsymbol{u}_i'\widehat{\boldsymbol{\beta}} = 0$ is *exactly* true for some column $\boldsymbol{u}_i$ of the matrix $\boldsymbol{U}$, usually the strict inequality $||\widehat{\boldsymbol{\beta}}^{(r-1)}||^2 < ||\widehat{\boldsymbol{\beta}}^{(r)}||^2$ will be valid, and the squared length of the principal components estimator will in fact become smaller for a decreasing value of r.

The Number of the Principal Components

When applying the principal components estimator $\widehat{\boldsymbol{\beta}}^{(r)}$, one is faced with problem of choosing an appropriate value r. If a certain number r is fixed, then under the additional model assumption (v) the corresponding restriction

$U_2'\beta = \mathbf{0}$ can be tested analogously to the pretest estimation procedure. The test statistic is given here as

$$F = \frac{(\widehat{\beta} - \widehat{\beta}^{(r)})' X' X (\widehat{\beta} - \widehat{\beta}^{(r)})}{y' M y} \frac{n-p}{p-r}, \qquad M = I_n - X(X'X)^{-1}X' .$$

If for a given level α the inequality

$$F \geq F_{p-r,n-p;1-\alpha}$$

is true, the hypothesis $H_0 : U_2'\beta = \mathbf{0}$ is rejected and the chosen r might be too small.

Basically it seems to be reasonable to restrict the estimate as little as possible and thus to choose a value for r as large as possible, but of course as small as necessary to obtain the desired effect of the estimator.

Example 3.9. In Example 2.6 the value of the test statistic F comes up to $F = 0.2918$, while the critical value $F_{p-r,n-p;1-\alpha}$ is computed as

$$F_{1,3;0.95} = 10.1280$$

when $\alpha = 0.05$. Hence, we cannot reject the hypothesis $H_0 : u_2'\beta = 0$. □

3.3.3 Drawbacks of the Principal Components Estimator

As noted before, the principal components estimator is a special case of the restricted least squares estimator with respect to restrictions

$$U_2'\beta = \mathbf{0} .$$

These restrictions are chosen in order to prevent the estimates to move into certain critical directions. This procedure does by no means assume that the unknown parameter vector β must satisfy the imposed restrictions.

Thus, the principal components estimator is considered as a biased estimator, where some bias is introduced (caused by restrictions which are not satisfied) in order to reduce the variance. As seen in the previous section, this appears to be a reasonable procedure only in those cases when the restrictions can be seen to be almost satisfied. The more the restrictions are violated, the less precise the estimator will be.

When applying the principal components estimator, without further ado we cannot assume that the imposed restrictions on the true parameter vector are true or almost true. For example if the unknown β happens to belong to the column space of U_2, then β can be written as $\beta = U_2 b$ for some $(p-r) \times 1$ vector b. In this case the restrictions are

$$b = \mathbf{0} .$$

Clearly, if the vector b happens to lie far away from the zero vector, then the restrictions cannot be said to be almost satisfied. See also [48] for some similar cautionary notes on the use of the principal components estimator.

3.3.4 The Marquardt Estimator

The principal components estimator

$$\widehat{\boldsymbol{\beta}}^{(r)} = \boldsymbol{U}_1(\boldsymbol{U}_1'\boldsymbol{X}'\boldsymbol{X}\boldsymbol{U}_1)^{-1}\boldsymbol{U}_1'\boldsymbol{X}'\boldsymbol{y} = \boldsymbol{U}_1\boldsymbol{\Lambda}_1^{-1}\boldsymbol{U}_1'\boldsymbol{X}'\boldsymbol{y}$$

can also be seen as a modification of the ordinary least squares estimator

$$\widehat{\boldsymbol{\beta}} = (\boldsymbol{X}'\boldsymbol{X})^{-1}\boldsymbol{X}'\boldsymbol{y} = (\boldsymbol{U}_1\boldsymbol{\Lambda}_1^{-1}\boldsymbol{U}_1' + \boldsymbol{U}_2\boldsymbol{\Lambda}_2^{-1}\boldsymbol{U}_2')\boldsymbol{X}'\boldsymbol{y} ,$$

which is simply achieved by deleting the matrix $\boldsymbol{U}_2\boldsymbol{\Lambda}_2^{-1}\boldsymbol{U}_2'$.

Now, instead of a full cancellation of the influence of $\boldsymbol{U}_2\boldsymbol{\Lambda}_2^{-1}\boldsymbol{U}_2'$, one may also think of only weakening this influence. The corresponding estimator is then of the form

$$\widehat{\boldsymbol{\beta}}^{(r,c)} = (\boldsymbol{U}_1\boldsymbol{\Lambda}_1^{-1}\boldsymbol{U}_1' + c\boldsymbol{U}_2\boldsymbol{\Lambda}_2^{-1}\boldsymbol{U}_2')\boldsymbol{X}'\boldsymbol{y}$$

for some c with $0 \le c \le 1$. In the special case $c = 1$, the estimator $\widehat{\boldsymbol{\beta}}^{(r,c)}$ coincides with the ordinary least squares estimator $\widehat{\boldsymbol{\beta}}$ (irrespective of the choice of r), while for the setting $c = 0$ the principal components estimator $\widehat{\boldsymbol{\beta}}^{(r)}$ is obtained. The estimator $\widehat{\boldsymbol{\beta}}^{(r,c)}$ is also called *Marquardt estimator*, see [83, p. 92]. Note that in the original paper by Marquardt [75] only the case $c = 0$ is discussed.

Although the principal components estimator $\widehat{\boldsymbol{\beta}}^{(r)}$ can be seen as a special case of the Marquardt estimator $\widehat{\boldsymbol{\beta}}^{(r,c)}$ for the choice $c = 0$, the two estimators have quite a different structure. The estimator $\widehat{\boldsymbol{\beta}}^{(r)}$ is a special restricted least squares estimator and satisfies the restrictions $\boldsymbol{U}_2'\widehat{\boldsymbol{\beta}}^{(r)} = \boldsymbol{0}$ for any possible realization $\boldsymbol{y} \in \mathbb{R}^n$. These restriction do not vanish whenever $r < p$, i.e. whenever the principal components estimator is different from the ordinary least squares estimator. In Example 3.7 this is confirmed by the fact that each principal components estimate lies *exactly* on a straight line determined by the restrictions, see Fig. 3.10. Dispersion of the estimates is only possible on this line.

In contrast to this behavior, the Marquardt estimator $\widehat{\boldsymbol{\beta}}^{(r,c)}$ for a choice $c \neq 0$ admits dispersion in all directions, so that this estimator (for $c > 0$) is *not* a special restricted estimator. Here the expansion of the estimates in the critical directions is reduced by some choice $c < 1$, but not forbidden.

This non-restrictiveness of the Marquardt estimator can also be derived from formal considerations. For this we assume that there exists a matrix $\boldsymbol{R}$ and a vector $\boldsymbol{r}$ such that $\boldsymbol{R}\widehat{\boldsymbol{\beta}}^{(r,c)} = \boldsymbol{r}$ is satisfied for any realization $\boldsymbol{y} \in \mathbb{R}^n$. Then this identity is also true for $\boldsymbol{y} = \boldsymbol{0}$, implying that $\boldsymbol{r}$ must be $\boldsymbol{0}$. Moreover, when $\boldsymbol{r} = \boldsymbol{0}$ we have

$$\boldsymbol{R}(\boldsymbol{U}_1\boldsymbol{\Lambda}_1^{-1}\boldsymbol{U}_1' + c\boldsymbol{U}_2\boldsymbol{\Lambda}_2^{-1}\boldsymbol{U}_2')\boldsymbol{X}' = \boldsymbol{0} .$$

Multiplying this equality from the right by the matrix

$$X(X'X)^{-1}(U_1\Lambda_1^{-1}U_1' + cU_2\Lambda_2^{-1}U_2')^{-1}$$

shows that also $R = 0$. Hence, there cannot exist a pair $(R, r) \neq (0, 0)$ such that $R\widehat{\beta}^{(r,c)} = r$ for any realization $y \in \mathbb{R}^n$.

Figures 3.11 and 3.12 show in the situation of Example 3.7 the already considered 30 ordinary least squares estimates and 30 Marquardt estimates for the case $r = 1$ and $c = 0.3$. Although from Fig. 3.12 one could get the impression that the Marquardt estimates scatter symmetrically around the true parameter vector $\beta = (2, 2)'$, it should be noted that the Marquardt estimator is not unbiased for β in the case $c < 1$.

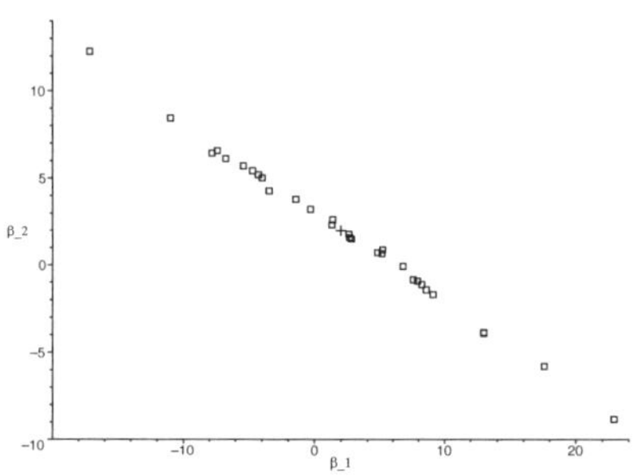

Fig. 3.11. 30 ordinary least squares estimates in Example 3.7

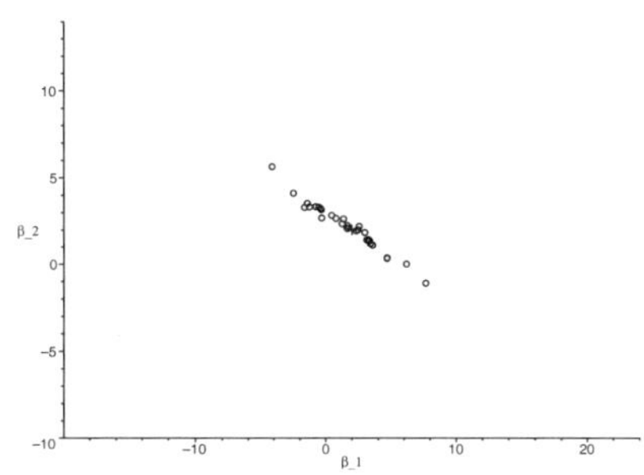

Fig. 3.12. 30 Marquardt estimates for $r = 1$ and $c = 0.3$ in Example 3.7

3.4 Ridge Estimator

The principal components estimation process aims to reduce the effect of collinearity by imposing certain linear restrictions on the unknown parameter vector. The ridge estimation procedure provides a different way to deal with the problem of collinearity. Here, the model matrix is altered in order to avoid the implications arising from its bad condition.

3.4.1 Preliminary Considerations

We have shown in Sect. 2.4 that the ridge estimator

$$\widehat{\beta}_k = (X'X + kI_p)^{-1}X'y, \quad k \geq 0\,,$$

which has been introduced by Hoerl and Kennard [51], is a biased linear estimator (for fixed $k > 0$), which in presence of collinearity can yield better estimates than the ordinary least squares estimator $\widehat{\beta}$. The use of this estimator is motivated from the fact that the matrix $\boldsymbol{X}'\boldsymbol{X}$ is stabilized when a positive number k is added to each of its main diagonal elements. This stabilization is understood in the sense that small relative changes in the matrix $\boldsymbol{X}'\boldsymbol{X} + k\boldsymbol{I}_p$ should involve also small relative changes in the inverse $(\boldsymbol{X}'\boldsymbol{X} + k\boldsymbol{I}_p)^{-1}$. This can be envisaged by noting that with increasing k the ridge of the matrix $\boldsymbol{X}'\boldsymbol{X} + k\boldsymbol{I}_p$ is raised, so that $\boldsymbol{X}'\boldsymbol{X} + k\boldsymbol{I}_p$ resembles more and more a diagonal matrix, which has an optimal condition. The condition can of course not be the only aspect when using the ridge estimator, since with increasing k the bias of the estimator grows as well. Hence the ultimate goal is to find some k which is large enough to reduce the variance compared to the ordinary least squares estimator, but which is small enough to produce some acceptable low bias.

The idea behind the ridge estimator is quite different from that behind the principal components estimator. In the following example we demonstrate the different behavior of both estimators.

Example 3.10. In Example 2.6 we have considered a linear regression model $\boldsymbol{y} = \boldsymbol{X}\boldsymbol{\beta} + \boldsymbol{\varepsilon}$ with

$$\boldsymbol{X}' = \begin{pmatrix} 1 & 1 & 1 & 1 & 1 \\ 1.9 & 2.1 & 2 & 2 & 1.8 \end{pmatrix}.$$

In Sect. 3.3 we have shown that for

$$\boldsymbol{y} = (6.0521, 7.0280, 7.1230, 4.4441, 5.0813)'$$

the value

$$\widehat{\boldsymbol{\beta}}^{(1)} = \boldsymbol{u}_1(\boldsymbol{u}_1'\boldsymbol{X}'\boldsymbol{X}\boldsymbol{u}_1)^{-1}\boldsymbol{u}_1'\boldsymbol{X}'\boldsymbol{y} = \begin{pmatrix} 1.2284 \\ 2.4128 \end{pmatrix}$$

of the principal components estimator is distinctly closer to the true parameter $\boldsymbol{\beta} = (2, 2)'$ than the value

$$\widehat{\boldsymbol{\beta}} = (\boldsymbol{X}'\boldsymbol{X})^{-1}\boldsymbol{X}'\boldsymbol{y} = \begin{pmatrix} -4.2489 \\ 5.2013 \end{pmatrix}$$

of the ordinary least squares estimator. Moreover, in Sect. 2.4.2 we have shown that the value

$$\widehat{\boldsymbol{\beta}}_k = (\boldsymbol{X}'\boldsymbol{X} + k\boldsymbol{I}_2)^{-1}\boldsymbol{X}'\boldsymbol{y} = \begin{pmatrix} 0.9396 \\ 2.5349 \end{pmatrix}, \quad k = 1/5\,,$$

of the ridge estimator is closer to the true $\boldsymbol{\beta} = (2, 2)'$ than the value of the ordinary least squares estimator.

Both estimators $\widehat{\boldsymbol{\beta}}^{(1)}$ and $\widehat{\boldsymbol{\beta}}_{1/5}$ are biased. For $\boldsymbol{\beta} = (2, 2)'$ we compute

$$\mathrm{E}(\widehat{\boldsymbol{\beta}}^{(1)}) = \begin{pmatrix} 1.2203 \\ 2.3969 \end{pmatrix} \quad \text{and} \quad \mathrm{E}(\widehat{\boldsymbol{\beta}}_{1/5}) = \begin{pmatrix} 1.25 \\ 2.3571 \end{pmatrix}.$$

If we have more than one realization of $\boldsymbol{y}$ available, we can compute more than one value of each of both estimators, and we should see that the estimates scatter somehow around the above computed expected values. To visualize this, we compute 30 estimates $\widehat{\boldsymbol{\beta}}_{1/5}$ for the 30 realizations of $\boldsymbol{y}$ which we have already generated in Sect. 3.3. Fig. 3.13 shows the result.

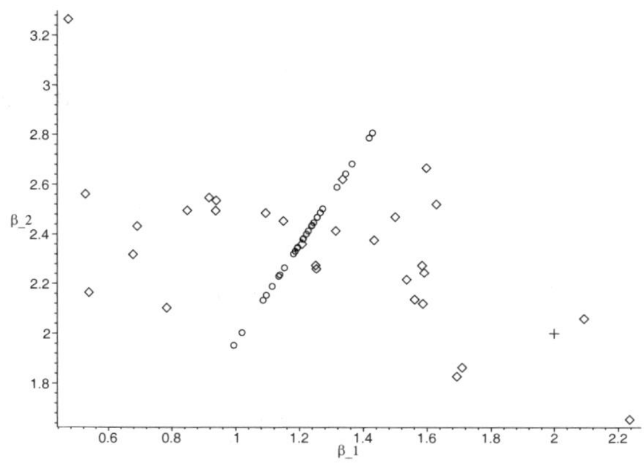

Fig. 3.13. True parameter (cross), 30 principal components estimates when $r = 1$ (circles), and 30 ridge estimates when $k = 1/5$ (diamonds) in Example 3.10

Note that although the expected values of $\widehat{\boldsymbol{\beta}}^{(1)}$ and $\widehat{\boldsymbol{\beta}}_{1/5}$ are rather similar, and the estimates do in fact scatter around these values, the behavior of the estimators is quite different. If we consider the observed unweighted squared error losses of both estimators, we can state the following. For the true $\boldsymbol{\beta} = (2, 2)'$ the minimum of observed losses of the ridge estimator is 0.0121, while the maximum is 3.9337. The average observed loss of the ridge estimator is 0.9533. For the principal components estimator the minimum of observed losses is 0.7655, while the maximum is 1.0144. This means that the range of observed losses of principal components estimates is smaller than the range of observed losses of the ridge estimator, which has a positive as well as a negative effect. An actual estimate of the principle components estimator is likely to be not too far away from the true parameter, but it can at the same time not come really close. The average observed loss of the principal components estimator is 0.8136. If we consider the ordinary least squares estimator, we can see that both estimators provide substantial better estimates with respect to the unweighted squared error loss. The minimum of observed losses of the ordinary least squares estimator is 0.3991, while the maximum is 553.4850. The average observed loss of the ordinary least squares estimator is 91.9358. □

The above example demonstrates the possibility for situations in which the application of the principal components estimator as well as the application of the ridge estimator can be appropriate methods for avoiding extremely large observed losses produced by the ordinary least squares estimator. In the following we will consider the ridge estimator in detail and try to find answers to the following questions:

- Under which conditions is the risk of the ridge estimator smaller than the risk of the ordinary least squares estimator?
- Does there always exist some positive k such that the risk of the corresponding ridge estimator is smaller than the risk of the ordinary least squares estimator?
- How can an appropriate ridge parameter k be found?

3.4.2 Properties of the Linear Ridge Estimator

As noted before, if k is non-stochastic, the ridge estimator

$$\widehat{\boldsymbol{\beta}}_k = (\boldsymbol{X}'\boldsymbol{X} + k\boldsymbol{I}_p)^{-1}\boldsymbol{X}'\boldsymbol{y}, \quad k \geq 0\,,$$

is a homogeneous linear estimator with

$$\mathrm{E}(\widehat{\boldsymbol{\beta}}_k) = \boldsymbol{\beta} - k(\boldsymbol{X}'\boldsymbol{X} + k\boldsymbol{I}_p)^{-1}\boldsymbol{\beta}$$

and

$$\mathrm{Cov}(\widehat{\boldsymbol{\beta}}_k) = \sigma^2\boldsymbol{X}'\boldsymbol{X}(\boldsymbol{X}'\boldsymbol{X} + k\boldsymbol{I}_p)^{-2}\,,$$

see Problem 2.15. We can then compare the risks of $\widehat{\boldsymbol{\beta}}_k$ and $\widehat{\boldsymbol{\beta}}$.

Risk Comparison of Ridge and Ordinary Least Squares Estimator

The following theorem gives conditions on the parameters $(\boldsymbol{\beta}, \sigma^2)$ and on the ridge parameter $k > 0$, being necessary and sufficient for $\widehat{\boldsymbol{\beta}}_k$ to be better than $\widehat{\boldsymbol{\beta}}$ with respect to $\mathrm{MSE}(\boldsymbol{\beta}, \widetilde{\boldsymbol{\beta}})$.

Theorem 3.8. *Under the linear regression model with assumptions (i) to (iv), let $k > 0$ be a (non-stochastic) number. Then the difference*

$$\boldsymbol{\Delta} = \mathrm{MSE}(\boldsymbol{\beta}, \widehat{\boldsymbol{\beta}}) - \mathrm{MSE}(\boldsymbol{\beta}, \widehat{\boldsymbol{\beta}}_k)$$

is nonnegative definite if and only if the inequality

$$\boldsymbol{\beta}'\left[\frac{2}{k}\boldsymbol{I}_p + (\boldsymbol{X}'\boldsymbol{X})^{-1}\right]^{-1}\boldsymbol{\beta} \leq \sigma^2$$

is satisfied. In addition, $\boldsymbol{\Delta} \neq \boldsymbol{0}$ whenever $k > 0$ and $p > 1$.

The above theorem turns out to be a special case of Theorem 3.19 in Sect. 3.6 and thus is not proved here. The following corollary gives a *sufficient* condition for the difference $\boldsymbol{\Delta}$ to be nonnegative definite.

Corollary 3.3. *Suppose that under the linear regression model with assumptions (i) to (iv) the parameters* $(\boldsymbol{\beta} \neq \mathbf{0}, \sigma^2)$ *satisfy the inequality*

$$0 < k \leq \frac{2\sigma^2}{\boldsymbol{\beta}'\boldsymbol{\beta}} \, .$$

Then the difference $\boldsymbol{\Delta} = \mathrm{MSE}(\boldsymbol{\beta}, \widehat{\boldsymbol{\beta}}) - \mathrm{MSE}(\boldsymbol{\beta}, \widehat{\boldsymbol{\beta}}_k)$ *is nonnegative definite.*

Proof. In the following let $k > 0$. By Theorem 3.8, the matrix $\boldsymbol{\Delta}$ is nonnegative definite if and only if

$$\boldsymbol{\beta}' \left[\frac{2}{k} \boldsymbol{I}_p + (\boldsymbol{X}'\boldsymbol{X})^{-1} \right]^{-1} \boldsymbol{\beta} \leq \sigma^2 \, ,$$

which in view of Theorem A.60 is satisfied if and only if the matrix

$$\sigma^2 \left[\frac{2}{k} \boldsymbol{I}_p + (\boldsymbol{X}'\boldsymbol{X})^{-1} \right] - \boldsymbol{\beta}\boldsymbol{\beta}'$$

is nonnegative definite. Since the matrix $\sigma^2(\boldsymbol{X}'\boldsymbol{X})^{-1}$ is always nonnegative definite (it is even positive definite), the latter holds when the symmetric matrix

$$\sigma^2 \frac{2}{k} \boldsymbol{I}_p - \boldsymbol{\beta}\boldsymbol{\beta}'$$

is nonnegative definite. If we set $\boldsymbol{A} = \frac{2}{k}\boldsymbol{I}_p$ (which is symmetric positive definite), $\alpha = \sigma^2$, and $\boldsymbol{a} = \boldsymbol{\beta}$, then by Theorem A.60 the matrix $\alpha\boldsymbol{A} - \boldsymbol{a}\boldsymbol{a}'$ is nonnegative definite if and only if $\boldsymbol{a}'\boldsymbol{A}^{-1}\boldsymbol{a} \leq \alpha$. The latter condition reads

$$\boldsymbol{\beta}' \frac{k}{2} \boldsymbol{I}_p \boldsymbol{\beta} \leq \sigma^2 \, ,$$

which is equivalent to $k\boldsymbol{\beta}'\boldsymbol{\beta} \leq 2\sigma^2$. Since $\boldsymbol{\beta} \neq \mathbf{0}$, this shows the assertion. □

The parametric function $\boldsymbol{\beta}'\boldsymbol{\beta}/\sigma^2$ is also called *signal to noise ratio.* For any $\boldsymbol{\beta} \neq \mathbf{0}$ and any $\sigma^2 > 0$ the number $2\sigma^2/\boldsymbol{\beta}'\boldsymbol{\beta}$ is strictly greater than 0. Then from Corollary 3.3 we can state the following:

Remark 3.2. For any $\boldsymbol{\beta} \neq \mathbf{0}$ with $\boldsymbol{\beta}'\boldsymbol{\beta} < \infty$ and any $\sigma^2 > 0$ there exists some $k > 0$ such that the matrix inequality

$$\mathrm{MSE}(\boldsymbol{\beta}, \widehat{\boldsymbol{\beta}}_k) \leq_{\mathrm{L}} \mathrm{MSE}(\boldsymbol{\beta}, \widehat{\boldsymbol{\beta}})$$

is satisfied, where in case $p > 1$ the two matrices involved in this inequality do not coincide. Thus it also follows

$$\mathrm{tr}[\mathrm{MSE}(\boldsymbol{\beta}, \widehat{\boldsymbol{\beta}}_k)] < \mathrm{tr}[\mathrm{MSE}(\boldsymbol{\beta}, \widehat{\boldsymbol{\beta}})]$$

for any such k.

The above noted existence of some $k > 0$ for which $\widehat{\boldsymbol{\beta}}_k$ is better than $\widehat{\boldsymbol{\beta}}$ does not depend on the degree of collinearity. Even if we cannot speak of the presence of collinearity, the existence of such a k is guaranteed. However, the expected improvement of $\widehat{\boldsymbol{\beta}}_k$ upon $\widehat{\boldsymbol{\beta}}$ will be more noticeable when collinearity is stronger.

The *existence* of an appropriate value $k > 0$, delivering a better estimator $\widehat{\boldsymbol{\beta}}_k$ than $\widehat{\boldsymbol{\beta}}$ for any configuration of the parameters, is thus guaranteed. This, however, does not provide us with a guideline which k to choose. From the above, a good choice of k would depend on $\boldsymbol{\beta}$ and σ^2, which, of course, are not known. We consider the problem of the choice of k in detail in Sect. 3.4.3.

The Improvement Region

Let us reconsider the necessary and sufficient condition for improvement of $\widehat{\boldsymbol{\beta}}_k$ upon $\widehat{\boldsymbol{\beta}}$ with respect to MSE$(\cdot)$ given in Theorem 3.8. The set of all vectors $\boldsymbol{\beta}$, satisfying the condition

$$\boldsymbol{\beta}' \left[\frac{2}{k} \boldsymbol{I}_p + (\boldsymbol{X}'\boldsymbol{X})^{-1} \right]^{-1} \boldsymbol{\beta} \leq \sigma^2$$

for an arbitrary but fixed $\sigma^2 \in (0, \infty)$ and an arbitrary but fixed $k \in (0, \infty)$, form an ellipsoid in IR^p, see also Problem 3.17.

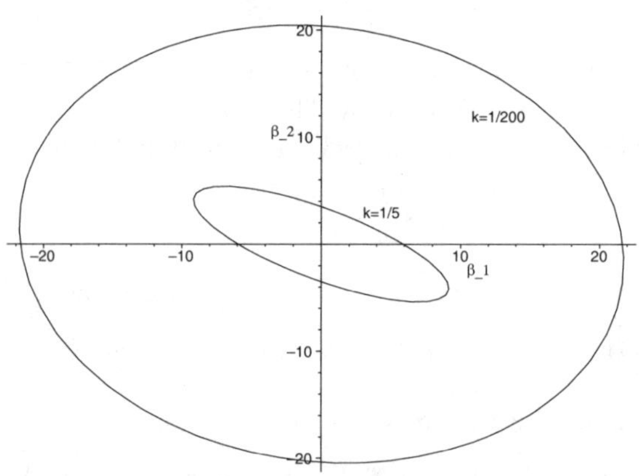

Fig. 3.14. Parameter regions in Example 2.6 for improvement of $\widehat{\boldsymbol{\beta}}_k$ upon $\widehat{\boldsymbol{\beta}}$ with respect to MSE$(\cdot)$, where $\sigma^2 = 1$

We already know that there always exists some $k > 0$ such that the unknown parameter $\boldsymbol{\beta}$ belongs to this ellipsoid, given an arbitrary but fixed $\sigma^2 > 0$. As a matter of fact, this ellipsoid becomes vaster when k becomes smaller, so that for arbitrary small k any vector $\boldsymbol{\beta}$ with finite elements can be made to belong to it. Fig. 3.14 shows two ellipsoids for $k = 1/5$ and $k = 1/200$ in Example 2.6 for the case $\sigma^2 = 1$. Although for small enough k the corresponding ridge estimator $\widehat{\boldsymbol{\beta}}_k$ will be not worse than the ordinary least squares estimator $\widehat{\boldsymbol{\beta}}$, it is clear that for very small k the achieved improvement can only be marginal and the respective estimates $\widehat{\boldsymbol{\beta}}_k$ and $\widehat{\boldsymbol{\beta}}$ will be almost the same.

The parameter regions in Fig. 3.14 are related to improvement of the estimator $\widehat{\boldsymbol{\beta}}_k$ upon $\widehat{\boldsymbol{\beta}}$ with respect to the matrix-valued risk $\mathrm{MSE}(\boldsymbol{\beta}, \widetilde{\boldsymbol{\beta}})$. If the true parameter $\boldsymbol{\beta}$ lies within this region, then $\widehat{\boldsymbol{\beta}}_k$ is advantageous over $\widehat{\boldsymbol{\beta}}$ with respect to this risk. But then, $\widehat{\boldsymbol{\beta}}_k$ is also advantageous over $\widehat{\boldsymbol{\beta}}$ with respect to the real-valued risk $\rho(\boldsymbol{\beta}, \widetilde{\boldsymbol{\beta}}) = \mathrm{tr}[\mathrm{MSE}(\boldsymbol{\beta}, \widetilde{\boldsymbol{\beta}})]$.

Usually the parameter region for improvement of the ridge estimator $\widehat{\boldsymbol{\beta}}_k$ upon $\widehat{\boldsymbol{\beta}}$ with respect to the real-valued risk $\rho(\cdot)$ will be vaster than with respect to the matrix-valued risk $\mathrm{MSE}(\cdot)$. This is illustrated by Fig. 3.15, which shows the two regions in Example 2.6 (for $\sigma^2 = 1$) where $\widehat{\boldsymbol{\beta}}_{1/5}$ is not worse than $\widehat{\boldsymbol{\beta}}$ with respect to $\rho(\cdot)$ and with respect to $\mathrm{MSE}(\cdot)$, see also Problem 3.18.

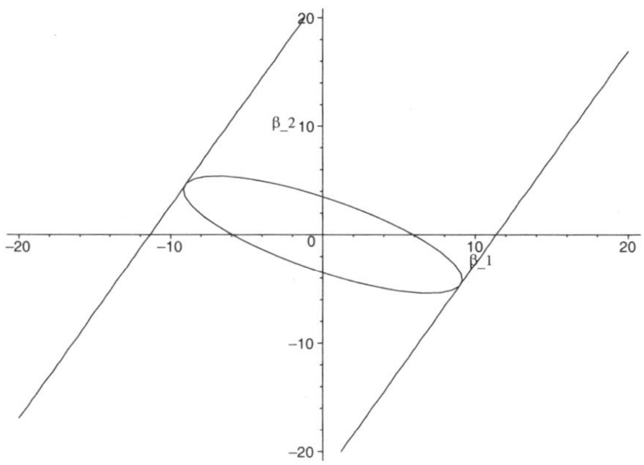

Fig. 3.15. Parameter regions in Example 2.6 for which $\widehat{\boldsymbol{\beta}}_{1/5}$ is not worse than $\widehat{\boldsymbol{\beta}}$ with respect to $\rho(\cdot)$ (almost degenerated ellipsoid enclosing small ellipsoid) and with respect to $\mathrm{MSE}(\cdot)$ (small ellipsoid), where $\sigma^2 = 1$

The Shrinkage Property of the Ridge Estimator

As mentioned before, when collinearity is present, an actual ordinary least squares estimate $\widehat{\boldsymbol{\beta}}$ will often have distinctly greater length than the true parameter vector $\boldsymbol{\beta}$. This effect of collinearity is softened by the ridge estimator $\widehat{\boldsymbol{\beta}}_k$, since the length of the ridge estimator becomes smaller for greater k.

Theorem 3.9. *Under the linear regression model with assumptions (i) to (iv), the inequality* $||\widehat{\beta}_{k_2}||^2 < ||\widehat{\beta}_{k_1}||^2$ *holds true for* $0 \leq k_1 < k_2$.

Proof. The squared length of the ridge estimator is given by

$$||\widehat{\beta}_k||^2 = \widehat{\beta}' \boldsymbol{H}_k \widehat{\beta}, \quad \boldsymbol{H}_k = \boldsymbol{X}'\boldsymbol{X}(\boldsymbol{X}'\boldsymbol{X} + k\boldsymbol{I}_p)^{-2}\boldsymbol{X}'\boldsymbol{X} .$$

If we spectrally decompose the matrix $\boldsymbol{X}'\boldsymbol{X}$ as $\boldsymbol{X}'\boldsymbol{X} = \boldsymbol{U}\boldsymbol{\Lambda}\boldsymbol{U}'$, then the matrix $\boldsymbol{H}_k$ can be written as

$$\boldsymbol{H}_k = \boldsymbol{U}\boldsymbol{\Gamma}\boldsymbol{U}', \quad \boldsymbol{\Gamma} = \operatorname{diag}\left(\frac{\lambda_1^2}{(\lambda_1 + k)^2}, \ldots, \frac{\lambda_p^2}{(\lambda_p + k)^2}\right),$$

where $\lambda_1, \ldots, \lambda_p$ are the main diagonal elements of $\boldsymbol{\Lambda}$, and thus the eigenvalues of $\boldsymbol{X}'\boldsymbol{X}$. If $0 \leq k_1 < k_2$, then the matrix $\boldsymbol{H}_{k_1} - \boldsymbol{H}_{k_2}$ is positive definite. Hence we have $\widehat{\beta}' \boldsymbol{H}_{k_2} \widehat{\beta} < \widehat{\beta}' \boldsymbol{H}_{k_1} \widehat{\beta}$ whenever $\widehat{\beta} \neq \mathbf{0}$. Note that we have already excluded the case $R^2 = 0$, being equivalent to $\widehat{\beta} = \mathbf{0}$, see also Assumption 2.1, p. 40. □

By the above theorem the squared length of the ridge estimator becomes smaller when k becomes greater. We can conclude $\lim_{k\to\infty} ||\widehat{\beta}_k||^2 = 0$ and therefore

$$\lim_{k\to\infty} \widehat{\beta}_k = \mathbf{0} .$$

Because of this, the ridge estimator is sometimes also called *shrinkage estimator*. Starting from $k = 0$, with increasing values of k the elements of the ordinary least squares estimator are *shrunken* towards the $p \times 1$ zero vector. It should be noted that this shrinking has a varying rate for the different elements.

The effect of shrinkage can also be seen to be achieved by multiplying the ordinary least squares estimator $\widehat{\beta}$ from the left by the matrix

$$\boldsymbol{W}(k) = (\boldsymbol{X}'\boldsymbol{X} + k\boldsymbol{I}_p)^{-1}\boldsymbol{X}'\boldsymbol{X} = (\boldsymbol{I}_p + k(\boldsymbol{X}'\boldsymbol{X})^{-1})^{-1}, \quad k \geq 0 .$$

The symmetric positive definite matrix $\boldsymbol{W}(k)$ can thus also be called the *ridge-shrinking* matrix.

Basically, it is also possible not only to shrink the ordinary least squares estimator $\widehat{\beta}$ but also other estimators $\widetilde{\beta}$ in the same way. Then one will obtain ridge-shrinkage estimators $\widetilde{\beta}_k = \boldsymbol{W}(k)\widetilde{\beta}$. Sarkar [94], for example, considers the ridge-shrinkage restricted least squares estimator $\boldsymbol{W}(k)\widehat{\beta}_{\mathrm{R}}$, while Silvapulle [102] investigates ridge-shrinkage robust estimators $\boldsymbol{W}(k)\widehat{\beta}^{\mathrm{rob}}$. See also Sect. 3.9.5 for a short discussion of the latter estimators.

3.4.3 The Choice of the Ridge Parameter

For any $(\boldsymbol{\beta} \neq \mathbf{0}, \sigma^2)$ there exists some optimal k_{opt} for which we can conclude

$$\rho(\boldsymbol{\beta}, \widehat{\boldsymbol{\beta}}_{k_{\text{opt}}}) \leq \rho(\boldsymbol{\beta}, \widehat{\boldsymbol{\beta}}_k)$$

for all $k \geq 0$. Here, $\rho(\boldsymbol{\beta}, \widetilde{\boldsymbol{\beta}})$ denotes the unweighted squared error risk of $\widetilde{\boldsymbol{\beta}}$, i.e. $\rho(\boldsymbol{\beta}, \widetilde{\boldsymbol{\beta}}) = \text{tr}[\text{MSE}(\boldsymbol{\beta}, \widetilde{\boldsymbol{\beta}})]$. We will not prove this fact, but illustrate its consequence in the following example.

Example 3.11. Consider the situation from Example 2.6. Since the data has been simulated, the true parameters are known to be

$$(\boldsymbol{\beta}, \sigma^2) = (\begin{pmatrix} 2 \\ 2 \end{pmatrix}, 1)\,.$$

For these specific values of $(\boldsymbol{\beta}, \sigma^2)$ we can plot the risk $\rho(\boldsymbol{\beta}, \widehat{\boldsymbol{\beta}}_k)$ as a function of k. Simultaneously, we can plot

$$v(k) = \sigma^2\, \text{tr}[\text{Cov}(\widehat{\boldsymbol{\beta}}_k)] \quad \text{(sum of variances)}$$

and

$$b(k) = (\text{E}(\widehat{\boldsymbol{\beta}}_k) - \boldsymbol{\beta})'(\text{E}(\widehat{\boldsymbol{\beta}}_k) - \boldsymbol{\beta}) \quad \text{(sum of squared biases)}$$

as functions of k. See Fig. 3.16 and note that $\rho(\boldsymbol{\beta}, \widehat{\boldsymbol{\beta}}_k) = v(k) + b(k)$.

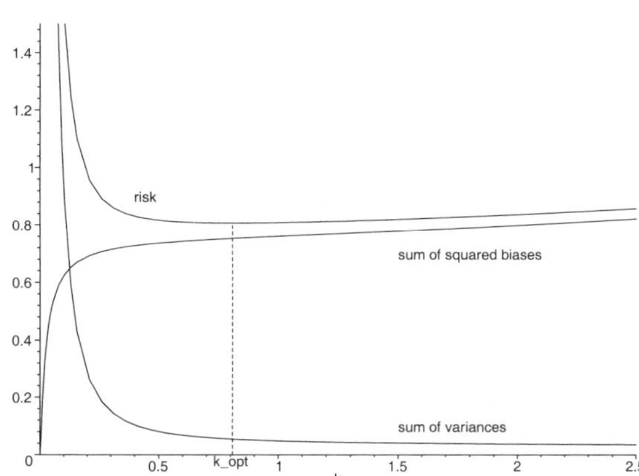

Fig. 3.16. Optimal choice of the ridge parameter k for the true parameters in Example 2.6

It can be seen that $v(k)$ decreases with increasing k, while $b(k)$ increases with increasing k. For $k = 0$ the risk of $\rho(\boldsymbol{\beta}, \widehat{\boldsymbol{\beta}}_k)$ coincides with the risk of

$\widehat{\beta}$, decreases with increasing k until $k = k_{\text{opt}}$, and increases for greater k. In this example we obtain

$$k_{\text{opt}} = 0.8078.$$

The risk of $\widehat{\beta}_{0.8078}$ is given by $\rho(\beta, \widehat{\beta}_{0.8078}) = 0.8077$, while the risk of $\widehat{\beta}$ is given by $\rho(\beta, \widehat{\beta}) = 93.3077$. This also shows that the previously employed choice $k = 1/5$ does not deliver the maximal improvement of $\widehat{\beta}_k$ upon $\widehat{\beta}$. On the other hand the risk $\rho(\beta, \widehat{\beta}_{1/5}) = 0.9721$ is only marginally greater than the risk $\rho(\beta, \widehat{\beta}_{0.8078})$, and thus it is also distinctly smaller than the risk $\rho(\beta, \widehat{\beta})$. □

The value k_{opt} has been obtained numerically in the above example for known parameter values (β, σ^2). In practice this is of course not possible, since we do not know the parameters. Hence it still remains open how to choose k. For determining an appropriate k we can distinguish between two methodical approaches.

The Ridge Trace (Subjective Method)

The ridge trace method considers the p elements of the ridge estimator $\widehat{\beta}_k$ as functions of k, all of them plotted into one figure. The user then chooses the value $k = k_{\text{tr}}$ at which the functions begin to stabilize.

Example 3.12. Under the model $\boldsymbol{y} = \boldsymbol{X}\beta + \varepsilon$, where

$$\boldsymbol{X}' = \begin{pmatrix} 1 & 1 & 1 & 1 & 1 \\ 1.9 & 2.1 & 2 & 2 & 1.8 \end{pmatrix}$$

and

$$\boldsymbol{y} = (6.0521, 7.0280, 7.1230, 4.4441, 5.0813)'$$

we can compute the elements of $\widehat{\beta}_k$ as functions of k and plot both elements into one figure, see Fig. 3.17. We can say that the ridge trace stabilizes somewhere between 0.2 and 0.4. A possible choice of the ridge parameter could be $k_{\text{tr}} = 0.3$. □

The judgement of the ridge trace depends on the user, but also on the range of values k for which the functions are drawn. The impression of stability will be quite different when the ridge trace is considered for different ranges. For instance, if in Example 3.12 the ridge trace is plotted for values k between 0 and 20, then one can easily get the impression that the ridge trace stabilizes for values of k which are greater than 1, see Fig. 3.18. To avoid this effect some proposals to use different scales exist, see [123, Sect. 7.2.5].

Remark 3.3. The estimator $\widehat{\beta}_{k_{\text{tr}}}$ is not a linear estimator, since k_{tr} depends on the actual value of $\boldsymbol{y}$. In addition, it is rather difficult to say something about the statistical properties of the non-linear estimator $\widehat{\beta}_{k_{\text{tr}}}$, since the choice of k_{tr} depends on the user.

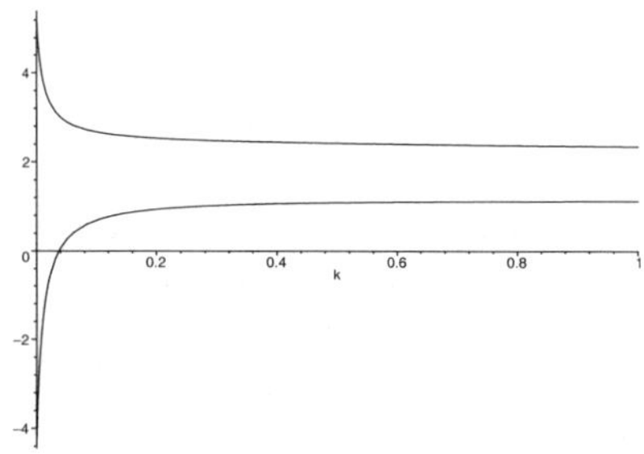

Fig. 3.17. Ridge trace in Example 3.12 for $k \in [0, 1]$

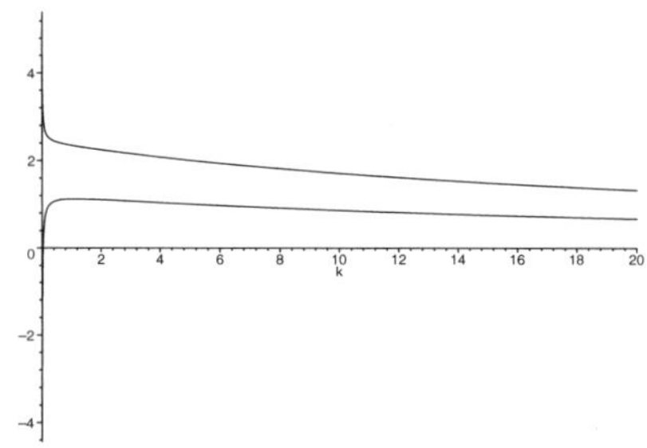

Fig. 3.18. Ridge trace in Example 3.12 for $k \in [0, 20]$

The Instability Function (Objectification of the Ridge Trace)

There are also approaches for finding an objective judgement of the ridge trace, see [120]. For this we consider the function

$$\phi(k) = (\widehat{\boldsymbol{\beta}}_k - \widehat{\boldsymbol{\beta}})'(\widehat{\boldsymbol{\beta}}_k - \widehat{\boldsymbol{\beta}}) = ||\widehat{\boldsymbol{\beta}}_k - \widehat{\boldsymbol{\beta}}||^2 .$$

The derivative $\partial\phi(k)/\partial k$ can be seen as the 'velocity of change' in ϕ under variation of k. If the maximum of $\partial\phi(k)/\partial k$ is unique and is attained for k^* (position of maximum instability of the ridge trace), the first inflection point k^{**} of $\partial\phi(k)/\partial k$ greater than k^* is defined to be the position of the beginning of a stabilize ridge trace. Fig. 3.19 shows the function $\partial\phi(k)/\partial k$ in Example 2.6. It can be computed that $k^{**} = 0.0107$, which, however, appears to be distinctly too small for the choice of ridge parameter k.

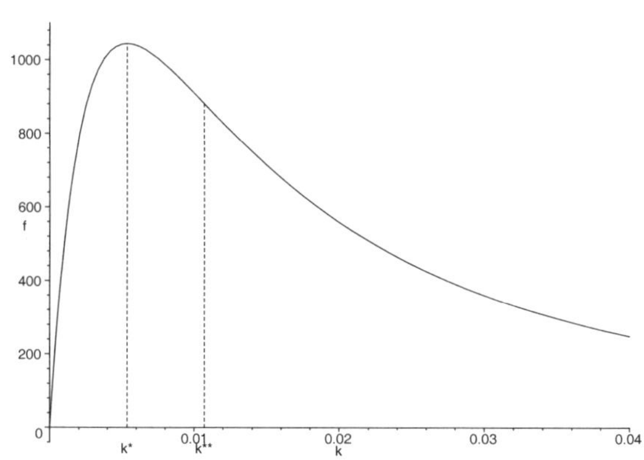

Fig. 3.19. Instability function $f = \partial\phi(k)/\partial k$ of the ridge trace in Example 2.6

Estimation of k (Objective Method)

The ridge trace method is subjective in the sense that the choice of k does not only depend on the actual value of $\boldsymbol{y}$ but also on the judgement of the user. If we consider an estimator $\widehat{k}$ as a choice of k, this can be seen as an objective method in the sense that the choice of k does only depend on $\boldsymbol{y}$. However, since it is hard to find an estimator for k satisfying some optimality criterion, the choice of estimator is again due to the user in the end.

Now, how can we obtain an appropriate estimator for k? As noted before, for every given parameter values $(\boldsymbol{\beta} \neq \mathbf{0}, \sigma^2)$ there exists some k_{opt} for which

$$\rho(\boldsymbol{\beta}, \widehat{\boldsymbol{\beta}}_{k_{\text{opt}}}) \leq \rho(\boldsymbol{\beta}, \widehat{\boldsymbol{\beta}}_k)$$

for any $k > 0$. In general it is not possible to find a closed form formula for k_{opt} (which of course would depend on $\boldsymbol{\beta}$ and σ^2). If, however, we consider model assumption (vii), $\boldsymbol{X}'\boldsymbol{X} = \boldsymbol{I}_p$, to be true, we can determine k_{opt} as a function of $\boldsymbol{\beta} \neq \mathbf{0}$ and σ^2, as shown by the following theorem.

Theorem 3.10. *Under the linear regression model with assumptions (i) to (iv) and (vii), the inequality*

$$\rho(\boldsymbol{\beta}, \widehat{\boldsymbol{\beta}}_{k_{\text{opt}}}) \leq \rho(\boldsymbol{\beta}, \widehat{\boldsymbol{\beta}}_k) \quad \forall\, k \geq 0$$

is true for $k_{\text{opt}} = p\,\sigma^2/\boldsymbol{\beta}'\boldsymbol{\beta}$.

Proof. If $\boldsymbol{X}'\boldsymbol{X} = \boldsymbol{I}_p$ is true, then

$$\rho(\boldsymbol{\beta}, \widehat{\boldsymbol{\beta}}_k) = \frac{p\,\sigma^2}{(1+k)^2} + \frac{k^2}{(1+k)^2}\boldsymbol{\beta}'\boldsymbol{\beta}\,.$$

We can now consider the risk as a function of k and determine its derivative with respect to k. If we put the derivative equal to zero and solve for k, then we obtain $k = p\,\sigma^2/\boldsymbol{\beta}'\boldsymbol{\beta}$. It is easy to check that the risk attains in fact its minimum for this value of k. □

Based on the above theorem we can consider some ad-hoc possibilities for determining an estimator for k, which then can also be used as an estimator under the linear regression model with assumptions (i) to (iv) but not necessarily (vii).

Estimator from Hoerl, Kennard and Baldwin [53]. Replace in $k_{\text{opt}} = p\,\sigma^2/\boldsymbol{\beta}'\boldsymbol{\beta}$ the parameter σ^2 by the estimator $\widehat{\sigma}^2$ and the parameter vector $\boldsymbol{\beta}$ by the estimator $\widehat{\boldsymbol{\beta}}$. Apply

$$\widehat{k}_{\text{HKB}} = \frac{p\,\widehat{\sigma}^2}{\widehat{\boldsymbol{\beta}}'\widehat{\boldsymbol{\beta}}}$$

as an estimator for k even in situations when model assumption (vii) is not satisfied.

Iterative Estimator from Hoerl and Kennard [52]. It has been experienced that the estimator $\widehat{k}_{\mathrm{HKB}}$ often yields a too small value for k. As a reason for this behavior one might take into account that in presence of collinearity the squared length $\widehat{\boldsymbol{\beta}}'\widehat{\boldsymbol{\beta}}$ of $\widehat{\boldsymbol{\beta}}$ is often distinctly greater than the squared length $\boldsymbol{\beta}'\boldsymbol{\beta}$ of $\boldsymbol{\beta}$. But on the other hand, if one considers $\widehat{k}_{\mathrm{HKB}}$ as an estimator for k_{opt}, then only in case $\boldsymbol{X}'\boldsymbol{X} = \boldsymbol{I}_p$, i.e. in absence of collinearity, the optimal value k_{opt} is given as $k_{\mathrm{opt}} = p\,\sigma^2/\boldsymbol{\beta}'\boldsymbol{\beta}$. Even in this case it is not clear whether the estimator $\widehat{k}_{\mathrm{HKB}}$ is a good estimator for k_{opt}.

A possibility for raising the actual estimate is provided by an iteration of the estimation process. For obtaining the estimate $\widehat{k}^{i+1}_{\mathrm{HKB}}$, the estimator $\widehat{\boldsymbol{\beta}}$ is replaced by $\widehat{\boldsymbol{\beta}}_{\widehat{k}^i_{\mathrm{HKB}}}$. Hence,

$$\widehat{k}^{i+1}_{\mathrm{HKB}} = \frac{p\,\widehat{\sigma}^2}{\widehat{\boldsymbol{\beta}}'_{\widehat{k}^i_{\mathrm{HKB}}}\widehat{\boldsymbol{\beta}}_{\widehat{k}^i_{\mathrm{HKB}}}}, \qquad \widehat{k}^0_{\mathrm{HKB}} := 0\,.$$

The iteration is continued until the difference between $\widehat{k}^{i+1}_{\mathrm{HKB}}$ and $\widehat{k}^{i}_{\mathrm{HKB}}$ is smaller than some given bound. According to Theorem 3.9 we can conclude that

$$\widehat{k}^{i+1}_{\mathrm{HKB}} > \widehat{k}^{i}_{\mathrm{HKB}}\,,$$

so that in fact the obtained estimate is greater than $\widehat{k}_{\mathrm{HKB}}$.

Estimator from Lawless and Wang [68]. If assumption (vii) is not satisfied, then one may consider the reparameterized model

$$\boldsymbol{y} = \boldsymbol{Z}\boldsymbol{\vartheta} + \boldsymbol{\varepsilon}, \quad \boldsymbol{Z} = \boldsymbol{X}(\boldsymbol{X}'\boldsymbol{X})^{-1/2}, \quad \boldsymbol{\vartheta} = (\boldsymbol{X}'\boldsymbol{X})^{1/2}\boldsymbol{\beta}\,,$$

under which $\boldsymbol{Z}'\boldsymbol{Z} = \boldsymbol{I}_p$ is satisfied, see Sect. 2.5.1. In this model the corresponding estimator $\widehat{k}_{\mathrm{HKB}}$ is given by

$$\widehat{k}_{\mathrm{LW}} = \frac{p\,\widehat{\sigma}^2}{\widehat{\boldsymbol{\vartheta}}'\widehat{\boldsymbol{\vartheta}}} = \frac{p\,\widehat{\sigma}^2}{\widehat{\boldsymbol{\beta}}'\boldsymbol{X}'\boldsymbol{X}\widehat{\boldsymbol{\beta}}}\,,$$

which may also be used in the original model even when assumption (vii) is not satisfied. Lawless and Wang [68] give a different, Bayesian oriented motivation for this estimator.

Estimator from Lindley and Smith [71]. Also motivated from Bayesian arguments, Lindley and Smith [71] derive a ridge estimator with ridge parameter

$$\widehat{k}_{\mathrm{LS}} = \frac{(n-p)(p+2)}{n+2}\,\frac{\widehat{\sigma}^2}{\widehat{\boldsymbol{\beta}}'\widehat{\boldsymbol{\beta}}}\,,$$

which differs from $\widehat{k}_{\mathrm{HKB}}$ by the factor by which $\widehat{\sigma}^2/\widehat{\boldsymbol{\beta}}'\widehat{\boldsymbol{\beta}}$ is multiplied.

Similar to $\widehat{k}_{\mathrm{HKB}}$ we can also use the estimates $\widehat{k}_{\mathrm{LW}}$ and $\widehat{k}_{\mathrm{LS}}$ as starting values for an iteration process.

Example 3.13. If in Example 3.12 we consider the estimator $\widehat{k} := \widehat{k}_{\mathrm{HKB}}$, then we obtain

$$\widehat{k} = 0.06154\,.$$

If we compare this value with the ridge trace in Fig. 3.17, the estimate would prove to be too small. If we now compute the corresponding iterative estimate, then

$$\widehat{k}_8 = 0.3895\,,$$

where the iteration process has been continued until $\widehat{k}_{i+1} - \widehat{k}_i \leq 0.0000001$. If we compare this value with the ridge trace, it appears to be an appropriate choice. □

If an estimator $\widehat{k}$ is used for the determination of the ridge parameter k, then it is rather difficult to derive exact statistical properties of the corresponding non-linear ridge estimator $\widehat{\beta}_{\widehat{k}}$, although there are some approaches. For example in [39] the moments of $\widehat{\beta}_{\widehat{k}_{\mathrm{LW}}}$ are derived under the additional model assumption (v). The respective formulas turn out to be complicated and a direct application to compare the risks of different non-linear ridge estimators is very difficult. Although simulation studies are not of general validity and cannot be taken as proofs, they appear to be more appropriate with regard to the desired risk comparisons. Gruber [47, Sect. 3.5] gives a review of these studies and their respective results in the literature.

3.4.4 Standardization

When we apply the linear regression model to real data, we see that the order of magnitude of the parameter value β_i is relative to the unit of measurement of the corresponding variable x_i. The following example illustrates this point.

Example 3.14. Consider the linear regression model described by

$$y = \beta_1 + \beta_2 x + \varepsilon\,,$$

where the $n = 9$ observations of y and x are given in Table 3.1. Suppose that the variable x has been measured in US$, while the unit of measurement of y depends on the specific meaning of y and is not of interest here.

Table 3.1. Observed values of the variables x and y

i	1	2	3	4	5	6	7	8	9
y_i	24.59	23.19	26.94	28.82	32.66	25.97	28.36	38.21	36.42
x_i	20.45	18.70	21.34	22.67	28.10	21.66	24.50	31.88	29.88

The true parameters in this model are known to be $\beta_1 = 0.5$ and $\beta_2 = 1.2$. Now, consider a second model

$$y = \delta_1 + \delta_2 z + \varepsilon \ ,$$

with the same variable y measured as before and with a variable z which is the same as x, but has been measured in a different currency. For simplicity it is assumed that this currency is computed as $2 \cdot$ US\$, so that we can compute the values of z from those of x by multiplying each value of x by the factor $d = 2$. Now, since both models are supposed to describe the same relationship, the parameters are necessarily related by $\delta_1 = \beta_1 = 0.5$ and $\delta_2 = d^{-1} \cdot 1.2 = 0.6$. These relationships are reflected by the respective ordinary least squares estimates under both models. We can compute $\widehat{\delta}_1 = \widehat{\beta}_1 = 2.0490$ and $\widehat{\delta}_2 = 0.5628 = d^{-1}\widehat{\beta}_2$. □

If we consider the model description $\boldsymbol{y} = \boldsymbol{X\beta} + \boldsymbol{\varepsilon}$, then a multiplicative change in the unit of measurement of the variable x_i corresponds to the multiplication of i-th column of the matrix $\boldsymbol{X}$ with a number $d_i \neq 0$. If we multiply each of the columns of $\boldsymbol{X}$ by some specific number, then this is equivalent to right-multiplication of $\boldsymbol{X}$ by some diagonal matrix $\boldsymbol{D}$ with all main diagonal elements different from 0. We then obtain two equivalent models

$$\text{(a)} \qquad \boldsymbol{y} = \boldsymbol{X\beta} + \boldsymbol{\varepsilon}, \quad \boldsymbol{\varepsilon} \sim (\boldsymbol{0}, \sigma^2 \boldsymbol{I}_n) \ ,$$

and

$$\text{(b)} \qquad \boldsymbol{y} = \boldsymbol{Z\delta} + \boldsymbol{\varepsilon}, \quad \boldsymbol{\varepsilon} \sim (\boldsymbol{0}, \sigma^2 \boldsymbol{I}_n) \ ,$$

where $\boldsymbol{Z} = \boldsymbol{XD}$ and $\boldsymbol{\delta} = \boldsymbol{D}^{-1}\boldsymbol{\beta}$. If we compute the ordinary least squares estimator $\widehat{\boldsymbol{\delta}}$ in model (b), we get

$$\widehat{\boldsymbol{\delta}} = (\boldsymbol{Z}'\boldsymbol{Z})^{-1}\boldsymbol{Z}'\boldsymbol{y} = (\boldsymbol{DX}'\boldsymbol{XD})^{-1}\boldsymbol{DX}'\boldsymbol{y} = \boldsymbol{D}^{-1}\widehat{\boldsymbol{\beta}} \ .$$

This shows that the ordinary least squares estimator $\widehat{\boldsymbol{\beta}}$ conforms to the multiplicative change in unit measurements, i.e. for the new parameter vector $\boldsymbol{D}^{-1}\boldsymbol{\beta}$ the ordinary least squares estimator is $\boldsymbol{D}^{-1}\widehat{\boldsymbol{\beta}}$.

The ridge estimator $\widehat{\boldsymbol{\beta}}_k$ does not have a similar property. If a specific value $k = k_1$ is chosen in model (a), while an appropriate value k_2 in model (b) (being possibly identical to k_1) is determined, then usually $\widehat{\boldsymbol{\delta}}_{k_2} \neq \boldsymbol{D}^{-1}\widehat{\boldsymbol{\beta}}_{k_1}$. This can easily be seen by considering

$$\begin{aligned} \widehat{\boldsymbol{\delta}}_{k_2} &= (\boldsymbol{DX}'\boldsymbol{XD} + k_2\boldsymbol{I}_p)^{-1}\boldsymbol{DX}'\boldsymbol{y} \\ &= (\boldsymbol{DX}'\boldsymbol{XD} + k_2\boldsymbol{DD}^{-2}\boldsymbol{D})^{-1}\boldsymbol{DX}'\boldsymbol{y} \\ &= \boldsymbol{D}^{-1}(\boldsymbol{X}'\boldsymbol{X} + k_2\boldsymbol{D}^{-2})^{-1}\boldsymbol{X}'\boldsymbol{y}. \end{aligned}$$

Thus, we have $\widehat{\boldsymbol{\delta}}_{k_2} = \boldsymbol{D}^{-1}\widehat{\boldsymbol{\beta}}_{k_1}$ for all $\boldsymbol{y} \in \mathbb{R}^n$ if and only if $\boldsymbol{D}^{-2} = (k_1/k_2)\boldsymbol{I}_p$, which usually will not be the case.

Example 3.15. In the situation of Example 3.14 we compute $\widehat{\boldsymbol{\beta}}_{k_1}$ and $\widehat{\boldsymbol{\delta}}_{k_2}$ with $k_1 = p\widehat{\sigma}^2/\widehat{\boldsymbol{\beta}}'\widehat{\boldsymbol{\beta}}$ and $k_2 = p\widehat{\sigma}^2/\widehat{\boldsymbol{\delta}}'\widehat{\boldsymbol{\delta}}$. We obtain $k_1 = 0.3157$ and $k_2 = 0.3822$ and thus

$$\widehat{\boldsymbol{\beta}}_{k_1} = \begin{pmatrix} 0.9770 \\ 1.1683 \end{pmatrix} \quad \text{and} \quad \widehat{\boldsymbol{\delta}}_{k_2} = \begin{pmatrix} 0.8623 \\ 0.5864 \end{pmatrix} .$$

In this case the relationship between the estimates corresponding to the relationship between elements of the parameter vectors in both models are at least approximately satisfied, i.e. $(\widehat{\boldsymbol{\delta}}_{k_2})_1 \approx (\widehat{\boldsymbol{\beta}}_{k_1})_1$ and $(\widehat{\boldsymbol{\delta}}_{k_2})_2 \approx d^{-1}(\widehat{\boldsymbol{\beta}}_{k_1})_2$. □

Remark 3.4. When computing two ridge estimates in equivalent models which only differ by some multiplicative changes in measurement units of the independent variables, estimates are obtained whose difference cannot be explained solely from these changes. The effect can even be more noticeable than in the above example.

Effect of Standardization

For the application of the ridge estimator in practice it is often recommended to standardize the non-constant independent variables beforehand. If some unified convention of standardization is used, the application of a specific ridge method yields the same result irrespective of different units of measurement of independent variables.

A Suggested Standardization Procedure. In the following an appropriate standardization procedure and its consequences are described. Let us assume that the linear regression model

$$y_i = \beta_1 x_{i,1} + \cdots + \beta_p x_{i,p} + \varepsilon_i, \qquad i = 1, \ldots, n$$

has an intercept, so that $x_{i,1} \equiv 1$ for $i = 1, \ldots, n$. In matrix notation we may write

$$\boldsymbol{y} = \mathbf{1}_n \beta_1 + \boldsymbol{X}_* \boldsymbol{\beta}_* + \boldsymbol{\varepsilon}, \quad \boldsymbol{\varepsilon} \sim (\mathbf{0}, \sigma^2 \boldsymbol{I}_n) .$$

Here $\mathbf{1}_n$ is the $n \times 1$ vector with each element equal to 1, and

$$\boldsymbol{X}_* = \begin{pmatrix} x_{1,2} & \ldots & x_{1,p} \\ \vdots & & \vdots \\ x_{n,2} & \ldots & x_{n,p} \end{pmatrix}, \quad \boldsymbol{\beta}_* = \begin{pmatrix} \beta_2 \\ \vdots \\ \beta_p \end{pmatrix} .$$

In the usual description

$$\boldsymbol{y} = \boldsymbol{X}\boldsymbol{\beta} + \boldsymbol{\varepsilon}$$

we thus have $\boldsymbol{X} = (\mathbf{1}_n, \boldsymbol{X}_*)$ and $\boldsymbol{\beta} = (\beta_1, \boldsymbol{\beta}'_*)'$.

Now, the standardization of the independent variables x_2 to x_p can be seen as transformations of their respective values such that the transformed values are related to fictive but appropriate units of measurement. Instead of the values $x_{i,j}$ we consider $z_{i,j}$ with

$$z_{i,1} = 1 \quad \text{for } i = 1, \ldots, n$$

and

$$z_{i,j} = \frac{x_{i,j} - \overline{x}_j}{\sqrt{\sum_{i=1}^{n}(x_{i,j} - \overline{x}_j)^2}}, \quad \overline{x}_j = \frac{1}{n}\sum_{i=1}^{n} x_{i,j} ,$$

for $i = 1, \ldots, n$ and $j = 2, \ldots, p$. Then $\boldsymbol{Z} = (\boldsymbol{1}_n, \boldsymbol{Z}_*)$, where $\boldsymbol{Z}_* = \boldsymbol{C}\boldsymbol{X}_*\boldsymbol{D}^{-1}$ with

$$\boldsymbol{C} = \boldsymbol{I}_n - \frac{1}{n}\boldsymbol{1}_n\boldsymbol{1}_n'$$

(the centering matrix), and

$$\boldsymbol{D} = \operatorname{diag}(d_2, \ldots, d_p), \quad d_j = \sqrt{\sum_{i=1}^{n}(x_{i,j} - \overline{x}_j)^2}, \quad j = 2, \ldots, p .$$

After the reparameterization we can consider the model

$$\boldsymbol{y} = \boldsymbol{Z}\boldsymbol{\delta} + \boldsymbol{\varepsilon} = \boldsymbol{1}_n\mu + \boldsymbol{Z}_*\boldsymbol{\gamma} + \boldsymbol{\varepsilon}, \quad \boldsymbol{\varepsilon} \sim (\boldsymbol{0}, \sigma^2\boldsymbol{I}_n) ,$$

with $\boldsymbol{\gamma} = (\gamma_1, \ldots, \gamma_{p-1})$ and $\boldsymbol{\delta} = (\mu, \boldsymbol{\gamma}')'$ (standardized model). The matrix $\boldsymbol{Z}$ is of full column rank (see problem 3.20) and for the matrix $\boldsymbol{Z}_*$ it follows that $\boldsymbol{R} = \boldsymbol{Z}_*'\boldsymbol{Z}_*$ is a correlation matrix. This is achieved by the above described determination of $z_{i,j}$ via the formula $z_{i,j} = (x_{i,j} - \overline{x}_j)/(\sqrt{n-1}\, s)$ and not via the formula $z_{i,j} = (x_{i,j} - \overline{x}_j)/s$, where

$$s = \sqrt{\frac{1}{n-1}\sum_{i=1}^{n}(x_{i,j} - \overline{x}_j)^2} .$$

Standardization has the additional advantage that the order of magnitude of the parameters $\gamma_1, \ldots, \gamma_{p-1}$ corresponding to the standardized model are comparable. Hence, also the estimates of these parameters are comparable. This means that if the estimate γ_3 is twice the estimate of γ_4 (say), then, *based on the estimates*, we can conclude that the variable x_3 explains twice as much of y as the variable x_4. Such a statement can only be reasonable when the estimates are based on standardized variables, since otherwise the order of magnitude of each estimate depends on the units of measurement of the corresponding variable.

Different Point of Views. Marquardt and Snee [76, p. 3] emphasize that the centering of the data (subtracting the respective means) has the additional implication that only those collinearity effects come into play, which are caused by the variables themselves. On the other hand, Belsley [11, Chap. 6] demonstrates that centering can be quite misleading in collinearity analysis and may only disguise conditioning problems. Moreover, Vinod and Ullah [123, p. 179] remind that there exist cases in which it is desired to work with specific units of measurement, so that standardization in theses cases is not an appropriate tool.

Ridge Estimation from Standardized Data. In the linear regression model described by $\boldsymbol{y} = \boldsymbol{Z}\boldsymbol{\delta} + \boldsymbol{\varepsilon}$ the ordinary least squares estimator of $\boldsymbol{\delta}$ is given as

$$\widehat{\boldsymbol{\delta}} = (\boldsymbol{Z}'\boldsymbol{Z})^{-1}\boldsymbol{Z}'\boldsymbol{y} = \begin{pmatrix} \mathbf{1}_n'\mathbf{1}_n & \mathbf{1}_n'\boldsymbol{Z}_* \\ \boldsymbol{Z}_*'\mathbf{1}_n & \boldsymbol{Z}_*'\boldsymbol{Z}_* \end{pmatrix}^{-1} \begin{pmatrix} \mathbf{1}_n' \\ \boldsymbol{Z}_*' \end{pmatrix} \boldsymbol{y} \,.$$

From the identity $\mathbf{1}_n'\boldsymbol{C} = \mathbf{0}$ it also follows $\mathbf{1}_n'\boldsymbol{Z}_* = \mathbf{0}$, so that

$$\begin{aligned} \widehat{\boldsymbol{\delta}} &= \begin{pmatrix} n & \mathbf{0} \\ \mathbf{0} & \boldsymbol{Z}_*'\boldsymbol{Z}_* \end{pmatrix}^{-1} \begin{pmatrix} \mathbf{1}_n' \\ \boldsymbol{Z}_*' \end{pmatrix} \boldsymbol{y} \\ &= \begin{pmatrix} \frac{1}{n} & \mathbf{0} \\ \mathbf{0} & (\boldsymbol{Z}_*'\boldsymbol{Z}_*)^{-1} \end{pmatrix} \begin{pmatrix} \mathbf{1}_n' \\ \boldsymbol{Z}_*' \end{pmatrix} \boldsymbol{y} \\ &= \begin{pmatrix} \frac{1}{n}\mathbf{1}_n'\boldsymbol{y} \\ (\boldsymbol{Z}_*'\boldsymbol{Z}_*)^{-1}\boldsymbol{Z}_*'\boldsymbol{y} \end{pmatrix} . \end{aligned}$$

This shows that the ordinary least squares estimators μ and γ are

$$\widehat{\mu} = \frac{1}{n}\mathbf{1}_n'\boldsymbol{y} = \overline{y} \quad \text{and} \quad \widehat{\gamma} = (\boldsymbol{Z}_*'\boldsymbol{Z}_*)^{-1}\boldsymbol{Z}_*'\boldsymbol{y} \,.$$

Let us now apply the ridge method in the model $\boldsymbol{y} = \boldsymbol{Z}\boldsymbol{\delta} + \boldsymbol{\varepsilon}$. When we consider the matrix

$$\boldsymbol{Z}'\boldsymbol{Z} = \begin{pmatrix} n & \mathbf{0} \\ \mathbf{0} & \boldsymbol{Z}_*'\boldsymbol{Z}_* \end{pmatrix} ,$$

we can see that a possible instability of the matrix $\boldsymbol{Z}'\boldsymbol{Z}$ can solely be accounted for a possible instability of the matrix $\boldsymbol{Z}_*'\boldsymbol{Z}_*$. Hence it appears to be reasonable not to replace the whole matrix $\boldsymbol{Z}'\boldsymbol{Z}$ by $k\boldsymbol{I}_p + \boldsymbol{Z}'\boldsymbol{Z}$, but only to replace $\boldsymbol{Z}_*'\boldsymbol{Z}_*$ by $k\boldsymbol{I}_{p-1} + \boldsymbol{Z}_*'\boldsymbol{Z}_*$. Then a modified ridge estimator for $\boldsymbol{\delta}$ is given by

$$\widehat{\boldsymbol{\delta}}_k = \begin{pmatrix} n & \mathbf{0} \\ \mathbf{0} & \boldsymbol{Z}_*'\boldsymbol{Z}_* + k\boldsymbol{I}_{p-1} \end{pmatrix}^{-1} \begin{pmatrix} \mathbf{1}_n' \\ \boldsymbol{Z}_*' \end{pmatrix} \boldsymbol{y} = \begin{pmatrix} \overline{y} \\ (\boldsymbol{Z}_*'\boldsymbol{Z}_* + k\boldsymbol{I}_{p-1})^{-1}\boldsymbol{Z}_*'\boldsymbol{y} \end{pmatrix} .$$

This modified ridge estimation approach thus implies that the ordinary least squares estimator $\widehat{\mu} = \overline{y}$ is still used for estimating μ, while the ridge estimator $\widehat{\gamma}_k = (\boldsymbol{Z}_*'\boldsymbol{Z}_* + k\boldsymbol{I}_{p-1})^{-1}\boldsymbol{Z}_*'\boldsymbol{y}$ is used for estimating γ.

We illustrate the ridge estimation procedure for standardized data by the following example, which is also discussed in detail in [16, Chap. 8].

Example 3.16. Consider the cement data given in Table 1.1, p. 4. We consider the linear regression model

$$y_i = \beta_1 + \beta_2 x_{i1} + \beta_3 x_{i2} + \beta_4 x_{i3} + \beta_5 x_{i4} + \varepsilon_i, \quad i = 1, \ldots, 13 \,,$$

and the related standardized model

$$y_i = \mu + \gamma_1 z_{i1} + \gamma_2 z_{i2} + \gamma_3 z_{i3} + \gamma_4 z_{i4} + \varepsilon_i, \quad i = 1, \ldots, 13 \,.$$

The matrix $\boldsymbol{R} = \boldsymbol{Z}_*'\boldsymbol{Z}_*$ is computed as

$$\boldsymbol{R} = \begin{pmatrix} 1.0 & 0.2286 & -0.8241 & -0.2454 \\ 0.2286 & 1.0 & -0.1392 & -0.9730 \\ -0.8241 & -0.1392 & 1.0 & 0.0295 \\ -0.2454 & -0.9730 & 0.0295 & 1.0 \end{pmatrix}.$$

The (i, j)th element of this matrix is the empirical correlation between the variables x_i and x_j. Obviously the variables x_2 and x_4 are highly negative correlated (the correlation coefficient is -0.9730), so that we can note an almost linear relationship between these two variables, giving a strong indication for the presence of collinearity in the data. As a matter of fact, if we add x_2 to x_4, then we obtain more or less 77 for each element (except for the batches 1 and 5). This can be explained by the fact that x_2 and x_4 are the only two variables which contain silicate, so that it seems that the overall percentage of silicate ingredients has been kept close to 77%. Also x_1 and x_3 have a high negative correlation of - 0.8241, the sum of their values is in most cases close to 20.

As pointed out in [11], correlation between independent variables can be seen as an indication for collinearity, but the converse is not necessarily true, i.e. collinearity involving two independent variables can also occur without any correlation among them.

The condition number of $\boldsymbol{Z}_*$ computes to $\kappa(\boldsymbol{Z}_*) = 37.1063$ and thus is relatively large, compare also p. 63. (The condition number of $\boldsymbol{Z}_*$ equals its scaled condition number.)

As a first step we compute the ordinary least squares estimates of μ and γ. We obtain $\widehat{\mu} = \overline{y} = 95.4231$ and

$$\widehat{\gamma} = (\boldsymbol{Z}_*'\boldsymbol{Z}_*)^{-1}\boldsymbol{Z}_*'\boldsymbol{y} = \begin{pmatrix} 31.6072 \\ 27.5003 \\ 2.2612 \\ -8.3530 \end{pmatrix}.$$

As noted before, if we apply the ridge estimation method to standardized data as above, then the ordinary least squares estimator $\widehat{\mu}$ is not altered, and it remains to determine $\widehat{\gamma}_k = (\boldsymbol{Z}_*'\boldsymbol{Z}_* + k\boldsymbol{I}_{p-1})^{-1}\boldsymbol{Z}_*'\boldsymbol{y}$ for an appropriate value k. The Hoerl, Kennard and Baldwin [53] estimator of k in this situation is given as

$$\widehat{k}_{\text{HKB}} = \frac{(p-1)\,\widehat{\sigma}^2}{\widehat{\gamma}'\widehat{\gamma}}.$$

Here $\widehat{\sigma}^2$ denotes the ordinary least squares variance estimator of σ^2 in the model $\boldsymbol{y} = \boldsymbol{X}\boldsymbol{\beta} + \varepsilon$, which coincides with the ordinary least squares variance estimator of σ^2 in the model $\boldsymbol{y} = \boldsymbol{Z}\boldsymbol{\delta} + \varepsilon$, see Problem 3.19, although there might be some difference due to rounding-off errors in the computations. With the estimate $\widehat{\sigma}^2 = 5.9830$ we compute

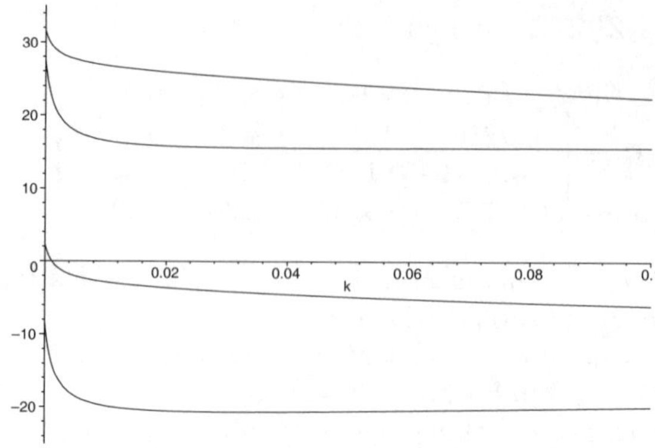

Fig. 3.20. Ridge trace of $\widehat{\gamma}_k$ in Example 3.16

$$\widehat{k}_{\mathrm{HKB}} = 0.0131 \,.$$

It has often been experienced that for standardized data an appropriate value for k is much smaller than 1. To apply a second tool, we also plot the ridge trace, see Fig. 3.20. We can see that there are no reasons to suppose that the choice $\widehat{k}_{\mathrm{HKB}} = 0.0131$ could be a bad one and we obtain

$$\widehat{\gamma}_{\widehat{k}_{\mathrm{HKB}}} = (\boldsymbol{Z}_*'\boldsymbol{Z}_* + \widehat{k}_{\mathrm{HKB}}\boldsymbol{I}_p)^{-1}\boldsymbol{Z}_*'\boldsymbol{y} = \begin{pmatrix} 26.4802 \\ 16.1675 \\ -3.1506 \\ -20.2136 \end{pmatrix}$$

as a ridge estimate of γ. As can be seen, the ridge estimates rate the influence of the amount of ingredient 2 on the heat lower and the influence of the amount of ingredient 4 distinctly higher than the ordinary least squares estimates do. Moreover, both estimators rate the influence of the amount of ingredient 3 as low, but with a different sign. □

3.4.5 Ridge and Restricted Least Squares Estimator

As already noted before, the ridge estimator satisfies

$$\lim_{k\to\infty} \widehat{\boldsymbol{\beta}}_k = \boldsymbol{0} \,,$$

and thus the estimates will come closer to the $p \times 1$ zero vector for increasing values of k, while for $k = 0$ the ridge estimator coincides with the ordinary least squares estimator $\widehat{\boldsymbol{\beta}}$. The $p \times 1$ zero vector can also be regarded as the restricted least squares estimator of $\boldsymbol{\beta}$ with respect to the restriction

$$\boldsymbol{R\beta} = \boldsymbol{r}, \quad \boldsymbol{R} = \boldsymbol{I}_p, \quad \boldsymbol{r} = \boldsymbol{0}.$$

Thus the ridge estimator can be seen as a weighted average between ordinary and restricted least squares estimator with respect to the aforementioned restrictions.

Now, it should also be possible to find an estimator which can be seen as a weighted average between the ordinary least squares estimator and the restricted least squares estimator with respect to *arbitrary* linear restrictions. Farebrother [37] considers the estimator

$$\widehat{\boldsymbol{\beta}}_{\mathrm{R}}(k) = (\boldsymbol{X}'\boldsymbol{X} + k\boldsymbol{R}'\boldsymbol{R})^{-1}(\boldsymbol{X}'\boldsymbol{y} + k\boldsymbol{R}'\boldsymbol{r}), \quad k \geq 0\,,$$

for some given $m \times p$ matrix $\boldsymbol{R}$ of full row rank and some given $m \times 1$ vector $\boldsymbol{r}$. It is obvious that $\widehat{\boldsymbol{\beta}}_{\mathrm{R}}(0) = \widehat{\boldsymbol{\beta}}$, and, as shown by the following theorem, $\widehat{\boldsymbol{\beta}}_{\mathrm{R}}(\infty) = \widehat{\boldsymbol{\beta}}_{\mathrm{R}}$ with respect to restrictions $\boldsymbol{R\beta} = \boldsymbol{r}$. Thus, the estimator $\widehat{\boldsymbol{\beta}}_{\mathrm{R}}(k)$ is in fact a weighted average between $\widehat{\boldsymbol{\beta}}$ and $\widehat{\boldsymbol{\beta}}_{\mathrm{R}}$.

Theorem 3.11. *Under the linear regression model with assumptions (i) to (iv),*

$$\lim_{k\to\infty} \widehat{\boldsymbol{\beta}}_{\mathrm{R}}(k) = \widehat{\boldsymbol{\beta}} - (\boldsymbol{X}'\boldsymbol{X})^{-1}\boldsymbol{R}'[\boldsymbol{R}(\boldsymbol{X}'\boldsymbol{X})^{-1}\boldsymbol{R}']^{-1}(\boldsymbol{R}\widehat{\boldsymbol{\beta}} - \boldsymbol{r})\,.$$

Proof. The inversion formula

$$\begin{aligned}&(\boldsymbol{X}'\boldsymbol{X} + k\boldsymbol{R}'\boldsymbol{R})^{-1}\\&= (\boldsymbol{X}'\boldsymbol{X})^{-1} - (\boldsymbol{X}'\boldsymbol{X})^{-1}\boldsymbol{R}'[\frac{1}{k}\boldsymbol{I}_m + \boldsymbol{R}(\boldsymbol{X}'\boldsymbol{X})^{-1}\boldsymbol{R}']^{-1}\boldsymbol{R}(\boldsymbol{X}'\boldsymbol{X})^{-1}\end{aligned}$$

holds true for $k > 0$. Then obviously

$$\begin{aligned}&\lim_{k\to\infty}(\boldsymbol{X}'\boldsymbol{X} + k\boldsymbol{R}'\boldsymbol{R})^{-1}\\&= (\boldsymbol{X}'\boldsymbol{X})^{-1} - (\boldsymbol{X}'\boldsymbol{X})^{-1}\boldsymbol{R}'[\boldsymbol{R}(\boldsymbol{X}'\boldsymbol{X})^{-1}\boldsymbol{R}']^{-1}\boldsymbol{R}(\boldsymbol{X}'\boldsymbol{X})^{-1}\,.\end{aligned}$$

Moreover, it follows

$$\begin{aligned}&(\boldsymbol{X}'\boldsymbol{X} + k\boldsymbol{R}'\boldsymbol{R})^{-1}k\boldsymbol{R}'\boldsymbol{r}\\&= (\boldsymbol{X}'\boldsymbol{X} + k\boldsymbol{R}'\boldsymbol{R})^{-1}k\boldsymbol{R}'\boldsymbol{R}\boldsymbol{R}'(\boldsymbol{R}\boldsymbol{R}')^{-1}\boldsymbol{r}\\&= (\boldsymbol{X}'\boldsymbol{X} + k\boldsymbol{R}'\boldsymbol{R})^{-1}(k\boldsymbol{R}'\boldsymbol{R} + \boldsymbol{X}'\boldsymbol{X} - \boldsymbol{X}'\boldsymbol{X})\boldsymbol{R}'(\boldsymbol{R}\boldsymbol{R}')^{-1}\boldsymbol{r}\\&= (\boldsymbol{I}_p - (\boldsymbol{X}'\boldsymbol{X} + k\boldsymbol{R}'\boldsymbol{R})^{-1}\boldsymbol{X}'\boldsymbol{X})\boldsymbol{R}'(\boldsymbol{R}\boldsymbol{R}')^{-1}\boldsymbol{r}\,.\end{aligned}$$

From the above formula for $\lim_{k\to\infty}(\boldsymbol{X}'\boldsymbol{X} + k\boldsymbol{R}'\boldsymbol{R})^{-1}$, it can thus be concluded that

$$\lim_{k\to\infty}(\boldsymbol{X}'\boldsymbol{X} + k\boldsymbol{R}'\boldsymbol{R})^{-1}k\boldsymbol{R}'\boldsymbol{r} = (\boldsymbol{X}'\boldsymbol{X})^{-1}\boldsymbol{R}'[\boldsymbol{R}(\boldsymbol{X}'\boldsymbol{X})^{-1}\boldsymbol{R}']^{-1}\boldsymbol{r}\,.$$

Hence, these formulas for $\lim_{k\to\infty}(\boldsymbol{X}'\boldsymbol{X} + k\boldsymbol{R}'\boldsymbol{R})^{-1}$ and $\lim_{k\to\infty}(\boldsymbol{X}'\boldsymbol{X} + k\boldsymbol{R}'\boldsymbol{R})^{-1}k\boldsymbol{R}'\boldsymbol{r}$ imply the assertion. □

If we choose $\boldsymbol{R} = \boldsymbol{I}_p$ but $\boldsymbol{r} \neq \boldsymbol{0}$, we obtain the estimator from Swindel [112]

$$\widehat{\boldsymbol{\beta}}_{k,\boldsymbol{r}} = (\boldsymbol{X}'\boldsymbol{X} + k\boldsymbol{I}_p)^{-1}(\boldsymbol{X}'\boldsymbol{y} + k\boldsymbol{r}), \quad k \geq 0 ,$$

which has the effect of shrinking the ordinary least squares estimator towards the vector $\boldsymbol{r}$. Hence it may also be called *direction modified ridge estimator.*

The application of $\widehat{\boldsymbol{\beta}}_{\mathrm{R}}(k)$ appears to be reasonable in cases when ridge estimation methods turn out to be appropriate for improving $\widehat{\boldsymbol{\beta}}$, while at the same time the estimate should not move towards the zero vector but to the restricted least squares estimate with respect to some given restrictions $\boldsymbol{R}\boldsymbol{\beta} = \boldsymbol{r}$. In such a situation, again the question of an appropriate choice of the ridge parameter k is crucial. To obtain some impression of the order of magnitude of k, the ridge trace is a practical device. For this, it seems reasonable not to plot the elements of the vector $\widehat{\boldsymbol{\beta}}_{\mathrm{R}}(k)$, but the elements of the difference

$$\widehat{\boldsymbol{\beta}}_{\mathrm{R}}(k) - \widehat{\boldsymbol{\beta}}_{\mathrm{R}}$$

as functions of k. The more the functions approach the zero line for increasing values of k, the less differs $\widehat{\boldsymbol{\beta}}_{\mathrm{R}}(k)$ from $\widehat{\boldsymbol{\beta}}_{\mathrm{R}}$.

Example 3.17. Let us again consider the linear regression model $\boldsymbol{y} = \boldsymbol{X}\boldsymbol{\beta} + \boldsymbol{\varepsilon}$ from Example 2.6, where

$$\boldsymbol{X}' = \begin{pmatrix} 1 & 1 & 1 & 1 & 1 \\ 1.9 & 2.1 & 2 & 2 & 1.8 \end{pmatrix}$$

and

$$\boldsymbol{y} = (6.0521, 7.0280, 7.1230, 4.4441, 5.0813)' .$$

In addition, let us consider the linear restrictions

$$\beta_1 + \beta_2 = 5 ,$$

which are only approximately met by the true parameter vector $\boldsymbol{\beta} = (2, 2)'$. We compute

$$\widehat{\boldsymbol{\beta}} = \begin{pmatrix} -4.2489 \\ 5.2013 \end{pmatrix} \quad \text{and} \quad \widehat{\boldsymbol{\beta}}_{\mathrm{R}} = \begin{pmatrix} 3.9678 \\ 1.0322 \end{pmatrix} .$$

If we now consider the individual elements of $\widehat{\boldsymbol{\beta}}_{\mathrm{R}}(k) - \widehat{\boldsymbol{\beta}}_{\mathrm{R}}$ as functions of k, then we can plot them as shown in Fig. 3.21.

Starting from the ordinary least squares estimator, the ridge trace visualizes how fast $\widehat{\boldsymbol{\beta}}_{\mathrm{R}}(k)$ approaches the restricted least squares estimator. Our choice of k can then be based on the stabilizing behavior of the ridge trace, but also on the magnitude of our prior believe in the restrictions. If we choose e.g. $k = 1$, then the resulting estimator already differs not much from the restricted least squares estimator, we obtain $\widehat{\boldsymbol{\beta}}_{\mathrm{R}}(1) = (3.5336, 1.2525)'$. This estimate has an observed unweighted squared error loss of $L(\boldsymbol{\beta}, \widehat{\boldsymbol{\beta}}_{\mathrm{R}}(1)) = 2.9108$ for the true $\boldsymbol{\beta} = (2, 2)'$, while the ordinary least squares estimator $\widehat{\boldsymbol{\beta}}_{\mathrm{R}}(0) = \widehat{\boldsymbol{\beta}}$ has an observed unweighted squared error loss of $L(\boldsymbol{\beta}, \widehat{\boldsymbol{\beta}}_{\mathrm{R}}(0)) = 49.2978$. □

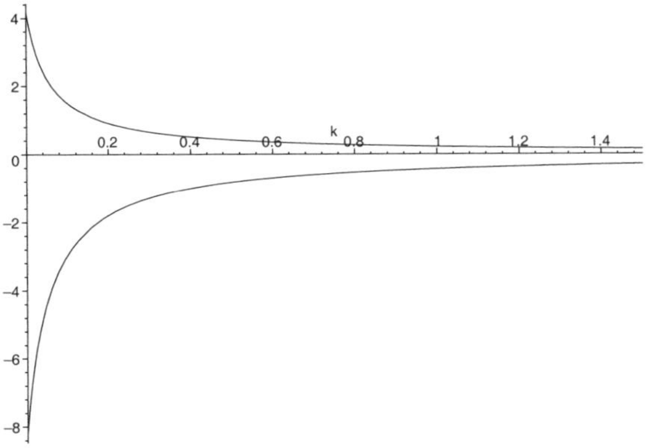

Fig. 3.21. Ridge trace of $\widehat{\beta}_{\mathrm{R}}(k) - \widehat{\beta}_{\mathrm{R}}$ in Example 3.17

If for the estimator $\widehat{\boldsymbol{\beta}}_{\mathrm{R}}(k)$ our choice of k is based on the ridge trace, we cannot say anything about the statistical properties of the estimator. Such a choice of k depends on the observed $\boldsymbol{y}$ as well as on the subjective judgement of the user. The situation is different, however, when we assume that k is fixed non-stochastic. In this case we can investigate under what conditions the estimator $\widehat{\boldsymbol{\beta}}_{\mathrm{R}}(k)$ is uniformly not worse than $\widehat{\boldsymbol{\beta}}$.

Theorem 3.12. *Under the linear regression model with assumptions (i) to (iv), the difference*

$$\boldsymbol{\Delta} = \mathrm{MSE}(\boldsymbol{\beta}, \widehat{\boldsymbol{\beta}}) - \mathrm{MSE}(\boldsymbol{\beta}, \widehat{\boldsymbol{\beta}}_{\mathrm{R}}(k))$$

is nonnegative definite for $k > 0$ if and only if the inequality

$$(\boldsymbol{R\beta} - \boldsymbol{r})' \left[\frac{2}{k} \boldsymbol{I}_m + \boldsymbol{R}(\boldsymbol{X}'\boldsymbol{X})^{-1}\boldsymbol{R}' \right]^{-1} (\boldsymbol{R\beta} - \boldsymbol{r}) \leq \sigma^2$$

is satisfied.

Proof. Let $\boldsymbol{T}_k = \boldsymbol{X}'\boldsymbol{X} + k\boldsymbol{R}'\boldsymbol{R}$. Then

$$\mathrm{Cov}(\widehat{\boldsymbol{\beta}}_{\mathrm{R}}(k)) = \sigma^2 \boldsymbol{T}_k^{-1} \boldsymbol{X}'\boldsymbol{X} \boldsymbol{T}_k^{-1}$$

and

$$\mathrm{bias}(\widehat{\boldsymbol{\beta}}_{\mathrm{R}}(k)) = \mathrm{E}(\widehat{\boldsymbol{\beta}}_{\mathrm{R}}(k)) - \boldsymbol{\beta} = -k\boldsymbol{T}_k^{-1}\boldsymbol{R}'(\boldsymbol{R\beta} - \boldsymbol{r}) \,.$$

This yields

$$\mathrm{MSE}(\boldsymbol{\beta}, \widehat{\boldsymbol{\beta}}_{\mathrm{R}}(k)) = \boldsymbol{T}_k^{-1} \left[\sigma^2 \boldsymbol{X}'\boldsymbol{X} + k^2 \boldsymbol{R}'\boldsymbol{\delta}\boldsymbol{\delta}'\boldsymbol{R} \right] \boldsymbol{T}_k^{-1} \,,$$

where $\boldsymbol{\delta} = \boldsymbol{R\beta} - \boldsymbol{r}$. The difference $\boldsymbol{\Delta} = \mathrm{MSE}(\boldsymbol{\beta}, \widehat{\boldsymbol{\beta}}) - \mathrm{MSE}(\boldsymbol{\beta}, \widehat{\boldsymbol{\beta}}_{\mathrm{R}}(k))$ is nonnegative definite if and only if $\boldsymbol{T}_k \boldsymbol{\Delta} \boldsymbol{T}_k$ is nonnegative definite. The latter can be written as

$$T_k \Delta T_k = \sigma^2 T_k (X'X)^{-1} T_k - \sigma^2 X'X - k^2 R' \delta\delta' R \, .$$

Since $T_k(X'X)^{-1}T_k = X'X + 2kR'R + k^2 R'R(X'X)^{-1}R'R$, it follows

$$T_k \Delta T_k = R' \left[\sigma^2 2k I_m + \sigma^2 k^2 R(X'X)^{-1} R' - k^2 \delta\delta' \right] R \, .$$

Now, the matrix $T_k \Delta T_k$ is nonnegative definite for $k > 0$ if and only if the matrix $\Phi = (1/k^2)(RR')^{-1} R T_k \Delta T_k R' (RR')^{-1}$ is nonnegative definite, where

$$\Phi = \sigma^2 \frac{2}{k} I_m + \sigma^2 R(X'X)^{-1} R' - \delta\delta' \, .$$

Since $\frac{2}{k} I_m + R(X'X)^{-1}R'$ is positive definite, Theorem A.60 shows that Φ is nonnegative definite if and only if the inequality

$$\delta' \left[\frac{2}{k} I_m + R(X'X)^{-1} R' \right]^{-1} \delta \leq \sigma^2$$

is satisfied. This confirms the assertion. □

The estimator $\widehat{\beta}_{\mathrm{R}}(k)$ has also the form of a mixed estimator under stochastic restrictions, see p. 103.

3.4.6 Ridge and Principal Components Estimator

Principal components and ridge estimator represent two different decision rules, whose application appears to be reasonable in presence of collinearity. Instead of using one or the other estimator, both of them may be appropriately combined.

Fine Tuning the Principal Components Estimator

Baye and Parker [9] propose the application of ridge methods to improve the principal components estimator. For this, the authors consider the estimator

$$\widehat{\beta}_k^{(r)} = U_1 (U_1' X' X U_1 + k I_r)^{-1} U_1' X' y, \quad k \geq 0 \, .$$

Here $X'X = U_1 \Lambda_1 U_1' + U_2 \Lambda_2 U_2'$ is a spectral decomposition of the matrix $X'X$, where Λ_1 and Λ_2 are diagonal matrices such that the main diagonal elements of the $r \times r$ matrix Λ_1 are the $r \leq p$ largest eigenvalues of $X'X$, while the main diagonal elements of the $(p-r) \times (p-r)$ matrix Λ_2 are the remaining $p - r$ eigenvalues. The $p \times p$ matrix $U = (U_1, U_2)$ is orthogonal, where $U_1 = (u_1, \ldots, u_r)$ consists of the first r columns of U and $U_2 = (u_{r+1}, \ldots, u_p)$ consists of the remaining $p - r$ columns.

We have already shown that in case $p > 1$, there always exists a $k > 0$ for any $\beta \neq 0$ and any $\sigma^2 > 0$, such that

$$\mathrm{tr}[\mathrm{MSE}(\boldsymbol{\beta}, \widehat{\boldsymbol{\beta}}_k)] < \mathrm{tr}[\mathrm{MSE}(\boldsymbol{\beta}, \widehat{\boldsymbol{\beta}})]$$

is satisfied, see p. 119. Baye and Parker [9] demonstrate a similar inequality for the comparison of $\widehat{\boldsymbol{\beta}}^{(r)}$ and $\widehat{\boldsymbol{\beta}}_k^{(r)}$. In case $p > 1$, there always exists a $k > 0$ for any $\boldsymbol{\beta} \neq \mathbf{0}$ and any $\sigma^2 > 0$, such that

$$\mathrm{tr}[\mathrm{MSE}(\boldsymbol{\beta}, \widehat{\boldsymbol{\beta}}_k^{(r)})] < \mathrm{tr}[\mathrm{MSE}(\boldsymbol{\beta}, \widehat{\boldsymbol{\beta}}^{(r)})]$$

is satisfied for any $0 < r \leq p$. This justifies the application of $\widehat{\boldsymbol{\beta}}_k^{(r)}$ for the purpose of improving upon the principal components estimator.

Nonetheless, we are also interested in conditions under which the estimator $\widehat{\boldsymbol{\beta}}_k^{(r)}$ improves upon the ordinary least squares estimator $\widehat{\boldsymbol{\beta}}$. If we compare the matrix-valued risks of $\widehat{\boldsymbol{\beta}}$ and $\widehat{\boldsymbol{\beta}}_k^{(r)}$, then we obtain the following result, compare also [95, Theorem 1].

Theorem 3.13. *Under the linear regression model with assumptions (i) to (iv), the difference* $\boldsymbol{\Delta} = \mathrm{MSE}(\boldsymbol{\beta}, \widehat{\boldsymbol{\beta}}) - \mathrm{MSE}(\boldsymbol{\beta}, \widehat{\boldsymbol{\beta}}_k^{(r)})$ *is nonnegative definite for* $k > 0$ *if and only if the inequality*

$$\boldsymbol{\beta}' \left[\boldsymbol{U}_1 (\frac{2}{k} \boldsymbol{I}_r + \boldsymbol{\Lambda}_1^{-1})^{-1} \boldsymbol{U}_1' + \boldsymbol{U}_2 \boldsymbol{\Lambda}_2 \boldsymbol{U}_2' \right] \boldsymbol{\beta} \leq \sigma^2$$

is satisfied.

Proof. In view of $\boldsymbol{U}_1' \boldsymbol{X}' \boldsymbol{X} \boldsymbol{U}_1 = \boldsymbol{\Lambda}_1$ it follows $\mathrm{Cov}(\widehat{\boldsymbol{\beta}}_k^{(r)}) = \sigma^2 \boldsymbol{U}_1 \boldsymbol{\Gamma} \boldsymbol{U}_1'$, where

$$\boldsymbol{\Gamma} = (\boldsymbol{\Lambda}_1 + k\boldsymbol{I}_r)^{-1} \boldsymbol{\Lambda}_1 (\boldsymbol{\Lambda}_1 + k\boldsymbol{I}_r)^{-1} = \mathrm{diag}\left(\frac{\lambda_1}{(\lambda_1 + k)^2}, \ldots, \frac{\lambda_r}{(\lambda_r + k)^2} \right) .$$

We can also write

$$\mathrm{Cov}(\widehat{\boldsymbol{\beta}}_k^{(r)}) = \sigma^2 \boldsymbol{U} \begin{pmatrix} \boldsymbol{\Gamma} & \mathbf{0} \\ \mathbf{0} & \mathbf{0} \end{pmatrix} \boldsymbol{U}' .$$

For $\mathrm{bias}(\widehat{\boldsymbol{\beta}}_k^{(r)}) = \mathrm{E}(\widehat{\boldsymbol{\beta}}_k^{(r)}) - \boldsymbol{\beta}$ we obtain $\mathrm{bias}(\widehat{\boldsymbol{\beta}}_k^{(r)}) = \boldsymbol{U}_1 \boldsymbol{\Phi} \boldsymbol{U}_1' \boldsymbol{\beta} - \boldsymbol{\beta}$, where

$$\boldsymbol{\Phi} = (\boldsymbol{\Lambda}_1 + k\boldsymbol{I}_r)^{-1} \boldsymbol{\Lambda}_1 = \mathrm{diag}\left(\frac{\lambda_1}{\lambda_1 + k}, \ldots, \frac{\lambda_r}{\lambda_r + k} \right) .$$

We now write

$$\mathrm{bias}(\widehat{\boldsymbol{\beta}}_k^{(r)}) = -\boldsymbol{U} \begin{pmatrix} \boldsymbol{I}_r - \boldsymbol{\Phi} & \mathbf{0} \\ \mathbf{0} & \boldsymbol{I}_{p-r} \end{pmatrix} \boldsymbol{U}' \boldsymbol{\beta} .$$

In view of

$$\mathrm{MSE}(\boldsymbol{\beta}, \widehat{\boldsymbol{\beta}}) = \mathrm{Cov}(\widehat{\boldsymbol{\beta}}) = \sigma^2 \boldsymbol{U} \begin{pmatrix} \boldsymbol{\Lambda}_1^{-1} & \mathbf{0} \\ \mathbf{0} & \boldsymbol{\Lambda}_2^{-1} \end{pmatrix} \boldsymbol{U}' ,$$

the difference $\boldsymbol{\Delta} = \mathrm{MSE}(\boldsymbol{\beta}, \widehat{\boldsymbol{\beta}}) - \mathrm{MSE}(\boldsymbol{\beta}, \widehat{\boldsymbol{\beta}}_k^{(r)})$ is given as $\boldsymbol{\Delta} = \sigma^2 \boldsymbol{U} \boldsymbol{A} \boldsymbol{U}' - \boldsymbol{U} \boldsymbol{a} \boldsymbol{a}' \boldsymbol{U}'$, where

$$\boldsymbol{A} = \begin{pmatrix} \boldsymbol{\Lambda}_1^{-1} - \boldsymbol{\Gamma} & \mathbf{0} \\ \mathbf{0} & \boldsymbol{\Lambda}_2^{-1} \end{pmatrix} \quad \text{and} \quad \boldsymbol{a} = \begin{pmatrix} \boldsymbol{I}_r - \boldsymbol{\Phi} & \mathbf{0} \\ \mathbf{0} & \boldsymbol{I}_{p-r} \end{pmatrix} \boldsymbol{U}'\boldsymbol{\beta}\,.$$

The matrix $\boldsymbol{\Delta}$ is nonnegative definite if and only if $\boldsymbol{U}'\boldsymbol{\Delta}\boldsymbol{U} = \sigma^2\boldsymbol{A} - \boldsymbol{a}\boldsymbol{a}'$ is nonnegative definite. Since the matrix

$$\boldsymbol{\Lambda}_1^{-1} - \boldsymbol{\Gamma} = \operatorname{diag}\left(\frac{2\lambda_1 k + k^2}{\lambda_1(\lambda_1 + k)^2}, \ldots, \frac{2\lambda_r k + k^2}{\lambda_r(\lambda_r + k)^2}\right)$$

is positive definite for $k > 0$, also the matrix $\boldsymbol{A}$ is positive definite. Then by Theorem A.60 the difference $\sigma^2\boldsymbol{A} - \boldsymbol{a}\boldsymbol{a}'$ is nonnegative definite if and only if $\boldsymbol{a}'\boldsymbol{A}^{-1}\boldsymbol{a} \leq \sigma^2$ is satisfied. From

$$\boldsymbol{A}^{-1} = \begin{pmatrix} (\boldsymbol{\Lambda}_1^{-1} - \boldsymbol{\Gamma})^{-1} & \mathbf{0} \\ \mathbf{0} & \boldsymbol{\Lambda}_2 \end{pmatrix},$$

the condition $\boldsymbol{a}'\boldsymbol{A}^{-1}\boldsymbol{a} \leq \sigma^2$ is equivalent to

$$\boldsymbol{\beta}'\boldsymbol{U} \begin{pmatrix} (\boldsymbol{I}_r - \boldsymbol{\Phi})(\boldsymbol{\Lambda}_1^{-1} - \boldsymbol{\Gamma})^{-1}(\boldsymbol{I}_r - \boldsymbol{\Phi}) & \mathbf{0} \\ \mathbf{0} & \boldsymbol{\Lambda}_2 \end{pmatrix} \boldsymbol{U}'\boldsymbol{\beta} \leq \sigma^2\,.$$

Then from the identity

$$\begin{aligned} &(\boldsymbol{I}_r - \boldsymbol{\Phi})(\boldsymbol{\Lambda}_1^{-1} - \boldsymbol{\Gamma})^{-1}(\boldsymbol{I}_r - \boldsymbol{\Phi}) \\ &= \operatorname{diag}\left(\frac{\lambda_1 k}{2\lambda_1 + k}, \ldots, \frac{\lambda_r k}{2\lambda_r + k}\right) = \left(\frac{2}{k}\boldsymbol{I}_r + \boldsymbol{\Lambda}_1^{-1}\right)^{-1} \end{aligned}$$

the assertion follows. □

Note that the estimator $\widehat{\boldsymbol{\beta}}_k^{(r)}$ satisfies the linear restrictions

$$\boldsymbol{U}_2'\widehat{\boldsymbol{\beta}}_k^{(r)} = \mathbf{0}$$

irrespective of the realization $\boldsymbol{y}$. In this sense, $\widehat{\boldsymbol{\beta}}_k^{(r)}$ is a restricted estimator, but it does not coincide with the restricted least squares estimator with respect to restrictions $\boldsymbol{U}_2'\boldsymbol{\beta} = \mathbf{0}$ unless $k = 0$.

Shrinking Towards the Principal Components Estimator

A second possibility for using ridge methods to modify principal components estimation is provided by the estimator

$$\widehat{\boldsymbol{\beta}}_{\mathrm{R}}(k) = (\boldsymbol{X}'\boldsymbol{X} + k\boldsymbol{R}'\boldsymbol{R})^{-1}(\boldsymbol{X}'\boldsymbol{y} + k\boldsymbol{R}'\boldsymbol{r}), \quad k \geq 0\,,$$

which we have considered in Sect. 3.4.5. If the matrix $\boldsymbol{R}$ is replaced by $\boldsymbol{U}_2'$ and the vector $\boldsymbol{r}$ is replaced by the zero vector, then we obtain the estimator

$$\widehat{\boldsymbol{\beta}}^{(r)}(k) = (\boldsymbol{X}'\boldsymbol{X} + k\boldsymbol{I}_p - k\boldsymbol{U}_1\boldsymbol{U}_1')^{-1}\boldsymbol{X}'\boldsymbol{y}, \quad k \geq 0 .$$

This estimator can be seen as a weighted average of the ordinary least squares and the principal components estimator. With increasing values of k the ordinary least squares estimator is shrunken towards the principal components estimator. A practical value of k can be obtained from an inspection of the ridge trace of $\widehat{\boldsymbol{\beta}}^{(r)}(k) - \widehat{\boldsymbol{\beta}}^{(r)}$, as described in Sect. 3.4.5.

Convex Combination

A third possibility for combining ridge and principal components estimation is the convex combination

$$\widehat{\boldsymbol{\beta}}_k^{(r)}(\alpha) = \alpha\widehat{\boldsymbol{\beta}}_k + (1-\alpha)\widehat{\boldsymbol{\beta}}^{(r)}, \quad 0 \leq \alpha \leq 1 ,$$

of the two estimators in question. Such a combination has the advantage that its actual loss cannot become greater than the maximum of the losses of $\widehat{\boldsymbol{\beta}}_k$ and $\widehat{\boldsymbol{\beta}}^{(r)}$. The parameter α can be seen as a control by which an increase of share of one estimator automatically yields a decrease of share of the other estimator. Of course, we can also consider convex combinations between other estimators, see Sect. 4.2.7.

3.4.7 Jackknife Modified Ridge Estimator

The jackknife method has been proposed by Quenouille [85] as an approach to reduce the bias of a point estimator. In the following we outline the jackknife method with regard to the ridge estimator $\widehat{\boldsymbol{\beta}}_k = (\boldsymbol{X}'\boldsymbol{X} + k\boldsymbol{I}_p)^{-1}\boldsymbol{X}'\boldsymbol{y}$.

A Jackknifed Ridge Estimator

Let in the sequel $\boldsymbol{E}_{(i)}$ denote the $(n-1) \times n$ matrix, obtained from the $n \times n$ identity matrix by deleting its i-th row. If in the observation matrix $(\boldsymbol{y}, \boldsymbol{X})$ the i-th row is deleted (so that the i-th observation vanishes), then the corresponding linear regression model can be written as

$$\boldsymbol{E}_{(i)}\boldsymbol{y} = \boldsymbol{E}_{(i)}\boldsymbol{X}\boldsymbol{\beta} + \boldsymbol{E}_{(i)}\boldsymbol{\varepsilon} .$$

Here, $\mathrm{E}(\boldsymbol{E}_{(i)}\boldsymbol{\varepsilon}) = 0$ and $\mathrm{Cov}(\boldsymbol{E}_{(i)}\boldsymbol{\varepsilon}) = \sigma^2\boldsymbol{I}_{n-1}$, but the matrix $\boldsymbol{E}_{(i)}\boldsymbol{X}$ is not necessarily of full column rank. Nonetheless, for the purpose of simplicity we will assume that $\boldsymbol{E}_{(i)}\boldsymbol{X}$ has full column rank for all $i = 1, \ldots, n$. Then the ridge estimator in this reduced model is given as

$$\widehat{\boldsymbol{\beta}}_{k,\,(i)} = (\boldsymbol{X}'\boldsymbol{E}_{(i)}'\boldsymbol{E}_{(i)}\boldsymbol{X} + k\boldsymbol{I}_p)^{-1}\boldsymbol{X}'\boldsymbol{E}_{(i)}'\boldsymbol{E}_{(i)}\boldsymbol{y} .$$

We assume that k is a non-stochastic number and the same for all $i = 1, \ldots, n$. If $\widehat{\boldsymbol{\beta}}_{k,\,(\cdot)}$ denotes the mean of these estimators, i.e.

$$\widehat{\boldsymbol{\beta}}_{k,(\cdot)} = \frac{1}{n}\sum_{i=1}^{n} \widehat{\boldsymbol{\beta}}_{k,(i)} ,$$

then the estimator for the bias $\mathrm{bias}(\widehat{\boldsymbol{\beta}}_k) = \mathrm{E}(\widehat{\boldsymbol{\beta}}_k) - \boldsymbol{\beta}$ due to Quenouille [85] is given as

$$\widehat{\boldsymbol{b}} = (n-1)(\widehat{\boldsymbol{\beta}}_{k,(\cdot)} - \widehat{\boldsymbol{\beta}}_k) .$$

The jackknifed ridge estimator is then

$$\breve{\boldsymbol{\beta}}_k^{\mathrm{J}} = \widehat{\boldsymbol{\beta}}_k - \widehat{\boldsymbol{b}} ,$$

i.e. the original estimator $\widehat{\boldsymbol{\beta}}_k$ minus the estimator for its bias. One can also think of obtaining similar estimators when different estimators for the bias are employed.

An Almost Unbiased Ridge Estimator

According to Problem 2.15, the bias of the ridge estimator $\widehat{\boldsymbol{\beta}}_k$ is given as

$$\mathrm{bias}(\widehat{\boldsymbol{\beta}}_k) = \mathrm{E}(\widehat{\boldsymbol{\beta}}_k) - \boldsymbol{\beta} = -k(\boldsymbol{X}'\boldsymbol{X} + k\boldsymbol{I}_p)^{-1}\boldsymbol{\beta} .$$

If we replace the unknown parameter vector $\boldsymbol{\beta}$ by $\widehat{\boldsymbol{\beta}}_k$, then

$$\widehat{\boldsymbol{b}} = -k(\boldsymbol{X}'\boldsymbol{X} + k\boldsymbol{I}_p)^{-1}\widehat{\boldsymbol{\beta}}_k$$

can be seen as an estimator for $\mathrm{bias}(\widehat{\boldsymbol{\beta}}_k) = \mathrm{E}(\widehat{\boldsymbol{\beta}}_k) - \boldsymbol{\beta}$. Then an estimator which is similar to the jackknifed ridge estimator is given as

$$\widehat{\boldsymbol{\beta}}_k^{\mathrm{J}} = \widehat{\boldsymbol{\beta}}_k - \widehat{\boldsymbol{b}} = \widehat{\boldsymbol{\beta}}_k + k(\boldsymbol{X}'\boldsymbol{X} + k\boldsymbol{I}_p)^{-1}\widehat{\boldsymbol{\beta}}_k .$$

If $\boldsymbol{U\Lambda U}'$ is a spectral decomposition of $\boldsymbol{X}'\boldsymbol{X}$, then $\widehat{\boldsymbol{\beta}}_k^{\mathrm{J}} = \boldsymbol{U}\widehat{\boldsymbol{\gamma}}_k^{\mathrm{J}}$, where

$$\widehat{\boldsymbol{\gamma}}_k^{\mathrm{J}} = (\boldsymbol{I}_p + k(\boldsymbol{\Lambda} + k\boldsymbol{I}_p)^{-1})(\boldsymbol{\Lambda} + k\boldsymbol{I}_p)^{-1}\boldsymbol{U}'\boldsymbol{X}'\boldsymbol{y} .$$

The estimator $\widehat{\boldsymbol{\gamma}}_k^{\mathrm{J}}$ is called *almost unbiased ordinary ridge estimator* for $\boldsymbol{\gamma} = \boldsymbol{U}'\boldsymbol{\beta}$ by Singh, Chaubey and Dwivedi [105], see also [80].

Gruber [47, p. 300] emphasizes that the estimator $\widehat{\boldsymbol{\gamma}}_k^{\mathrm{J}}$ is not well motivated in the literature and, in order to meet this shortcoming, gives an illustration in the light of a simple model, see also Problem 3.23. Our above derivation of the estimator, however, appears to be a rather obvious one.

For the application of the jackknifed ridge estimator or the almost unbiased ridge estimator it seems to be appropriate to determine the ridge parameter k in advance by using known methods and then compute the modified ridge estimator for this value of k.

Example 3.18. In Example 3.17 we have considered the linear regression model $\boldsymbol{y} = \boldsymbol{X}\boldsymbol{\beta} + \varepsilon$, where

$$\boldsymbol{X}' = \begin{pmatrix} 1 & 1 & 1 & 1 & 1 \\ 1.9 & 2.1 & 2 & 2 & 1.8 \end{pmatrix} .$$

If we compare the expectations of $\widehat{\boldsymbol{\beta}}_k$ and $\widehat{\boldsymbol{\beta}}_k^{\mathrm{J}}$ for $k = 1/5$, then

$$\mathrm{E}(\widehat{\boldsymbol{\beta}}_{1/5}) = \begin{pmatrix} 1.25 \\ 2.3571 \end{pmatrix} \quad \text{and} \quad \mathrm{E}(\widehat{\boldsymbol{\beta}}_{1/5}^{\mathrm{J}}) = \begin{pmatrix} 1.2976 \\ 2.3574 \end{pmatrix} ,$$

where $\boldsymbol{\beta} = (2, 2)'$. The expectation vector of the almost unbiased ridge estimator $\widehat{\boldsymbol{\beta}}_{1/5}^{\mathrm{J}}$ is a little closer to the true parameter vector $\boldsymbol{\beta}$ than the expectation vector of the ridge estimator $\widehat{\boldsymbol{\beta}}_{1/5}$. To illustrate the behavior of the two estimators, we generate 30 realizations of $\boldsymbol{y}$ and compute the respective estimates, see Fig. 3.22 and 3.23. For the true $\boldsymbol{\beta} = (2, 2)'$ the average observed unweighted squared error loss of $\widehat{\boldsymbol{\beta}}_{k=1/5}$ is 0.9533, while the average observed unweighted squared error loss of $\widehat{\boldsymbol{\beta}}_{k=1/5}^{\mathrm{J}}$ is 1.5315. This shows that, although $\widehat{\boldsymbol{\beta}}_{k=1/5}^{\mathrm{J}}$ has a smaller bias compared to $\widehat{\boldsymbol{\beta}}_{k=1/5}$, the estimator $\widehat{\boldsymbol{\beta}}_{k=1/5}^{\mathrm{J}}$ is disadvantageous here with respect to the average loss. □

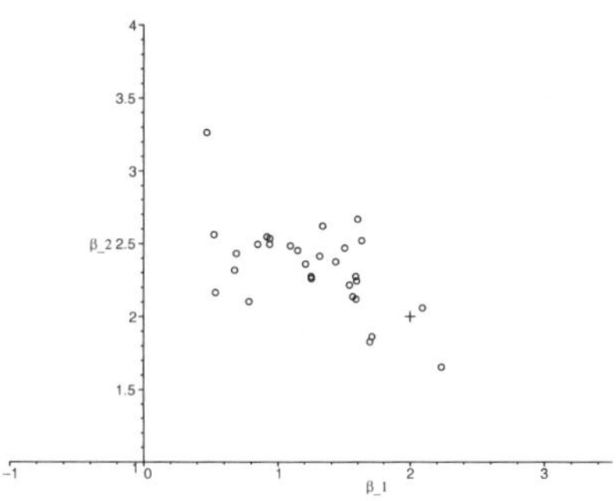

Fig. 3.22. 30 ridge estimates in Example 3.18

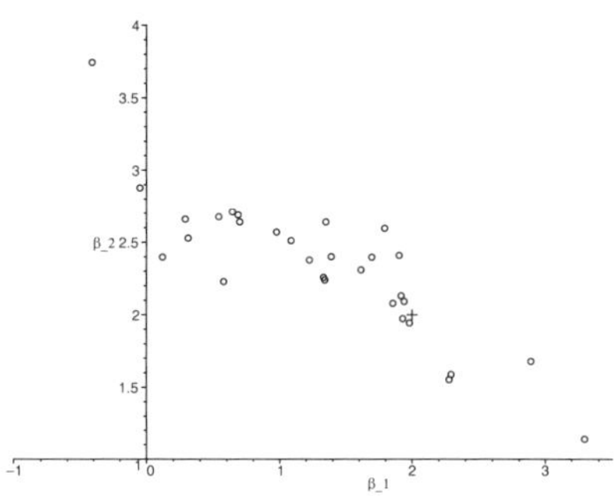

Fig. 3.23. 30 almost unbiased ridge estimates in Example 3.18

Eventually we can show that the almost unbiased ridge estimator has in fact a shorter bias vector than the ordinary ridge estimator, where the bias vector of an estimator $\widetilde{\boldsymbol{\beta}}$ is given as $\mathrm{bias}(\widetilde{\boldsymbol{\beta}}) = \mathrm{E}(\widetilde{\boldsymbol{\beta}}) - \boldsymbol{\beta}$.

Theorem 3.14. *Under the linear regression model with assumptions (i) to (iv), the inequality*

$$\| \, \mathrm{bias}(\widehat{\boldsymbol{\beta}}_k^{\mathrm{J}})\|^2 < \| \, \mathrm{bias}(\widehat{\boldsymbol{\beta}}_k)\|^2.$$

holds true for $\boldsymbol{\beta} \neq \mathbf{0}$ *and* $k > 0$.

Proof. We have $\text{bias}(\widehat{\boldsymbol{\beta}}_k) = -k(\boldsymbol{X}'\boldsymbol{X} + k\boldsymbol{I}_p)^{-1}\boldsymbol{\beta}$ and

$$\text{bias}(\widehat{\boldsymbol{\beta}}^{\text{J}}_k) = -k^2(\boldsymbol{X}'\boldsymbol{X} + k\boldsymbol{I}_p)^{-2}\boldsymbol{\beta} \,.$$

Thus

$$\|\,\text{bias}(\widehat{\boldsymbol{\beta}}_k)\|^2 - \|\,\text{bias}(\widehat{\boldsymbol{\beta}}^{\text{J}}_k)\|^2 = \boldsymbol{\beta}'\boldsymbol{G}\boldsymbol{\beta} \,,$$

where

$$\boldsymbol{G} = k^2(\boldsymbol{X}'\boldsymbol{X} + k\boldsymbol{I}_p)^{-2} - k^4(\boldsymbol{X}'\boldsymbol{X} + k\boldsymbol{I}_p)^{-4} \,.$$

By using the spectral decomposition of $\boldsymbol{X}'\boldsymbol{X}$, it can easily be shown that the symmetric matrix $\boldsymbol{G}$ is positive definite. This gives the assertion. □

3.4.8 Iteration Estimator

If we consider a spectral decomposition $\boldsymbol{U}\boldsymbol{\Lambda}\boldsymbol{U}'$ of $\boldsymbol{X}'\boldsymbol{X}$, then we can write

$$(\boldsymbol{X}'\boldsymbol{X} + k\boldsymbol{I}_p)^{-1} = \boldsymbol{U}\boldsymbol{\Gamma}\boldsymbol{U}', \quad \boldsymbol{\Gamma} = (\boldsymbol{\Lambda} + k\boldsymbol{I}_p)^{-1} \,.$$

If $k > 0$, then the main diagonal elements $\gamma_i = 1/(\lambda_i + k)$ of the matrix $\boldsymbol{\Gamma}$ satisfy

$$0 < \gamma_i < \frac{1}{\lambda_i} \le \frac{1}{\lambda_{\max}} \,.$$

Here $\lambda_{\max}$ denotes the largest eigenvalue of the matrix $\boldsymbol{X}'\boldsymbol{X}$. We can obtain a simpler estimator than the ridge estimator if instead of the different γ_i we employ only a single number δ satisfying the inequalities

$$0 < \delta < \frac{1}{\lambda_{\max}} \,.$$

We then replace $\boldsymbol{\Gamma}$ by the matrix $\delta\boldsymbol{I}_p$ and obtain the simplified ridge estimator

$$\widehat{\boldsymbol{\beta}}_{\delta,0} = \boldsymbol{U}\delta\boldsymbol{I}_p\boldsymbol{U}'\boldsymbol{X}'\boldsymbol{y} = \delta\boldsymbol{X}'\boldsymbol{y} \,.$$

The reason for the notation $\widehat{\boldsymbol{\beta}}_{\delta,0}$ will later be explained. In case $p > 1$ we can choose

$$\delta = \frac{1}{\text{tr}(\boldsymbol{X}'\boldsymbol{X})} \,,$$

and the so obtained simplified ridge estimator

$$\widehat{\boldsymbol{\beta}}_{1/\text{tr}(\boldsymbol{X}'\boldsymbol{X}),0} = \frac{1}{\text{tr}(\boldsymbol{X}'\boldsymbol{X})}\boldsymbol{X}'\boldsymbol{y}$$

can also be seen as a strong simplification of the ordinary least squares estimator $\widehat{\boldsymbol{\beta}} = (\boldsymbol{X}'\boldsymbol{X})^{-1}\boldsymbol{X}'\boldsymbol{y}$, where instead of the inverse of the complete matrix $\boldsymbol{X}'\boldsymbol{X}$ only the inverse of its trace is involved.

The Iteration Estimator

Trenkler [119] defines the *iteration estimator* as

$$\widehat{\boldsymbol{\beta}}_{\delta,m} = \delta \sum_{j=0}^{m} (\boldsymbol{I}_p - \delta \boldsymbol{X}'\boldsymbol{X})^j \boldsymbol{X}'\boldsymbol{y}, \quad 0 < \delta < \frac{1}{\lambda_{\max}}, \quad m = 0, 1, \ldots .$$

Since we will show that $\widehat{\boldsymbol{\beta}}_{\delta,0} = \delta \boldsymbol{X}'\boldsymbol{y}$ and $\widehat{\boldsymbol{\beta}}_{\delta,\infty} = \widehat{\boldsymbol{\beta}}$, the iteration estimator can be seen as a weighted average between the simplified ridge estimator $\delta \boldsymbol{X}'\boldsymbol{y}$ and the ordinary least squares estimator $\widehat{\boldsymbol{\beta}}$.

Now, let us consider the iteration estimator in more detail.

Theorem 3.15. *Under the linear regression model with assumptions (i) to (iv), let $\boldsymbol{U\Lambda U}'$ denote a spectral decomposition of the matrix $\boldsymbol{X}'\boldsymbol{X}$. Then the iteration estimator $\widehat{\boldsymbol{\beta}}_{\delta,m}$ can be written in the form*

$$\widehat{\boldsymbol{\beta}}_{\delta,m} = \boldsymbol{U\Gamma U}'\widehat{\boldsymbol{\beta}} ,$$

where

$$\boldsymbol{\Gamma} = \delta \boldsymbol{\Lambda} \sum_{j=0}^{m} (\boldsymbol{I}_p - \delta \boldsymbol{\Lambda})^j = \boldsymbol{I}_p - (\boldsymbol{I}_p - \delta \boldsymbol{\Lambda})^{m+1}$$

for $0 < \delta < 1/\lambda_{\max}$ and $m = 0, 1, \ldots$.

Proof. If $\boldsymbol{U\Lambda U}'$ is a spectral decomposition of $\boldsymbol{X}'\boldsymbol{X}$, then we can write

$$\begin{aligned}
\widehat{\boldsymbol{\beta}}_{\delta,m} &= \delta \sum_{j=0}^{m} (\boldsymbol{I}_p - \delta \boldsymbol{X}'\boldsymbol{X})^j \boldsymbol{X}'\boldsymbol{X}(\boldsymbol{X}'\boldsymbol{X})^{-1}\boldsymbol{X}'\boldsymbol{y} \\
&= \delta \sum_{j=0}^{m} \boldsymbol{U}(\boldsymbol{I}_p - \delta \boldsymbol{\Lambda})^j \boldsymbol{\Lambda U}'\widehat{\boldsymbol{\beta}} \\
&= \boldsymbol{U\Gamma U}'\widehat{\boldsymbol{\beta}} ,
\end{aligned}$$

where $\boldsymbol{\Gamma} = \delta \boldsymbol{\Lambda} \sum_{j=0}^{m} (\boldsymbol{I}_p - \delta \boldsymbol{\Lambda})^j$. Let us consider the i-th main diagonal element

$$\gamma_i = \delta \lambda_i \sum_{j=0}^{m} (1 - \delta \lambda_i)^j$$

of the matrix $\boldsymbol{\Gamma}$. From $0 < \delta < 1/\lambda_{\max}$ it follows $0 < \delta\lambda_i < 1$ and thus $0 < 1 - \delta\lambda_i < 1$. This means that $1 - \delta\lambda_i \neq 1$, so that we can apply the geometric sum formula

$$\sum_{j=0}^{m} (1 - \delta\lambda_i)^j = \frac{1 - (1 - \delta\lambda_i)^{m+1}}{1 - (1 - \delta\lambda_i)} .$$

Therefore,

$$\gamma_i = 1 - (1 - \delta\lambda_i)^{m+1} ,$$

showing that the matrix $\boldsymbol{\Gamma}$ can also be written in the form $\boldsymbol{\Gamma} = \boldsymbol{I}_p - (\boldsymbol{I}_p - \delta\boldsymbol{\Lambda})^{m+1}$. □

In view of $\boldsymbol{U}(\boldsymbol{I}_p - \delta\boldsymbol{\Lambda})^{m+1}\boldsymbol{U}' = [\boldsymbol{U}(\boldsymbol{I}_p - \delta\boldsymbol{\Lambda})\boldsymbol{U}']^{m+1}$ and the above theorem, it is easy to see that the iteration estimator can also be written in a form which does not involve the sum, namely,

$$\widehat{\boldsymbol{\beta}}_{\delta,m} = \widehat{\boldsymbol{\beta}} - (\boldsymbol{I}_p - \delta\boldsymbol{X}'\boldsymbol{X})^{m+1}\widehat{\boldsymbol{\beta}}, \quad 0 < \delta < \frac{1}{\lambda_{\max}}, \quad m = 0, 1, \ldots .$$

It is obvious that for $m = 0$ we obtain $\widehat{\boldsymbol{\beta}}_{\delta,0} = \delta\boldsymbol{X}'\boldsymbol{y}$, while the following corollary shows that for $m = \infty$ we obtain $\widehat{\boldsymbol{\beta}}_{\delta,\infty} = \widehat{\boldsymbol{\beta}}$.

Corollary 3.4. *Under the linear regression model with assumptions (i) to (iv), the identity*

$$\widehat{\boldsymbol{\beta}}_{\delta,\infty} = \widehat{\boldsymbol{\beta}}$$

holds true for any $0 < \delta < 1/\lambda_{\max}$.

Proof. From Theorem 3.15 we have $\widehat{\boldsymbol{\beta}}_{\delta,\infty} = \boldsymbol{U}\boldsymbol{\Gamma}\boldsymbol{U}'\widehat{\boldsymbol{\beta}}$, where the i-th main diagonal element γ_i of the matrix $\boldsymbol{\Gamma}$ is given as

$$\gamma_i = \delta\lambda_i \sum_{j=0}^{\infty}(1 - \delta\lambda_i)^j .$$

Here λ_i denotes the i-th eigenvalue of the matrix $\boldsymbol{X}'\boldsymbol{X}$. As in the proof of Theorem 3.15, we can conclude that $0 < 1 - \delta\lambda_i < 1$, so that

$$\sum_{j=0}^{\infty}(1 - \delta\lambda_i)^j = \frac{1}{1 - (1 - \delta\lambda_i)} .$$

This shows $\gamma_i = 1$ and thus completes the proof. □

For the application of the iteration estimator in practice it is necessary to know how to choose the parameters δ and m. Different choices of δ will yield different choices of m. One possibility for determining the parameters is to fix δ in advance and then to plot the elements of the estimator as functions of m. This is similar to the ridge trace procedure. While, however, in case of the ridge trace the elements of the ridge estimator move towards zero with increasing k, the elements of the iteration estimator start from the elements of $\delta\boldsymbol{X}'\boldsymbol{y}$ and move towards the elements of the ordinary least squares estimator $\widehat{\boldsymbol{\beta}}$ with increasing values of the integer m. In the following section this behavior is exemplarily illustrated.

3.4.9 An Example for Ridge Estimation

Again we consider the introductory example from Sect. 1.1.4 with $n = 12$, $p = 3$ and observations from Table 1.2, p. 8. The true parameters in the model are

$$(\boldsymbol{\beta}, \sigma^2) = (\begin{pmatrix} 0.2 \\ 0.4 \\ 0.7 \end{pmatrix}, 1) .$$

Determination of the Ridge Parameter. Since we know the true parameters in the model, we can derive the optimal ridge parameter k_{opt} by minimizing the function $\rho(\boldsymbol{\beta}, \widehat{\boldsymbol{\beta}}_k) = \text{tr}[\text{MSE}(\boldsymbol{\beta}, \widehat{\boldsymbol{\beta}}_k)]$ with respect to k. A numerical computer algorithm yields

$$k_{\text{opt}} = 7.2558 .$$

Note that k_{opt} is not equal to $p\sigma^2/\boldsymbol{\beta}'\boldsymbol{\beta}$ here, since model assumption (vii) is not satisfied.

We can now apply methods for determining the ridge parameter k and compare our efforts with the optimal choice in order to illustrate how these methods work. Among the possible estimators for k we confine to $\widehat{k}_{\text{HKB}} = p\widehat{\sigma}^2/\widehat{\boldsymbol{\beta}}'\widehat{\boldsymbol{\beta}}$ and its iterative version $\widehat{k}^i_{\text{HKB}}$. To get an impression of the behavior of these estimators in this example, we generate 50 realizations of $\boldsymbol{y}$ and compute the respective estimates for k.

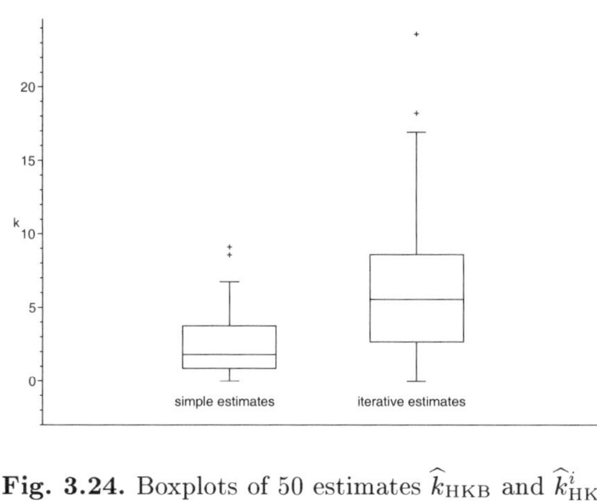

Fig. 3.24. Boxplots of 50 estimates $\widehat{k}_{\text{HKB}}$ and $\widehat{k}^i_{\text{HKB}}$

Fig. 3.24 shows the boxplots of the 50 estimates. If we compare the estimates with the value for k_{opt}, then it is confirmed here that the estimates $\widehat{k}_{\text{HKB}}$ tend to be too small, while the estimates $\widehat{k}^i_{\text{HKB}}$ appear to be more appropriate. Nonetheless, it can be seen that some of the iterative estimates are distinctly too large, implying that some caution is necessary when using this estimation procedure. The ridge trace as an additional aid for decision-making should also be taken into account.

Let us now try to determine some value for k for the realization of $\boldsymbol{y}$ given in Table 1.2, irrespective of our knowledge of k_{opt}. The estimator $\widehat{k}_{\text{HKB}}$ yields $\widehat{k}_{\text{HKB}} = 1.2288$.

If we compare this value with the ridge trace in Fig. 3.25, then it appears to be too small. The iterative estimator yields $\widehat{k}^{i}_{\text{HKB}} = 4.4201$. Compared to the ridge trace, this value still seems to be a little too small, but since we know that an increase of k comes along with an increase of the bias of the ridge estimator, we decide to stick to this choice of k. By this procedure we have not found a value of k which is very close to k_{opt}, but nonetheless we can be satisfied with having found some nonzero value of k being more or less in accordance with the ridge trace.

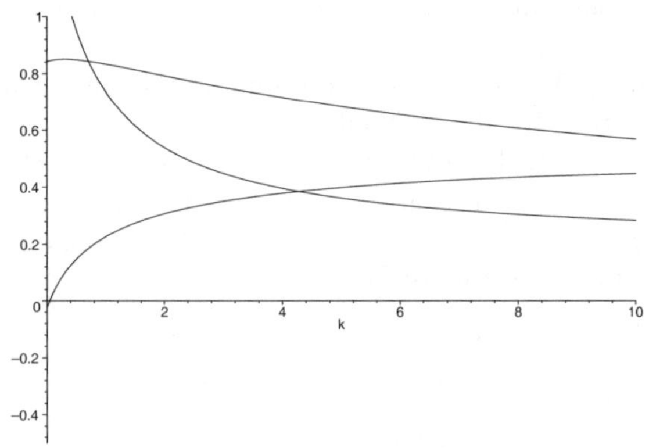

Fig. 3.25. Ridge trace in the introductory example from Sect. 1.1.4

Different Estimators. Having determined the ridge parameter as $k = 4.4201$, we can compute estimates $\widehat{\boldsymbol{\beta}}_{k=4.4201}$ and $\widehat{\boldsymbol{\beta}}^{\text{J}}_{k=4.4201}$ together with their observed unweighted squared error losses when $\boldsymbol{\beta} = (0.2, 0.4, 0.7)'$ is the true parameter vector.

We can also compare the performance of these estimators with the performance of the principal components estimator. The eigenvalues of the matrix $\boldsymbol{X}'\boldsymbol{X}$ are 122.5058, 8.5255, and 0.7389, so that $\widehat{\boldsymbol{\beta}}^{(2)}$ appears to be an appropriate choice. Moreover, we can consider the estimator $\widehat{\boldsymbol{\beta}}^{(2)}_{k}$ from Baye and Parker, being a fine tuning of the principal components estimator with the aid of ridge methods. The quality of this estimator does of course depend on the choice of k. We determine this parameter similar to the above and obtain a value of $k = 4.7054$.

If we want to incorporate the ridge and the principal components estimator into one estimation process, then we can consider a convex combination $\alpha\widehat{\boldsymbol{\beta}}_k + (1-\alpha)\widehat{\boldsymbol{\beta}}^{(2)}$. When one estimator should not be weighted more more than the other, then the appropriate choice is $\alpha = 0.5$.

Table 3.2 shows the values of the considered estimators and their observed losses.

Table 3.2. Different estimates in the introductory example from Sect. 1.1.4

	$\boldsymbol{\beta}$	$\widehat{\boldsymbol{\beta}}$	$\widehat{\boldsymbol{\beta}}_{k=4.4201}$	$\widehat{\boldsymbol{\beta}}^{\mathrm{J}}_{k=4.4201}$	$\widehat{\boldsymbol{\beta}}^{(2)}$	$\widehat{\boldsymbol{\beta}}^{(2)}_{k=4.7054}$	$0.5\widehat{\boldsymbol{\beta}}_{k=4.4201} + 0.5\widehat{\boldsymbol{\beta}}^{(2)}$
β_1	0.2	1.4570	0.3801	0.5349	0.1959	0.1994	0.2880
β_2	0.4	- 0.0233	0.3883	0.2979	0.3701	0.4472	0.3792
β_3	0.7	0.8423	0.7016	0.8446	0.9491	0.7070	0.8254
loss	0	1.7794	0.0326	0.1435	0.0630	0.0023	0.0239

As can be seen, all considered estimators perform better than the ordinary least squares estimator here. Although the degree of collinearity is not very high, see Table 2.6, p. 63, the ordinary least squares estimator tends to yield estimates which especially for the parameter β_1 distinctly differ from the true value.

When we compare the observed losses of all considered estimators, then $\widehat{\boldsymbol{\beta}}^{(2)}_{k=4.7054}$ performs best, so that applying the ridge method to the principal components estimator is successful here. As a reason for the good performance of this estimator one might also see the fact that the principal components estimator itself performs quite well.

It is also noteworthy that the arithmetic mean of principal components and ridge estimator yields an observed loss, which is smaller than each of the two losses of the two estimators.

Eventually we consider the iteration estimator

$$\widehat{\boldsymbol{\beta}}_{\delta,m} = \widehat{\boldsymbol{\beta}} - (\boldsymbol{I}_p - \delta\boldsymbol{X}'\boldsymbol{X})^{m+1}\widehat{\boldsymbol{\beta}}, \quad 0 < \delta < \frac{1}{\lambda_{\max}}, \qquad m = 0, 1, \ldots .$$

We choose $\delta = 1/\operatorname{tr}(\boldsymbol{X}'\boldsymbol{X})$, compute the estimates for $m = 0, 1, \ldots$ and plot the estimates, see Fig. 3.26 and Fig. 3.27.

We decide in favor of $m = 30$, since for greater values of m the changes in the estimates appear to be rather slow. One should, however, acknowledge that similar to the ridge trace the impression of stabilization depends on the range of m values. This is illustrated by Figs. 3.26 and 3.27 for which the estimates are plotted.

It is clear that for increasing values of m one comes closer to the ordinary least squares estimate. The iteration estimate for $m = 30$ is given as

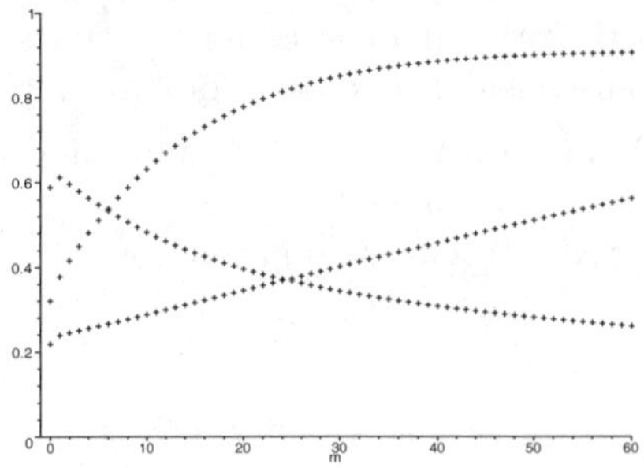

Fig. 3.26. Iteration estimates $\widehat{\boldsymbol{\beta}}_{\delta,m}$ for the case $\delta = 1/\operatorname{tr}(\boldsymbol{X}'\boldsymbol{X})$ and $m = 0,\dots,60$

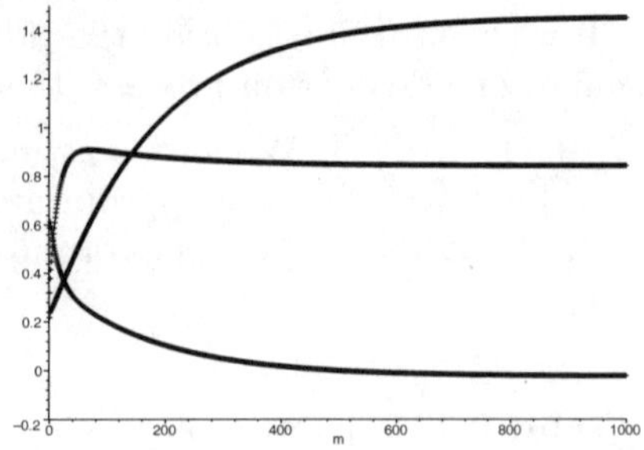

Fig. 3.27. Iteration estimates $\widehat{\boldsymbol{\beta}}_{\delta,m}$ for the case $\delta = 1/\operatorname{tr}(\boldsymbol{X}'\boldsymbol{X})$ and $m = 0,\dots,1000$

$$\widehat{\boldsymbol{\beta}}_{1/\operatorname{tr}(\boldsymbol{X}'\boldsymbol{X}),30} = \begin{pmatrix} 0.4019 \\ 0.3430 \\ 0.8504 \end{pmatrix}$$

with an observed unweighted squared error loss of 0.0666. Again, the alternative estimate is closer to the true $\boldsymbol{\beta}$ than the ordinary least squares estimate $\widehat{\boldsymbol{\beta}}$.

It should be noted that the above results concerning the performance of alternatives to the ordinary least squares estimator depend on the actual value $\boldsymbol{y}$. Hence, the results are not of general validity but illustrate the possibility for improvement upon $\widehat{\boldsymbol{\beta}}$. For other realizations $\boldsymbol{y}$ the situation can be quite different.

3.5 Shrinkage Estimator

The ridge estimator $\widehat{\boldsymbol{\beta}}_k$ discussed in the previous section is sometimes also called shrinkage estimator, since its squared length is smaller than the squared length of the ordinary least squares estimator (see Theorem 3.9). More often, however, a shrinkage estimator is simply understood as the multiplication of the ordinary least squares estimator with a number from the interval $[0,1]$.

3.5.1 Preliminary Considerations

The shrinkage estimator (also called *contraction* estimator)

$$\widehat{\boldsymbol{\beta}}(\varrho) = \frac{1}{1+\varrho}\widehat{\boldsymbol{\beta}}, \quad \varrho \geq 0\,,$$

has been introduced by Mayer and Willke [77] as a further alternative to the ordinary least squares estimator. If ϱ is considered to be a fixed number, then $\widehat{\boldsymbol{\beta}}(\varrho)$ is a homogeneous linear estimator satisfying

$$\mathrm{E}(\widehat{\beta}(\varrho)) = \frac{1}{1+\varrho}\beta \quad \text{and} \quad \mathrm{Cov}(\widehat{\beta}(\varrho)) = \frac{\sigma^2}{(1+\varrho)^2}(\boldsymbol{X}'\boldsymbol{X})^{-1} .$$

The shrinkage estimator is designed such that each element of the ordinary least squares estimator is multiplied by the same factor $1/(1+\varrho)$. Hence, this estimator can be advantageous when the sign and the proportion of the individual ordinary least squares estimates should be retained, while at the same time it is desired to obtain some reduction of variance at the cost of some small bias.

We can illustrate the behavior of the shrinkage estimator by computing some estimates.

Example 3.19. Consider the linear regression model $\boldsymbol{y} = \boldsymbol{X}\boldsymbol{\beta} + \boldsymbol{\varepsilon}$ with

$$\boldsymbol{X}' = \begin{pmatrix} 1 & 1 & 1 & 1 & 1 \\ 1.9 & 2.1 & 2 & 2 & 1.8 \end{pmatrix} .$$

We generate 30 realizations of the random vector $\boldsymbol{y}$ and compute the respective estimates $\widehat{\boldsymbol{\beta}}$ as well as $\widehat{\boldsymbol{\beta}}(7/3)$. Fig. 3.28 now gives an impression on how the shrinkage estimator works. □

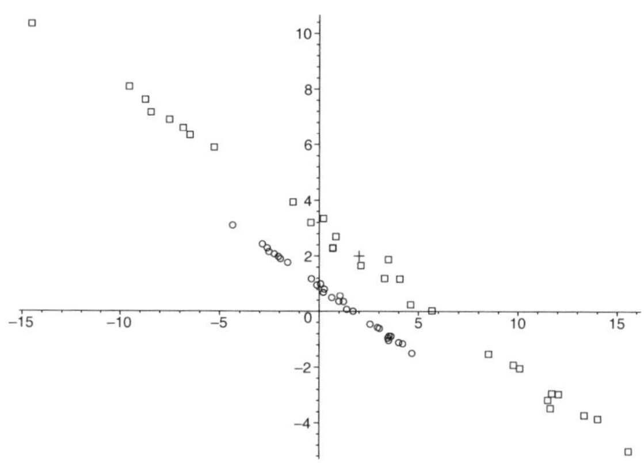

Fig. 3.28. 30 estimates $\widehat{\boldsymbol{\beta}}$ (boxes) and $\widehat{\boldsymbol{\beta}}(7/3)$ (circles) in Example 3.19

The factor $\gamma := 1/(1+\varrho)$ lies in the interval $(0,1)$ whenever $\varrho > 0$, and one might wonder why we introduce ϱ as the shrinkage parameter and not γ. A reason for this is, that for the determination of the parameter ϱ we only have to ensure its positiveness, similar to the ridge parameter k. Moreover, in the special case $\boldsymbol{X}'\boldsymbol{X} = \boldsymbol{I}_p$, i.e. when model assumption (i) to (iv) and (vii) are satisfied, then the identity

$$\widehat{\boldsymbol{\beta}}_k = \widehat{\boldsymbol{\beta}}(k)$$

holds true. In this case the ridge estimator with parameter k coincides with the shrinkage estimator with parameter $\varrho = k$.

3.5.2 Risk Comparison to Ordinary Least Squares

Similar to the case of ridge estimation we can investigate under what conditions the risk of the shrinkage estimator $\widehat{\boldsymbol{\beta}}(\varrho)$ is smaller than the risk of the ordinary least squares estimator $\widehat{\boldsymbol{\beta}}$.

Theorem 3.16. *Under the linear regression model with assumptions (i) to (iv), let $\varrho > 0$ be a (non-stochastic) number. Then the difference*

$$\boldsymbol{\Delta} = \mathrm{MSE}(\boldsymbol{\beta}, \widehat{\boldsymbol{\beta}}) - \mathrm{MSE}(\boldsymbol{\beta}, \widehat{\boldsymbol{\beta}}(\varrho))$$

is nonnegative definite if and only if the inequality

$$\boldsymbol{\beta}'\boldsymbol{X}'\boldsymbol{X}\boldsymbol{\beta} \leq \frac{\varrho+2}{\varrho}\sigma^2$$

is satisfied. In addition, $\boldsymbol{\Delta} \neq \boldsymbol{0}$ whenever $\varrho > 0$ and $p > 1$.

The above theorem is a special case of Theorem 3.19 in Sect. 3.6 and thus is not proved here. The set of all vectors $\boldsymbol{\beta}$ satisfying the condition

$$\boldsymbol{\beta}'\boldsymbol{X}'\boldsymbol{X}\boldsymbol{\beta} \leq \frac{\varrho+2}{\varrho}\sigma^2$$

for some $\sigma^2 \in (0, \infty)$ and some $\varrho \in (0, \infty)$ can be seen as an ellipsoid in $\mathbb{R}^p$. Similar to the ellipsoid corresponding to the ridge estimator, for an arbitrary fixed σ^2 we can always find some $\varrho > 0$ such that $\boldsymbol{\beta}$ belongs to the above ellipsoid. This follows from the fact that the ellipsoid becomes vaster for increasing values ϱ.

Corollary 3.5. *Suppose that under the linear regression model with assumptions (i) to (iv) the parameters $(\boldsymbol{\beta} \neq \boldsymbol{0}, \sigma^2)$ satisfy the inequality*

$$0 < \varrho \leq \frac{2\sigma^2}{\boldsymbol{\beta}'\boldsymbol{X}'\boldsymbol{X}\boldsymbol{\beta}} \,.$$

Then the difference $\boldsymbol{\Delta} = \mathrm{MSE}(\boldsymbol{\beta}, \widehat{\boldsymbol{\beta}}) - \mathrm{MSE}(\boldsymbol{\beta}, \widehat{\boldsymbol{\beta}}(\varrho))$ is nonnegative definite.

Proof. Let $\varrho > 0$ in the sequel. From Theorem 3.16 the matrix $\boldsymbol{\Delta}$ is nonnegative definite, if the condition

$$\boldsymbol{\beta}'\boldsymbol{X}'\boldsymbol{X}\boldsymbol{\beta} \leq \frac{\varrho+2}{\varrho}\sigma^2$$

is satisfied. For any $\boldsymbol{\beta} \neq \boldsymbol{0}$ the identity $\boldsymbol{\beta}'\boldsymbol{X}'\boldsymbol{X}\boldsymbol{\beta} \neq 0$ holds true in view of the positive definiteness of the matrix $\boldsymbol{X}'\boldsymbol{X}$. If the parameters $\boldsymbol{\beta} \neq \boldsymbol{0}$ and σ^2 satisfy $\boldsymbol{\beta}'\boldsymbol{X}'\boldsymbol{X}\boldsymbol{\beta} \leq \sigma^2$, then also the condition $\boldsymbol{\beta}'\boldsymbol{X}'\boldsymbol{X}\boldsymbol{\beta} \leq \frac{\varrho+2}{\varrho}\sigma^2$ holds true for any $0 < \varrho$ and thus for any $0 < \varrho \leq 2\sigma^2/\boldsymbol{\beta}'\boldsymbol{X}'\boldsymbol{X}\boldsymbol{\beta}$. If the parameters

$\boldsymbol{\beta} \neq \mathbf{0}$ and σ^2 satisfy $\boldsymbol{\beta}'\boldsymbol{X}'\boldsymbol{X}\boldsymbol{\beta} > \sigma^2$, then the inequality $\boldsymbol{\beta}'\boldsymbol{X}'\boldsymbol{X}\boldsymbol{\beta} \leq \frac{\varrho+2}{\varrho}\sigma^2$ can be written as

$$\varrho \leq \frac{2\sigma^2}{\boldsymbol{\beta}'\boldsymbol{X}'\boldsymbol{X}\boldsymbol{\beta} - \sigma^2} \, .$$

But this inequality is satisfied for any $0 < \varrho \leq 2\sigma^2/\boldsymbol{\beta}'\boldsymbol{X}'\boldsymbol{X}\boldsymbol{\beta}$ whenever $\boldsymbol{\beta}'\boldsymbol{X}'\boldsymbol{X}\boldsymbol{\beta} > \sigma^2$ holds true. □

We have already noted that in presence of collinearity the (squared) length of $\boldsymbol{\beta}$ is often overestimated by the ordinary least squares estimator $\widehat{\boldsymbol{\beta}}$. The estimators $\widehat{\boldsymbol{\beta}}(\varrho)$ and $\widehat{\boldsymbol{\beta}}_k$ reduce the (squared) length of $\widehat{\boldsymbol{\beta}}$, i.e.

$$||\widehat{\boldsymbol{\beta}}(\varrho)||^2 < ||\widehat{\boldsymbol{\beta}}||^2 \quad \text{and} \quad ||\widehat{\boldsymbol{\beta}}_k||^2 < ||\widehat{\boldsymbol{\beta}}||^2$$

whenever $\widehat{\boldsymbol{\beta}} \neq \mathbf{0}$ and $\varrho, k > 0$. The difference is, that the velocity of reduction is the same for all elements of the vector $\widehat{\boldsymbol{\beta}}(\varrho)$, while it is not the same for different elements of the ridge estimator $\widehat{\boldsymbol{\beta}}_k$.

3.5.3 The Choice of the Shrinkage Parameter

Similar to the case of ridge estimation, for any $(\boldsymbol{\beta} \neq \mathbf{0}, \sigma^2)$ there is an optimal value ϱ_{opt} such that $\rho(\boldsymbol{\beta}, \widehat{\boldsymbol{\beta}}(\varrho_{\text{opt}})) \leq \rho(\boldsymbol{\beta}, \widehat{\boldsymbol{\beta}}(\varrho))$ for every $\varrho \geq 0$. Here $\rho(\boldsymbol{\beta}, \widetilde{\boldsymbol{\beta}})$ denotes the unweighted squared error risk of $\widetilde{\boldsymbol{\beta}}$, i.e. $\rho(\boldsymbol{\beta}, \widetilde{\boldsymbol{\beta}}) = \text{tr}[\text{MSE}(\boldsymbol{\beta}, \widetilde{\boldsymbol{\beta}})]$. Unlike in the case of ridge estimation, this optimal value can explicitly be determined.

Theorem 3.17. *Under the linear regression model with assumptions (i) to (iv), for any* $(\boldsymbol{\beta} \neq \mathbf{0}, \sigma^2)$ *the inequality*

$$\rho(\boldsymbol{\beta}, \widehat{\boldsymbol{\beta}}(\varrho_{\text{opt}})) \leq \rho(\boldsymbol{\beta}, \widehat{\boldsymbol{\beta}}(\varrho)) \quad \forall \varrho \geq 0$$

holds true, where $\varrho_{\text{opt}} = \sigma^2 \, \text{tr}[(\boldsymbol{X}'\boldsymbol{X})^{-1}]/\boldsymbol{\beta}'\boldsymbol{\beta}$.

Proof. The risk of the shrinkage estimator $\widehat{\boldsymbol{\beta}}(\varrho)$ is given as

$$\rho(\boldsymbol{\beta}, \widehat{\boldsymbol{\beta}}(\varrho)) = \frac{\sigma^2 \, \text{tr}[(\boldsymbol{X}'\boldsymbol{X})^{-1}]}{(1+\varrho)^2} + \frac{\varrho^2}{(1+\varrho)^2}\boldsymbol{\beta}'\boldsymbol{\beta} \, .$$

Then the assertion follows analogously to Theorem 3.10. □

According to the above theorem, the quantity

$$\widehat{\varrho} = \frac{\widehat{\sigma}^2 \, \text{tr}[(\boldsymbol{X}'\boldsymbol{X})^{-1}]}{\widehat{\boldsymbol{\beta}}'\widehat{\boldsymbol{\beta}}}$$

can be seen as an obvious ad-hoc estimator for ϱ. The resulting shrinkage estimator $\widehat{\boldsymbol{\beta}}(\widehat{\varrho})$ is of course a non-linear estimator. In the literature there are also other non-linear shrinkage estimators for $\boldsymbol{\beta}$ known. We present some of them in the following, by introducing the corresponding estimator for ϱ.

The Farebrother Estimator

If the quantity

$$\widehat{\varrho}_{\mathrm{M}} = \frac{\widehat{\sigma}^2}{\widehat{\boldsymbol{\beta}}'\boldsymbol{X}'\boldsymbol{X}\widehat{\boldsymbol{\beta}}}$$

is applied for the estimation of ϱ, then

$$\widehat{\boldsymbol{\beta}}_{\mathrm{M}} = \widehat{\boldsymbol{\beta}}(\widehat{\varrho}_{\mathrm{M}}) = \frac{\widehat{\boldsymbol{\beta}}'\boldsymbol{X}'\boldsymbol{y}}{\widehat{\sigma}^2 + \widehat{\boldsymbol{\beta}}'\boldsymbol{X}'\boldsymbol{X}\widehat{\boldsymbol{\beta}}}\widehat{\boldsymbol{\beta}}$$

is the corresponding non-linear shrinkage estimator for $\boldsymbol{\beta}$ with stochastic shrinkage parameter $\widehat{\varrho}_{\mathrm{M}}$, compare also Problem 3.26. This estimator can also be obtained via an approach which initially is not related to the shrinkage method.

For this, consider a homogeneous linear estimator $\boldsymbol{C}\boldsymbol{y}$ for $\boldsymbol{\beta}$ such that the matrix-valued risk

$$\mathrm{MSE}(\boldsymbol{\beta}, \boldsymbol{C}\boldsymbol{y}) = \mathrm{E}[(\boldsymbol{C}\boldsymbol{y} - \boldsymbol{\beta})(\boldsymbol{C}\boldsymbol{y} - \boldsymbol{\beta})']$$

is minimized with respect to $\boldsymbol{C}$. The solution is given by

$$\boldsymbol{C} = \boldsymbol{\beta}\boldsymbol{\beta}'\boldsymbol{X}'(\boldsymbol{X}\boldsymbol{\beta}\boldsymbol{\beta}'\boldsymbol{X}' + \sigma^2\boldsymbol{I}_n)^{-1},$$

see also [123, Sect. 6.4.1]. By some transformations it can be shown that $\boldsymbol{C}\boldsymbol{y} =$ '$\widehat{\boldsymbol{\beta}}_{\mathrm{M}}$', where

$$\text{‘}\widehat{\boldsymbol{\beta}}_{\mathrm{M}}\text{’} = \frac{\boldsymbol{\beta}'\boldsymbol{X}'\boldsymbol{y}}{\sigma^2 + \boldsymbol{\beta}'\boldsymbol{X}'\boldsymbol{X}\boldsymbol{\beta}}\boldsymbol{\beta}\,,$$

see also Theil [113, p. 125]. But clearly this is not an estimator, since it contains the unknowns $\boldsymbol{\beta}$ and σ^2. Farebrother [35] proposes to replace $\boldsymbol{\beta}$ by $\widehat{\boldsymbol{\beta}}$ and σ^2 by $\widehat{\sigma}^2$, which just gives $\widehat{\boldsymbol{\beta}}_{\mathrm{M}}$.

Remark 3.5. In the literature, the vector '$\widehat{\boldsymbol{\beta}}_{\mathrm{M}}$' as well as $\widehat{\boldsymbol{\beta}}_{\mathrm{M}}$ are often called *minimum mean squared error estimator*. This is, however, a misnomer in both cases, since '$\widehat{\boldsymbol{\beta}}_{\mathrm{M}}$' is definitely not an estimator, while $\widehat{\boldsymbol{\beta}}_{\mathrm{M}}$ does neither minimize $\mathrm{MSE}(\cdot)$ nor $\rho(\cdot)$.

The Ohtani Estimator

Ohtani [82] investigates the estimator $\widehat{\boldsymbol{\beta}}(\widehat{\varrho}_{\mathrm{AM}})$, where

$$\widehat{\varrho}_{\mathrm{AM}} = \frac{\widehat{\sigma}^2\, p}{\widehat{\boldsymbol{\beta}}'\boldsymbol{X}'\boldsymbol{X}\widehat{\boldsymbol{\beta}}}\,.$$

The estimator $\widehat{\boldsymbol{\beta}}(\widehat{\varrho}_{\mathrm{AM}})$ is called *adjusted minimum mean squared error estimator*, since it has the representation

$$\widehat{\beta}_{\mathrm{AM}} = \widehat{\beta}(\widehat{\varrho}_{\mathrm{AM}}) = \frac{\widehat{\beta}' X' y / p}{\widehat{\sigma}^2 + \widehat{\beta}' X' X \widehat{\beta} / p} \widehat{\beta} \,,$$

and thus can be seen as an adjustment of the minimum mean squared error estimator

$$\widehat{\beta}_{\mathrm{M}} = \frac{\widehat{\beta}' X' y}{\widehat{\sigma}^2 + \widehat{\beta}' X' X \widehat{\beta}} \widehat{\beta} \,.$$

It is noteworthy that here an estimator for the shrinkage parameter ϱ is employed, which is also used for the estimation of the ridge parameter k, that is

$$\widehat{\varrho}_{\mathrm{AM}} = \widehat{k}_{\mathrm{LW}} \,,$$

compare also p. 127.

The estimator $\widehat{\varrho}_{\mathrm{AM}}$ can also be seen as an obvious ad-hoc estimator for ϱ when the predictive squared error risk $\rho(\beta, \widehat{\beta}(\varrho)) = \mathrm{tr}[X'X \,\mathrm{MSE}(\beta, \widehat{\beta}(\varrho))]$ instead of the unweighted squared error risk $\rho(\beta, \widehat{\beta}(\varrho)) = \mathrm{tr}[\mathrm{MSE}(\beta, \widehat{\beta}(\varrho))]$ is minimized with respect to ϱ.

Theorem 3.18. *Under the linear regression model with assumptions (i) to (iv), let $\rho(\beta, \widetilde{\beta}) = \mathrm{tr}[X'X \,\mathrm{MSE}(\beta, \widetilde{\beta})]$. Then for any $(\beta \neq \mathbf{0}, \sigma^2)$ the inequality*

$$\rho(\beta, \widehat{\beta}(\varrho_{\mathrm{opt}})) \leq \rho(\beta, \widehat{\beta}(\varrho)) \quad \forall \varrho \geq 0$$

holds true, where $\varrho_{\mathrm{opt}} = p\sigma^2 / \beta' X' X \beta$.

Proof. The predictive squared error risk of the shrinkage estimator $\widehat{\beta}(\varrho)$ is given as

$$\rho(\beta, \widehat{\beta}(\varrho)) = \mathrm{tr}[X'X \,\mathrm{MSE}(\beta, \widehat{\beta}(\varrho))] = \frac{p\sigma^2}{(1+\varrho)^2} + \frac{\varrho^2}{(1+\varrho)^2} \beta' X' X \beta \,.$$

Then the assertion follows analogously to Theorem 3.10. □

Now, the estimator $\widehat{\varrho}_{\mathrm{AM}}$ can be obtained by replacing the parameter σ^2 by $\widehat{\sigma}^2$ and the parameter vector β by $\widehat{\beta}$ in $\varrho_{\mathrm{opt}} = p\sigma^2 / \beta' X' X \beta$.

Example 3.20. In the linear regression model $y = X\beta + \varepsilon$ with

$$X' = \begin{pmatrix} 1 & 1 & 1 & 1 & 1 \\ 1.9 & 2.1 & 2 & 2 & 1.8 \end{pmatrix}$$

and

$$y = (6.0521, 7.0280, 7.1230, 4.4441, 5.0813)'$$

we compute

$$\widehat{\varrho}_{\mathrm{M}} = \frac{\widehat{\sigma}^2}{\widehat{\beta}' X' X \widehat{\beta}} = 0.0078 \quad \text{and} \quad \widehat{\varrho}_{\mathrm{AM}} = \frac{\widehat{\sigma}^2 p}{\widehat{\beta}' X' X \widehat{\beta}} = 0.0156 \,.$$

Then $1/(1+\widehat{\varrho}_{\mathrm{M}}) = 0.9922$ and $1/(1+\widehat{\varrho}_{\mathrm{AM}}) = 0.9847$, showing that the estimates $\widehat{\boldsymbol{\beta}}$, $\widehat{\boldsymbol{\beta}}_{\mathrm{M}}$, and $\widehat{\boldsymbol{\beta}}_{\mathrm{AM}}$ will differ only marginally in this example. On the other hand, we compute

$$\widehat{\varrho} = \frac{\widehat{\sigma}^2 \operatorname{tr}[(\boldsymbol{X}'\boldsymbol{X})^{-1}]}{\widehat{\boldsymbol{\beta}}'\widehat{\boldsymbol{\beta}}} = 2.8711$$

and thus $1/(1+\widehat{\varrho}) = 0.2583$. By using this value for the shrinkage parameter, the ordinary least squares estimator is shrunken rather strongly. The observed unweighted squared error loss is

$$L(\boldsymbol{\beta}, \widehat{\boldsymbol{\beta}}(\widehat{\varrho})) = (\widehat{\boldsymbol{\beta}}(\widehat{\varrho}) - \boldsymbol{\beta})'(\widehat{\boldsymbol{\beta}}(\widehat{\varrho}) - \boldsymbol{\beta}) = 10.0260$$

when $\boldsymbol{\beta} = (2,2)'$ is the true parameter vector. This is distinctly smaller than the observed loss $L(\boldsymbol{\beta}, \widehat{\boldsymbol{\beta}}) = 49.2978$ of the ordinary least squares estimator. □

The Stein Estimator

Let us consider again the quantity $\varrho_{\mathrm{opt}} = p\sigma^2/\boldsymbol{\beta}'\boldsymbol{X}'\boldsymbol{X}\boldsymbol{\beta}$, minimizing the predictive squared error risk $\rho(\boldsymbol{\beta}, \widehat{\boldsymbol{\beta}}(\varrho)) = \operatorname{tr}[\boldsymbol{X}'\boldsymbol{X}\,\mathrm{MSE}(\boldsymbol{\beta}, \widehat{\boldsymbol{\beta}}(\varrho))]$ with respect to $\varrho \geq 0$. We can also write

$$\varrho_{\mathrm{opt}} = \frac{p\sigma^2}{\boldsymbol{\beta}'\boldsymbol{X}'\boldsymbol{X}\boldsymbol{\beta} + p\sigma^2 - p\sigma^2} .$$

Now, a further ad-hoc estimator for ϱ can be constructed by replacing the function $f(\boldsymbol{\beta}, \sigma^2) := \boldsymbol{\beta}'\boldsymbol{X}'\boldsymbol{X}\boldsymbol{\beta} + p\sigma^2$ by $\widehat{\boldsymbol{\beta}}'\boldsymbol{X}'\boldsymbol{X}\widehat{\boldsymbol{\beta}}$, being unbiased for $f(\boldsymbol{\beta}, \sigma^2)$, and by replacing $p\sigma^2$ by $c\widehat{\boldsymbol{\varepsilon}}'\widehat{\boldsymbol{\varepsilon}}$. Here $\widehat{\boldsymbol{\varepsilon}} = \boldsymbol{y} - \boldsymbol{X}\widehat{\boldsymbol{\beta}}$, and $c \geq 0$ is a number such that the denominator $\widehat{\boldsymbol{\beta}}'\boldsymbol{X}'\boldsymbol{X}\widehat{\boldsymbol{\beta}} - c\widehat{\boldsymbol{\varepsilon}}'\widehat{\boldsymbol{\varepsilon}}$ is positive. Then we can write

$$\widehat{\varrho}_{\mathrm{S}}(c) = \frac{c\widehat{\boldsymbol{\varepsilon}}'\widehat{\boldsymbol{\varepsilon}}}{\widehat{\boldsymbol{\beta}}'\boldsymbol{X}'\boldsymbol{X}\widehat{\boldsymbol{\beta}} - c\widehat{\boldsymbol{\varepsilon}}'\widehat{\boldsymbol{\varepsilon}}}, \quad 0 \leq c < \frac{R^2}{1-R^2} ,$$

where it is assumed that $0 < R^2 < 1$, see also Problem 3.30. The corresponding non-linear shrinkage estimator for $\boldsymbol{\beta}$ is

$$\widehat{\boldsymbol{\beta}}_{\mathrm{S}}(c) = \frac{1}{1+\widehat{\varrho}_{\mathrm{S}}(c)}\widehat{\boldsymbol{\beta}} = \left[1 - \frac{c\widehat{\boldsymbol{\varepsilon}}'\widehat{\boldsymbol{\varepsilon}}}{\widehat{\boldsymbol{\beta}}'\boldsymbol{X}'\boldsymbol{X}\widehat{\boldsymbol{\beta}}}\right]\widehat{\boldsymbol{\beta}}, \quad 0 \leq c < \frac{R^2}{1-R^2} .$$

If $p \geq 3$, the choice

$$c = \frac{p-2}{n-p+2}$$

yields the usual Stein estimator $\widehat{\boldsymbol{\beta}}_{\mathrm{S}}$ as already considered in Sect. 2.5. If the estimator $\widehat{\boldsymbol{\beta}}_{\mathrm{S}}(c)$ is applied for some non-stochastic $c \geq 0$, then it could happen that $c \geq R^2/(1-R^2)$ for some realization $\boldsymbol{y}$. In that case the estimator $\widehat{\boldsymbol{\beta}}_{\mathrm{S}}(c)$ cannot be interpreted as a shrinkage estimator. Since

$$1 - \frac{c\,\widehat{\boldsymbol{\varepsilon}}'\widehat{\boldsymbol{\varepsilon}}}{\widehat{\boldsymbol{\beta}}'\boldsymbol{X}'\boldsymbol{X}\widehat{\boldsymbol{\beta}}} \begin{cases} = 0 & \text{for } c = \frac{R^2}{1-R^2} \\ < 0 & \text{for } c > \frac{R^2}{1-R^2} \end{cases}$$

holds true, the estimator

$$\widehat{\boldsymbol{\beta}}_{\mathrm{S}}^{+}(c) = \left[1 - \frac{c\,\widehat{\boldsymbol{\varepsilon}}'\widehat{\boldsymbol{\varepsilon}}}{\widehat{\boldsymbol{\beta}}'\boldsymbol{X}'\boldsymbol{X}\widehat{\boldsymbol{\beta}}}\right]_{+} \widehat{\boldsymbol{\beta}}, \quad 0 \leq c\,,$$

where $[\gamma]_+ = \max\{\gamma, 0\}$, can be considered instead of the estimator $\widehat{\boldsymbol{\beta}}_{\mathrm{S}}(c)$. Obviously, for any $c \geq 0$ the estimator

$$\widehat{\boldsymbol{\beta}}_{\mathrm{S}}^{+}(c) = \begin{cases} \widehat{\boldsymbol{\beta}}_{\mathrm{S}}(c) & \text{if } c < R^2/(1-R^2) \\ \mathbf{0} & \text{otherwise} \end{cases}$$

can be interpreted as a non-linear shrinkage estimator for $\boldsymbol{\beta}$. If $p \geq 3$, the choice

$$c = \frac{p-2}{n-p+2}$$

yields the positive part Stein estimator $\widehat{\boldsymbol{\beta}}_{\mathrm{S}}^{+}$ as already considered in Sect. 2.5.4. The positive part Stein estimator can (for $p \geq 3$) even be applied when the additional model assumption (v) is not satisfied. Then, however, the estimator $\widehat{\boldsymbol{\beta}}_{\mathrm{S}}^{+}$ will not be *uniformly* better than $\widehat{\boldsymbol{\beta}}$ with respect to the predictive squared error risk $\rho(\boldsymbol{\beta}, \widetilde{\boldsymbol{\beta}}) = \mathrm{tr}[\boldsymbol{X}'\boldsymbol{X}\,\mathrm{MSE}(\boldsymbol{\beta}, \widetilde{\boldsymbol{\beta}})]$.

3.5.4 Direction Modified Shrinkage Estimators

The estimator $\widehat{\boldsymbol{\beta}}(\varrho)$ shrinks the ordinary least squares estimator towards the zero vector. On the other hand, if one had prior knowledge that the true parameter vector $\boldsymbol{\beta}$ is close to some known vector $\boldsymbol{\beta}_0$, then one would appreciate to shrink in the direction of this vector. Such a shrinking behavior can be achieved by considering the shrinkage estimator $\widehat{\overline{\boldsymbol{\beta}}}(\varrho)$ for the modified parameter vector $\overline{\boldsymbol{\beta}} = \boldsymbol{\beta} - \boldsymbol{\beta}_0$ in advance, and then apply

$$\widehat{\boldsymbol{\beta}}_{\boldsymbol{\beta}_0}(\varrho) = \boldsymbol{\beta}_0 + \widehat{\overline{\boldsymbol{\beta}}}(\varrho)$$

for the estimation of the original parameter vector $\boldsymbol{\beta}$. For deriving this estimator, we note that the model equation $\boldsymbol{y} = \boldsymbol{X}\boldsymbol{\beta} + \boldsymbol{\varepsilon}$ can equivalently be written in the form

$$\overline{\boldsymbol{y}} = \boldsymbol{X}\overline{\boldsymbol{\beta}} + \boldsymbol{\varepsilon}, \quad \overline{\boldsymbol{y}} = \boldsymbol{y} - \boldsymbol{X}\boldsymbol{\beta}_0, \quad \overline{\boldsymbol{\beta}} = \boldsymbol{\beta} - \boldsymbol{\beta}_0\,.$$

The ordinary least squares estimator $\widehat{\overline{\beta}}$ for $\overline{\beta}$ in this model is given as

$$\widehat{\overline{\beta}} = (\boldsymbol{X}'\boldsymbol{X})^{-1}\boldsymbol{X}'\overline{\boldsymbol{y}} = \widehat{\beta} - \beta_0 \,.$$

Hence the corresponding shrinkage estimator $\widehat{\beta}_{\beta_0}(\varrho)$ for β is

$$\widehat{\beta}_{\beta_0}(\varrho) = \beta_0 + \frac{1}{1+\varrho}\widehat{\overline{\beta}} = \beta_0 + \frac{1}{1+\varrho}(\widehat{\beta} - \beta_0)$$

(*direction modified shrinkage estimator*). This estimator shrinks the ordinary least squares estimator in the direction of β_0 and can also be written as the convex combination

$$\widehat{\beta}_{\beta_0}(\varrho) = (1-\alpha)\beta_0 + \alpha\widehat{\beta}, \quad \alpha = \frac{1}{1+\varrho},$$

of β_0 and $\widehat{\beta}$.

For an appropriate estimator of the shrinkage parameter ϱ it is reasonable to start from the model $\overline{\boldsymbol{y}} = \boldsymbol{X}\overline{\beta} + \varepsilon$. Then a possible ad-hoc estimator for ϱ could be

$$\widehat{\varrho}_{\beta_0} = \frac{\widehat{\sigma}^2 \operatorname{tr}[(\boldsymbol{X}'\boldsymbol{X})^{-1}]}{(\widehat{\beta} - \beta_0)'(\widehat{\beta} - \beta_0)},$$

for example.

Instead of a fixed vector β_0 one can also consider a further estimator $\widetilde{\beta}$, so that

$$\widehat{\beta}_{\widetilde{\beta}}(\widehat{\varrho}_{\widetilde{\beta}}) = (1-\alpha)\widetilde{\beta} + \alpha\widehat{\beta}, \quad \alpha = \frac{1}{1+\widehat{\varrho}_{\widetilde{\beta}}}$$

is the convex combination of the two estimators $\widetilde{\beta}$ and $\widehat{\beta}$. A possible choice for $\widetilde{\beta}$ is the restricted least squares estimator.

The Bock estimator

Consider the restricted least squares estimator

$$\widehat{\beta}_{\mathrm{R}} = \widehat{\beta} - (\boldsymbol{X}'\boldsymbol{X})^{-1}\boldsymbol{R}'[\boldsymbol{R}(\boldsymbol{X}'\boldsymbol{X})^{-1}\boldsymbol{R}']^{-1}(\boldsymbol{R}\widehat{\beta} - \boldsymbol{r})$$

with corresponding restrictions $\boldsymbol{R}\beta = \boldsymbol{r}$, where $\boldsymbol{R}$ is a $m \times p$ matrix of full row rank. The statistic for testing the linear hypothesis $H_0 : \boldsymbol{R}\beta = \boldsymbol{r}$ is given as

$$F = \frac{(\widehat{\beta} - \widehat{\beta}_{\mathrm{R}})'\boldsymbol{X}'\boldsymbol{X}(\widehat{\beta} - \widehat{\beta}_{\mathrm{R}})}{\widehat{\varepsilon}'\widehat{\varepsilon}}\,\frac{n-p}{m}, \qquad \widehat{\varepsilon} = \boldsymbol{y} - \boldsymbol{X}\widehat{\beta}\,.$$

In the following we assume $F > 0$.

Bock [17] considers the estimator

$$\widehat{\beta}_{\mathrm{Bk}}(d) = \widehat{\beta}_{\mathrm{R}} + \frac{1}{1+\widehat{\varrho}_{\mathrm{R}}}(\widehat{\beta} - \widehat{\beta}_{\mathrm{R}})$$

for $\boldsymbol{\beta}$, where

$$\widehat{\varrho}_{\mathrm{R}} = \frac{d}{F-d}, \quad 0 \leq d < F .$$

This estimator $\widehat{\boldsymbol{\beta}}_{\mathrm{Bk}}(d)$ is of the form of a non-linear direction modified shrinkage estimator. It can also be written as

$$\widehat{\boldsymbol{\beta}}_{\mathrm{Bk}}(d) = \widehat{\boldsymbol{\beta}}_{\mathrm{R}} + \left[1 - \frac{d}{F}\right](\widehat{\boldsymbol{\beta}} - \widehat{\boldsymbol{\beta}}_{\mathrm{R}}), \qquad 0 \leq d < F .$$

The greater F, the less the ordinary least squares estimator is shrunken in the direction of the restricted least squares estimator. This is somehow similar to the behavior of the pretest estimator, where large values of F imply the use of the ordinary instead of the restricted least squares estimator. The difference is that the Bock estimator does not choose between one of the estimators $\widehat{\boldsymbol{\beta}}_{\mathrm{R}}$ and $\widehat{\boldsymbol{\beta}}$. Instead both estimators have a part in $\widehat{\boldsymbol{\beta}}_{\mathrm{Bk}}(d)$, where the respective weight is essentially determined by the actual value F.

If the Bock estimator $\widehat{\boldsymbol{\beta}}_{\mathrm{Bk}}(d)$ is applied for some non-stochastic $d \geq 0$, then it might happen that $F \leq d$. In that case the Bock estimator cannot be interpreted as a direction modified shrinkage estimator with stochastic shrinkage parameter as described above. One can, however, turn to a nearby variant of the estimator, given as

$$\widehat{\boldsymbol{\beta}}^{+}_{\mathrm{Bk}}(d) = \widehat{\boldsymbol{\beta}}_{\mathrm{R}} + \left[1 - \frac{d}{F}\right]_{+}(\widehat{\boldsymbol{\beta}} - \widehat{\boldsymbol{\beta}}_{\mathrm{R}}), \qquad 0 \leq d$$

(*modified Bock estimator*). Obviously, when $F \leq d$, then $\widehat{\boldsymbol{\beta}}^{+}_{\mathrm{Bk}}(d) = \widehat{\boldsymbol{\beta}}_{\mathrm{R}}$, while otherwise $\widehat{\boldsymbol{\beta}}^{+}_{\mathrm{Bk}}(d)$ shrinks $\widehat{\boldsymbol{\beta}}$ towards $\widehat{\boldsymbol{\beta}}_{\mathrm{R}}$.

Example 3.21. We reconsider Example 3.2 from Sect. 3.1, i.e. the linear regression model $\boldsymbol{y} = \boldsymbol{X\beta} + \boldsymbol{\varepsilon}$, $\boldsymbol{\varepsilon} \sim N(\mathbf{0}, \sigma^2 \boldsymbol{I}_5)$, with

$$\boldsymbol{X}' = \begin{pmatrix} 1 & 1 & 1 & 1 & 1 \\ 1 & 2 & 3 & 4 & 5 \end{pmatrix} .$$

It is assumed that the unknown parameter vector $\boldsymbol{\beta}$ satisfies the linear restriction

$$\boldsymbol{R\beta} = \boldsymbol{r}, \qquad \boldsymbol{R} = (1, 1), \quad \boldsymbol{r} = (1) .$$

To illustrate the behavior of the modified variant of the Bock estimator $\widehat{\boldsymbol{\beta}}^{+}_{\mathrm{Bk}}(d)$, we consider two different cases.

- Case 1. The true parameters are given as $\sigma^2 = 1$ and $\boldsymbol{\beta} = (0.2, 0.9)'$, so that the assumed linear restriction is not exactly but nearly satisfied by the parameter vector $\boldsymbol{\beta}$. For this parameter configuration we generate 50 realizations of $\boldsymbol{y}$ and compute the respective estimates $\widehat{\boldsymbol{\beta}}$ and $\widehat{\boldsymbol{\beta}}^{+}_{\mathrm{Bk}}(d)$ for the choice $d = (n-p)/n$. Figures 3.29 and 3.30 show the results.

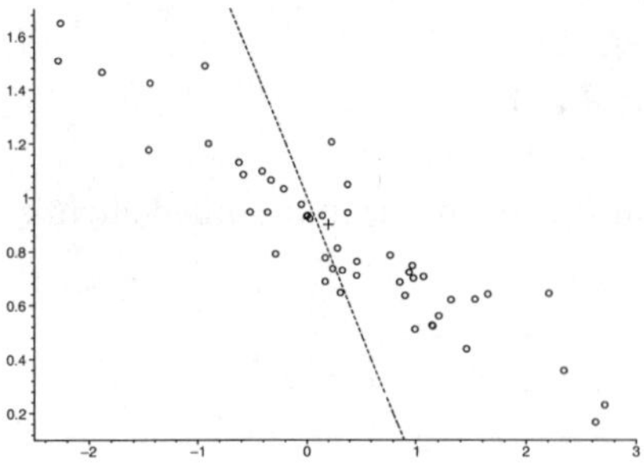

Fig. 3.29. 50 estimates $\widehat{\boldsymbol{\beta}}$ in Case 1 of Example 3.21

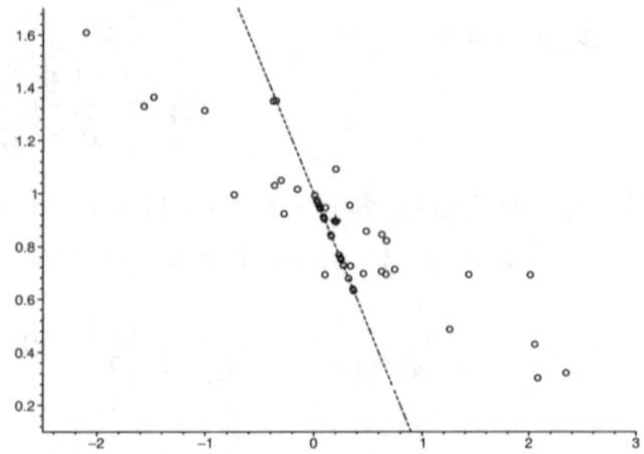

Fig. 3.30. 50 estimates $\widehat{\boldsymbol{\beta}}^+_{\mathrm{Bk}}((n-p)/n)$ in Case 1 of Example 3.21

As can be seen, quite a few of the Bock estimates lie exactly on the line determined by the restriction, while on the other hand a number of estimates are rather far away. Nonetheless the average observed unweighted squared error loss of the 50 Bock estimates of 0.7511 is distinctly smaller than the average observed loss of the ordinary least squares estimates of 1.3601. (The expected unweighted squared error loss of $\widehat{\boldsymbol{\beta}}$ is 1.2, so that the average observed loss of 1.3601 is greater than the expected loss, here.)

– Case 2. The true parameters are given as $\sigma^2 = 1$ and $\boldsymbol{\beta} = (1.8, 1.9)'$, so that the linear restriction is not even nearly satisfied by the parameter vector $\boldsymbol{\beta}$. Again, for this new parameter configuration we generate 50 realizations of $\boldsymbol{y}$ and compute the respective ordinary least squares and modified Bock estimates for the choice $d = (n-p)/n$. Note that we have already considered this case in Sect. 3.1.5 for an illustration of the behavior of the pretest estimator see also Fig. 3.4, p. 101. Figures 3.31 and 3.32 show the estimates.

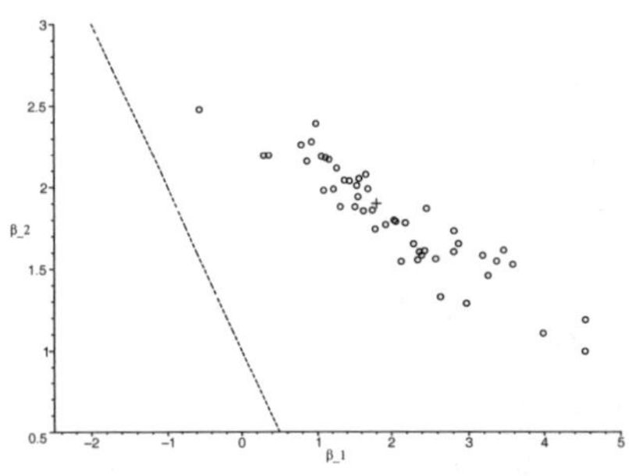

Fig. 3.31. 50 estimates $\widehat{\boldsymbol{\beta}}$ in Case 2 of Example 3.21

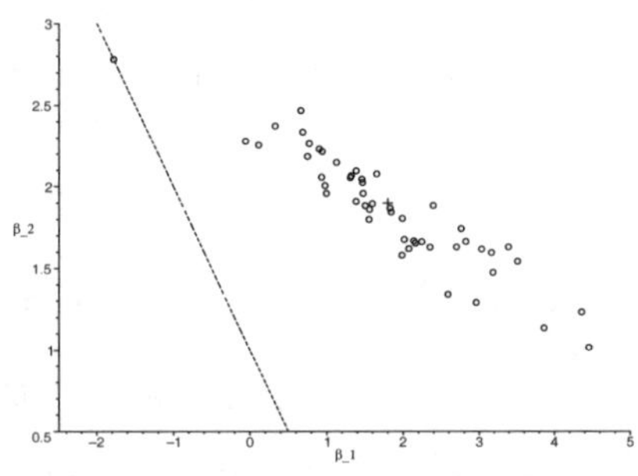

Fig. 3.32. 50 estimates $\widehat{\boldsymbol{\beta}}^+_{\mathrm{Bk}}((n-p)/n)$ in Case 2 of Example 3.21

As can be seen, only one out of the 50 Bock estimates lies exactly on the line which is described by the restriction. The average observed loss of the

50 Bock estimates of 1.4451 is not much greater than the average observed loss of the ordinary least squares estimates of 1.2562.

As follows from Case 2, if the assumed restriction is strongly violated, then the Bock estimator does not behave very differently from the ordinary least squares estimator in this example. Thus, it is quite a better estimator than the pretest estimator. If the restriction is nearly satisfied, then the Bock estimator turns out to be preferable to the ordinary least squares estimator. □

As special cases of the modified variant of the Bock estimator the estimators by Lindley [70], Sclove [98], and the above considered Stein type estimator $\widehat{\beta}_{\mathrm{S}}^{+}(c)$ can be obtained. We describe these three cases in the following.

The Lindley Estimator

If we consider the linear restrictions $\boldsymbol{R}\beta = \boldsymbol{r}$ such that $\boldsymbol{R} = \boldsymbol{I}_p$, then the restricted least squares estimator $\widehat{\beta}_{\mathrm{R}}$ is given by $\widehat{\beta}_{\mathrm{R}} = \boldsymbol{r}$. If in addition, we choose

$$d = \frac{n-p}{n} \, ,$$

then the modified Bock estimator $\widehat{\beta}_{\mathrm{Bk}}^{+}(d)$ becomes

$$\widehat{\beta}_{\mathrm{L}}^{+}(\boldsymbol{r}) = \boldsymbol{r} + \left[1 - \frac{p}{n}\frac{\widehat{\varepsilon}'\widehat{\varepsilon}}{(\widehat{\beta}-\boldsymbol{r})'\boldsymbol{X}'\boldsymbol{X}(\widehat{\beta}-\boldsymbol{r})}\right]_{+}(\widehat{\beta}-\boldsymbol{r}) \, ,$$

which is the estimator considered by Lindley [70]. The estimator $\widehat{\beta}_{\mathrm{L}}^{+}(\boldsymbol{r})$ shrinks the ordinary least squares estimator in direction of a given vector $\boldsymbol{r}$. If we have no prior idea about an appropriate destination vector $\boldsymbol{r}$ to shrink towards, then we might use

$$\boldsymbol{r} = \widehat{\mu}\mathbf{1}_p, \quad \widehat{\mu} = \frac{\mathbf{1}_p'\boldsymbol{X}'\boldsymbol{y}}{\mathbf{1}_p'\boldsymbol{X}'\boldsymbol{X}\mathbf{1}_p} \, ,$$

in which case, however, $\boldsymbol{r}$ is not fixed but stochastic. The symbol $\mathbf{1}_p$ denotes the $p \times 1$ vector whose elements are all equal to 1. The motivation for this choice of $\boldsymbol{r}$ is, that under the additional model assumption (vii), $\boldsymbol{X}'\boldsymbol{X} = \boldsymbol{I}_p$, the quantity $\widehat{\mu}$ is the mean of the ordinary least squares estimates, i.e. $\widehat{\mu} = (1/p)\sum_{i=1}^{p}\widehat{\beta}_i$. The estimator

$$\widehat{\beta}_{\mathrm{L}}^{+}(\widehat{\mu}\mathbf{1}_p) = \widehat{\mu}\mathbf{1}_p + \left[1 - \frac{p}{n}\frac{\widehat{\varepsilon}'\widehat{\varepsilon}}{(\widehat{\beta}-\widehat{\mu}\mathbf{1}_p)'\boldsymbol{X}'\boldsymbol{X}(\widehat{\beta}-\widehat{\mu}\mathbf{1}_p)}\right]_{+}(\widehat{\beta}-\widehat{\mu}\mathbf{1}_p)$$

can be seen as an operational variant of Lindley's estimator $\widehat{\beta}_{\mathrm{L}}^{+}(\boldsymbol{r})$ when $\boldsymbol{r}$ is unknown, see also [123, p. 146].

The Stein Estimator

If we consider the linear restrictions $\boldsymbol{R\beta} = \boldsymbol{r}$ such that $\boldsymbol{R} = \boldsymbol{I}_p$ and in addition $\boldsymbol{r} = \boldsymbol{0}$, then $\widehat{\boldsymbol{\beta}}_{\mathrm{R}}$ becomes $\widehat{\boldsymbol{\beta}}_{\mathrm{R}} = \boldsymbol{0}$, and the statistic F becomes

$$F = \frac{n-p}{p} \frac{R^2}{1-R^2} \,.$$

If in addition we choose the parameter d of the modified Bock estimator as

$$d = c\frac{n-p}{p}, \quad c \geq 0 \,,$$

then we obtain the already considered estimator

$$\widehat{\boldsymbol{\beta}}^{+}_{\mathrm{S}}(c) = \left[1 - \frac{c\widehat{\boldsymbol{\varepsilon}}'\widehat{\boldsymbol{\varepsilon}}}{\widehat{\boldsymbol{\beta}}'\boldsymbol{X}'\boldsymbol{X}\widehat{\boldsymbol{\beta}}}\right]_{+} \widehat{\boldsymbol{\beta}}, \quad c \geq 0 \,,$$

as a special case of the modified Bock estimator for $\boldsymbol{\beta}$.

The Sclove Estimator

A further special case of the Bock estimator can be obtained by employing the principal components estimator as a specific restricted least squares estimator. For this let us consider the spectral decomposition $\boldsymbol{U}_1\boldsymbol{\Lambda}_1\boldsymbol{U}_1' + \boldsymbol{U}_2\boldsymbol{\Lambda}_2\boldsymbol{U}_2'$ of the matrix $\boldsymbol{X}'\boldsymbol{X}$, where the $r \times r$ diagonal matrix $\boldsymbol{\Lambda}_1$ contains the r largest eigenvalues of $\boldsymbol{X}'\boldsymbol{X}$ on its main diagonal. In the sequel we assume $r < p$ so that the trivial case $r = p$ is excluded. The principal components estimator

$$\widehat{\boldsymbol{\beta}}^{(r)} = \boldsymbol{U}_1(\boldsymbol{U}_1'\boldsymbol{X}'\boldsymbol{X}\boldsymbol{U}_1)^{-1}\boldsymbol{U}_1'\boldsymbol{X}'\boldsymbol{y}$$

is a specific restricted least squares estimator with respect to the restrictions $\boldsymbol{R\beta} = \boldsymbol{r}$ where $\boldsymbol{R} = \boldsymbol{U}_2'$ and $\boldsymbol{r} = \boldsymbol{0}$. In view of

$$(\widehat{\boldsymbol{\beta}}^{(r)})'\boldsymbol{X}'\boldsymbol{X}\widehat{\boldsymbol{\beta}} = (\widehat{\boldsymbol{\beta}}^{(r)})'\boldsymbol{X}'\boldsymbol{X}\widehat{\boldsymbol{\beta}}^{(r)} \,,$$

the test statistic F can be written in the form

$$F = \frac{\widehat{\boldsymbol{\beta}}'\boldsymbol{X}'\boldsymbol{X}\widehat{\boldsymbol{\beta}} - (\widehat{\boldsymbol{\beta}}^{(r)})'\boldsymbol{X}'\boldsymbol{X}\widehat{\boldsymbol{\beta}}^{(r)}}{\widehat{\boldsymbol{\varepsilon}}'\widehat{\boldsymbol{\varepsilon}}} \frac{n-p}{n-r} \,.$$

If we choose the parameter d in the modified Bock estimator as $d = (p-r-1)/(n-r)$, then we obtain

$$\widehat{\boldsymbol{\beta}}^{+(r)}_{\mathrm{Sc}} = \widehat{\boldsymbol{\beta}}^{(r)} + \left[1 - \frac{p-r-1}{n-p} \frac{\widehat{\boldsymbol{\varepsilon}}'\widehat{\boldsymbol{\varepsilon}}}{\widehat{\boldsymbol{\beta}}'\boldsymbol{X}'\boldsymbol{X}\widehat{\boldsymbol{\beta}} - (\widehat{\boldsymbol{\beta}}^{(r)})'\boldsymbol{X}'\boldsymbol{X}\widehat{\boldsymbol{\beta}}^{(r)}}\right]_{+} (\widehat{\boldsymbol{\beta}} - \widehat{\boldsymbol{\beta}}^{(r)})$$

as a modification of the positive part Stein estimator, which has been developed by Sclove [98], see also [43].

3.6 General Ridge Estimator

The ridge estimator $\widehat{\boldsymbol{\beta}}_k$ as well as the shrinkage estimator $\widehat{\boldsymbol{\beta}}(\varrho)$ can be seen as special cases of a specific homogeneous linear estimator of $\boldsymbol{\beta}$, also known as *general ridge estimator*.

3.6.1 A Class of Estimators

A possible justification for the application of the ridge estimator is, that the addition of $k\boldsymbol{I}_p$ and $\boldsymbol{X}'\boldsymbol{X}$ yields a more stable matrix

$$\boldsymbol{X}'\boldsymbol{X} + k\boldsymbol{I}_p$$

than the matrix $\boldsymbol{X}'\boldsymbol{X}$, and the estimator $(\boldsymbol{X}'\boldsymbol{X} + k\boldsymbol{I}_p)^{-1}\boldsymbol{X}'\boldsymbol{y}$ should have a smaller dispersion than the estimator $(\boldsymbol{X}'\boldsymbol{X})^{-1}\boldsymbol{X}'\boldsymbol{y}$. If $\boldsymbol{U\Lambda U}'$ is a spectral decomposition of $\boldsymbol{X}'\boldsymbol{X}$, then

$$\boldsymbol{X}'\boldsymbol{X} + k\boldsymbol{I}_p = \boldsymbol{U}(\boldsymbol{\Lambda} + k\boldsymbol{I}_p)\boldsymbol{U}' \; .$$

Thus, summing the matrices $k\boldsymbol{I}_p$ and $\boldsymbol{X}'\boldsymbol{X}$ is nothing else but adding the number k to each of the eigenvalues of the matrix $\boldsymbol{X}'\boldsymbol{X}$. When some of the eigenvalues of the matrix $\boldsymbol{X}'\boldsymbol{X}$ are rather small, as it is usually the case in presence of collinearity, then clearly the corresponding eigenvalues of the matrix $\boldsymbol{X}'\boldsymbol{X} + k\boldsymbol{I}_p$ are greater, and this matrix is more stable. One can, however, criticize that all eigenvalues of the matrix $\boldsymbol{X}'\boldsymbol{X}$ are equally treated by this procedure, while for the purpose of stabilization it is reasonable to add rather large values to small eigenvalues of $\boldsymbol{X}'\boldsymbol{X}$ but rather small values to large eigenvalues of $\boldsymbol{X}'\boldsymbol{X}$. The course of action is then to add a specific number k_i to each eigenvalue of $\boldsymbol{X}'\boldsymbol{X}$, which can also be expressed as the addition of the matrix

$$\boldsymbol{K} = \boldsymbol{U}\mathrm{diag}(k_1, \ldots, k_p)\boldsymbol{U}', \quad k_i \geq 0, \quad i = 1, \ldots, p \, ,$$

to the matrix $\boldsymbol{X}'\boldsymbol{X}$. The corresponding estimator $(\boldsymbol{X}'\boldsymbol{X} + \boldsymbol{K})^{-1}\boldsymbol{X}'\boldsymbol{y}$ is then

$$\widehat{\boldsymbol{\beta}}_{k_1,\ldots,k_p} = \boldsymbol{U}\mathrm{diag}\left(\frac{1}{\lambda_1 + k_1}, \ldots, \frac{1}{\lambda_p + k_p}\right)\boldsymbol{U}'\boldsymbol{X}'\boldsymbol{y} \, .$$

The disadvantage of this procedure is that instead of only a single ridge parameter k, now, p appropriate ridge parameters $k_1, \ldots, k_p$ must be determined.

Let us turn our attention to the form of the derived estimator. As can be seen, the above matrix $\boldsymbol{K}$ is symmetric nonnegative definite. For any such matrix $\boldsymbol{K}$ we can call the estimator

$$\widehat{\boldsymbol{\beta}}_{\boldsymbol{K}} = (\boldsymbol{X}'\boldsymbol{X} + \boldsymbol{K})^{-1}\boldsymbol{X}'\boldsymbol{y}, \quad \boldsymbol{0} \leq_{\mathrm{L}} \boldsymbol{K} \, ,$$

a *general ridge estimator.*

It is quite obvious that the estimators $\widehat{\beta}_k$, $\widehat{\beta}_{k_1,\ldots,k_p}$, and $\widehat{\beta}(\varrho)$ are special cases of $\widehat{\beta}_K$ with $K = kI_p$, $K = U\text{diag}(k_1,\ldots,k_p)U'$, and $K = \varrho X'X$, respectively. In addition, the Marquardt estimator, the almost unbiased ridge estimator, the iteration estimator, and the estimator by Sommers [106] are also of the form of a general ridge estimator for special choices of K, see Problems 3.33 to 3.36. Further application of the general ridge estimator in case of elliptical restrictions is discussed in Sect. 3.7.

3.6.2 Risk Comparison of General Ridge and Ordinary Least Squares Estimator

In the special cases $K = kI_p$ for $k > 0$ and $K = \varrho X'X$ for $\varrho > 0$, the matrix K is not only symmetric nonnegative definite but also positive definite. For such a matrix K the following risk comparison between general ridge and ordinary least squares estimator holds true.

Theorem 3.19. *Under the linear regression model with assumptions (i) to (iv), let K be a $p \times p$ symmetric positive definite matrix. Then the difference*

$$\Delta = \text{MSE}(\beta, \widehat{\beta}) - \text{MSE}(\beta, \widehat{\beta}_K)$$

is nonnegative definite if and only if the inequality

$$\beta'[2K^{-1} + (X'X)^{-1}]^{-1}\beta \leq \sigma^2$$

is satisfied. In addition $\Delta \neq 0$ whenever $p > 1$.

Proof. For the estimator $\widehat{\beta}_K = (X'X + K)^{-1}X'y$ we have

$$\text{Cov}(\widehat{\beta}_K) = \sigma^2(X'X + K)^{-1}X'X(X'X + K)^{-1}$$

and

$$\text{bias}(\widehat{\beta}_K) = \text{E}(\widehat{\beta}_K) - \beta = (X'X + K)^{-1}X'X\beta - \beta\,.$$

If we write $\beta = (X'X + K)^{-1}(X'X + K)\beta$, then we obtain $\text{bias}(\widehat{\beta}_K) = -(X'X + K)^{-1}K\beta$. Then

$$\begin{aligned}\text{MSE}(\beta, \widehat{\beta}_K) &= \text{Cov}(\widehat{\beta}_K) + \text{bias}(\widehat{\beta}_K)\,\text{bias}(\widehat{\beta}_K)' \\ &= L[\sigma^2 X'X + K\beta\beta'K]L\,,\end{aligned}$$

where $L = (X'X + K)^{-1}$. The difference $\Delta = \text{MSE}(\beta, \widehat{\beta}) - \text{MSE}(\beta, \widehat{\beta}_K)$ is nonnegative definite if and only if $L^{-1}\Delta L^{-1}$ is nonnegative definite. The matrix $L^{-1}\Delta L^{-1}$ can be written as

$$L^{-1}\Delta L^{-1} = \sigma^2 L^{-1}(X'X)^{-1}L^{-1} - \sigma^2 X'X - K\beta\beta'K\,.$$

In view of $\boldsymbol{L}^{-1}(\boldsymbol{X}'\boldsymbol{X})^{-1}\boldsymbol{L}^{-1} = \boldsymbol{X}'\boldsymbol{X} + 2\boldsymbol{K} + \boldsymbol{K}(\boldsymbol{X}'\boldsymbol{X})^{-1}\boldsymbol{K}$, the identity

$$\boldsymbol{L}^{-1}\boldsymbol{\Delta}\boldsymbol{L}^{-1} = \sigma^2[2\boldsymbol{K} + \boldsymbol{K}(\boldsymbol{X}'\boldsymbol{X})^{-1}\boldsymbol{K}] - \boldsymbol{K}\boldsymbol{\beta}\boldsymbol{\beta}'\boldsymbol{K}$$

holds true. Now, the matrix $\boldsymbol{L}^{-1}\boldsymbol{\Delta}\boldsymbol{L}^{-1}$ is nonnegative definite if and only if $\boldsymbol{K}^{-1}\boldsymbol{L}^{-1}\boldsymbol{\Delta}\boldsymbol{L}^{-1}\boldsymbol{K}^{-1}$ is nonnegative definite, where

$$\boldsymbol{K}^{-1}\boldsymbol{L}^{-1}\boldsymbol{\Delta}\boldsymbol{L}^{-1}\boldsymbol{K}^{-1} = \sigma^2[2\boldsymbol{K}^{-1} + (\boldsymbol{X}'\boldsymbol{X})^{-1}] - \boldsymbol{\beta}\boldsymbol{\beta}' .$$

Since the matrix $2\boldsymbol{K}^{-1} + (\boldsymbol{X}'\boldsymbol{X})^{-1}$ is symmetric positive definite, Theorem A.60 can be applied, and $\boldsymbol{K}^{-1}\boldsymbol{L}^{-1}\boldsymbol{\Delta}\boldsymbol{L}^{-1}\boldsymbol{K}^{-1}$ is nonnegative definite if and only if the inequality

$$\boldsymbol{\beta}'[2\boldsymbol{K}^{-1} + (\boldsymbol{X}'\boldsymbol{X})^{-1}]^{-1}\boldsymbol{\beta} \le \sigma^2$$

is satisfied. Moreover, $\boldsymbol{\Delta} = \boldsymbol{0}$ if and only $\boldsymbol{K}^{-1}\boldsymbol{L}^{-1}\boldsymbol{\Delta}\boldsymbol{L}^{-1}\boldsymbol{K}^{-1} = \boldsymbol{0}$, that is

$$\sigma^2[2\boldsymbol{K}^{-1} + (\boldsymbol{X}'\boldsymbol{X})^{-1}] = \boldsymbol{\beta}\boldsymbol{\beta}' .$$

The rank of the left-hand matrix is p, while the rank of the right-hand matrix is either 0 or 1. Therefore $\boldsymbol{\Delta} = \boldsymbol{0}$ cannot hold true whenever $p > 1$. □

For the special case $\boldsymbol{K} = k\boldsymbol{I}_p$ the inequality

$$\boldsymbol{\beta}'[2\boldsymbol{K}^{-1} + (\boldsymbol{X}'\boldsymbol{X})^{-1}]^{-1}\boldsymbol{\beta} \le \sigma^2$$

becomes

$$\boldsymbol{\beta}'[\frac{2}{k}\boldsymbol{I}_p + (\boldsymbol{X}'\boldsymbol{X})^{-1}]^{-1}\boldsymbol{\beta} \le \sigma^2 ,$$

see Theorem 3.8, while for the case $\boldsymbol{K} = \varrho\boldsymbol{X}'\boldsymbol{X}$ this inequality can equivalently be written as

$$\boldsymbol{\beta}'\boldsymbol{X}'\boldsymbol{X}\boldsymbol{\beta} \le \frac{\varrho + 2}{\varrho}\sigma^2 ,$$

see Theorem 3.16.

3.7 Linear Minimax Estimator

In the previous sections the parameter space $\Theta_{\boldsymbol{\beta}}$ of the vector $\boldsymbol{\beta}$ is assumed to be either the whole space $\Theta_{\boldsymbol{\beta}} = \mathbb{R}^p$ or the affine subspace $\Theta_{\boldsymbol{\beta}} = \{\boldsymbol{\beta} \in \mathbb{R}^p : \boldsymbol{R}\boldsymbol{\beta} = \boldsymbol{r}\}$ of $\mathbb{R}^p$, both spaces being unbounded. When the parameter space $\Theta_{\boldsymbol{\beta}} \subseteq \mathbb{R}^p$ is a compact set, then the application of the minimax principle suggests itself for deriving estimators of $\boldsymbol{\beta}$.

3.7.1 Preliminary Considerations

If in the linear regression model

$$\boldsymbol{y} = \boldsymbol{X}\boldsymbol{\beta} + \varepsilon, \quad \varepsilon \sim (\boldsymbol{0}, \sigma^2 \boldsymbol{I}_n) ,$$

with assumptions (i) to (iv), additional prior information about the parameters $(\boldsymbol{\beta}, \sigma^2)$ of the form

$$\boldsymbol{\beta}'\boldsymbol{T}\boldsymbol{\beta} \leq \sigma^2$$

is given, where $\boldsymbol{T}$ is a $p \times p$ symmetric positive definite matrix, then for the estimation of $\boldsymbol{\beta}$ one can apply the estimator

$$\widehat{\boldsymbol{\beta}}_{\boldsymbol{T}} = (\boldsymbol{X}'\boldsymbol{X} + \boldsymbol{T})^{-1}\boldsymbol{X}'\boldsymbol{y} ,$$

being of the form of a generalized ridge estimator. As described in the following, this estimator can be interpreted as a linear minimax estimator for $\boldsymbol{\beta}$ in a certain sense. However, the estimator $\widehat{\boldsymbol{\beta}}_{\boldsymbol{T}}$ does *not* necessarily satisfy the inequality $\widehat{\boldsymbol{\beta}}'_{\boldsymbol{T}}\boldsymbol{T}\widehat{\boldsymbol{\beta}}_{\boldsymbol{T}} \leq \sigma^2$ and thus, contrary to the restricted least squares estimator $\widehat{\boldsymbol{\beta}}_{\mathrm{R}}$, the estimator $\widehat{\boldsymbol{\beta}}_{\boldsymbol{T}}$ does not obey the imposed restrictions on $\boldsymbol{\beta}$. Such a behavior is not as surprising as it seems to be on first sight, since there simply does not exist a linear estimator $\boldsymbol{A}\boldsymbol{y}$ satisfying the condition $\boldsymbol{y}'\boldsymbol{A}'\boldsymbol{T}\boldsymbol{A}\boldsymbol{y} \leq \sigma^2$ for any possible realization $\boldsymbol{y}$. Hence, if we confine to linear estimators, we have to abandon our legitimate demand for an estimate obeying the same restrictions as the parameter vector $\boldsymbol{\beta}$.

For an arbitrary but fixed σ^2 the set of vectors $\boldsymbol{\beta}$ satisfying the condition $\boldsymbol{\beta}'\boldsymbol{T}\boldsymbol{\beta} \leq \sigma^2$ is an ellipsoid with the zero vector as its center. Quite often, however, the center vector $\boldsymbol{\beta}_0$ of the ellipsoid describing the prior information is not zero. Hence, we have restrictions of the form

$$(\boldsymbol{\beta} - \boldsymbol{\beta}_0)'\boldsymbol{T}(\boldsymbol{\beta} - \boldsymbol{\beta}_0) \leq \sigma^2$$

for some known vector $\boldsymbol{\beta}_0 \in \mathbb{R}^p$. Setting $\overline{\boldsymbol{\beta}} = \boldsymbol{\beta} - \boldsymbol{\beta}_0$ and rewriting the model equation as

$$\overline{\boldsymbol{y}} = \boldsymbol{X}\overline{\boldsymbol{\beta}} + \varepsilon, \quad \overline{\boldsymbol{y}} = \boldsymbol{y} - \boldsymbol{X}\boldsymbol{\beta}_0, \quad \overline{\boldsymbol{\beta}}'\boldsymbol{T}\overline{\boldsymbol{\beta}} \leq \sigma^2 ,$$

we can consider

$$\widehat{\overline{\boldsymbol{\beta}}}_{\boldsymbol{T}} = (\boldsymbol{X}'\boldsymbol{X} + \boldsymbol{T})^{-1}\boldsymbol{X}'\overline{\boldsymbol{y}}$$

as an estimator for $\overline{\boldsymbol{\beta}}$. Then

$$\widehat{\boldsymbol{\beta}}_{\boldsymbol{T},\boldsymbol{\beta}_0} = \widehat{\overline{\boldsymbol{\beta}}}_{\boldsymbol{T}} + \boldsymbol{\beta}_0 = \boldsymbol{\beta}_0 + (\boldsymbol{X}'\boldsymbol{X} + \boldsymbol{T})^{-1}\boldsymbol{X}'(\boldsymbol{y} - \boldsymbol{X}\boldsymbol{\beta}_0)$$

can be considered as an estimator for $\boldsymbol{\beta}$.

Obviously the estimator $\widehat{\boldsymbol{\beta}}_{\boldsymbol{T},\boldsymbol{\beta}_0}$ belongs to the set of non-homogeneous linear estimators for $\boldsymbol{\beta}$.

3.7.2 Inequality Restrictions

If we have prior information about the elements β_i of the parameter vector $\boldsymbol{\beta}$ in the form of inequalities

$$a_i \leq \beta_i \leq b_i, \quad i = 1, \ldots, p\,,$$

where $a_i, b_i \in \mathbb{R}$ are known, then we can apply the non-linear inequality restricted least squares estimator $\widehat{\boldsymbol{\beta}}_{\mathrm{IR}}$ described in Sec 3.2.2.

A different course of action encloses these inequalities by an ellipsoid

$$(\boldsymbol{\beta} - \boldsymbol{\beta}_0)'\boldsymbol{T}(\boldsymbol{\beta} - \boldsymbol{\beta}_0) \leq 1\,.$$

For this, $\boldsymbol{T}$ and $\boldsymbol{\beta}_0$ are given as

$$\boldsymbol{T} = \frac{4}{p}\mathrm{diag}\left((b_1 - a_1)^{-2}, \ldots, (b_p - a_p)^{-2}\right), \quad \boldsymbol{\beta}_0 = \frac{1}{2}\begin{pmatrix} a_1 + b_1 \\ \vdots \\ a_p + b_p \end{pmatrix},$$

see [118]. If in addition we have prior information

$$\sigma_0^2 \leq \sigma^2\,,$$

where $\sigma_0^2 > 0$ is known, then

$$(\boldsymbol{\beta} - \boldsymbol{\beta}_0)'(\sigma_0^2\boldsymbol{T})(\boldsymbol{\beta} - \boldsymbol{\beta}_0) \leq \sigma_0^2 \leq \sigma^2\,,$$

and according to our preliminary considerations, we can estimate $\boldsymbol{\beta}$ by

$$\widehat{\boldsymbol{\beta}}_{\sigma_0^2\boldsymbol{T},\boldsymbol{\beta}_0} = \boldsymbol{\beta}_0 + (\boldsymbol{X}'\boldsymbol{X} + \sigma_0^2\boldsymbol{T})^{-1}\boldsymbol{X}'(\boldsymbol{y} - \boldsymbol{X}\boldsymbol{\beta}_0)\,.$$

This specific procedure of incorporating additional prior information about σ^2, however, appears to be reasonable only when σ_0^2 is close to the unknown σ^2, since otherwise the considered restriction

$$(\boldsymbol{\beta} - \boldsymbol{\beta}_0)'(\sigma_0^2\boldsymbol{T})(\boldsymbol{\beta} - \boldsymbol{\beta}_0) \leq \sigma^2$$

looses too much information about $\boldsymbol{\beta}$, compared to the original information given by the inequality

$$(\boldsymbol{\beta} - \boldsymbol{\beta}_0)'(\sigma_0^2\boldsymbol{T})(\boldsymbol{\beta} - \boldsymbol{\beta}_0) \leq \sigma_0^2\,,$$

i.e. $(\boldsymbol{\beta} - \boldsymbol{\beta}_0)'\boldsymbol{T}(\boldsymbol{\beta} - \boldsymbol{\beta}_0) \leq 1$. We will later discuss a practical estimator for $\boldsymbol{\beta}$, which can be used in those cases where only elliptical restrictions about $\boldsymbol{\beta}$ are given but no additional information about σ^2 is available.

Example 3.22. In Sect. 3.1.4 we have considered a linear regression model $\boldsymbol{y} = \boldsymbol{X}\boldsymbol{\beta} + \varepsilon$, where

$$\boldsymbol{y} = \begin{pmatrix} 2.6262 \\ 2.1640 \\ 2.3589 \\ 3.9844 \\ 4.2513 \end{pmatrix}, \quad \boldsymbol{X} = \begin{pmatrix} 1 & 1 \\ 1 & 2 \\ 1 & 3 \\ 1 & 4 \\ 1 & 5 \end{pmatrix}, \quad \boldsymbol{\beta} = \begin{pmatrix} \beta_1 \\ \beta_2 \end{pmatrix}.$$

We now assume that we have prior information

$$(1)\ 0 \le \beta_1 \le 2, \quad (2)\ 0.75 \le \beta_2 \le 1.5, \quad (3)\ 0.5 \le \sigma^2$$

about the parameters. This information is correct in the sense that it is satisfied by the true parameters $\boldsymbol{\beta} = (0.2, 0.9)$ and $\sigma^2 = 1$.

The parallelepiped described by (1) and (2) can be enclosed by an ellipsoid

$$(\boldsymbol{\beta} - \boldsymbol{\beta}_0)'\boldsymbol{T}(\boldsymbol{\beta} - \boldsymbol{\beta}_0) \le 1\,,$$

where

$$\boldsymbol{T} = \begin{pmatrix} 1/2 & 0 \\ 0 & 32/9 \end{pmatrix} \quad \text{and} \quad \boldsymbol{\beta}_0 = \begin{pmatrix} 1 \\ 9/8 \end{pmatrix},$$

see Fig. 3.33.

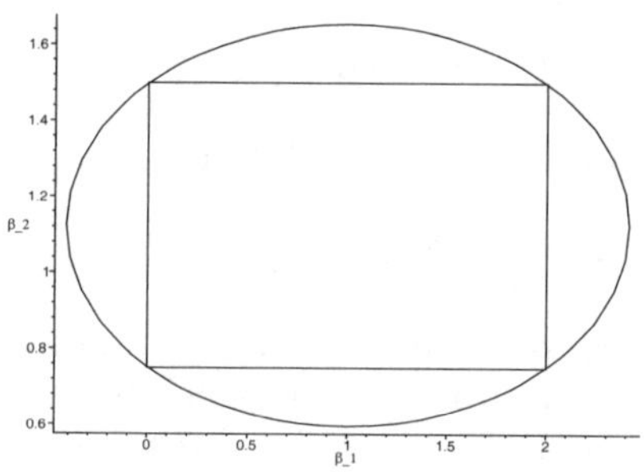

Fig. 3.33. Enclosure of inequality restrictions by an ellipsoid in Example 3.22

It is clear that the ellipsoid encircles a bigger area than the parallelepiped given by (1) and (2), so that advantage of the prior information is not wholly taken. On the other hand, the above ellipsoid has minimal volume compared to other possible ellipsoids enclosing the parallelepiped, and, in addition, is mathematically easier to handle.

If we estimate $\boldsymbol{\beta}$ by $\widehat{\boldsymbol{\beta}}_{\sigma_0^2\boldsymbol{T},\boldsymbol{\beta}_0}$ with $\sigma_0^2 = 0.5$, then we obtain

$$\widehat{\boldsymbol{\beta}}_{\sigma_0^2\boldsymbol{T},\boldsymbol{\beta}_0} = \begin{pmatrix} 1.2224 \\ 0.6145 \end{pmatrix}.$$

The ordinary least squares estimator, not taking any of the restrictions (1) to (3) into consideration, yields

$$\widehat{\beta} = \begin{pmatrix} 1.5558 \\ 0.5070 \end{pmatrix} .$$

Clearly, the estimate $\widehat{\beta}_{\sigma_0^2 T, \beta_0}$ is a little closer to the true parameter $\beta = (0.2, 0.9)'$ than the ordinary least squares estimate.

To obtain a better impression of the behavior of the estimators $\widehat{\beta}_{\sigma_0^2 T, \beta_0}$ and $\widehat{\beta}$, we generate 30 new realizations of y and compute the respective estimates of β. Figures 3.34 and 3.35 show the results.

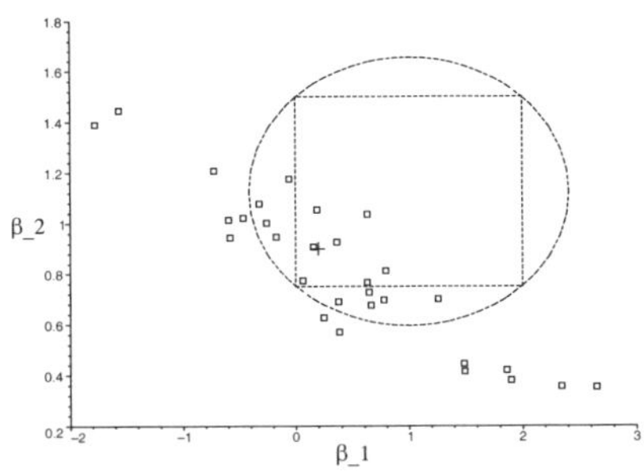

Fig. 3.34. 30 estimates $\widehat{\beta}$ in Example 3.22

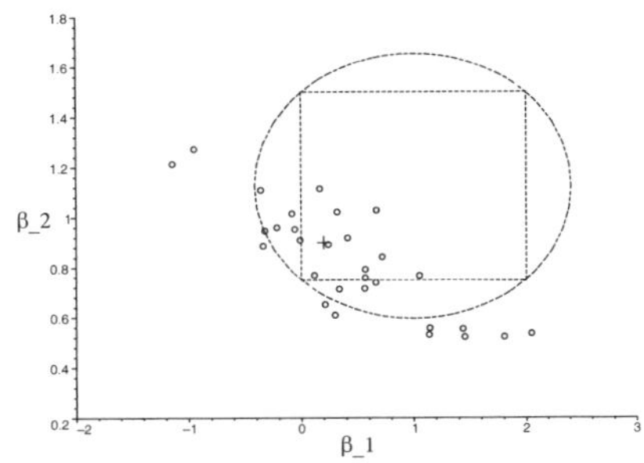

Fig. 3.35. 30 estimates $\widehat{\beta}_{\sigma_0^2 T, \beta_0}$ in Example 3.22

It can be seen that on the average the estimates $\widehat{\beta}_{\sigma_0^2 T, \beta_0}$ are closer to the true parameter vector than the estimates $\widehat{\beta}$. The unweighted squared error loss $L(\beta, \widetilde{\beta}) = (\widetilde{\beta} - \beta)'(\widetilde{\beta} - \beta)$ yields an average observed loss of 0.6208 for $\widehat{\beta}_{\sigma_0^2 T, \beta_0}$, compared to an average observed loss of 1.1765 for $\widehat{\beta}$ when $\beta = (0.2, 0.9)'$ is the true parameter. □

3.7.3 Linear Minimax Solutions

We have claimed above that in case of restrictions of the form

$$(\beta - \beta_0)' T (\beta - \beta_0) \leq \sigma^2$$

the application of the non-homogeneous linear ridge estimator $\widehat{\beta}_{T, \beta_0}$ can be useful. We have not explained so far why this one, of all linear estimators, should be used as the alternative to least squares estimation. A possible motivation is, that under elliptical restrictions the corresponding ridge estimator can be interpreted as a linear minimax estimator in a certain sense.

The Minimax Principle

In the following we consider the set of homogeneous linear estimators $\mathcal{L}(\boldsymbol{\beta})$ for $\boldsymbol{\beta}$, that is

$$\mathcal{L}(\boldsymbol{\beta}) = \{\boldsymbol{A}\boldsymbol{y} : \boldsymbol{A} \in \mathbb{R}^{p \times n}\} ,$$

where $\boldsymbol{A}$ is a non-stochastic matrix. To assess the performance of an element $\widetilde{\boldsymbol{\beta}} \in \mathcal{L}(\boldsymbol{\beta})$, we consider the weighted squared error risk

$$\rho(\boldsymbol{\beta}, \widetilde{\boldsymbol{\beta}}) = \mathrm{E}\left[(\widetilde{\boldsymbol{\beta}} - \boldsymbol{\beta})'\boldsymbol{W}(\widetilde{\boldsymbol{\beta}} - \boldsymbol{\beta})\right] ,$$

where $\boldsymbol{W}$ is a known $p \times p$ symmetric nonnegative definite matrix.

Definition 3.1. *A vector* $\breve{\boldsymbol{\beta}} \in \mathcal{L}(\boldsymbol{\beta})$ *is called* linear minimax solution *for* $\boldsymbol{\beta}$, *if*

$$\sup_{\boldsymbol{\beta} \in \Theta_{\boldsymbol{\beta}}} \rho(\boldsymbol{\beta}, \breve{\boldsymbol{\beta}}) = \inf_{\widetilde{\boldsymbol{\beta}} \in \mathcal{L}(\boldsymbol{\beta})} \sup_{\boldsymbol{\beta} \in \Theta_{\boldsymbol{\beta}}} \rho(\boldsymbol{\beta}, \widetilde{\boldsymbol{\beta}})$$

is satisfied.

As will be seen in the following, linear minimax solutions for $\boldsymbol{\beta}$ can be functions of the unknown parameter σ^2. This is the reason why we use the term 'solution' and not the term 'estimator'. For the application of a minimax solution as an estimator for $\boldsymbol{\beta}$ it can be necessary to replace the unknown σ^2 by an appropriate estimator.

From the above definition it follows that linear minimax solutions initially assume a most unfavorable parameter vector $\boldsymbol{\beta}$ with respect to the given risk, since for an arbitrary linear (possibly not completely known) decision rule $\widetilde{\boldsymbol{\beta}} \in \mathcal{L}(\boldsymbol{\beta})$ the greatest possible risk under variation of all possible $\boldsymbol{\beta} \in \Theta_{\boldsymbol{\beta}}$ is determined at first. This is, of course, only possible when the maximum or at least the supremum exists, i.e. when the risk is appropriately bounded. In a second step, among all solutions $\widetilde{\boldsymbol{\beta}}$ from $\mathcal{L}(\boldsymbol{\beta})$ the one is chosen, which yields the smallest of all greatest risks $\sup_{\boldsymbol{\beta} \in \Theta_{\boldsymbol{\beta}}} \rho(\boldsymbol{\beta}, \widetilde{\boldsymbol{\beta}})$.

In this sense minimax solutions represent a guard against the most unfavorable case, which nevertheless must not necessarily occur. This is the reason why minimax solutions are often identified with a *pessimistic* point of view.

Linear Minimax Solutions for Elliptical Parameter Spaces

We start with considering the set

$$\Theta_{\boldsymbol{\beta}} = \{\boldsymbol{\beta} \in \mathbb{R}^p : \boldsymbol{\beta}'\boldsymbol{T}\boldsymbol{\beta} \leq t^2\} ,$$

where $\boldsymbol{T}$ is a $p \times p$ symmetric positive definite matrix and t^2 is a finite positive number. As shown by the following theorem, we can determine the maximal weighted squared error risk of an arbitrary $\boldsymbol{A}\boldsymbol{y} \in \mathcal{L}(\boldsymbol{\beta})$ with respect to all parameter vectors $\boldsymbol{\beta} \in \Theta_{\boldsymbol{\beta}}$.

Theorem 3.20. *Under the linear regression model with assumptions (i) to (iv), for* $\boldsymbol{Ay} \in \mathcal{L}(\boldsymbol{\beta})$ *the identity*

$$\max_{\boldsymbol{\beta} \in \Theta_{\boldsymbol{\beta}}} \rho(\boldsymbol{\beta}, \boldsymbol{Ay}) = \sigma^2 \operatorname{tr}(\boldsymbol{WAA}') + t^2 \lambda_{\max}(\boldsymbol{B})$$

holds true, where $\boldsymbol{B} = \boldsymbol{T}^{-1/2}(\boldsymbol{AX} - \boldsymbol{I}_p)'\boldsymbol{W}(\boldsymbol{AX} - \boldsymbol{I}_p)\boldsymbol{T}^{-1/2}$.

Proof. The weighted squared error risk of $\boldsymbol{Ay}$ is given as

$$\rho(\boldsymbol{\beta}, \boldsymbol{Ay}) = \sigma^2 \operatorname{tr}(\boldsymbol{WAA}') + \boldsymbol{\beta}'(\boldsymbol{AX} - \boldsymbol{I}_p)'\boldsymbol{W}(\boldsymbol{AX} - \boldsymbol{I}_p)\boldsymbol{\beta} ,$$

and therefore

$$\max_{\boldsymbol{\beta} \in \Theta_{\boldsymbol{\beta}}} \rho(\boldsymbol{\beta}, \boldsymbol{Ay}) = \sigma^2 \operatorname{tr}(\boldsymbol{WAA}') + \max_{\boldsymbol{\beta} \in \Theta_{\boldsymbol{\beta}}} \boldsymbol{\beta}'\boldsymbol{T}^{1/2}\boldsymbol{B}\boldsymbol{T}^{1/2}\boldsymbol{\beta} ,$$

where $\boldsymbol{B} = \boldsymbol{T}^{-1/2}(\boldsymbol{AX} - \boldsymbol{I}_p)'\boldsymbol{W}(\boldsymbol{AX} - \boldsymbol{I}_p)\boldsymbol{T}^{-1/2}$. According to Theorem A.48, the inequality

$$\boldsymbol{\beta}'\boldsymbol{T}^{1/2}\boldsymbol{B}\boldsymbol{T}^{1/2}\boldsymbol{\beta} \leq \lambda_{\max}(\boldsymbol{B})\, \boldsymbol{\beta}'\boldsymbol{T}\boldsymbol{\beta}$$

is valid for any vector $\boldsymbol{\beta} \in \mathbb{R}^p$. Thus

$$\boldsymbol{\beta}'\boldsymbol{T}^{1/2}\boldsymbol{B}\boldsymbol{T}^{1/2}\boldsymbol{\beta} \leq \lambda_{\max}(\boldsymbol{B})\, t^2$$

for any vector $\boldsymbol{\beta} \in \mathbb{R}^p$ satisfying $\boldsymbol{\beta}'\boldsymbol{T}\boldsymbol{\beta} \leq t^2$.

It remains to show that there always exists a vector $\boldsymbol{\beta} \in \mathbb{R}^p$ satisfying $\boldsymbol{\beta}'\boldsymbol{T}\boldsymbol{\beta} \leq t^2$ and $\boldsymbol{\beta}'\boldsymbol{T}^{1/2}\boldsymbol{B}\boldsymbol{T}^{1/2}\boldsymbol{\beta} = \lambda_{\max}(\boldsymbol{B})t^2$. For this, we consider an eigenvector $\boldsymbol{x}$ of $\boldsymbol{B}$ with $\boldsymbol{x}'\boldsymbol{x} = 1$ corresponding to the greatest eigenvalue $\lambda_{\max}(\boldsymbol{B})$ of the matrix $\boldsymbol{B}$ (such an eigenvector does always exist). If we now choose $\boldsymbol{\beta} = t\boldsymbol{T}^{-1/2}\boldsymbol{x}$, then this vector $\boldsymbol{\beta}$ satisfies the required identities $\boldsymbol{\beta}'\boldsymbol{T}\boldsymbol{\beta} = t^2$ and $\boldsymbol{\beta}'\boldsymbol{T}^{1/2}\boldsymbol{B}\boldsymbol{T}^{1/2}\boldsymbol{\beta} = \lambda_{\max}(\boldsymbol{B})t^2$. This completes the proof. □

The Minimization Problem

According to the above theorem, we can obtain a linear minimax solution by minimizing the expression

$$\sigma^2 \operatorname{tr}(\boldsymbol{WAA}') + t^2 \lambda_{\max}\left(\boldsymbol{T}^{-1/2}(\boldsymbol{AX} - \boldsymbol{I}_p)'\boldsymbol{W}(\boldsymbol{AX} - \boldsymbol{I}_p)\boldsymbol{T}^{-1/2}\right)$$

with respect to the $p \times n$ matrix $\boldsymbol{A}$. Unfortunately, the explicit solution is rather complicated in general, see also [69]. An exception is given when the symmetric nonnegative definite weight matrix $\boldsymbol{W}$ is of rank 1, so that $\boldsymbol{W}$ can be written in the form $\boldsymbol{W} = \boldsymbol{w}\boldsymbol{w}'$ for some vector $\boldsymbol{w}$. Then the linear minimax solution is given by

$$\breve{\boldsymbol{\beta}} = (\boldsymbol{X}'\boldsymbol{X} + \frac{\sigma^2}{t^2}\boldsymbol{T})^{-1}\boldsymbol{X}'\boldsymbol{y} .$$

This solution is of the form of a general ridge estimator $\widehat{\boldsymbol{\beta}}_{\boldsymbol{K}}$, where $\boldsymbol{K} = (\sigma^2/t^2)\boldsymbol{T}$ is symmetric positive definite, but depends on the unknown σ^2. It is remarkable that the linear minimax solution $\breve{\boldsymbol{\beta}}$ does not depend on the vector $\boldsymbol{w}$, so that any choice of $\boldsymbol{w}$ yields the same result. The minimax solution for this special case has been obtained by Kuks and Olman [66, 67], see also [117, Chap. 4].

Since $\breve{\boldsymbol{\beta}}$ depends on σ^2 it is not an estimator of $\boldsymbol{\beta}$, but of course can be made a (non-linear) estimator by replacing the unknown parameter σ^2 by the estimator $\widehat{\sigma}^2$.

If the parameter space $\Theta_{\boldsymbol{\beta}}$ is given as

$$\Theta_{\boldsymbol{\beta}} = \{\boldsymbol{\beta} \in \mathbb{R}^p : (\boldsymbol{\beta} - \boldsymbol{\beta}_0)'\boldsymbol{T}(\boldsymbol{\beta} - \boldsymbol{\beta}_0) \leq t^2\}$$

for some known vector $\boldsymbol{\beta}_0$, then the corresponding minimax solution is

$$\breve{\boldsymbol{\beta}} = \boldsymbol{\beta}_0 + (\boldsymbol{X}'\boldsymbol{X} + \frac{\sigma^2}{t^2}\boldsymbol{T})^{-1}\boldsymbol{X}'(\boldsymbol{y} - \boldsymbol{X}\boldsymbol{\beta}_0) ,$$

which, again, can be made practical by replacing σ^2 by $\widehat{\sigma}^2$.

The Choice of The Weight Matrix

The application of a weight matrix $\boldsymbol{W} = \boldsymbol{w}\boldsymbol{w}'$ appears to be rather unsuitable when estimation of the parameter vector $\boldsymbol{\beta}$ is the matter of interest. On the other hand, if one wishes to estimate a linear combination $\boldsymbol{w}'\boldsymbol{\beta}$ via $\boldsymbol{w}'\boldsymbol{A}\boldsymbol{y}$ for some given vector $\boldsymbol{w} \in \mathbb{R}^p$, then

$$\begin{aligned}\rho(\boldsymbol{w}'\boldsymbol{\beta}, \boldsymbol{w}'\boldsymbol{A}\boldsymbol{y}) &= \mathrm{E}\left[(\boldsymbol{y}'\boldsymbol{A}'\boldsymbol{w} - \boldsymbol{\beta}'\boldsymbol{w})(\boldsymbol{w}'\boldsymbol{A}\boldsymbol{y} - \boldsymbol{w}'\boldsymbol{\beta})\right] \\ &= \mathrm{E}\left[(\boldsymbol{A}\boldsymbol{y} - \boldsymbol{\beta})'\boldsymbol{w}\boldsymbol{w}'(\boldsymbol{A}\boldsymbol{y} - \boldsymbol{\beta})\right] .\end{aligned}$$

Hence, in this situation a weight matrix of rank 1 comes naturally into play.

As noted before, the above linear minimax solution $\breve{\boldsymbol{\beta}}$ does not depend on the actual choice of the vector $\boldsymbol{w}$. If we choose $\boldsymbol{w} = \boldsymbol{e}_i$, where $\boldsymbol{e}_i$ is the i-th unit vector, then $\boldsymbol{e}_i'\breve{\boldsymbol{\beta}}$ can be interpreted as the linear minimax solution for the i-th element $\boldsymbol{e}_i'\boldsymbol{\beta}$ of the vector $\boldsymbol{\beta}$.

Operational Minimax Solutions

If the parameter space is given by

$$\Theta_{\boldsymbol{\beta}} = \{\boldsymbol{\beta} \in \mathbb{R}^p : (\boldsymbol{\beta} - \boldsymbol{\beta}_0)'\boldsymbol{T}(\boldsymbol{\beta} - \boldsymbol{\beta}_0) \leq t^2\,\sigma^2\} ,$$

then analogously to the above

$$\breve{\boldsymbol{\beta}} = \boldsymbol{\beta}_0 + (\boldsymbol{X}'\boldsymbol{X} + t^{-2}\boldsymbol{T})^{-1}\boldsymbol{X}'(\boldsymbol{y} - \boldsymbol{X}\boldsymbol{\beta}_0)$$

is a linear minimax solution when the weight matrix is given by $\boldsymbol{W} = \boldsymbol{w}\boldsymbol{w}'$. Contrary to the above, the linear minimax solution is operational and thus can directly be applied as an estimator.

Let us reconsider Example 3.22 from Sect. 3.7.2, where inequality restrictions with respect to the elements of $\boldsymbol{\beta}$ are enclosed by an ellipsoid of the form $(\boldsymbol{\beta} - \boldsymbol{\beta}_0)'\boldsymbol{T}(\boldsymbol{\beta} - \boldsymbol{\beta}_0) \leq 1$. The additional prior information $\sigma_0^2 \leq \sigma^2$ can then be used to form a parameter space

$$\Theta_{\boldsymbol{\beta}} = \{\boldsymbol{\beta} \in \mathrm{I\!R}^p : (\boldsymbol{\beta} - \boldsymbol{\beta}_0)'(\sigma_0^2 \boldsymbol{T})(\boldsymbol{\beta} - \boldsymbol{\beta}_0) \leq \sigma^2\} .$$

The corresponding linear minimax solution

$$\widehat{\boldsymbol{\beta}}_{\sigma_0^2 \boldsymbol{T}, \boldsymbol{\beta}_0} = \boldsymbol{\beta}_0 + (\boldsymbol{X}'\boldsymbol{X} + \sigma_0^2 \boldsymbol{T})^{-1}\boldsymbol{X}'(\boldsymbol{y} - \boldsymbol{X}\boldsymbol{\beta}_0) ,$$

can be used an an estimator for $\boldsymbol{\beta}$. The same estimator can be obtained by assuming the parameter space

$$\Theta_{\boldsymbol{\beta}} = \{\boldsymbol{\beta} \in \mathrm{I\!R}^p : (\boldsymbol{\beta} - \boldsymbol{\beta}_0)'\boldsymbol{T}(\boldsymbol{\beta} - \boldsymbol{\beta}_0) \leq 1\} ,$$

and replacing σ^2 by σ_0^2 in the corresponding linear minimax solution

$$\widehat{\boldsymbol{\beta}}_{\sigma^2 \boldsymbol{T}, \boldsymbol{\beta}_0} = \boldsymbol{\beta}_0 + (\boldsymbol{X}'\boldsymbol{X} + \sigma^2 \boldsymbol{T})^{-1}\boldsymbol{X}'(\boldsymbol{y} - \boldsymbol{X}\boldsymbol{\beta}_0) .$$

This shows that applying the restriction $\sigma_0^2 \leq \sigma^2$ as above, is somehow equivalent to using σ_0^2 as an 'estimator' for σ^2. It also confirms our previous judgement that such a procedure is likely to be beneficial only when σ_0^2 is close to σ^2, since otherwise σ_0^2 is a bad 'estimator' for σ^2.

If we have no prior knowledge about some $\sigma_0^2 \leq \sigma^2$ being close to σ^2, then it is more advisable to use the non-linear estimator

$$\widehat{\boldsymbol{\beta}}_{\widehat{\sigma}^2 \boldsymbol{T}, \boldsymbol{\beta}_0} = \boldsymbol{\beta}_0 + (\boldsymbol{X}'\boldsymbol{X} + \widehat{\sigma}^2 \boldsymbol{T})^{-1}\boldsymbol{X}'(\boldsymbol{y} - \boldsymbol{X}\boldsymbol{\beta}_0) ,$$

which is obtained from the non-operational linear minimax solution by replacing σ^2 by the estimate $\widehat{\sigma}^2$.

Example 3.23. In Example 3.22 we have generated 30 realizations $\boldsymbol{y}$ and computed the corresponding ordinary least squares estimates $\widehat{\boldsymbol{\beta}}$ as well as the ridge estimates $\widehat{\boldsymbol{\beta}}_{\sigma_0^2 \boldsymbol{T}, \boldsymbol{\beta}_0}$ for $\sigma_0^2 = 0.5$, see Figs. 3.34 and 3.35. The average observed unweighted squared error loss of $\widehat{\boldsymbol{\beta}}_{\sigma_0^2 \boldsymbol{T}, \boldsymbol{\beta}_0}$ is 0.6208, while the average observed loss of $\widehat{\boldsymbol{\beta}}$ is 1.1765 when $\boldsymbol{\beta} = (0.2, 0.9)'$ is the true parameter.

We can now also compute the corresponding estimates $\widehat{\boldsymbol{\beta}}_{\widehat{\sigma}^2 \boldsymbol{T}, \boldsymbol{\beta}_0}$ and compare them with the already obtained estimates $\widehat{\boldsymbol{\beta}}_{\sigma_0^2 \boldsymbol{T}, \boldsymbol{\beta}_0}$. It turns out that the average observed loss of $\widehat{\boldsymbol{\beta}}_{\widehat{\sigma}^2 \boldsymbol{T}, \boldsymbol{\beta}_0}$ is 0.3783. Thus, $\widehat{\boldsymbol{\beta}}_{\widehat{\sigma}^2 \boldsymbol{T}, \boldsymbol{\beta}_0}$ yields a stronger improvement upon the ordinary least squares estimator than the estimator $\widehat{\boldsymbol{\beta}}_{\sigma_0^2 \boldsymbol{T}, \boldsymbol{\beta}_0}$, compare also Figs. 3.36 and 3.37. Here, the application of $\widehat{\sigma}^2$ as an estimator for σ^2 turns out to be a better choice than the application of the 'estimator' $\sigma_0^2 = 0.5$ (the true σ^2 is $\sigma^2 = 1$). □

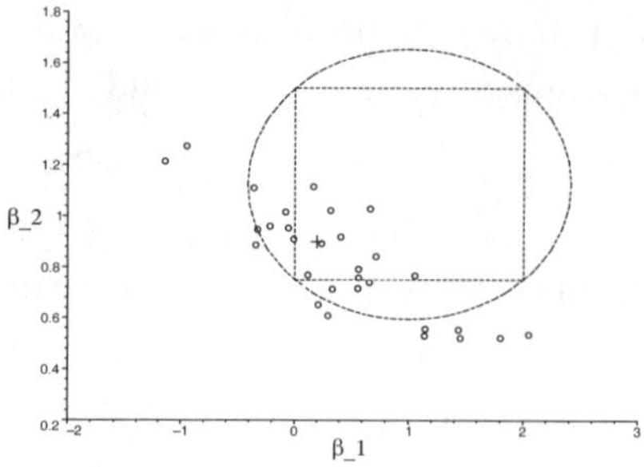

Fig. 3.36. 30 estimates $\widehat{\boldsymbol{\beta}}_{\sigma_0^2 \boldsymbol{T},\boldsymbol{\beta}_0}$ in Example 3.23

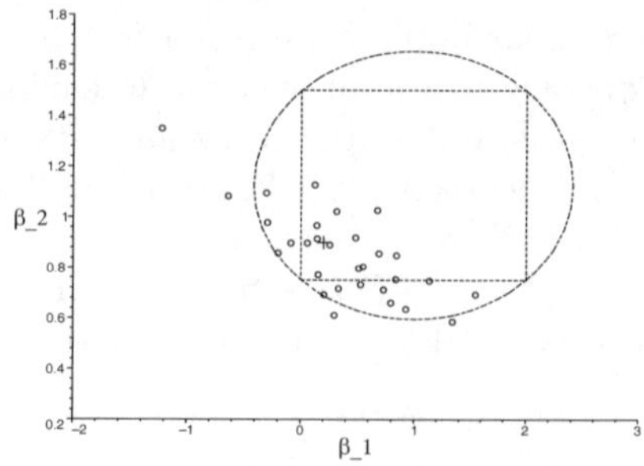

Fig. 3.37. 30 estimates $\widehat{\boldsymbol{\beta}}_{\widehat{\sigma^2} \boldsymbol{T},\boldsymbol{\beta}_0}$ in Example 3.23

3.7.4 Alternative Approaches

As in the first paragraph of Sect. 3.7.3, let us consider the parameter space

$$\Theta_{\boldsymbol{\beta}} = \{\boldsymbol{\beta} \in \mathbb{R}^p : \boldsymbol{\beta}'\boldsymbol{T}\boldsymbol{\beta} \leq t^2\} ,$$

where $\boldsymbol{T}$ is a $p \times p$ symmetric positive definite matrix and $0 < t^2 < \infty$. Then $\breve{\boldsymbol{A}}\boldsymbol{y}$ is a linear minimax solution for $\boldsymbol{\beta}$ if $\breve{\boldsymbol{A}}$ minimizes the maximal weighted squared error risk

$$\max_{\boldsymbol{\beta} \in \Theta_{\boldsymbol{\beta}}} \rho(\boldsymbol{\beta}, \boldsymbol{A}\boldsymbol{y}) = \sigma^2 \operatorname{tr}(\boldsymbol{W}\boldsymbol{A}\boldsymbol{A}') + t^2 \lambda_{\max}(\boldsymbol{B})$$

with respect to $\boldsymbol{A}$, where $\boldsymbol{B} = \boldsymbol{T}^{-1/2}(\boldsymbol{A}\boldsymbol{X} - \boldsymbol{I}_p)'\boldsymbol{W}(\boldsymbol{A}\boldsymbol{X} - \boldsymbol{I}_p)\boldsymbol{T}^{-1/2}$. An explicit solution $\breve{\boldsymbol{A}}$ can be given in two special cases, one of which we have already considered above.

- Case 1. The symmetric nonnegative definite weight matrix $\boldsymbol{W}$ has rank 1, i.e. $\boldsymbol{W} = \boldsymbol{w}\boldsymbol{w}'$ for some vector $\boldsymbol{w}$. In this case the matrix

$$\breve{\boldsymbol{A}} = (\boldsymbol{X}'\boldsymbol{X} + \frac{\sigma^2}{t^2}\boldsymbol{T})^{-1}\boldsymbol{X}'$$

  yields a linear minimax solution for $\boldsymbol{\beta}$.
- Case 2. The symmetric nonnegative definite weight matrix $\boldsymbol{W}$ and the matrix $\boldsymbol{T}$ are both identical to the identity matrix, i.e. $\boldsymbol{W} = \boldsymbol{T} = \boldsymbol{I}_p$. In this case the matrix

$$\breve{\boldsymbol{A}} = \frac{1}{1 + (\sigma^2/t^2)\operatorname{tr}[(\boldsymbol{X}'\boldsymbol{X})^{-1}]}(\boldsymbol{X}'\boldsymbol{X})^{-1}\boldsymbol{X}'$$

  yields a linear minimax solution for $\boldsymbol{\beta}$, see [69]. Thus, the linear minimax solution is of the form of a shrinkage estimator $\widehat{\boldsymbol{\beta}}(\breve{\varrho}) = \frac{1}{1+\breve{\varrho}}\widehat{\boldsymbol{\beta}}$ with shrinkage parameter $\breve{\varrho} = (\sigma^2/t^2)\operatorname{tr}[(\boldsymbol{X}'\boldsymbol{X})^{-1}]$.

Besides the two cases above, no other situation is known under which the matrix $\breve{\boldsymbol{A}}$ is of similar simplicity. As a fundamental reason for the difficulties in deriving explicit solutions, one may see the fact that the maximal eigenvalue of a matrix is not an easy function of the elements of this matrix.

Quasi Minimax Solutions

Since the above matrix $\boldsymbol{B}$ is symmetric nonnegative definite, the maximal eigenvalue of $\boldsymbol{B}$ must lie in the interval

$$\frac{1}{p}\,\mathrm{tr}(\boldsymbol{B}) \le \lambda_{\max}(\boldsymbol{B}) \le \mathrm{tr}(\boldsymbol{B}) \, .$$

Now, the idea of the quasi minimax approach is to replace $\lambda_{\max}(\boldsymbol{B})$ in the formula for the maximal risk either by the upper bound $\mathrm{tr}(\boldsymbol{B})$ (upper quasi minimax) or by the lower bound $\frac{1}{p}\,\mathrm{tr}(\boldsymbol{B})$ (lower quasi minimax) and then minimize the new function. One may expect some benefit from this procedure, since contrary to $\lambda_{\max}(\boldsymbol{B})$ the trace $\mathrm{tr}(\boldsymbol{B})$ is an easy function of the elements of $\boldsymbol{B}$, and thus one can hope to obtain explicit solutions to the new minimization problem. As a matter of fact, replacing $\lambda_{\max}(\boldsymbol{B})$ by $\mathrm{tr}(\boldsymbol{B})$ and minimizing the function

$$\sigma^2\,\mathrm{tr}(\boldsymbol{W}\boldsymbol{A}\boldsymbol{A}') + t^2\,\mathrm{tr}(\boldsymbol{B})$$

with respect to $\boldsymbol{A}$ yields the matrix

$$\breve{\boldsymbol{A}} = (\boldsymbol{X}'\boldsymbol{X} + \frac{\sigma^2}{t^2}\boldsymbol{T})^{-1}\boldsymbol{X}'$$

as the solution. The upper quasi minimax approach thus gives the same result as the linear minimax approach with respect to a rank 1 weight matrix $\boldsymbol{W}$.

Analogously, replacing $\lambda_{\max}(\boldsymbol{B})$ by $(1/p)\,\mathrm{tr}(\boldsymbol{B})$ yields

$$\breve{\boldsymbol{A}} = (\boldsymbol{X}'\boldsymbol{X} + \frac{p\,\sigma^2}{t^2}\boldsymbol{T})^{-1}\boldsymbol{X}'$$

as the lower quasi minimax solution, see e.g. [110, 96, 97]. If

$$\Theta_{\boldsymbol{\beta}} = \{\boldsymbol{\beta} \in \mathrm{IR}^p : (\boldsymbol{\beta} - \boldsymbol{\beta}_0)'\boldsymbol{T}(\boldsymbol{\beta} - \boldsymbol{\beta}_0) \le t^2\} \, ,$$

then the non-linear estimators $\widehat{\boldsymbol{\beta}}_{(\widehat{\sigma^2}/t^2)\boldsymbol{T},\boldsymbol{\beta}_0}$ and $\widehat{\boldsymbol{\beta}}_{p\,(\widehat{\sigma^2}/t^2)\boldsymbol{T},\boldsymbol{\beta}_0}$ can be called the (operational) *upper quasi minimax estimator* and the *lower quasi minimax estimator* for $\boldsymbol{\beta}$. If the parameter space is

$$\Theta_{\boldsymbol{\beta}} = \{\boldsymbol{\beta} \in \mathrm{IR}^p : (\boldsymbol{\beta} - \boldsymbol{\beta}_0)'\boldsymbol{T}(\boldsymbol{\beta} - \boldsymbol{\beta}_0) \le t^2\sigma^2\},$$

then accordingly $\widehat{\boldsymbol{\beta}}_{t^{-2}\boldsymbol{T},\boldsymbol{\beta}_0}$ and $\widehat{\boldsymbol{\beta}}_{p\,t^{-2}\boldsymbol{T},\boldsymbol{\beta}_0}$ are the linear upper quasi minimax estimator and the linear lower quasi minimax estimator for $\boldsymbol{\beta}$.

The main difference in the behavior of upper and lower quasi minimax estimator is, that the lower quasi minimax estimator causes a stronger shrinking towards the center vector $\boldsymbol{\beta}_0$ of the parametric ellipsoid than the upper quasi minimax estimator.

Example 3.24. In Example 3.23 we have seen that for 30 generated realizations $\boldsymbol{y}$ in the linear regression model from Example 3.22, the average observed loss of 1.1765 of the ordinary least squares estimator is distinctly greater than the average observed loss 0.3783 of the upper quasi minimax estimator $\widehat{\boldsymbol{\beta}}_{\widehat{\sigma^2}\boldsymbol{T},\boldsymbol{\beta}_0}$. If, in addition, we compute the corresponding lower quasi minimax estimates $\widehat{\boldsymbol{\beta}}_{p\widehat{\sigma^2}\boldsymbol{T},\boldsymbol{\beta}_0}$, then their average observed loss is 0.2433, so that again an improvement is obtained, see also Figs. 3.38 and 3.39. □

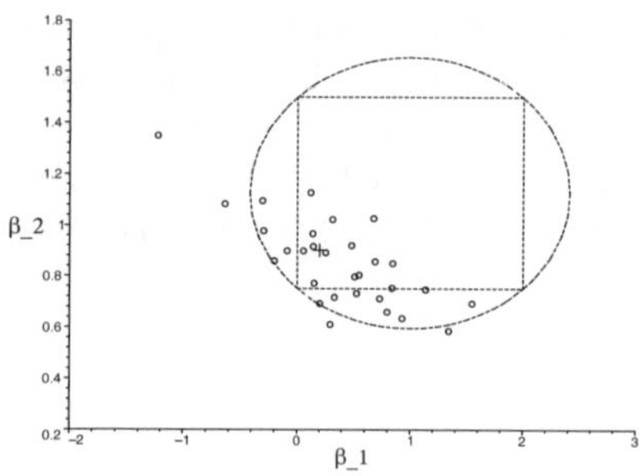

Fig. 3.38. 30 estimates $\widehat{\boldsymbol{\beta}}_{\widehat{\sigma^2}\boldsymbol{T},\boldsymbol{\beta}_0}$ in Example 3.24

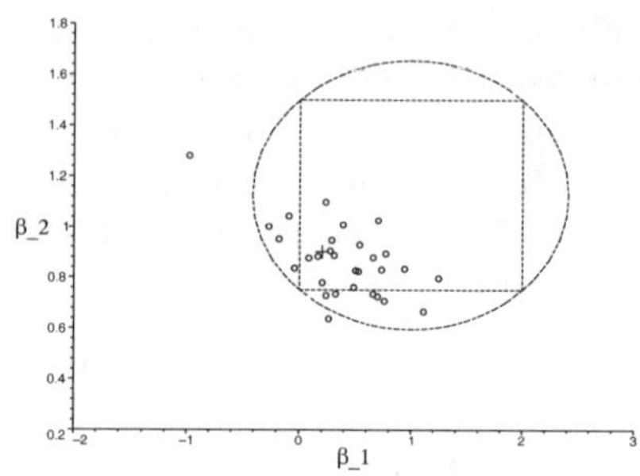

Fig. 3.39. 30 estimates $\widehat{\boldsymbol{\beta}}_{p\widehat{\sigma^2}\boldsymbol{T},\boldsymbol{\beta}_0}$ in Example 3.24

Inequality Restrictions

In case of inequality restrictions

$$a_i \leq \beta_i \leq b_i, \quad i = 1, \ldots, p,$$

for known $a_i, b_i \in \mathbb{R}$, Pilz [83, p. 244 ff.] proposes the application of a general ridge estimator with $p\boldsymbol{T}$ instead of the matrix

$$\boldsymbol{T} = \frac{4}{p}\operatorname{diag}\left((b_1 - a_1)^{-2}, \ldots, (b_p - a_p)^{-2}\right).$$

This means that instead of an operational upper quasi minimax estimator $\widehat{\boldsymbol{\beta}}_{\widehat{\sigma^2}\boldsymbol{T},\boldsymbol{\beta}_0}$, as suggested before, one should use an operational lower quasi minimax estimator $\widehat{\boldsymbol{\beta}}_{p\widehat{\sigma^2}\boldsymbol{T},\boldsymbol{\beta}_0}$ for estimating $\boldsymbol{\beta}$. A good reason for this is provided by the fact that $\widehat{\boldsymbol{\beta}}_{p\sigma^2\boldsymbol{T},\boldsymbol{\beta}_0}$ is indeed the actual linear minimax solution with respect to the inequality restrictions themselves, as opposed to $\widehat{\boldsymbol{\beta}}_{\sigma^2\boldsymbol{T},\boldsymbol{\beta}_0}$, which is the linear minimax solution with respect to the enclosing ellipsoid and a weight matrix $\boldsymbol{W}$ of rank 1, see [83, p. 245] for the details.

One may, of course, also apply the non-linear inequality restricted least squares estimator $\widehat{\boldsymbol{\beta}}_{\mathrm{IR}}$ described in Sect. 3.2.2, guaranteeing that the actual estimate satisfies the inequalities.

Minimax Shrinkage Solutions

A further possibility for simplification of the above described minimization problem is offered, when the minimax solutions are only sought within some subset of $\mathcal{L}(\boldsymbol{\beta})$. If we consider the set

$$\mathcal{S}(\boldsymbol{\beta}) = \{\boldsymbol{A}\boldsymbol{y} : \boldsymbol{A} = \frac{1}{1+\varrho}(\boldsymbol{X}'\boldsymbol{X})^{-1}\boldsymbol{X}',\ \varrho > 0\}$$

of linear shrinkage estimators for $\boldsymbol{\beta}$, then it follows

$$\max_{\boldsymbol{\beta}\in\Theta_{\boldsymbol{\beta}}} \rho(\boldsymbol{\beta}, \frac{1}{1+\varrho}\widehat{\boldsymbol{\beta}})$$
$$= \frac{1}{(1+\varrho)^2}\sigma^2 \operatorname{tr}[\boldsymbol{W}(\boldsymbol{X}'\boldsymbol{X})^{-1}] + \frac{\varrho^2}{(1+\varrho)^2}t^2\lambda_{\max}(\boldsymbol{W}\boldsymbol{T}^{-1}) .$$

We have $\operatorname{tr}[\boldsymbol{W}(\boldsymbol{X}'\boldsymbol{X})^{-1}] > 0$ and $\operatorname{tr}[\boldsymbol{W}(\boldsymbol{X}'\boldsymbol{X})^{-1}] > 0$ whenever $\boldsymbol{W} \neq \boldsymbol{0}$, so that minimization of the maximal risk with respect to ϱ yields

$$\breve{\varrho} = \frac{\sigma^2 \operatorname{tr}[\boldsymbol{W}(\boldsymbol{X}'\boldsymbol{X})^{-1}]}{t^2\,\lambda_{\max}(\boldsymbol{W}\boldsymbol{T}^{-1})} .$$

Note that we have repeatedly considered the minimization of a function of the form $f_{a,b}(\varrho) = a/(1+\varrho)^2 + b\varrho^2/(1+\varrho)^2$, where $a, b > 0$, with respect to ϱ, see Theorems 3.10, 3.17 and 3.18.

The linear *minimax shrinkage* solution is therefore

$$\widehat{\boldsymbol{\beta}}(\breve{\varrho}) = \frac{1}{1+\breve{\varrho}}\widehat{\boldsymbol{\beta}} .$$

For the special case $\boldsymbol{W} = \boldsymbol{T} = \boldsymbol{I}_p$, this solution is not only minimax within the set $\mathcal{S}(\boldsymbol{\beta})$ but also within the set $\mathcal{L}(\boldsymbol{\beta})$, see [69]. Minimax shrinkage solutions are also considered in [109, p. 94] and [107]. When the parameter space is

$$\Theta_{\boldsymbol{\beta}} = \{\boldsymbol{\beta} \in \mathbb{R}^p : (\boldsymbol{\beta} - \boldsymbol{\beta}_0)'\boldsymbol{T}(\boldsymbol{\beta} - \boldsymbol{\beta}_0) \le t^2\} ,$$

then

$$\widehat{\boldsymbol{\beta}}_{\boldsymbol{\beta}_0}(\breve{\varrho}) = \boldsymbol{\beta}_0 + \frac{1}{1+\breve{\varrho}}(\boldsymbol{X}'\boldsymbol{X})^{-1}\boldsymbol{X}'(\boldsymbol{y} - \boldsymbol{X}\boldsymbol{\beta}_0)$$

is the corresponding minimax shrinkage solution.

Example 3.25. The previous Examples 3.23 and 3.24 are concerned with estimation of $\boldsymbol{\beta}$ for the situation from Example 3.22. We can also use the 30 generated realizations $\boldsymbol{y}$ to compute minimax shrinkage estimates $\widehat{\boldsymbol{\beta}}_{\boldsymbol{\beta}_0}(\breve{\varrho})$ with shrinkage parameter

$$\breve{\varrho} = \frac{\widehat{\sigma}^2 \operatorname{tr}[\boldsymbol{W}(\boldsymbol{X}'\boldsymbol{X})^{-1}]}{t^2\,\lambda_{\max}(\boldsymbol{W}\boldsymbol{T}^{-1})} ,$$

where $\boldsymbol{W} = \boldsymbol{I}_p$ and $t^2 = 1$. From $\mathrm{tr}[(\boldsymbol{X}'\boldsymbol{X})^{-1}] = 6/5$ and $\lambda_{\max}(\boldsymbol{T}^{-1}) = 2$ the shrinkage parameter is $\breve{\varrho} = (3/5)\widehat{\sigma}^2$. If we compute the average observed unweighted squared error loss of $\widehat{\boldsymbol{\beta}}_{\boldsymbol{\beta}_0}(\breve{\varrho})$, then we obtain a value of 0.5971, which is greater than the average observed loss of 0.2433 of the estimator $\widehat{\boldsymbol{\beta}}_{p\widehat{\sigma}^2,\boldsymbol{\beta}_0}$. Figures 3.40 and 3.41 show the estimates of both estimators. □

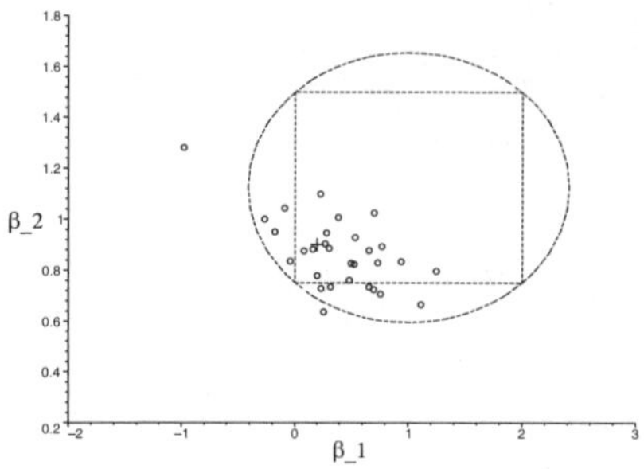

Fig. 3.40. 30 estimates $\widehat{\boldsymbol{\beta}}_{p\widehat{\sigma}^2\boldsymbol{T},\boldsymbol{\beta}_0}$ in Example 3.25

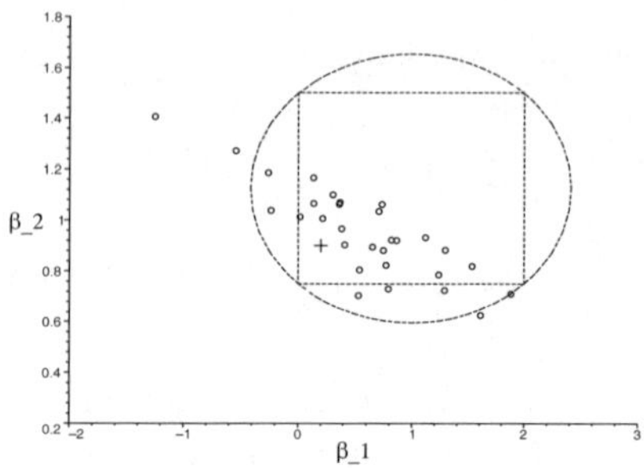

Fig. 3.41. 30 estimates $\widehat{\boldsymbol{\beta}}_{\boldsymbol{\beta}_0}(\breve{\varrho})$ in Example 3.25

Projection Onto an Ellipsoid

As shown by the Examples 3.22 to 3.25, none of the previously considered estimators must take its values within the ellipsoid encircling the given parallelepiped. This behavior does not stem from replacing the unknown σ^2 by the estimator $\widehat{\sigma}^2$. Even if σ^2 was known, the upper and lower quasi minimax estimates as well as the minimax shrinkage estimates could lie outside the ellipsoid.

Aa a remedy, Hoffmann [56] considers the modified estimator

$$\widehat{\boldsymbol{\beta}}_{\lambda\boldsymbol{T}} = (\boldsymbol{X}'\boldsymbol{X} + \lambda\boldsymbol{T})^{-1}\boldsymbol{X}'\boldsymbol{y}$$

for some number $\lambda \geq 0$. Since the continuous function

$$g(\lambda) = \widehat{\boldsymbol{\beta}}'_{\lambda\boldsymbol{T}}\boldsymbol{T}\widehat{\boldsymbol{\beta}}_{\lambda\boldsymbol{T}}$$

is monotonically decreasing for $\lambda \geq 0$, a given positive value t^2 either satisfies the inequality $g(0) \leq t^2$, or there exists some $\lambda_t > 0$ such that the condition $g(\lambda_t) = t^2$ is met. For this λ_t, the estimate $\widehat{\boldsymbol{\beta}}_{\lambda_t\boldsymbol{T}}$ lies on the edge of the ellipsoid

$$\mathcal{E} = \{\boldsymbol{\beta} : \boldsymbol{\beta}'\boldsymbol{T}\boldsymbol{\beta} \leq t^2\}\,.$$

Based on this observation, Hoffmann [56] proposes the estimator

$$\widehat{\beta}_{\mathcal{E}} = \begin{cases} \widehat{\beta} & \text{if } \widehat{\beta}'T\widehat{\beta} \leq t^2 \\ \widehat{\beta}_{\lambda_t T} & \text{otherwise} \end{cases},$$

which can be interpreted as a projection of the ordinary least squares estimator onto the ellipsoid $\mathcal{E}$. Analogously, the corresponding projection onto some ellipsoid

$$\mathcal{E} = \{\beta : (\beta - \beta_0)'T(\beta - \beta_0) \leq t^2\}$$

with center vector $\beta_0 \neq \mathbf{0}$ can be obtained by considering the function

$$\begin{aligned} g(\lambda) &= \widehat{\overline{\beta}}'_{\lambda T} T \widehat{\overline{\beta}}_{\lambda T} \\ &= (y - X\beta_0)'X(X'X + \lambda T)^{-1}T(X'X + \lambda T)^{-1}X'(y - X\beta_0) . \end{aligned}$$

Then

$$\widehat{\beta}_{\mathcal{E}} = \begin{cases} \widehat{\beta} & \text{if } g(0) \leq t^2 \\ \widehat{\beta}_{\lambda_t T, \beta_0} & \text{otherwise} \end{cases}$$

is the projection of $\widehat{\beta}$ onto the considered ellipsoid.

Example 3.26. We consider again the situation from Example 3.22 and use the 30 generated realizations y to compute the estimates $\widehat{\beta}_{\mathcal{E}}$, where $\mathcal{E}$ stands for the ellipsoid which encloses the parallelepiped, i.e.

$$\mathcal{E} = \{\beta : (\beta - \beta_0)'T(\beta - \beta_0) \leq 1\}$$

with

$$T = \begin{pmatrix} 1/2 & 0 \\ 0 & 32/9 \end{pmatrix} \quad \text{and} \quad \beta_0 = \begin{pmatrix} 1 \\ 9/8 \end{pmatrix} .$$

The average observed unweighted squared error loss of $\widehat{\beta}_{\mathcal{E}}$ is then 0.4541. This is a greater value than those obtained from the upper and lower quasi minimax estimators, but it is a smaller value than that obtained from the minimax shrinkage estimator. Compare also Figs. 3.42 and 3.43. □

It is of course also possible to apply this method to any estimator $\widetilde{\beta}$ other than $\widehat{\beta}$. One can in a first step check whether the actual estimate lies within some specified ellipsoid. If this is not the case, then in a second step one can compute an appropriate estimate $\widehat{\beta}_{\lambda_t T, \beta_0}$ which lies on the edge of the ellipsoid.

3.7.5 Admissibility

If we assume that under the linear regression model with assumptions (i) to (iv) restrictions of the form $\beta'T\beta \leq \sigma^2$ are satisfied, then the ridge estimator $\widehat{\beta}_T$ is an easy alternative to the ordinary least squares estimator, the latter ignoring any type of additional information about the parameters. Consequently, a direct comparison between the risks of $\widehat{\beta}$ and $\widehat{\beta}_T$ shows that the ridge estimator is preferable to the ordinary least squares estimator.

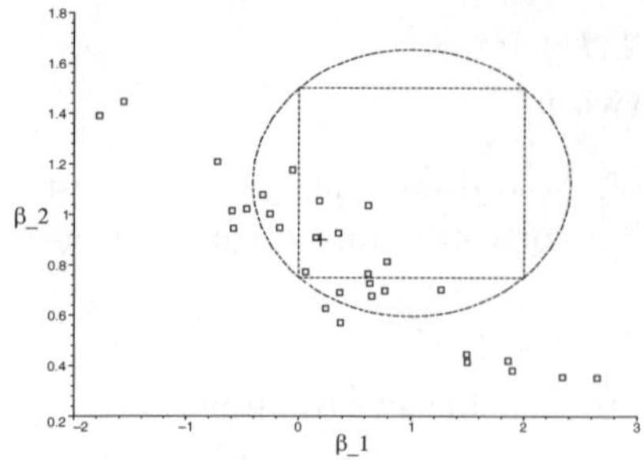

Fig. 3.42. 30 estimates $\widehat{\boldsymbol{\beta}}$ in Example 3.26

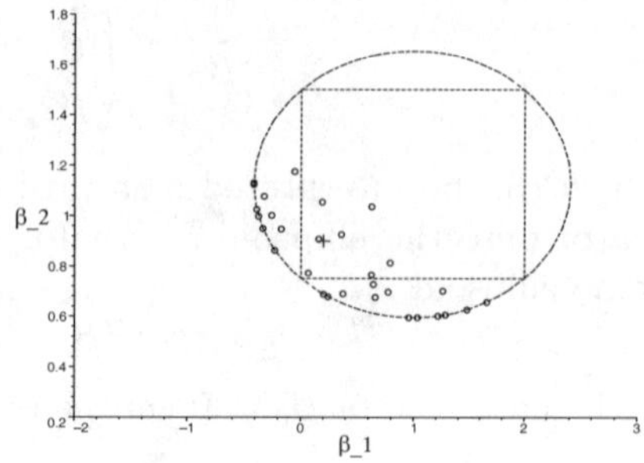

Fig. 3.43. 30 estimates $\widehat{\boldsymbol{\beta}}_{\mathcal{E}}$ in Example 3.26

Theorem 3.21. *If under the linear regression model with assumptions (i) to (iv), the inequality $\boldsymbol{\beta}'\boldsymbol{T}\boldsymbol{\beta} \leq \sigma^2$ is true for some given $p\times p$ symmetric positive definite matrix $\boldsymbol{T}$, then the difference*

$$\boldsymbol{\Delta} = \mathrm{MSE}(\boldsymbol{\beta}, \widehat{\boldsymbol{\beta}}) - \mathrm{MSE}(\boldsymbol{\beta}, \widehat{\boldsymbol{\beta}}_{\boldsymbol{T}})$$

is nonnegative definite.

Proof. Since the matrix $\boldsymbol{T}^{-1} + (\boldsymbol{X}'\boldsymbol{X})^{-1}$ is nonnegative definite, it follows $\boldsymbol{T}^{-1} \leq_{\mathrm{L}} 2\boldsymbol{T}^{-1} + (\boldsymbol{X}'\boldsymbol{X})^{-1}$. Due to Theorem A.67, the latter inequality is equivalent to $[2\boldsymbol{T}^{-1} + (\boldsymbol{X}'\boldsymbol{X})^{-1}]^{-1} \leq_{\mathrm{L}} \boldsymbol{T}$. Then

$$\boldsymbol{\beta}'[2\boldsymbol{T}^{-1} + (\boldsymbol{X}'\boldsymbol{X})^{-1}]^{-1}\boldsymbol{\beta} \leq \boldsymbol{\beta}'\boldsymbol{T}\boldsymbol{\beta}$$

holds true for all vectors $\boldsymbol{\beta} \in \mathbb{R}^p$. In view of the condition $\boldsymbol{\beta}'\boldsymbol{T}\boldsymbol{\beta} \leq \sigma^2$ the assertion follows from Theorem 3.19. □

According to the above theorem, the inequality $\mathrm{MSE}(\boldsymbol{\beta}, \widehat{\boldsymbol{\beta}}_{\boldsymbol{T}}) \leq_{\mathrm{L}} \mathrm{MSE}(\boldsymbol{\beta}, \widehat{\boldsymbol{\beta}})$ holds true for all vectors $\boldsymbol{\beta} \in \mathbb{R}^p$ satisfying the condition $\boldsymbol{\beta}'\boldsymbol{T}\boldsymbol{\beta} \leq \sigma^2$. Since in case $p > 1$ the two matrices $\mathrm{MSE}(\boldsymbol{\beta}, \widehat{\boldsymbol{\beta}}_{\boldsymbol{T}})$ and $\mathrm{MSE}(\boldsymbol{\beta}, \widehat{\boldsymbol{\beta}})$ do not coincide, it follows that

$$\rho(\boldsymbol{\beta}, \widehat{\boldsymbol{\beta}}_{\boldsymbol{T}}) < \rho(\boldsymbol{\beta}, \widehat{\boldsymbol{\beta}}) \quad \text{for all } \boldsymbol{\beta} \in \mathbb{R}^p \text{ such that } \boldsymbol{\beta}'\boldsymbol{T}\boldsymbol{\beta} \leq \sigma^2 ,$$

where $\rho(\boldsymbol{\beta}, \widetilde{\boldsymbol{\beta}}) = \mathrm{tr}[\mathrm{MSE}(\boldsymbol{\beta}, \widetilde{\boldsymbol{\beta}})]$. This means that the ridge estimator $\widehat{\boldsymbol{\beta}}_{\boldsymbol{T}}$ is *strictly better* than the ordinary least squares estimator $\widehat{\boldsymbol{\beta}}$ with respect to the unweighted squared error risk for all vectors $\boldsymbol{\beta}$ satisfying the condition $\boldsymbol{\beta}'\boldsymbol{T}\boldsymbol{\beta} \leq \sigma^2$. Hence, in this situation the ordinary least squares estimator is *inadmissible* for estimating $\boldsymbol{\beta}$ with respect to the unweighted squared error risk.

The ridge estimator $\widehat{\boldsymbol{\beta}}_{\boldsymbol{T}}$ is preferable to the ordinary least squares estimator $\widehat{\boldsymbol{\beta}}$ whenever the condition $\boldsymbol{\beta}'\boldsymbol{T}\boldsymbol{\beta} \leq \sigma^2$ is satisfied. Nonetheless, in this case, the ridge estimator $\widehat{\boldsymbol{\beta}}_{\boldsymbol{T}}$ is itself inadmissible with respect to the unweighted squared error risk when the assumed parameter space is $\Theta = \Theta_\beta = \{\boldsymbol{\beta} : \boldsymbol{\beta}'\boldsymbol{T}\boldsymbol{\beta} \leq \sigma^2\}$. See also the remarks following Theorem 4.22, p. 250.

Hoffmann [54] notes that $\widehat{\boldsymbol{\beta}}_{p\boldsymbol{T}}$ is a better linear estimator than $\widehat{\boldsymbol{\beta}}_{\boldsymbol{T}}$ when $\boldsymbol{X}'\boldsymbol{X} = \boldsymbol{I}_p$, $\boldsymbol{T} = \boldsymbol{I}_p$, and $p > 1$, see Problem 3.38.

When $\boldsymbol{X}'\boldsymbol{X} \neq \boldsymbol{I}_p$ and/or $\boldsymbol{T} \neq \boldsymbol{I}_p$, then the dominance of $\widehat{\boldsymbol{\beta}}_{p\boldsymbol{T}}$ over $\widehat{\boldsymbol{\beta}}_{\boldsymbol{T}}$ cannot be confirmed, but it can be shown that

$$\widehat{\boldsymbol{\beta}}_{p\boldsymbol{T}} = (\boldsymbol{X}'\boldsymbol{X} + p\boldsymbol{T})^{-1}\boldsymbol{X}'\boldsymbol{y}$$

is admissible within the set of homogeneous linear estimators of $\boldsymbol{\beta}$ when $\boldsymbol{\beta}'\boldsymbol{T}\boldsymbol{\beta} \leq \sigma^2$, see again the remarks following Thorem 4.22. On the other hand, the estimator $\widehat{\boldsymbol{\beta}}_{p\boldsymbol{T}}$ is not necessarily better than the ordinary least squares estimator $\widehat{\boldsymbol{\beta}}$ for all vectors $\boldsymbol{\beta}$ such that $\boldsymbol{\beta}'\boldsymbol{T}\boldsymbol{\beta} \leq \sigma^2$ when $\boldsymbol{X}$ can be any matrix of full column rank and $\boldsymbol{T}$ can be any symmetric positive definite matrix.

A necessary and sufficient condition for

$$\mathrm{MSE}(\boldsymbol{\beta}, \widehat{\boldsymbol{\beta}}_{p\boldsymbol{T}}) \leq_{\mathrm{L}} \mathrm{MSE}(\boldsymbol{\beta}, \widehat{\boldsymbol{\beta}}) \forall \boldsymbol{\beta} \in \{\boldsymbol{\beta} : \boldsymbol{\beta}'\boldsymbol{T}\boldsymbol{\beta} \leq \sigma^2\}$$

is given by

$$0 \leq_{\mathrm{L}} \frac{2-p}{p}\boldsymbol{T}^{-1} + (\boldsymbol{X}'\boldsymbol{X})^{-1} ,$$

see also Problem 4.17. This condition is always satisfied when $p = 1, 2$, while in the case $p \geq 3$ it is equivalent to

$$\boldsymbol{X}'\boldsymbol{X} \leq_{\mathrm{L}} \frac{p}{p-2}\boldsymbol{T} ,$$

being not necessarily fulfilled. In Sect. 4.4 we consider the admissibility of linear estimators for a parameter space $\Theta = \Theta_\beta = \{\boldsymbol{\beta} : \boldsymbol{\beta}'\boldsymbol{T}\boldsymbol{\beta} \leq \sigma^2\}$ in more detail.

3.8 Linear Bayes Estimator

When the unknown parameter vector $\boldsymbol{\beta}$ is not assumed to be fixed, but the realization of some random vector $\boldsymbol{b}$ for which a-priori information about its distribution is available, then Bayes methods suggest themselves for creating appropriate estimators for $\boldsymbol{\beta}$.

3.8.1 Preliminary Considerations

Usually, in the linear regression model with assumptions (i) to (iv), we assume that the true parameter vector $\boldsymbol{\beta}$ is some specific but unknown element from $\mathbb{R}^p$. If we have no further a-priori information about it, then any value from $\mathbb{R}^p$ could be the unknown. If we adopt the assumption of a Bayesian that $\boldsymbol{\beta}$ is the realization of a random vector $\boldsymbol{b}$ and we have some a-priori information about its distribution available, then it is reasonable to mobilize these information in order to find some better estimator for $\boldsymbol{\beta}$.

If we are mainly interested in *linear* estimation of $\boldsymbol{\beta}$, then it is not necessary to have knowledge about the distribution of $\boldsymbol{b}$ as a whole, but it suffices to have information about expectation vector and covariance matrix

$$\mathrm{E}(\boldsymbol{b}) = \boldsymbol{\mu} \quad \text{and} \quad \mathrm{Cov}(\boldsymbol{b}) = \sigma^2 \boldsymbol{T} .$$

Here $\boldsymbol{\mu}$ is a known $p \times 1$ vector and $\boldsymbol{T}$ is a known $p \times p$ symmetric nonnegative definite matrix. Now, how can knowledge of $\boldsymbol{\mu}$ and $\boldsymbol{T}$ be employed for estimation of $\boldsymbol{\beta}$?

To find an answer, let us consider the usual unweighted squared error risk of a linear estimator $\boldsymbol{Ay} + \boldsymbol{a}$ for $\boldsymbol{\beta}$, given as

$$\begin{aligned}\rho(\boldsymbol{\beta}, \boldsymbol{Ay} + \boldsymbol{a}) &= \mathrm{E}[(\boldsymbol{Ay} + \boldsymbol{a} - \boldsymbol{\beta})'(\boldsymbol{Ay} + \boldsymbol{a} - \boldsymbol{\beta})] \\ &= \sigma^2 \operatorname{tr}(\boldsymbol{AA}') + \operatorname{tr}[((\boldsymbol{AX} - \boldsymbol{I}_p)\boldsymbol{\beta} + \boldsymbol{a})((\boldsymbol{AX} - \boldsymbol{I}_p)\boldsymbol{\beta} + \boldsymbol{a})'] .\end{aligned}$$

This risk can be considered as a function $R_{\boldsymbol{\beta},\sigma^2}(\boldsymbol{A}, \boldsymbol{a})$ of $\boldsymbol{A}$ and $\boldsymbol{a}$, which in addition depends on the unknown parameters $\boldsymbol{\beta}$ and σ^2. A rather natural way to remove the dependency of this function on $\boldsymbol{\beta}$ is, to apply the information about $\boldsymbol{b}$ by considering the *Bayes risk*

$$\rho^{\mathrm{B}}(\boldsymbol{\beta}, \boldsymbol{Ay} + \boldsymbol{a}) = \mathrm{E}[R_{\boldsymbol{b},\sigma^2}(\boldsymbol{A}, \boldsymbol{a})]$$

instead of the usual risk $\rho(\boldsymbol{\beta}, \boldsymbol{Ay} + \boldsymbol{a}) = R_{\boldsymbol{\beta},\sigma^2}(\boldsymbol{A}, \boldsymbol{a})$. The Bayes risk can be interpreted as averaging the usual risk with respect to all possible realizations $\boldsymbol{\beta}$ of the random vector $\boldsymbol{b}$. It is easy to see that knowledge about $\mathrm{E}(\boldsymbol{b}) = \boldsymbol{\mu}$ and $\mathrm{Cov}(\boldsymbol{b}) = \sigma^2 \boldsymbol{T}$ are sufficient for evaluating $\mathrm{E}[R_{\boldsymbol{b},\sigma^2}(\boldsymbol{A}, \boldsymbol{a})]$.

Then a linear estimator $\boldsymbol{A}_*\boldsymbol{y} + \boldsymbol{a}_*$ is called *linear Bayes estimator* for $\boldsymbol{\beta}$, if

$$\rho^{\mathrm{B}}(\boldsymbol{\beta}, \boldsymbol{A}_*\boldsymbol{y} + \boldsymbol{a}_*) \leq \rho^{\mathrm{B}}(\boldsymbol{\beta}, \boldsymbol{Ay} + \boldsymbol{a})$$

holds true for all linear estimator $\boldsymbol{Ay}+\boldsymbol{a}$ of $\boldsymbol{\beta}$. This simply means that a linear Bayes estimator minimizes the Bayes risk within the set of linear estimators.

3.8.2 Characterization of Linear Bayes Estimators

The following theorem may be deduced from formula (4.11) in [90]. We omit its proof here. The reader might get an impression of the structure of the proof from Problems 3.39 and 3.40.

Theorem 3.22. *Under the linear regression model with assumptions (i) to (iv), let* $\boldsymbol{\beta} \in \mathbb{R}^p$ *be the realization of a* $p \times 1$ *random vector* $\boldsymbol{b}$ *with* $\mathrm{E}(\boldsymbol{b}) = \boldsymbol{\mu}$ *and* $\mathrm{Cov}(\boldsymbol{b}) = \sigma^2 \boldsymbol{T}$. *Then*

$$\widehat{\boldsymbol{\beta}}^{\mathrm{B}} = \boldsymbol{\mu} + \boldsymbol{TX}'(\boldsymbol{XTX}' + \boldsymbol{I}_n)^{-1}(\boldsymbol{y} - \boldsymbol{X\mu})$$

is the (uniquely determined) linear Bayes estimator of $\boldsymbol{\beta}$.

The estimator $\widehat{\boldsymbol{\beta}}^{\mathrm{B}}$ from the above theorem is the uniquely determined linear estimator $\boldsymbol{A}_*\boldsymbol{y} + \boldsymbol{a}_*$, which minimizes the Bayes risk as described in the previous subsection. It can also be written in alternative forms described in the following.

Corollary 3.6. *Under the linear regression model with assumptions (i) to (iv), let $\boldsymbol{\beta} \in \mathbb{R}^p$ be the realization of a $p \times 1$ random vector $\boldsymbol{b}$ with $\mathrm{E}(\boldsymbol{b}) = \boldsymbol{\mu}$ and $\mathrm{Cov}(\boldsymbol{b}) = \sigma^2\boldsymbol{T}$. Then the linear Bayes estimator $\widehat{\boldsymbol{\beta}}^{\mathrm{B}}$ can be written in the form*

$$\widehat{\boldsymbol{\beta}}^{\mathrm{B}} = \boldsymbol{\mu} + \boldsymbol{T}[\boldsymbol{T} + (\boldsymbol{X}'\boldsymbol{X})^{-1}]^{-1}(\widehat{\boldsymbol{\beta}} - \boldsymbol{\mu}) \, ,$$

where $\widehat{\boldsymbol{\beta}} = (\boldsymbol{X}'\boldsymbol{X})^{-1}\boldsymbol{X}'\boldsymbol{y}$ is the ordinary least squares estimator.

Proof. If we write $\boldsymbol{S} = \boldsymbol{X}'\boldsymbol{X}$, then with Problem 3.41 it can easily be shown that

$$\boldsymbol{T}(\boldsymbol{T} + \boldsymbol{S}^{-1})^{-1}\boldsymbol{S}^{-1}\boldsymbol{X}'(\boldsymbol{I}_n + \boldsymbol{X}\boldsymbol{T}\boldsymbol{X}') = \boldsymbol{T}\boldsymbol{X}' \, ,$$

giving

$$\boldsymbol{T}(\boldsymbol{T} + \boldsymbol{S}^{-1})^{-1}\boldsymbol{S}^{-1}\boldsymbol{X}' = \boldsymbol{T}\boldsymbol{X}'(\boldsymbol{I}_n + \boldsymbol{X}\boldsymbol{T}\boldsymbol{X}')^{-1} \, .$$

Then the assertion follows from Theorem 3.22. □

The representation of $\widehat{\boldsymbol{\beta}}^{\mathrm{B}}$ from Corollary 3.6 can also be found in [83, p. 93]. In this context, it should be noted that the uniqueness of the linear Bayes estimator $\widehat{\boldsymbol{\beta}}^{\mathrm{B}}$ cannot be postulated when a weighted Bayes risk

$$\rho^{\mathrm{B}}(\boldsymbol{\beta}, \boldsymbol{A}\boldsymbol{y} + \boldsymbol{a}) = \mathrm{E}[(\boldsymbol{A}\boldsymbol{X}\boldsymbol{b} + \boldsymbol{a} - \boldsymbol{b})'\boldsymbol{W}(\boldsymbol{A}\boldsymbol{X}\boldsymbol{b} + \boldsymbol{a} - \boldsymbol{b})]$$

is considered, and $\boldsymbol{W}$ is a singular $p \times p$ symmetric nonnegative definite matrix. Nevertheless, if $\boldsymbol{W}$ is nonsingular and hence symmetric positive definite, then minimization of the weighted Bayes risk gives the same linear Bayes estimator for any such $\boldsymbol{W}$, see also Sect. 5.3 in [83].

Corollary 3.7. *Under the linear regression model with assumptions (i) to (iv), let $\boldsymbol{\beta} \in \mathbb{R}^p$ be the realization of a $p \times 1$ random vector $\boldsymbol{b}$ with $\mathrm{E}(\boldsymbol{b}) = \boldsymbol{\mu}$ and $\mathrm{Cov}(\boldsymbol{b}) = \sigma^2\boldsymbol{T}$. If $\boldsymbol{T}$ is nonsingular, then the linear Bayes estimator $\widehat{\boldsymbol{\beta}}^{\mathrm{B}}$ can be written in the form*

$$\widehat{\boldsymbol{\beta}}^{\mathrm{B}} = \boldsymbol{\mu} + (\boldsymbol{X}'\boldsymbol{X} + \boldsymbol{T}^{-1})^{-1}\boldsymbol{X}'(\boldsymbol{y} - \boldsymbol{X}\boldsymbol{\mu}) \, .$$

Proof. If we write $\boldsymbol{S} = \boldsymbol{X}'\boldsymbol{X}$, then the assertion follows in view of

$$\boldsymbol{T}(\boldsymbol{T} + \boldsymbol{S}^{-1})^{-1}\boldsymbol{S}^{-1}\boldsymbol{X}' = \boldsymbol{T}\boldsymbol{X}'(\boldsymbol{I}_n + \boldsymbol{X}\boldsymbol{T}\boldsymbol{X}')^{-1}$$

and

$$\boldsymbol{T}(\boldsymbol{T} + \boldsymbol{S}^{-1})^{-1}\boldsymbol{S}^{-1} = (\boldsymbol{S} + \boldsymbol{T}^{-1})^{-1}$$

from Theorem 3.22. □

According to the above corollary

$$\widehat{\boldsymbol{\beta}}^{\mathrm{B}} = \widehat{\boldsymbol{\beta}}_{\boldsymbol{T}^{-1},\boldsymbol{\mu}}$$

when the matrix $\boldsymbol{T}$ is nonsingular. In this case the linear Bayes estimator coincides with the upper quasi minimax estimator obtained under elliptical restrictions

$$\Theta_{\boldsymbol{\beta}} = \{\boldsymbol{\beta} \in \mathrm{I\!R}^p : (\boldsymbol{\beta} - \boldsymbol{\mu})'\boldsymbol{T}^{-1}(\boldsymbol{\beta} - \boldsymbol{\mu}) \leq \sigma^2\} .$$

3.8.3 Non-Operational Bayes Solutions

For the derivation of the linear Bayes estimator we have assumed that $\mathrm{Cov}(\boldsymbol{b}) = \sigma^2\boldsymbol{T}$ depends on the unknown parameter σ^2. This assumption has the effect that we can obtain an *operational* linear Bayes estimator $\widehat{\boldsymbol{\beta}}^{\mathrm{B}}$. Nonetheless, it is more realistic to assume prior information of the form

$$\mathrm{E}(\boldsymbol{b}) = \boldsymbol{\mu} \quad \text{and} \quad \mathrm{Cov}(\boldsymbol{b}) = \boldsymbol{W}$$

for some known $p \times 1$ vector $\boldsymbol{\mu}$ and some known $p \times p$ symmetric nonnegative definite matrix $\boldsymbol{W}$. In this case minimization of the Bayes risk yields the *non-operational* Bayes solution

$$\widehat{\boldsymbol{\beta}}^{\mathrm{B}} = \boldsymbol{\mu} + \boldsymbol{W}(\boldsymbol{W} + \sigma^2(\boldsymbol{X}'\boldsymbol{X})^{-1})^{-1}(\widehat{\boldsymbol{\beta}} - \boldsymbol{\mu}) .$$

As before, we may obtain a practical estimator by replacing the unknown σ^2 by the estimator $\widehat{\sigma}^2$. The result can be interpreted as a practical Bayes estimator for $\boldsymbol{\beta}$, which, however, is not a linear estimator any longer.

For the case that $\boldsymbol{W}$ is non-singular, the non-operational Bayes solution can also be written as

$$\widehat{\boldsymbol{\beta}}^{\mathrm{B}} = \boldsymbol{\mu} + (\boldsymbol{X}'\boldsymbol{X} + \sigma^2\boldsymbol{W}^{-1})^{-1}\boldsymbol{X}'(\boldsymbol{y} - \boldsymbol{X}\boldsymbol{\mu}) ,$$

being identical to

$$\widehat{\boldsymbol{\beta}}^{\mathrm{B}} = (\sigma^{-2}\boldsymbol{X}'\boldsymbol{X} + \boldsymbol{W}^{-1})^{-1}(\sigma^{-2}\boldsymbol{X}'\boldsymbol{y} + \boldsymbol{W}^{-1}\boldsymbol{\mu}) .$$

The latter has the form of the generalized least squares estimator under the model

$$\begin{pmatrix} \boldsymbol{y} \\ \boldsymbol{\mu} \end{pmatrix} = \begin{pmatrix} \boldsymbol{X} \\ \boldsymbol{I}_p \end{pmatrix} \boldsymbol{\beta} + \begin{pmatrix} \boldsymbol{\varepsilon} \\ \boldsymbol{v} \end{pmatrix} ,$$

where

$$\mathrm{E}\begin{pmatrix} \boldsymbol{\varepsilon} \\ \boldsymbol{v} \end{pmatrix} = \begin{pmatrix} \mathbf{0} \\ \mathbf{0} \end{pmatrix} \quad \text{and} \quad \mathrm{Cov}\begin{pmatrix} \boldsymbol{\varepsilon} \\ \boldsymbol{v} \end{pmatrix} = \begin{pmatrix} \sigma^2\boldsymbol{I}_n & \mathbf{0} \\ \mathbf{0} & \boldsymbol{W} \end{pmatrix} .$$

Thus, this Bayes solution can also be interpreted as a non-operational mixed estimator, see Sect. 3.2.1.

3.8.4 A-priori Assumptions

The previous subsection introduces the linear Bayes estimator $\widehat{\boldsymbol{\beta}}^{\mathrm{B}}$ for $\boldsymbol{\beta}$ with respect to a-priori information $\mathrm{E}(\boldsymbol{b}) = \boldsymbol{\mu}$ and $\mathrm{Cov}(\boldsymbol{b}) = \sigma^2\boldsymbol{T}$ or $\mathrm{Cov}(\boldsymbol{b}) = \boldsymbol{W}$. But under what situations is such information available?

Noninformative Prior Knowledge

Before discussing two examples for possible prior knowledge, we assume that we have no idea what $\mathrm{E}(\boldsymbol{b}) = \boldsymbol{\mu}$ and $\mathrm{Cov}(\boldsymbol{b}) = \boldsymbol{W}$ could be. This means that we have no prior knowledge about the distribution of the individual elements b_i of the vector $\boldsymbol{b}$.

If nonetheless we want to specify an a-priori distribution, we may e.g. assume that the individual elements are pairwise uncorrelated and distributed as $N(0,\tau^2)$ with $\tau^2 = \infty$. This may be justified from the fact that the density of $N(0,\tau^2)$ becomes flatter for increasing values of τ^2, and for $\tau^2 = \infty$ the density may be interpreted as the density of a uniform distribution over the interval $(-\infty,\infty)$, being of course not a genuine density.

Since the variance of such a quasi-uniformly distributed random variable is not finite, the covariance matrix of the vector $\boldsymbol{b}$ is the matrix $\mathrm{Cov}(\boldsymbol{b}) = \lim_{t\to\infty} t\boldsymbol{I}_p$. This matrix is not altered when we multiply any finite σ^2 to it, so we may as well assume $\mathrm{Cov}(\boldsymbol{b}) = \sigma^2 \lim_{t\to\infty} t\boldsymbol{I}_p$ in the following. In addition, the expectation $\boldsymbol{\mu}$ can be the zero vector but also any other vector from $\mathbb{R}^p$. Hence, our *noninformative* prior knowledge about $\boldsymbol{b}$ can be represented by

$$\mathrm{E}(\boldsymbol{b}) = \boldsymbol{\mu} \quad \text{and} \quad \mathrm{Cov}(\boldsymbol{b}) = \sigma^2\boldsymbol{T}, \quad \boldsymbol{T} = \lim_{t\to\infty} t\boldsymbol{I}_p\,,$$

where $\boldsymbol{\mu}$ is an arbitrary vector from $\mathbb{R}^p$.

What is the linear Bayes estimator $\widehat{\boldsymbol{\beta}}^{\mathrm{B}}$ in this case? If $\boldsymbol{U}\boldsymbol{\Lambda}\boldsymbol{U}'$ is a spectral decomposition of the matrix $\boldsymbol{X}'\boldsymbol{X}$, then we can write

$$\begin{aligned}
\widehat{\boldsymbol{\beta}}^{\mathrm{B}} &= \boldsymbol{\mu} + \boldsymbol{T}[\boldsymbol{T} + (\boldsymbol{X}'\boldsymbol{X})^{-1}]^{-1}(\widehat{\boldsymbol{\beta}} - \boldsymbol{\mu}) \\
&= \lim_{t\to\infty} \boldsymbol{\mu} + t[t\boldsymbol{I}_p + (\boldsymbol{X}'\boldsymbol{X})^{-1}]^{-1}(\widehat{\boldsymbol{\beta}} - \boldsymbol{\mu}) \\
&= \lim_{t\to\infty} \boldsymbol{\mu} + t\boldsymbol{U}(t\boldsymbol{I}_p + \boldsymbol{\Lambda}^{-1})^{-1}\boldsymbol{U}'(\widehat{\boldsymbol{\beta}} - \boldsymbol{\mu}) \\
&= \lim_{t\to\infty} \boldsymbol{\mu} + \boldsymbol{U}\boldsymbol{\Gamma}\boldsymbol{U}'(\widehat{\boldsymbol{\beta}} - \boldsymbol{\mu})\,,
\end{aligned}$$

where

$$\boldsymbol{\Gamma} = \mathrm{diag}\left(\frac{t}{t+\frac{1}{\lambda_1}}, \ldots, \frac{t}{t+\frac{1}{\lambda_p}}\right).$$

In view of $\lim_{t\to\infty} \boldsymbol{\Gamma} = \boldsymbol{I}_p$, it follows

$$\widehat{\boldsymbol{\beta}}^{\mathrm{B}} = \boldsymbol{\mu} + \boldsymbol{I}_p(\widehat{\boldsymbol{\beta}} - \boldsymbol{\mu}) = \widehat{\boldsymbol{\beta}}\,.$$

Hence, if we specify noninformative a-priori assumptions on $\mathrm{E}(\boldsymbol{b})$ and $\mathrm{Cov}(\boldsymbol{b})$ as above, then the corresponding linear Bayes estimator is nothing else but the ordinary least squares estimator for $\boldsymbol{\beta}$.

Different Prior Knowledge

In the following we give two examples for the possible specification of prior assumptions on the distribution of the vector $\boldsymbol{b}$. See also [83, Sect. 3.1], where some additional ad-hoc proposals are discussed.

Earlier Estimates. Suppose that under a given linear regression model we have already obtained a number of samples (say s) of the dependent variable for a fixed set of values of the independent variables. If not the samples but the s estimates $\widehat{\boldsymbol{\beta}}_1, \dots, \widehat{\boldsymbol{\beta}}_s$ and $\widehat{\sigma}_1^2, \dots, \widehat{\sigma}_s^2$ of $\boldsymbol{\beta}$ and σ^2 are still available, then we can compute

$$\widetilde{\sigma}^2 = \frac{1}{s}\sum_{j=1}^{s}\widehat{\sigma}_j^2, \quad \widetilde{\boldsymbol{\mu}} = \frac{1}{s}\sum_{j=1}^{s}\widehat{\boldsymbol{\beta}}_j, \quad \widetilde{\boldsymbol{T}} = \frac{1}{(s-1)\widetilde{\sigma}^2}\sum_{j=1}^{s}(\widehat{\boldsymbol{\beta}}_j - \widetilde{\boldsymbol{\mu}})(\widehat{\boldsymbol{\beta}}_j - \widetilde{\boldsymbol{\mu}})' .$$

These estimates may be interpreted as prior knowledge about $\boldsymbol{b}$ and thus can be applied for the determination of the linear Bayes estimate $\widehat{\boldsymbol{\beta}}^{\mathrm{B}}$ for $\boldsymbol{\beta}$ when putting $\boldsymbol{\mu} = \widetilde{\boldsymbol{\mu}}$ and $\boldsymbol{T} = \widetilde{\boldsymbol{T}}$.

Example 3.27. We consider again the linear regression model $\boldsymbol{y} = \boldsymbol{X}\boldsymbol{\beta} + \boldsymbol{\varepsilon}$ with

$$\boldsymbol{y} = \begin{pmatrix} 2.6262 \\ 2.1640 \\ 2.3589 \\ 3.9844 \\ 4.2513 \end{pmatrix} \quad \text{and} \quad \boldsymbol{X} = \begin{pmatrix} 1 & 1 \\ 1 & 2 \\ 1 & 3 \\ 1 & 4 \\ 1 & 5 \end{pmatrix} .$$

The ordinary least squares estimator $\widehat{\boldsymbol{\beta}}$ yields the estimate

$$\widehat{\boldsymbol{\beta}} = \begin{pmatrix} 1.5558 \\ 0.5070 \end{pmatrix}$$

with an observed unweighted squared error loss of $L(\boldsymbol{\beta}, \widehat{\boldsymbol{\beta}}) = 1.9925$ for the true parameter $\boldsymbol{\beta} = (0.2, 0.9)'$.

Now assume that from earlier investigations we have $s = 3$ additional least squares estimates of $\boldsymbol{\beta}$ and σ^2 available, see Table 3.27. From these estimates we compute

$$\widetilde{\sigma}^2 = 1.1551, \quad \widetilde{\boldsymbol{\mu}} = \begin{pmatrix} 0.0503 \\ 0.8833 \end{pmatrix}, \quad \text{and} \quad \widetilde{\boldsymbol{T}} = \begin{pmatrix} 0.4025 & -0.1750 \\ -0.1750 & 0.0791 \end{pmatrix} .$$

Then the linear Bayes estimator $\widehat{\boldsymbol{\beta}}^{\mathrm{B}}$ yields the estimate

Table 3.3. Three least squares estimates for $\boldsymbol{\beta}$ and σ^2 in Example 3.27

j	1	2	3
$(\widehat{\beta}_1)_j$	-0.0900	0.7914	-0.5505
$(\widehat{\beta}_2)_j$	0.8782	0.5837	1.1880
$\widehat{\sigma}_j^2$	0.6295	2.3259	0.5097

$$\widehat{\boldsymbol{\beta}}^{\mathrm{B}} = \widetilde{\boldsymbol{\mu}} + \widetilde{\boldsymbol{T}}[\widetilde{\boldsymbol{T}} + (\boldsymbol{X}'\boldsymbol{X})^{-1}]^{-1}(\widehat{\boldsymbol{\beta}} - \widetilde{\boldsymbol{\mu}}) = \begin{pmatrix} 0.2884 \\ 0.7899 \end{pmatrix}$$

with an observed unweighted squared error loss of $L(\boldsymbol{\beta}, \widehat{\boldsymbol{\beta}}^{\mathrm{B}}) = 0.0199$. Hence, the incorporation of prior knowledge via the linear Bayes estimator gives a strong improvement upon the ordinary least squares estimator here. □

If in a given linear regression model different estimates for the vector $\boldsymbol{\beta}$ are available, as it is the case in the above example, then an easy alternative approach to the application of the linear Bayes estimator is the computation of the mean of *all* ordinary least squares estimates. Sometimes, this mean will even produce a smaller loss than the linear Bayes estimator. In the above example

$$\widetilde{\boldsymbol{\beta}} = \frac{3}{4}\widetilde{\boldsymbol{\mu}} + \frac{1}{4}(\boldsymbol{X}'\boldsymbol{X})^{-1}\boldsymbol{X}'\boldsymbol{y} = \begin{pmatrix} 0.4267 \\ 0.7892 \end{pmatrix}$$

with an observed unweighted squared error loss of $L(\boldsymbol{\beta}, \widetilde{\boldsymbol{\beta}}) = 0.0637$ for the true $\boldsymbol{\beta} = (0.2, 0.9)'$. In this example the mean of all ordinary least squares estimates has a slightly greater loss than the linear Bayes estimator.

The estimator $\widetilde{\boldsymbol{\beta}}$ can also be seen as a convex combination of $\widetilde{\boldsymbol{\mu}}$ and $\widehat{\boldsymbol{\beta}}$, i.e.

$$\widetilde{\boldsymbol{\beta}} = \widehat{\boldsymbol{\beta}}_{\widetilde{\boldsymbol{\mu}}}(\varrho) = (1-\alpha)\widetilde{\boldsymbol{\mu}} + \alpha\widehat{\boldsymbol{\beta}}, \quad \alpha = \frac{1}{1+\varrho}, \quad \varrho = s\,,$$

which may as well be interpreted as shrinking $\widehat{\boldsymbol{\beta}}$ into the direction of $\widetilde{\boldsymbol{\mu}}$.

Further possibilities for choosing ϱ are discussed in Sect. 3.5.4. The estimator $\widehat{\boldsymbol{\beta}}_{\widetilde{\boldsymbol{\mu}}}(\varrho)$ is also a linear Bayes estimator with respect to the prior assumptions

$$\mathrm{E}(\boldsymbol{b}) = \widetilde{\boldsymbol{\mu}} \quad \text{and} \quad \mathrm{Cov}(\boldsymbol{b}) = \varrho\, \sigma^2 (\boldsymbol{X}'\boldsymbol{X})^{-1}, \quad \varrho > 0\,,$$

see Problem 3.42.

Another possibility for shrinking towards $\widetilde{\boldsymbol{\mu}}$ is the application of the direction modified ridge estimator estimator from Swindel [112]

$$\widehat{\boldsymbol{\beta}}_{k,\widetilde{\boldsymbol{\mu}}} = (\boldsymbol{X}'\boldsymbol{X} + k\boldsymbol{I}_p)^{-1}(\boldsymbol{X}'\boldsymbol{y} + k\widetilde{\boldsymbol{\mu}})$$

for some appropriate positive k, see also Sect. 3.4.5. Again, this estimator is a linear Bayes estimator, but now with respect to the prior assumptions

$$\mathrm{E}(\boldsymbol{b}) = \widetilde{\boldsymbol{\mu}} \quad \text{and} \quad \mathrm{Cov}(\boldsymbol{b}) = k\,\sigma^2 \boldsymbol{I}_p, \quad k > 0\,,$$

see also Problem 3.43.

Known Parameter Intervals. If we know that the individual parameters β_i lie within specific intervals $c_i \leq \beta_i \leq d_i$, then we may for example assume that the corresponding random variables b_i are uniformly distributed over $[c_i, d_i]$, respectively. Thus $\mathrm{E}(b_i) = \frac{1}{2}(c_i + d_i)$ and $\mathrm{Var}(b_i) = \frac{1}{12}(d_i - c_i)^2$. If in addition we assume that $\mathrm{Cov}(b_i, b_j) = 0$ for $i \neq j$, then we obtain the prior assumptions

$$\mathrm{E}(\boldsymbol{b}) = \frac{1}{2}(\boldsymbol{c} + \boldsymbol{d}) \quad \text{and} \quad \mathrm{Cov}(\boldsymbol{b}) = \frac{1}{12}\,\mathrm{diag}\left((d_1 - c_1)^2, \ldots, (d_p - c_p)^2\right)\,,$$

which can be employed for computing the corresponding linear Bayes estimator.

Example 3.28. For the linear regression model from Example 3.27 we now assume that the random variable b_1 corresponding to the parameter β_1 is uniformly distributed over $[0, 2]$. Moreover, we assume that b_2 is uniformly distribute over $[0.75, 1.5]$, and b_1 and b_2 are uncorrelated. Then our prior assumptions are

$$\mathrm{E}(\boldsymbol{b}) = \boldsymbol{\mu} = \begin{pmatrix} 1 \\ 9/8 \end{pmatrix} \quad \text{and} \quad \mathrm{Cov}(\boldsymbol{b}) = \boldsymbol{W} = \begin{pmatrix} 1/3 & 0 \\ 0 & 3/64 \end{pmatrix}.$$

When we apply the practical linear Bayes estimator

$$\widehat{\boldsymbol{\beta}}^{\mathrm{B}} = \boldsymbol{\mu} + \boldsymbol{W}(\boldsymbol{W} + \widehat{\sigma}^2(\boldsymbol{X}'\boldsymbol{X})^{-1})^{-1}(\widehat{\boldsymbol{\beta}} - \boldsymbol{\mu})\,,$$

then we obtain $\widehat{\sigma}^2 = 0.3946$ and

$$\widehat{\boldsymbol{\beta}}^{\mathrm{B}} = \begin{pmatrix} 0.8394 \\ 0.7585 \end{pmatrix}.$$

The observed squared error loss is $L(\boldsymbol{\beta}, \widehat{\boldsymbol{\beta}}^{\mathrm{B}}) = 0.4289$ for the true parameter $\boldsymbol{\beta} = (0.2, 0.9)'$. As in Example 3.27, incorporating prior knowledge via the linear Bayes estimator yields an improvement upon the ordinary least squares estimator. □

3.9 Robust Estimator

In this section we turn our attention to alternatives to the ordinary least squares estimator, meant to yield better estimates for $\boldsymbol{\beta}$ when outliers are present.

3.9.1 Preliminary Considerations

In the previous sections we have considered alternative estimators for $\boldsymbol{\beta}$, which can yield better estimates than the ordinary least squares estimator $\widehat{\boldsymbol{\beta}}$ in presence of collinearity, or when additional prior information about the parameters is available.

A further situation when the ordinary least squares can produce unsatisfactory results is the presence of outliers being caused by (unobservable) realizations of individual errors ε_i, which are quite far away from the majority of the other errors. To have a practical identification method, an outlier may also be defined as an observation with a corresponding large ordinary least squares residual, see Definition 6.1.

Example 3.29. Consider the linear regression model

$$y_i = \beta_1 + \beta_2 x_i + \varepsilon_i, \quad i = 1, \ldots, n,$$

with $n = 12$ observations given in Table 3.4.

Table 3.4. Observations in Example 3.29

i	1	2	3	4	5	6	7	8	9	10	11	12
y_i	4.0	4.5	5.1	5.2	6.0	5.3	6.3	7.5	6.0	5.8	6.9	7.0
x_i	2.9	3.6	3.9	3.7	4.4	4.1	4.7	3.7	4.5	5.1	5.7	5.5

If we plot the points (x_i, y_i), then we can see that one of these data points seems to be an outlier, see Fig. 3.44.

The ordinary least squares estimate is $\widehat{\boldsymbol{\beta}} = (2.1785, 0.8389)'$. Since the data is simulated, we know the true parameter vector given as $\boldsymbol{\beta} = (0.5, 1.2)'$. Obviously, the ordinary least squares estimate is rather far away from the true value. If we plot the line $y = \beta_1 + \beta_2 x$ for $\beta_1 = 0.5$ and $\beta_2 = 1.2$ (undisturbed relationship between x and y), as well as $y = \widehat{\beta}_1 + \widehat{\beta}_2 x$ (estimated relationship between x and y), see Fig. 3.45, then we may conclude:

- The aforementioned data point is rather far away from both, the undisturbed and the estimated relationship between x and y. Hence, it may in fact be qualified as an outlier (with respect to the corresponding true error as well as with respect to the corresponding ordinary least squares residual).
- The estimated relationship would have been much closer to the undisturbed relationship if this data point had been absent. The outlier has thus a strong influence on the ordinary least squares fit.

If eventually we compute the observed unweighted squared error loss, then we obtain $L(\boldsymbol{\beta}, \widehat{\boldsymbol{\beta}}) = (\widehat{\boldsymbol{\beta}} - \boldsymbol{\beta})'(\widehat{\boldsymbol{\beta}} - \boldsymbol{\beta}) = 2.9478$. □

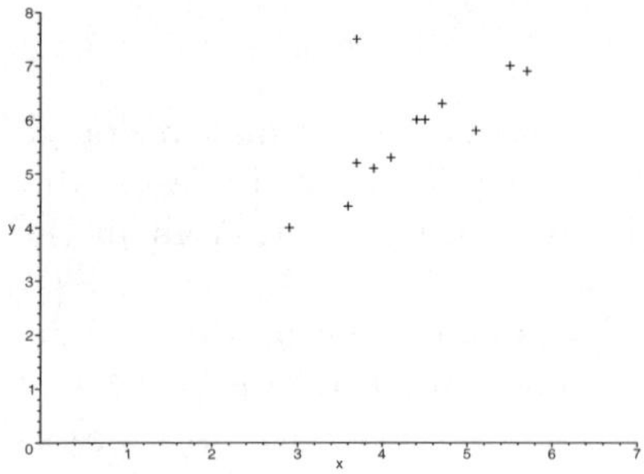

Fig. 3.44. Data points (x_i, y_i) in Example 3.29

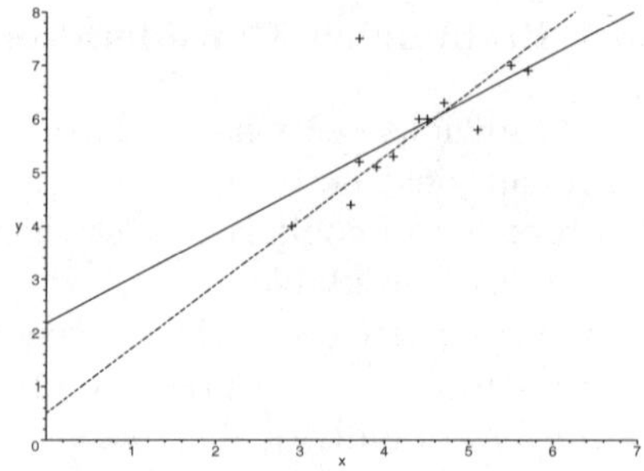

Fig. 3.45. Undisturbed relationship (dotted line) and estimated relationship (continuous line) in Example 3.29

Non-Normality

If model assumption (v) is satisfied, i.e. $\varepsilon \sim N(\mathbf{0}, \sigma^2 \boldsymbol{I}_n)$, then we have $\text{Prob}(-3\sigma \leq \varepsilon_i \leq 3\sigma) = 0.9973$ for any element ε_i of the vector ε. Therefore, extreme realizations of ε_i outside the interval $[-3\sigma, 3\sigma]$ are unlikely to occur. If model assumption (v) is not satisfied and the errors have a distribution whose tails have more probability mass than the tails of the normal distribution (*heavy-tail distribution*), then extreme individual errors are more likely to occur, producing outliers in the data.

Non-normality of the random errors is considered here as a reason for the presence of outliers in the data. These outliers can essentially be characterized as strong deviations of individual observations of the dependent variable y from the remaining observations of y (see also Definition 6.1, p. 309). They are sometimes also called *vertical outliers*. If, on the other hand, there are outliers in the data which can rather be characterized as deviations from the rest of values of the independent variables (*high-leverage points*), then these points cannot be seen as being caused by some heavy-tail distribution of the errors. For additional notes on the actual identification of outliers and high-leverage points, see Sects. 6.4 and 6.5.

Now, what are the implications of a distribution with heavier tails than the normal distribution? The following example illustrates this point.

Example 3.30. Let us consider a random variable X with cumulative distribution function

$$F_X(x) = (1-\alpha)\Phi\left(\frac{x-\mu}{\tau}\right) + \alpha\Phi\left(\frac{x-\mu}{c\,\tau}\right), \quad 0 \leq \alpha \leq 1, \quad 0 < c, \quad 0 < \tau,$$

where

$$\Phi(x) = \frac{1}{\sqrt{2\pi}} \int_{-\infty}^{x} \mathrm{e}^{-y^2/2}\, dy$$

is the cumulative distribution function of the standard normal distribution. It can be shown that the identities $\mathrm{E}(X) = \mu$ and $\mathrm{Var}(X) = (1-\alpha)\tau^2 + \alpha c^2 \tau^2$ are true.

We now choose $\mu = 0$, $\tau = 1$, $\alpha = 0.2$, and $c = 5$. Then

$$\sigma^2 := \mathrm{Var}(X) = (1-\alpha)\tau^2 + \alpha c^2 \tau^2 = 5.8 \, .$$

Fig. 3.46 shows the density of X compared to the density of a $N(0, \sigma^2)$ distributed random variable Y.

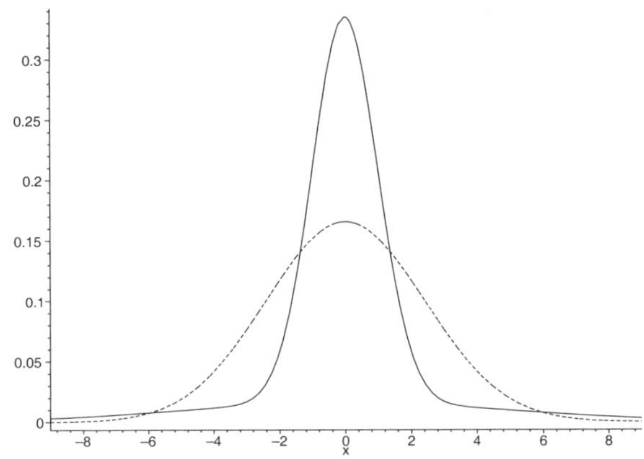

Fig. 3.46. Density of $X \sim$ '$0.8\Phi(x) + 0.2\Phi(x/5)$' (continuous graph) compared to the density of $Y \sim N(0, \sigma^2 = \mathrm{Var}(X))$ (dotted graph)

How will an actual sample of n independent '$0.8\Phi(x) + 0.2\Phi(x/5)$' distributed random variables look like, compared to a sample of n independent $N(0, \sigma^2)$ distributed random variables? From Fig. 3.46 it can be deduced that under the '$0.8\Phi(x)+0.2\Phi(x/5)$' distribution there will occur more realizations close to 0 than under $N(0, \sigma^2)$, while on the other hand, some realizations will be farther away from the majority of realizations.

For the 1σ intervals of X and Y we have $P(-\sigma \leq X \leq \sigma) = 0.8612$ and $P(-\sigma \leq Y \leq \sigma) = 0.6827$ ('more realizations close to 0'), while for the 3σ interval we have $P(-3\sigma \leq X \leq 3\sigma) = 0.9703$ and $P(-3\sigma \leq Y \leq 3\sigma) = 0.9973$ ('some realizations farther away from the majority').

Figures 3.47 and 3.48 respectively show $n = 50$ data points (x_i, y_i), observed in the linear regression model

$$y_i = \beta_1 + \beta_2 x_i + \varepsilon_i, \quad i = 1, \ldots, n \, .$$

In Fig. 3.47 the errors are independent $N(0, \sigma^2)$ distributed (where $\sigma^2 = 5.8$), while in Fig. 3.48 the errors are independent '$0.8\Phi(x)+0.2\Phi(x/5)$' distributed. In addition, the true relationship between the variables y and x is plotted as a dotted straight line, respectively. The described behavior of the '$0.8\Phi(x) + 0.2\Phi(x/5)$' distribution compared to the behavior of the $N(0, \sigma^2)$ distribution is resembled by the data points here. □

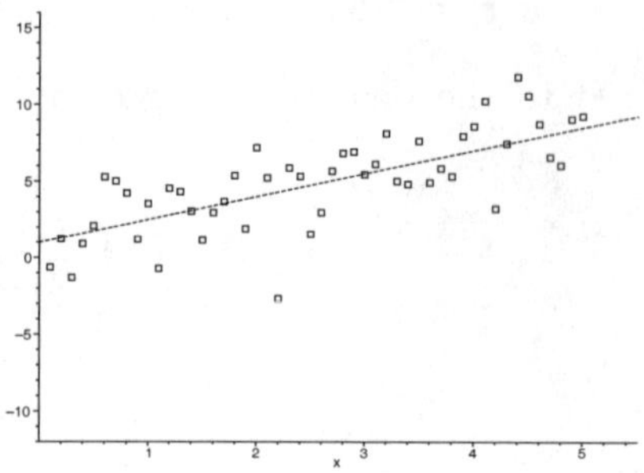

Fig. 3.47. 50 data points in a linear regression model with independent $N(0, \sigma^2)$ distributed errors

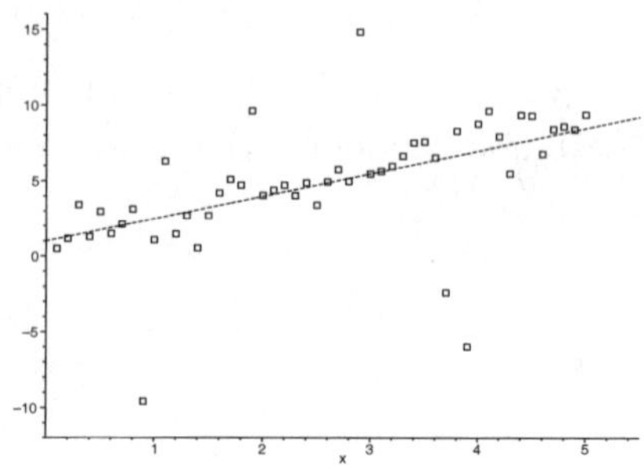

Fig. 3.48. 50 data points in a linear regression model with independent '$0.8\Phi(x) + 0.2\Phi(x/5)$' distributed errors

Quite often the Laplace distribution (also called double exponential distribution) with density

$$f(x) = \frac{1}{2\beta} e^{-\frac{|x-\alpha|}{\beta}}, \qquad -\infty < \alpha < \infty, \quad 0 < \beta,$$

is cited as an example for a possible distribution of the errors ε_i which can produce outliers. Fig. 3.49 shows the density of the Laplace distribution compared to the densities from Fig. 3.46. Here, $\alpha = 0$ and $\beta = \sqrt{\sigma^2/2}$, where $\sigma^2 = 5.8$ as above. The expectation of random variable with such an Laplace distribution is therefore $\alpha = 0$ and its variance is $2\beta^2 = \sigma^2 = 5.8$.

3.9.2 Weighted Least Squares Estimation

An easy way to deal with outliers removes them from the data set before carrying out the analysis. This is, however, quite often not a justified procedure, since outliers, though extreme in a certain sense, are not necessarily useless. As a matter of fact, they can provide important information for the analysis of a linear regression model. Therefore, instead of simply neglecting the outlier and thus weighting it as zero, it could be better to find some method which can rate an outlier and then give an appropriate weight to it.

For the estimation of the vector $\boldsymbol{\beta}$, the application of a weighted least squares estimator

$$\widehat{\boldsymbol{\beta}}_{\boldsymbol{W}} = (\boldsymbol{X}'\boldsymbol{W}\boldsymbol{X})^{-1}\boldsymbol{X}'\boldsymbol{W}\boldsymbol{y}$$

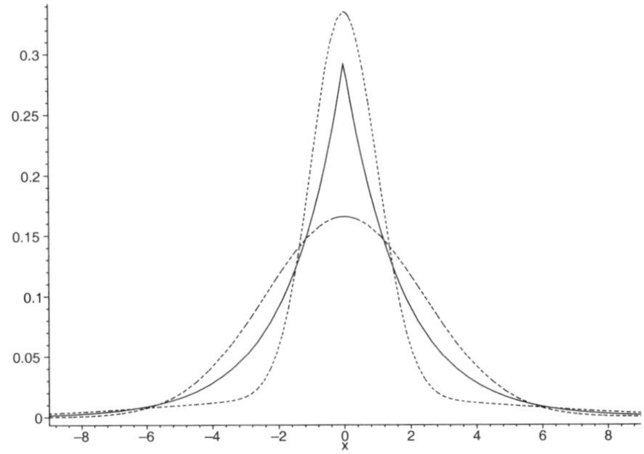

Fig. 3.49. Densities of $X \sim$ '$0.8\Phi(x) + 0.2\Phi(x/5)$' and $Y \sim N(0, \sigma^2 = \text{Var}(X))$ (dotted graphs) compared to the density of a Laplace($\alpha = 0$, $\beta = \sqrt{\text{Var}(X)/2}$) distributed random variable (continuous graph)

instead of the ordinary least squares estimator can be the basis for a such a method. The matrix $\boldsymbol{W}$ is diagonal with diagonal elements w_{ii} depending on the i-th element $\widehat{\varepsilon}_i$ of the vector of least squares residuals $\widehat{\boldsymbol{\varepsilon}} = \boldsymbol{y} - \boldsymbol{X}\widehat{\boldsymbol{\beta}}$. To be more precise, the weights are computed as

$$w_{ii} = \begin{cases} \dfrac{\psi(\widehat{\varepsilon}_i/\widehat{\sigma}_{\text{rob}})}{\widehat{\varepsilon}_i/\widehat{\sigma}_{\text{rob}}} & \text{for } \widehat{\varepsilon}_i \neq 0 \\ 1 & \text{for } \widehat{\varepsilon}_i = 0 \end{cases} .$$

Here, $\widehat{\sigma}_{\text{rob}}$ is a robust estimator for σ (not for σ^2) and $\psi(\cdot)$ is a function yielding a value with the same sign as $\widehat{\varepsilon}_i/\widehat{\sigma}_{\text{rob}}$, but with an absolute value not greater but only smaller than $\widehat{\varepsilon}_i/\widehat{\sigma}_{\text{rob}}$. This ensures that the weights satisfy $0 \leq w_{ii} \leq 1$.

An example for such a function $\psi(\cdot)$ is Huber's t-function, see [60], given by

$$\psi(r) = \begin{cases} r & \text{for } |r| \leq t \\ t\,\text{sign}(r) & \text{for } |r| > t \end{cases} .$$

If this function is applied to some $\widehat{\varepsilon}_i/\widehat{\sigma}_{\text{rob}}$ which has an absolute value much greater than a given number $t > 0$, then the computed weight w_{ii} is rather small, where, however, the case $w_{ii} = 0$ can be excluded. If on the other hand the absolute value of $\widehat{\varepsilon}_i/\widehat{\sigma}_{\text{rob}}$ is not greater than t, then $w_{ii} = 1$.

The so determined weight matrix $\boldsymbol{W}$ can assumed to be positive definite, since none of its diagonal elements is exactly zero. Then we can consider the minimization of the function

$$f_{\boldsymbol{W}}(\boldsymbol{\beta}_*) = (\boldsymbol{y} - \boldsymbol{X}\boldsymbol{\beta}_*)'\boldsymbol{W}(\boldsymbol{y} - \boldsymbol{X}\boldsymbol{\beta}_*)$$

with respect to $\boldsymbol{\beta}_*$, yielding $\min_{\boldsymbol{\beta}_*} f_{\boldsymbol{W}}(\boldsymbol{\beta}_*) = f_{\boldsymbol{W}}(\widehat{\boldsymbol{\beta}}_{\boldsymbol{W}})$, where

$$\widehat{\boldsymbol{\beta}}_{\boldsymbol{W}} = (\boldsymbol{X}'\boldsymbol{W}\boldsymbol{X})^{-1}\boldsymbol{X}'\boldsymbol{W}\boldsymbol{y} \,,$$

see also Problem 2.13. It should be noted that the diagonal elements of the matrix $\boldsymbol{W}$ are stochastic, so that a more precise notation is $\boldsymbol{W}(\boldsymbol{y})$ instead of $\boldsymbol{W}$. The above weighted least squares estimator $\widehat{\boldsymbol{\beta}}_{\boldsymbol{W}(\boldsymbol{y})}$ is therefore a non-linear estimator whose exact statistical properties are difficult to determine. Even the derivation of expectation of $\widehat{\boldsymbol{\beta}}_{\boldsymbol{W}(\boldsymbol{y})}$ appears to be complicated.

Remark 3.6. A weighted least squares estimator with a robust determination of the weights is a *non-linear* estimator of $\boldsymbol{\beta}$.

Example 3.31. We can compute a weighted least squares estimator of $\boldsymbol{\beta}$ for the data from Example 3.29. To start with, we need a robust estimate $\widehat{\sigma}_{\mathrm{rob}}$ for σ. As in [123, p. 324] and [4, p. 335] we employ

$$\widehat{\sigma}_{\mathrm{rob}} = \frac{1}{0.6745}\,\mathrm{med}(|\widehat{\varepsilon}_i - \mathrm{med}(\widehat{\varepsilon}_i)|) \,,$$

where $\widehat{\varepsilon}_i$ denotes the i-th element of the residual vector $\widehat{\boldsymbol{\varepsilon}} = \boldsymbol{y} - \boldsymbol{X}\widehat{\boldsymbol{\beta}}$, and med$(\cdot)$ denotes the median of a set of values. The computed estimate for the given data is $\widehat{\sigma}_{\mathrm{rob}} = 0.3920$. As a next step, we compute the weights $w_{ii} = \psi(\widehat{\varepsilon}_i/\widehat{\sigma}_{\mathrm{rob}})/(\widehat{\varepsilon}_i/\widehat{\sigma}_{\mathrm{rob}})$, where $\psi(\cdot)$ stands for Huber's t-function with $t = 1.345$, see also [123]. The computed weights w_{ii} are given in Table 3.5.

Table 3.5. Computed weights based on Huber's t-function with $t = 1.345$ in Example 3.31

i	1	2	3	4	5	6	7	8	9	10	11	12
w_{ii}	0.8623	0.7546	1.0	1.0	1.0	1.0	1.0	0.2378	1.0	0.8023	1.0	1.0

The smallest weight is $w_{88} = 0.2378$. The data point (x_8, y_8) is the one which we have described as a presumed outlier in Example 3.29.

As the last step, we can now compute $\widehat{\boldsymbol{\beta}}_{\boldsymbol{W}}$, yielding

$$\widehat{\boldsymbol{\beta}}_{\boldsymbol{W}} = (1.4172, 0.9848)'$$

with an observed unweighted squared error loss of $L(\boldsymbol{\beta}, \widehat{\boldsymbol{\beta}}_{\boldsymbol{W}}) = 0.8875$ for the true $\boldsymbol{\beta} = (0.5, 1.2)'$. Fig. 3.50 shows the estimated relationship between y and x compared to the undisturbed relationship. □

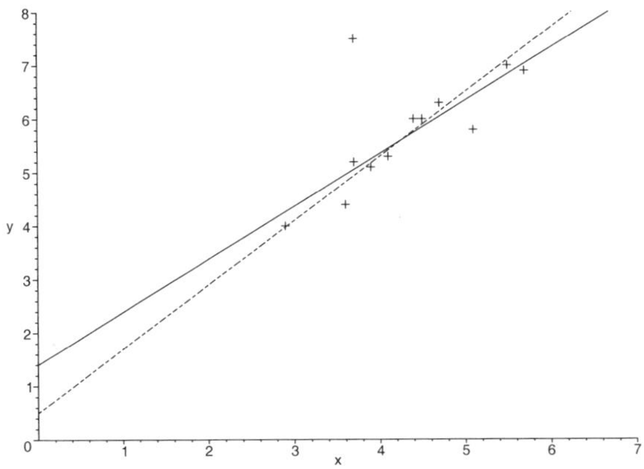

Fig. 3.50. Estimated relationship (continuous line) via the weighted least squares estimator and undisturbed relationship (dotted line) in Example 3.31

Robust Estimation of σ

In the above Example 3.31 we estimate σ by the median of the absolute deviations of the least squares residuals from their median (MADM) divided by 0.6745. As an alternative robust estimator for σ, Birkes and Dodge [16, p. 79] apply $\widehat{\sigma}_{\text{rob}} = \text{MAD}/0.6745$, where MAD is the median of the absolute values of the least squares residuals which are different from 0. In the above Example 3.31 both estimators yield almost the same estimate.

The application of these two estimators can be justified by their behavior under normality. If $Z \sim N(0, \sigma^2)$, then

$$\text{med}(|Z - \text{med}(Z)|) = \text{med}(|Z|) = \sigma\, u_{0.75},$$

where $u_{0.75} = 0.6745$ is the 0.75- quantile of the standard normal distribution. When $z_1, \ldots, z_n$ are realizations of n independent $N(0, \sigma^2)$ distributed random variables, then $\text{med}(|z_i - \text{med}(z_i)|)/u_{0.75}$ as well as $\text{med}(|z_i|)/u_{0.75}$ are likely to be close to σ for large n.

The application of the estimator $\widehat{\sigma}_{\text{rob}} = \text{MAD}/0.6745$ is especially reasonable when it is already known that the elements of the vector ε are *symmetrically* distributed around 0.

Note that when $\widehat{\sigma}_{\text{rob}} = \text{MADM}/0.6745$ is used for the estimation of σ, it is sometimes referred as 'MAD' instead of 'MADM' in the literature.

Iteratively Reweighted Least Squares Estimator

The above described estimation process can be iterated by using the residuals of the weighted least squares estimator $\widehat{\varepsilon}_{\boldsymbol{W}} = \boldsymbol{y} - \boldsymbol{X}\widehat{\boldsymbol{\beta}}_{\boldsymbol{W}}$ to compute a new estimate $\widehat{\sigma}_{\text{rob}}$ and new weights w_{ii}. Then the residuals of the so obtained weighted least squares estimator can again be used to compute new estimates and so on. The resulting estimator is then called *iteratively reweighted least squares* (IRLS) estimator. Askin and Montgomery [4] compute IRLS estimates without recomputing $\widehat{\sigma}_{\text{rob}}$ in each step.

Different Weights

The actual weights w_{ii} in $\widehat{\boldsymbol{\beta}}_{\boldsymbol{W}}$ depend on the choice of the estimator $\widehat{\sigma}_{\text{rob}}$ as well as on the function $\psi(\cdot)$. Further possible choices for $\psi(\cdot)$ are for example Hampel's (a, b, c)-function

$$\psi(r) = \begin{cases} r & \text{for } |r| \le a \\ a\,\text{sign}(r) & \text{for } a < |r| \le b \\ a\,\text{sign}(r)(\dfrac{c-|r|}{c-b}) & \text{for } b < |r| \le c \\ 0 & \text{for } |r| > c \end{cases},$$

Ramsay's E_a-function

$$\psi(r) = \mathrm{e}^{-a|r|}r\,,$$

or Andrews' $\sin(a)$-function

$$\psi(r) = \begin{cases} \sin(r/a) & \text{for } |r| \le \pi a \\ 0 & \text{for } |r| > \pi a \end{cases},$$

compare [3] and [87]. If one of the computed weights w_{ii} is exactly 0, then the weighted least squares estimator $\widehat{\boldsymbol{\beta}}_{\boldsymbol{W}} = (\boldsymbol{X}'\boldsymbol{W}\boldsymbol{X})^{-1}\boldsymbol{X}'\boldsymbol{W}\boldsymbol{y}$ does not necessarily exist, since in that case the matrix $\boldsymbol{X}'\boldsymbol{W}\boldsymbol{X}$ can (but must not) be singular .

Each of the above functions $\psi(\cdot)$ depends on one or more parameters being differently chosen in the literature, and then yielding different estimates. The t in Huber's t-function is often $t = 1.345$, but also $t = 1.5$ and $t = 2.0$ are proposed. The usual choice in Hampel's (a, b, c)-function is $a = 1.7$, $b = 3.4$, and $c = 8.5$. Ramsay's E_a-function is often applied with $a = 0.3$, while some common choice in Andrews' $\sin(a)$-function is $a = 1.0$ or $a = 1.339$. Fig. 3.51 displays the four functions for given parameters.

Example 3.32. In Example 3.31 a weighted least squares estimator for $\boldsymbol{\beta}$ based on Huber's t-function for $t = 1.345$ is computed.

We can compute a further weighted least squares estimator based on Andrews' $\sin(a)$-function with $a = 1.0$. As an estimator for σ we stick to the same $\widehat{\sigma}_{\text{rob}}$ as before with an actual value of $\widehat{\sigma}_{\text{rob}} = 0.3920$. The new weights are given in Table 3.6.

In this case the weight w_{88} is exactly 0. Nonetheless the weighted least squares estimator $\widehat{\boldsymbol{\beta}}_{\boldsymbol{W}} = (\boldsymbol{X}'\boldsymbol{W}\boldsymbol{X})^{-1}\boldsymbol{X}'\boldsymbol{W}\boldsymbol{y}$ can be computed and yields an estimate of $\widehat{\boldsymbol{\beta}}_{\boldsymbol{W}} = (1.1441, 1.0398)'$ with an observed unweighted squared

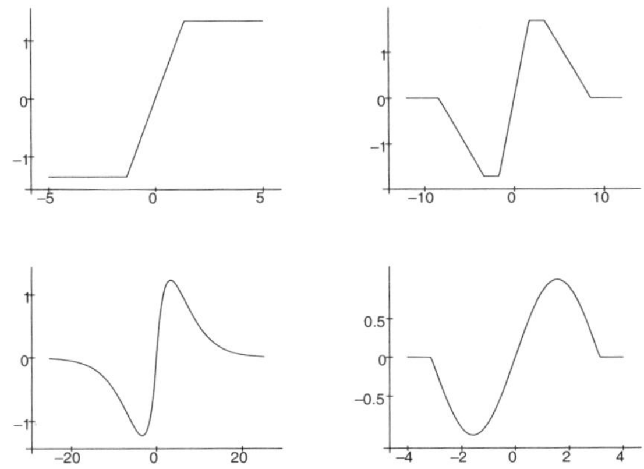

Fig. 3.51. Huber's 1.345-, Hampel's (1.7, 3.4, 8.5)-, Ramsey's $E_{0.3}$- and Andrews' sin(1.0)-function

error loss of $L(\boldsymbol{\beta}, \widehat{\boldsymbol{\beta}}_{\boldsymbol{W}}) = 0.4405$ for the true $\boldsymbol{\beta} = (0.5, 1.2)'$. This estimate almost coincides with the ordinary least squares estimate, being obtained when x_8 and y_8 are deleted from the data set, see Problem 3.44. □

Table 3.6. Computed weights based on Andrews' sin(a)-function with $a = 1.0$ in Example 3.32

i	1	2	3	4	5	6	7	8	9	10	11	12
w_{ii}	.6411	.5485	.8720	.9926	.9817	.8937	.9658	.0000	.9977	.5932	.9960	.9541

3.9.3 The l_1 Estimator

The least squares approach aims at minimizing the function

$$f(\boldsymbol{\beta}_*) = \|\boldsymbol{y} - \boldsymbol{X}\boldsymbol{\beta}_*\|_2^2 = (\boldsymbol{y} - \boldsymbol{X}\boldsymbol{\beta}_*)'(\boldsymbol{y} - \boldsymbol{X}\boldsymbol{\beta}_*) = \sum_{i=1}^{n}(y_i - \boldsymbol{x}'_{[i]}\boldsymbol{\beta}_*)^2 ,$$

where $\boldsymbol{x}'_{[i]}$ is the i-th row of $\boldsymbol{X}$ and $\|\boldsymbol{x}\|_2 = (\sum_{i=1}^n x_i^2)^{1/2}$ is the usual Euclidian norm (also called l_2-norm). The function $f(\boldsymbol{\beta}_*)$ attains its minimum at

$$\widehat{\boldsymbol{\beta}} = (\boldsymbol{X}'\boldsymbol{X})^{-1}\boldsymbol{X}'\boldsymbol{y} ,$$

meaning that $\min_{\boldsymbol{\beta}_*} f(\boldsymbol{\beta}_*) = f(\widehat{\boldsymbol{\beta}})$.

In contrast to this, the *least absolute deviations* (LAD) approach considers $\|\cdot\|_1$ instead of $\|\cdot\|_2^2$, where $\|\boldsymbol{x}\|_1 = \sum_{i=1}^n |x_i|$ denotes the l_1-norm. It is desired to find a vector $\widehat{\boldsymbol{\beta}}_{\mathrm{LAD}}$, such that the function

$$f(\boldsymbol{\beta}_*) = \|\boldsymbol{y} - \boldsymbol{X}\boldsymbol{\beta}_*\|_1 = \sum_{i=1}^{n} |y_i - (\boldsymbol{X}\boldsymbol{\beta}_*)_i|$$

attains its minimum at $\widehat{\boldsymbol{\beta}}_{\text{LAD}}$. Such a vector will be called l_1 estimator for $\boldsymbol{\beta}$. The method of least absolute deviations is older and not more difficult to propose than the method of least squares. However, it is more difficult to obtain a solution $\widehat{\boldsymbol{\beta}}_{\text{LAD}}$ to the minimization problem. As a matter fact, no explicit formula can be derived, see also Chap. 4 in [16].

Robustness

Let us consider $\widetilde{\varepsilon}_i = y_i - \boldsymbol{x}'_{[i]}\widetilde{\boldsymbol{\beta}}$, the i-th residual of an estimate $\widetilde{\boldsymbol{\beta}}$. If an individual $\widetilde{\varepsilon}_i$ has an absolute value which is relatively large compared to the absolute values of the other residuals, then this difference will even be greater when squared residuals are considered. Since the ordinary least squares estimator $\widehat{\boldsymbol{\beta}}$ minimizes the sum of squared residuals, the resulting estimate will often avoid large squared residuals $\widehat{\varepsilon}_i^2$. But this also implies that an individual outlier, which would have caused a large residual $\widetilde{\varepsilon}_i^2$ if some other estimator $\widetilde{\boldsymbol{\beta}}$ had been applied, strongly influences the ordinary least squares estimate. As a matter of fact it can often be observed that the ordinary least squares estimate moves into the direction of the outlier, which is not surprising since this decreases the value of corresponding squared residual. (In such cases the minimization of the *sum* of squared residuals strongly depends on the order of magnitude of *individual* residuals.)

When the l_1 estimator $\widehat{\boldsymbol{\beta}}_{\text{LAD}}$ instead of the ordinary least squares estimator $\widehat{\boldsymbol{\beta}}$ is applied, then the above described reaction on outliers will be less pronounced, since not the squares of the residuals but only the absolute values are considered. In view of this, the l_1 estimator is less liable to outliers and thus it is more robust than the ordinary least squares estimator.

This robustness property holds true with respect to vertical outliers, which may essentially be characterized as strong deviations of individual observations in the dependent variable y. The robustness property of the l_1 estimator does not hold true with respect to high-leverage points (strong deviations of individual observations in the independent variables), see [92, p. 10 ff] or [30]. The following example illustrates this point.

Example 3.33. Consider the linear regression model

$$y_i = \beta_1 + \beta_2 x_i + \varepsilon_i, \quad i = 1, \ldots, n\,,$$

with $n = 9$ observations given in Table 3.7.

Table 3.7 provides the same data as Example 2.1, p. 38, with the only distinction that instead of $x_4 = 4.148$ the observation here is $x_4 = 34.15$. Thus, the fourth observation can be identified as an outlier in the direction of the independent variable. The l_1 estimator yields $\widehat{\boldsymbol{\beta}}_{\text{LAD}} = (4.3664, -0.0152)'$,

Table 3.7. Observations in Example 3.33

i	1	2	3	4	5	6	7	8	9
y_i	3.545	3.307	12.48	3.848	9.796	2.619	4.284	22.20	15.02
x_i	3.575	7.153	13.58	34.15	3.141	6.830	5.430	13.48	10.50

which is obviously strongly influenced by the fourth observation, see Fig. 3.52. □

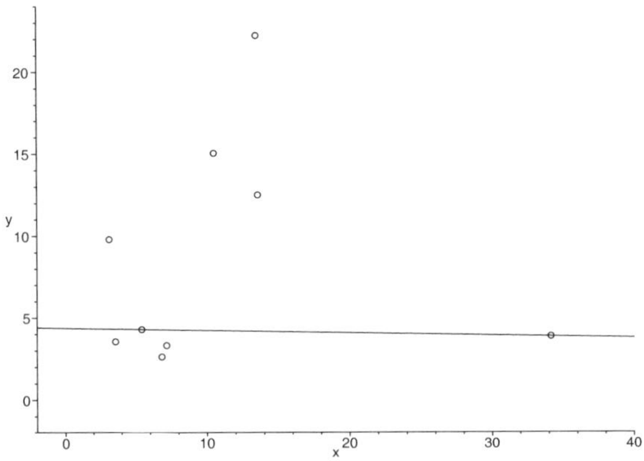

Fig. 3.52. Data points (x_i, y_i) with high-leverage point (x_4, y_4) and corresponding l_1 estimation fit

Properties of the l_1 Estimator

In the following, we shortly discuss some of the properties of the l_1 estimator.

Computation. As noted before, we cannot derive an explicit formula for $\widehat{\beta}_{\text{LAD}}$. The computation of $\widehat{\beta}_{\text{LAD}}$ can be carried out via the methods of linear programming. Most of the statistical software packages offer the possibility for l_1 regression, e.g. S-PLUS (function `l1fit`).

Non-Uniqueness. The least absolute deviations approach does not guarantee a unique solution $\widehat{\beta}_{\text{LAD}}$. The following simple example is given by Birkes and Dodge [16, p. 80].

Example 3.34. Consider the three points $(1, 1)$, $(2, 1)$, and $(2, 2)$. It is desired to draw a line $y = (\widehat{\beta}_{\text{LAD}})_1 + (\widehat{\beta}_{\text{LAD}})_2 x$ through the 'cloud' of the three points (x_i, y_i), $i = 1, 2, 3$, such that

$$\sum_{i=1}^{3} |y_i - (\widehat{\beta}_{\text{LAD}})_1 - (\widehat{\beta}_{\text{LAD}})_2 x_i| = \min_{\beta_{*1},\beta_{*2}} \sum_{i=1}^{3} |y_i - \beta_{*1} - \beta_{*2} x_i|$$

is satisfied. This means that the line is chosen in such a way that the sum of vertical absolute deviations of y_i from the corresponding values on the line is minimal, compared to the sum of absolute deviations caused by any other line (LAD approach). When we plot the three points into one system of coordinates, then we can see that any line which covers the point $(1,1)$ and in addition an arbitrary point on the line between $(2,1)$ and $(2,2)$ is a LAD solution in the above sense, see Fig. 3.53. Hence, the LAD approach does not yield a unique solution here, we have $0 \leq (\widehat{\beta}_{\text{LAD}})_1 \leq 1$ and $(\widehat{\beta}_{\text{LAD}})_2 = 1 - (\widehat{\beta}_{\text{LAD}})_1$. □

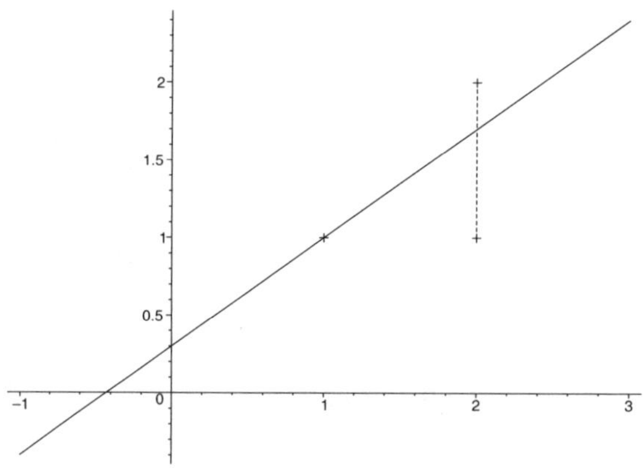

Fig. 3.53. l_1 estimation fit in Example 3.34

LAD Equation and *p* Data Points. Consider the linear regression model *with* intercept, described by

$$\boldsymbol{y} = \boldsymbol{X}\boldsymbol{\beta} + \varepsilon, \quad \varepsilon \sim (\boldsymbol{0}, \sigma^2 \boldsymbol{I}_n)\ .$$

Since each element of the first column of the matrix $\boldsymbol{X}$ equals 1, the actual observation matrix consists of the n rows

$$(x_{i2}, \ldots, x_{ip}, y_i), \quad i = 1, \ldots, n\ ,$$

each of them being a p-dimensional observation.

For the derivation of a solution $\widehat{\beta}_{\text{LAD}}$, we can always assume that at least p out of n of these p-dimensional observations exactly lie on the hyperplane which is generated by the solution, see [16, Chap. 4]. For an illustration of the notion of a hyperplane, let us consider the vector

$$\boldsymbol{v}'_* = (1, v_2, \ldots, v_p) \,.$$

Then the set of p-dimensional points

$$\mathcal{H}_{\text{LAD}} = \{(v_2, \ldots, v_p, u) : u = \boldsymbol{v}'_* \widehat{\boldsymbol{\beta}}_{\text{LAD}}\}$$

is called the hyperplane corresponding to $\widehat{\boldsymbol{\beta}}_{\text{LAD}}$. When $p = 2$ the hyperplane is a line, while in case $p = 3$ the hyperplane is a plane in $\mathbb{R}^3$. Then,

$$(x_{i2}, \ldots, x_{ip}, y_i) \in \mathcal{H}_{\text{LAD}}$$

holds true for at least p of the given p-dimensional observations.

If we consider the vector $\widehat{\boldsymbol{y}}_{\text{LAD}} = \boldsymbol{X}\widehat{\boldsymbol{\beta}}_{\text{LAD}}$, then at least p elements of $\widehat{\boldsymbol{y}}_{\text{LAD}}$ and $\boldsymbol{y}$ are identical.

Example 3.35. We compute the l_1 estimator of $\boldsymbol{\beta}$ for the data from Example 3.29, p. 189. The estimator is computed as $\widehat{\boldsymbol{\beta}}_{\text{LAD}} = (0.6538, 1.1538)'$ via the S-PLUS function `l1fit`. The observed unweighted squared error loss is $L(\boldsymbol{\beta}, \widehat{\boldsymbol{\beta}}_{\text{LAD}}) = 0.0258$ for the true $\boldsymbol{\beta} = (0.5, 1.2)'$. The hyperplane $\mathcal{H}_{\text{LAD}}$ corresponding to $\widehat{\boldsymbol{\beta}}_{\text{LAD}}$ is the set of all points (x, y) with $y = 0.6538 + 1.1538x$, i.e. the line which describes the estimated relationship between the variables y and x, similar to the ordinary least squares and the weighted least squares regression lines computed in Examples 3.29 and 3.31, respectively. Fig. 3.54 shows the line, which exactly covers the $p = 2$ data points (x_1, y_1) and (x_{12}, y_{12}). □

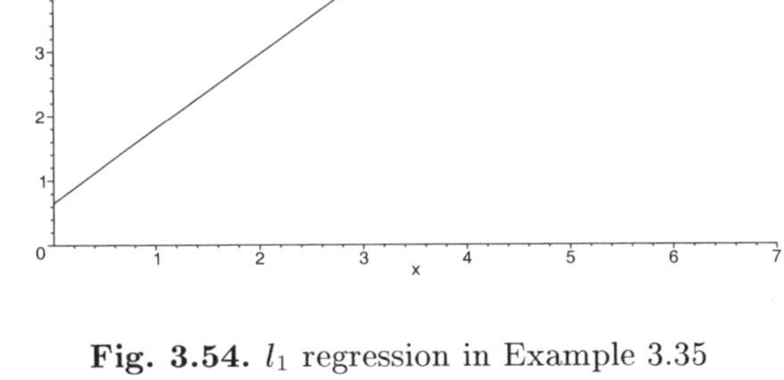

Fig. 3.54. l_1 regression in Example 3.35

3.9.4 M Estimator

The least squares estimator $\widehat{\boldsymbol{\beta}}$ is the solution to the minimization problem

$$\min_{\boldsymbol{\beta}_*} \sum_{i=1}^{n} \varepsilon_{*i}^2, \quad \varepsilon_{*i} = y_i - \boldsymbol{x}'_{[i]}\boldsymbol{\beta}_* ,$$

while the l_1 estimator $\widehat{\boldsymbol{\beta}}_{\text{LAD}}$ is a solution to $\min_{\boldsymbol{\beta}_*} \sum_{i=1}^{n} |\varepsilon_{*i}|$. Instead of considering the two approaches separately, one can take an appropriate combination into account, as provided by the minimization problem

$$\min_{\boldsymbol{\beta}_*} \sum_{i=1}^{n} \varrho(\varepsilon_{*i}) .$$

The function $\varrho(r)$ can be chosen such that it is smooth and in addition 'essentially' coincides with r^2 when the absolute value of r is smaller than a given number t while it 'essentially' coincides with $|r|$ otherwise. (The term 'essentially' means that the minimization of the function is then equivalent to the minimization of r^2 and $|r|$, respectively.) A function which satisfies such a requirement is

$$\varrho(r) = \begin{cases} r^2/2 & \text{for } |r| \le t \\ t|r| - t^2/2 & \text{for } |r| > t \end{cases}$$

due to Huber [60], see also Fig. 3.55.

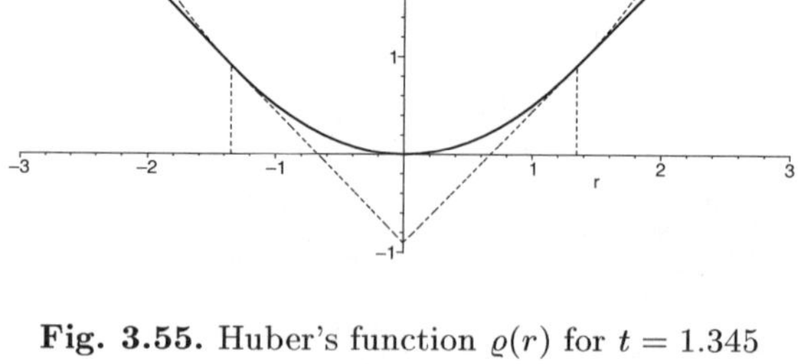

Fig. 3.55. Huber's function $\varrho(r)$ for $t = 1.345$

The derivative of $\varrho(r)$ is exactly Huber's t-function as considered before. The corresponding IRLS estimator can be regarded as an algorithm for the (approximate) solution to the problem $\min_{\boldsymbol{\beta}_*} \sum_{i=1}^{n} \varrho(\varepsilon_{*i})$.

For any of the other functions $\psi(\cdot)$, which we have considered before, it is also possible to find a corresponding function $\varrho(\cdot)$. The solution $\widehat{\boldsymbol{\beta}}_{\varrho(\cdot)}$ to the minimization problem $\min_{\boldsymbol{\beta}_*} \sum_{i=1}^n \varrho(\varepsilon_{*i})$ for some given appropriate function $\varrho(\cdot)$ is also called *M estimator* for $\boldsymbol{\beta}$. In this sense $\widehat{\boldsymbol{\beta}}$ and $\widehat{\boldsymbol{\beta}}_{\mathrm{LAD}}$ are also specific M estimators with $\varrho(r) = r^2$ and $\varrho(r) = |r|$, respectively.

The letter 'M' refers to Maximum Likelihood estimation. Some (but not all) M estimators are also Maximum Likelihood estimators under a specific distribution of the error vector $\boldsymbol{\varepsilon}$. Birkes and Dodge [16, p. 99] note that this property is of less importance, since it is aimed to find robust estimators for a wide range of distributions of $\boldsymbol{\varepsilon}$.

Similar to the l_1 estimator, also M estimators are rather unreliable with respect to high-leverage points. If evidence of such data points is given, then often generalized M estimators are applied, which are designed to reweight the influence of outliers in the direction of the independent variables, see e.g. [104].

3.9.5 Robust Ridge Estimator

If a given data set comprises the presence of collinearity and in addition the presence of outliers, then the combination of ridge and robust methods can be an appropriate procedure to deal with both problems simultaneously, see also [58]. The ridge estimator

$$\widehat{\boldsymbol{\beta}}_k = (\boldsymbol{X}'\boldsymbol{X} + k\boldsymbol{I}_p)^{-1}\boldsymbol{X}'\boldsymbol{y}$$

shrinks the ordinary least squares estimator $\widehat{\boldsymbol{\beta}}$ towards the zero vector by multiplying $\widehat{\boldsymbol{\beta}}$ with the matrix $(\boldsymbol{X}'\boldsymbol{X} + k\boldsymbol{I}_p)^{-1}\boldsymbol{X}'\boldsymbol{X}$ from the left, i.e.

$$\widehat{\boldsymbol{\beta}}_k = (\boldsymbol{X}'\boldsymbol{X} + k\boldsymbol{I}_p)^{-1}\boldsymbol{X}'\boldsymbol{X}\widehat{\boldsymbol{\beta}} \, .$$

Silvapulle [102] proposes not to shrink the ordinary least squares estimator $\widehat{\boldsymbol{\beta}}$ but some robust estimator $\widehat{\boldsymbol{\beta}}^{\mathrm{rob}}$ for $\boldsymbol{\beta}$, giving a robust ridge estimator

$$\widehat{\boldsymbol{\beta}}_k^{\mathrm{rob}} = (\boldsymbol{X}'\boldsymbol{X} + k\boldsymbol{I}_p)^{-1}\boldsymbol{X}'\boldsymbol{X}\widehat{\boldsymbol{\beta}}^{\mathrm{rob}} \, .$$

This is similar to the approach by Holland [58], which implies the determination of a robust weighted least squares estimator in a first step, and then the application of ridge estimation methods in a second step. The opposite procedure is proposed by Askin and Montgomery [4], who propose the determination of the ridge parameter in a first step and then the application of a robust estimator in a second step. Simpson and Montgomery [103, p. 6] observe that the method in [4] is rather sensitive to high-leverage points, i.e. outliers which can be characterized as deviations into the direction of the independent variables rather than into the direction of the dependent variable y. This is, however, also the case for the approach [102], which aims at improved estimation in the presence of collinearity and vertical outliers.

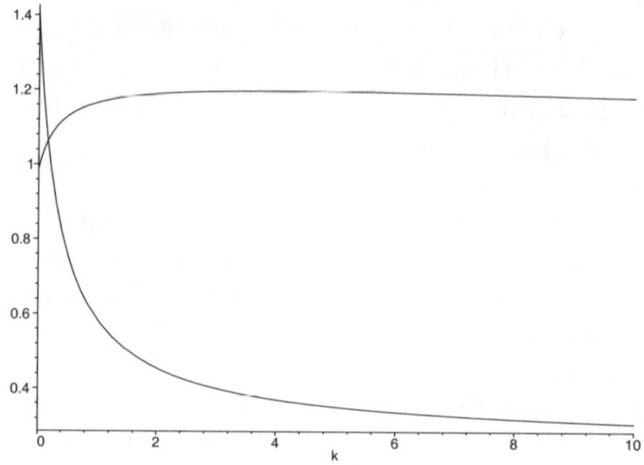

Fig. 3.56. Ridge trace of $\widehat{\boldsymbol{\beta}}_k^{\mathrm{rob}}$ in Example 3.36

Example 3.36. We can determine a robust ridge estimator $\widehat{\boldsymbol{\beta}}_k^{\mathrm{rob}}$ for the data in Example 3.29, based on the weighted least squares estimator which is computed from Huber's t-function, see Example 3.31. The robust ridge estimator is thus given as

$$\widehat{\boldsymbol{\beta}}_k^{\mathrm{rob}} = (\boldsymbol{X}'\boldsymbol{X} + k\boldsymbol{I}_2)^{-1}\boldsymbol{X}'\boldsymbol{X}\begin{pmatrix}1.4172\\0.9848\end{pmatrix},$$

and it remains to determine the ridge parameter. Silvapulle [102] proposes some estimators for the ridge parameter k, which are similar to the estimators from Sect. 3.4.3. Here, we confine to the interpretation of the ridge trace which is a plot of the elements of the vector $\widehat{\boldsymbol{\beta}}_k^{\mathrm{rob}}$ as functions of k, see Fig. 3.56.

We choose $k = 3$ and obtain $\widehat{\boldsymbol{\beta}}_3^{\mathrm{rob}} = (0.4019, 1.1967)'$ as an estimate of $\boldsymbol{\beta}$. This is very close to the true parameter vector $\boldsymbol{\beta} = (0.5, 1.2)'$, the observed unweighted squared error loss is $L(\boldsymbol{\beta}, \widehat{\boldsymbol{\beta}}_k^{\mathrm{rob}}) = 0.0096$. □

3.10 Problems

3.1. Show that the restricted least squares estimator $\widehat{\boldsymbol{\beta}}_{\mathrm{R}}$ satisfies the identity $\boldsymbol{R}\widehat{\boldsymbol{\beta}}_{\mathrm{R}} = \boldsymbol{r}$.

3.2. (See Theorem 3.1) Show that under the linear regression model with assumptions (i) to (iv)

$$f(\boldsymbol{\beta}_*) = f(\widehat{\boldsymbol{\beta}}_{\mathrm{R}}) + (\widehat{\boldsymbol{\beta}}_{\mathrm{R}} - \boldsymbol{\beta}_*)'\boldsymbol{X}'\boldsymbol{X}(\widehat{\boldsymbol{\beta}}_{\mathrm{R}} - \boldsymbol{\beta}_*)$$

is satisfied for every vector $\boldsymbol{\beta}_* \in \mathbb{R}^p$ with $\boldsymbol{R}\boldsymbol{\beta}_* = \boldsymbol{r}$.

3.3. Show that under the linear regression model with assumptions (i) to (iv) and (viii) an estimator $\boldsymbol{Ay}+\boldsymbol{a}$ is unbiased for $\boldsymbol{\beta}$, if and only if the conditions

$$\boldsymbol{AX}(\boldsymbol{I}_p - \boldsymbol{R}'(\boldsymbol{RR}')^{-1}\boldsymbol{R}) + \boldsymbol{R}'(\boldsymbol{RR}')^{-1}\boldsymbol{R} = \boldsymbol{I}_p$$

and

$$\boldsymbol{a} = (\boldsymbol{I}_p - \boldsymbol{AX})\boldsymbol{R}'(\boldsymbol{RR}')^{-1}\boldsymbol{r}$$

are met. [Hint: The set of all $\boldsymbol{\beta}$ such that $\boldsymbol{R\beta} = \boldsymbol{r}$ coincides with the set of all $\boldsymbol{\beta}$ satisfying $\boldsymbol{\beta} = \boldsymbol{R}'(\boldsymbol{RR}')^{-1}\boldsymbol{r} + (\boldsymbol{I}_p - \boldsymbol{R}'(\boldsymbol{RR}')^{-1}\boldsymbol{R})\boldsymbol{z}$, $\boldsymbol{z} \in \mathbb{R}^p$.]

3.4. Consider a linear regression model with

$$\boldsymbol{y} = \begin{pmatrix} y_1 \\ y_2 \\ y_3 \\ y_4 \end{pmatrix}, \quad \boldsymbol{X} = \begin{pmatrix} 1\,0 \\ 1\,0 \\ 0\,1 \\ 0\,1 \end{pmatrix}, \quad \boldsymbol{\beta} = \begin{pmatrix} \beta_1 \\ \beta_2 \end{pmatrix}$$

and linear restrictions $\beta_1 + \beta_2 = 0$.

(a) Derive the restricted least squares estimator $\widehat{\boldsymbol{\beta}}_{\mathrm{R}}$ of $\boldsymbol{\beta}$ as a function of y_i, $i = 1, \ldots, 4$.

(b) Check whether

$$\widetilde{\boldsymbol{\beta}} = \begin{pmatrix} 2y_1 + y_2 + y_3 + y_4 \\ y_1 + y_2 + y_3 + 2y_4 \end{pmatrix}$$

is unbiased for $\boldsymbol{\beta}$ when the given linear restrictions are satisfied (model assumptions (i) to (iv) and (viii)).

3.5. Determine $\mathrm{MSE}(\boldsymbol{\beta}, \widehat{\boldsymbol{\beta}}_{\mathrm{R}})$ and $\rho(\boldsymbol{\beta}, \widehat{\boldsymbol{\beta}}_{\mathrm{R}})$ for $\widehat{\boldsymbol{\beta}}_{\mathrm{R}}$ from Problem 3.4. Suppose that model assumptions (i) to (iv) but not necessarily (viii) are satisfied.

3.6. Explain why under the linear regression model with assumptions (i) to (iv) none of the two estimators $\widehat{\boldsymbol{\beta}}$ and $\widehat{\boldsymbol{\beta}}_{\mathrm{R}}$ can be uniformly better than the other with respect to $\mathrm{MSE}(\boldsymbol{\beta}, \widetilde{\boldsymbol{\beta}})$.

3.7. Consider the regression model from Problem 3.4, where it is only known that $(\beta_1 + \beta_2)^2 \leq \sigma^2$. Is it nonetheless reasonable to use the restricted least squares estimator with corresponding restrictions $\beta_1 + \beta_2 = 0$?

3.8. Plot $\rho(\boldsymbol{\beta}, \widehat{\boldsymbol{\beta}}_{\mathrm{R}})$ and $\rho(\boldsymbol{\beta}, \widehat{\boldsymbol{\beta}})$ from Problem 3.4 as functions of $\lambda = \boldsymbol{\delta}'(\boldsymbol{RS}^{-1}\boldsymbol{R}')^{-1}\boldsymbol{\delta}/2\sigma^2$ with $\sigma^2 = 1$.

3.9. Consider a linear regression model with $n = 9$, $p = 3$ and normally distributed errors, so that assumptions (i) to (vi) are satisfied. The model matrix is given as

$$\boldsymbol{X}' = \begin{pmatrix} 1\,1\,1\,0\,0\,0\,0\,0\,0 \\ 0\,0\,0\,1\,1\,1\,0\,0\,0 \\ 0\,0\,0\,0\,0\,0\,1\,1\,1 \end{pmatrix}.$$

Moreover, the observed vector $\boldsymbol{y}$ is given as

$$\boldsymbol{y} = (-.8946, -.8882, -.0525, -.5289, .6568, .0842, 1.072, -.0459, .4267)'.$$

(a) Compute the level $\alpha = 0.05$ pretest estimator $\widehat{\boldsymbol{\beta}}_{\mathrm{P}}$ for $\boldsymbol{\beta}$ under the imposed restriction $\beta_1 + \beta_2 + \beta_3 = 0$..
(b) Compute the observed loss of $\widehat{\boldsymbol{\beta}}_{\mathrm{P}}$ when $\boldsymbol{\beta} = (0, 0.1, 0.2)'$ is the true parameter and the usual unweighted squared error loss function is used. Compare the result with the loss of $\widehat{\boldsymbol{\beta}}$ (Problem 2.25 (c)).

3.10. Under the situation of Problem 3.9 compute

$$\lambda = \boldsymbol{\delta}'(\boldsymbol{R}\boldsymbol{S}^{-1}\boldsymbol{R}')^{-1}\boldsymbol{\delta}/2\sigma^2$$

when $(\boldsymbol{\beta}, \sigma^2) = (\begin{pmatrix} 0 \\ 0.1 \\ 0.2 \end{pmatrix}, 0.5)$ are the true parameters.

3.11. Compute in the situation of Example 3.7 the risks $\rho(\boldsymbol{\beta}, \widehat{\boldsymbol{\beta}})$ and $\rho(\boldsymbol{\beta}, \widehat{\boldsymbol{\beta}}^{(1)})$ when $(\boldsymbol{\beta}, \sigma^2) = (\begin{pmatrix} 2 \\ 2 \end{pmatrix}, 1)$ is the true parameter. Compare the obtained values with the observed average losses given in Example 3.7.

3.12. Prove Theorem 3.6.

3.13. Show that if $\boldsymbol{\beta} \in \mathcal{C}(\boldsymbol{U}_2)$, i.e. $\boldsymbol{\beta} = \boldsymbol{U}_2\boldsymbol{b}$ for some vector $\boldsymbol{b}$, then $\mathrm{MSE}(\boldsymbol{\beta}, \boldsymbol{0}) \leq_{\mathrm{L}} \mathrm{MSE}(\boldsymbol{\beta}, \widehat{\boldsymbol{\beta}}^{(r)})$.

3.14. Consider the linear regression model with assumptions (i) to (iv), where $\boldsymbol{X}'\boldsymbol{X} = \lambda\boldsymbol{I}_p$ for some number $\lambda > 0$. Show that for any fixed $k > 0$ the inequality $\rho(\boldsymbol{\beta}, \widehat{\boldsymbol{\beta}}_k) < \rho(\boldsymbol{\beta}, \widehat{\boldsymbol{\beta}})$ holds true for all parameters $(\boldsymbol{\beta}, \sigma^2)$ satisfying the condition

$$\frac{\boldsymbol{\beta}'\boldsymbol{\beta}}{\sigma^2} < p\frac{2\lambda + k}{\lambda k} \,.$$

Here $\rho(\cdot)$ stands for the unweighted squared error risk.

3.15. Show that for an $n \times p$ matrix $\boldsymbol{X}$ of full column rank the identity

$$\boldsymbol{X}'\boldsymbol{X}(\boldsymbol{X}'\boldsymbol{X} + k\boldsymbol{I}_p)^{-1} = (\boldsymbol{X}'\boldsymbol{X} + k\boldsymbol{I}_p)^{-1}\boldsymbol{X}'\boldsymbol{X}$$

is satisfied for any $k \in \mathbb{R}$.

3.16. Show that for an $n \times p$ matrix $\boldsymbol{X}$ of full column rank the identity

$$\boldsymbol{X}'\boldsymbol{X}(\boldsymbol{X}'\boldsymbol{X} + k\boldsymbol{I}_p)^{-1} = (k(\boldsymbol{X}'\boldsymbol{X})^{-1} + \boldsymbol{I}_p)^{-1}$$

is satisfied for any $k \geq 0$.

3.17. The set of all vectors $\{\boldsymbol{\beta} \in \mathbb{R}^p : \boldsymbol{\beta}'\boldsymbol{H}\boldsymbol{\beta} \leq r^2\}$, where $\boldsymbol{H}$ is an $p \times p$ symmetric positive definite matrix and $r^2 > 0$, is called a p-dimensional ellipsoid. The p semi-axes have length

$$r/\sqrt{\mu_i}, \qquad i = 1, \ldots, p \,,$$

where $\mu_1, \ldots, \mu_p$ are the p eigenvalues of the matrix $\boldsymbol{H}$.

(a) Under the linear regression model with assumptions (i) to (iv), let $\boldsymbol{H} = \left(\frac{2}{k}\boldsymbol{I}_p + (\boldsymbol{X}'\boldsymbol{X})^{-1}\right)^{-1}$ and $r^2 = \sigma^2$. Show that the semi-axes of the corresponding ellipsoid are given as

$$\sigma\sqrt{\frac{2}{k} + \frac{1}{\lambda_i}}, \qquad i = 1, \ldots, p\,,$$

where $\lambda_1, \ldots, \lambda_p$ are the eigenvalues of $\boldsymbol{X}'\boldsymbol{X}$. [Hint: Use the spectral decomposition $\boldsymbol{X}'\boldsymbol{X} = \boldsymbol{U}\boldsymbol{\Lambda}\boldsymbol{U}'$ with $\boldsymbol{U}\boldsymbol{U}' = \boldsymbol{U}'\boldsymbol{U} = \boldsymbol{I}_p$ and $\boldsymbol{\Lambda} = \text{diag}(\lambda_1, \ldots, \lambda_p)$.]

(b) Compute the length of the semi-axes of the ellipsoid from part (a) for $k = 1/5$, $\sigma^2 = 1$, and $\boldsymbol{X}$ as in Example 2.6. Compare the results with Fig. 3.14.

3.18. The set of all vector $\boldsymbol{\beta}$ for which the inequality $\rho(\boldsymbol{\beta}, \widehat{\boldsymbol{\beta}}_k) \leq \rho(\boldsymbol{\beta}, \widehat{\boldsymbol{\beta}})$ hold true can be written as $\{\boldsymbol{\beta} \in \mathrm{IR}^p : \boldsymbol{\beta}'\boldsymbol{H}\boldsymbol{\beta} \leq \sigma^2\}$.

(a) Determine the matrix $\boldsymbol{H}$.

(b) Compute the length of the semi-axes of the corresponding ellipsoid for $k = 1/5$ and $\sigma^2 = 1$, and $\boldsymbol{X}$ as in Example 2.6. Compare the results with Fig. 3.15.

3.19. Let $\boldsymbol{X} = (\boldsymbol{1}_n, \boldsymbol{X}_*)$ be an $n \times p$ matrix of full column rank. Let $\boldsymbol{Z} = (\boldsymbol{1}_n, \boldsymbol{Z}_*)$ where $\boldsymbol{Z}_* = \boldsymbol{C}\boldsymbol{X}_*\boldsymbol{D}^{-1}$ with $\boldsymbol{C} = \boldsymbol{I}_n - \frac{1}{n}\boldsymbol{1}_n\boldsymbol{1}_n'$ and $\boldsymbol{D}^{-1}$ being a $(p-1) \times (p-1)$ diagonal matrix with positive main diagonal elements.

(a) Show that $\mathcal{C}(\boldsymbol{X}) = \mathcal{C}(\boldsymbol{Z})$. [Hint: The condition $\boldsymbol{a} \in \mathcal{C}(\boldsymbol{X})$ is satisfied if and only if there exists some number α and some vector $\boldsymbol{u}$ such that $\boldsymbol{a} = \boldsymbol{1}_n\alpha + \boldsymbol{X}_*\boldsymbol{u}$. Find a number γ and a vector $\boldsymbol{v}$ such that $\boldsymbol{a} = \boldsymbol{1}_n\gamma + \boldsymbol{Z}_*\boldsymbol{v}$. This shows that $\mathcal{C}(\boldsymbol{X}) \subseteq \mathcal{C}(\boldsymbol{Z})$. It then remains to demonstrate the reverse inclusion.]

(b) Show that the least squares variance estimator for σ^2 is the same in the two models $\boldsymbol{y} = \boldsymbol{X}\boldsymbol{\beta} + \boldsymbol{\varepsilon}$ and $\boldsymbol{y} = \boldsymbol{Z}\boldsymbol{\delta} + \boldsymbol{\varepsilon}$, where , $\boldsymbol{\varepsilon} \sim (\boldsymbol{0}, \sigma^2\boldsymbol{I}_n)$. [Hint: The equality of two column spaces is equivalent to the equality of the orthogonal projectors onto these column spaces.]

3.20. Consider the situation from Problem 3.19.

(a) Show that

$$\text{rk}(\boldsymbol{Z}) = \text{rk}(\boldsymbol{I}_n - \boldsymbol{C}, \boldsymbol{C}\boldsymbol{X}_*) = \text{rk}(\boldsymbol{I}_n - \boldsymbol{C}) + \text{rk}(\boldsymbol{C}\boldsymbol{X}_*)$$

holds true. [Hint: For suitable matrices $\boldsymbol{A}$ and $\boldsymbol{B}$ the formula $\text{rk}(\boldsymbol{A}, \boldsymbol{B}) = \text{rk}(\boldsymbol{A}) + \text{rk}(\boldsymbol{B}) - \dim[\mathcal{C}(\boldsymbol{A}) \cap \mathcal{C}(\boldsymbol{B})]$ is valid.]

(b) Show that

$$\text{rk}(\boldsymbol{C}\boldsymbol{X}_*) = \text{rk}(\boldsymbol{X}_*)\,.$$

[Hint: For suitable matrices $\boldsymbol{A}$ and $\boldsymbol{B}$ the formula $\text{rk}(\boldsymbol{A}\boldsymbol{B}) = \text{rk}(\boldsymbol{B}) - \dim[\mathcal{N}(\boldsymbol{A}) \cap \mathcal{C}(\boldsymbol{B})]$ holds true. If $\boldsymbol{A}$ is an orthogonal projector, then $\mathcal{N}(\boldsymbol{A}) = \mathcal{C}(\boldsymbol{I} - \boldsymbol{A})$.]

(c) Show that $\boldsymbol{Z}$ is of full column rank.

3.21. Consider the linear regression model with $n = 9$ observations given as

i	1	2	3	4	5	6	7	8	9
y_i	41.9	45	49.2	50.6	52.6	55.1	56.2	57.3	57.8
$x_{i,1}$	12.4	16.9	18.4	19.4	20.1	19.6	19.8	21.1	21.7
$x_{i,2}$	28.2	32.2	37	37	38.6	40.7	41.5	42.9	45.3

(Here y is the total of consumption in the USA, x_1 is the total income from profits and x_2 is the total income from wages. The variables have been observed in the years 1921 to 1929.)

Assume a linear regression model with intercept, standardize the non-constant independent variables, and compute the ordinary least squares and ridge estimates.

3.22. Consider the estimator

$$\widetilde{\boldsymbol{\beta}}(d) = (\boldsymbol{X}'\boldsymbol{X} + \boldsymbol{I}_p)^{-1}(\boldsymbol{X}'\boldsymbol{y} + d\widehat{\boldsymbol{\beta}}), \quad 0 \le d \le 1 ,$$

see [64].

(a) Show that $||\widetilde{\boldsymbol{\beta}}(d)||^2 \le ||\widehat{\boldsymbol{\beta}}||^2$ is satisfied.
(b) Show that $\widetilde{\boldsymbol{\beta}}(0) = \widehat{\boldsymbol{\beta}}_{k=1}$ and $\widetilde{\boldsymbol{\beta}}(1) = \widehat{\boldsymbol{\beta}}$.

(As follows from (b), the estimator $\widetilde{\boldsymbol{\beta}}(d)$ can be seen as a weighted average between the ordinary least squares and the ridge estimator with ridge parameter $k = 1$.)

3.23. Consider the linear regression model described by $\boldsymbol{y} = \boldsymbol{1}_n\mu + \boldsymbol{\varepsilon}$, $\boldsymbol{\varepsilon} \sim (\boldsymbol{0}, \sigma^2\boldsymbol{I}_n)$, where $\boldsymbol{1}_n$ stands for the $n \times 1$ vector whose every element is 1. Show the following statements:

(a) The ridge estimator for μ is given as $\widehat{\mu}_k = \dfrac{n}{n+k}\overline{\boldsymbol{y}}$.
(b) The jackknifed ridge estimator for μ is given as $\breve{\mu}_k^{\mathrm{J}} = \widehat{\mu}_k + \dfrac{(n-1)k}{n(n-1+k)}\widehat{\mu}_k$.
(c) The almost unbiased ridge estimator is given as $\widehat{\mu}_k^{\mathrm{J}} = \widehat{\mu}_k + \dfrac{k}{n+k}\widehat{\mu}_k$.

3.24. Consider the situation from Problem 3.23. Show that (for the case $p > 1$ and $k > 0$) the absolute value of the bias of the jackknifed ridge estimator $\breve{\mu}_k^{\mathrm{J}}$ is smaller than the absolute value of the bias of the ridge estimator $\widehat{\mu}_k$.

3.25. Consider the situation from Example 3.20, where the true parameters are $(\boldsymbol{\beta}, \sigma^2) = (\begin{pmatrix} 2 \\ 2 \end{pmatrix}, 1)$.

(a) Compute $\rho(\boldsymbol{\beta}, \widehat{\boldsymbol{\beta}}(7/3))$ and $\rho(\boldsymbol{\beta}, \widehat{\boldsymbol{\beta}})$.

(b) Compute $\mathbf{\Delta} = \text{MSE}(\boldsymbol{\beta}, \widehat{\boldsymbol{\beta}}) - \text{MSE}(\boldsymbol{\beta}, \widehat{\boldsymbol{\beta}}(7/3))$. Is the matrix $\mathbf{\Delta}$ nonnegative definite?

3.26. Show that $\widehat{\boldsymbol{\beta}}(\widehat{\varrho}_{\text{M}})$ can be written as

$$\widehat{\boldsymbol{\beta}}(\widehat{\varrho}_{\text{M}}) = \frac{\widehat{\boldsymbol{\beta}}'\boldsymbol{X}'\boldsymbol{y}}{\widehat{\sigma}^2 + \widehat{\boldsymbol{\beta}}'\boldsymbol{X}'\boldsymbol{X}\widehat{\boldsymbol{\beta}}}\widehat{\boldsymbol{\beta}} = \frac{(n-p)R^2}{1+(n-p-1)R^2}\widehat{\boldsymbol{\beta}} .$$

3.27. Consider the situation from Example 3.20, where the true parameters are $(\boldsymbol{\beta}, \sigma^2) = (\binom{2}{2}, 1)$. Compute the matrix $\boldsymbol{C}$ in '$\widehat{\boldsymbol{\beta}}_{\text{M}}$' $= \boldsymbol{Cy}$ as well as the estimate $\boldsymbol{Cy}$.

3.28. Find a non-homogeneous 'estimator' $\boldsymbol{Cy} + \boldsymbol{c}$ such that the risk

$$\text{MSE}(\boldsymbol{\beta}, \boldsymbol{Cy} + \boldsymbol{c})$$

is minimized with respect to $\boldsymbol{C}$ and $\boldsymbol{c}$. The matrix $\boldsymbol{C}$ and the vector $\boldsymbol{c}$ can be functions of $(\boldsymbol{\beta}, \sigma^2)$.

3.29. Show that $\widehat{\boldsymbol{\beta}}(\widehat{\varrho}_{\text{AM}})$ can be written as

$$\widehat{\boldsymbol{\beta}}(\widehat{\varrho}_{\text{AM}}) = \frac{\widehat{\boldsymbol{\beta}}'\boldsymbol{X}'\boldsymbol{y}/p}{\widehat{\sigma}^2 + \widehat{\boldsymbol{\beta}}'\boldsymbol{X}'\boldsymbol{X}\widehat{\boldsymbol{\beta}}/p}\widehat{\boldsymbol{\beta}} = \frac{(n-p)R^2}{p+(n-2p)R^2}\widehat{\boldsymbol{\beta}} .$$

3.30. Consider the linear regression model with assumptions (i) to (iv) and $0 < R^2 < 1$. Show that $\widehat{\boldsymbol{\beta}}'\boldsymbol{X}'\boldsymbol{X}\widehat{\boldsymbol{\beta}} - c\,\widehat{\boldsymbol{\varepsilon}}'\widehat{\boldsymbol{\varepsilon}}$ is positive if and only if $c < R^2/(1-R^2)$.

3.31. Find a matrix $\boldsymbol{X}_*$ and a vector $\boldsymbol{y}_*$ such that the general ridge estimator $\widehat{\boldsymbol{\beta}}_{\boldsymbol{K}} = (\boldsymbol{X}'\boldsymbol{X} + \boldsymbol{K})^{-1}\boldsymbol{X}'\boldsymbol{y}$ can be written in the form

$$\widehat{\boldsymbol{\beta}}_{\boldsymbol{K}} = (\boldsymbol{X}_*'\boldsymbol{X}_*)^{-1}\boldsymbol{X}_*'\boldsymbol{y}_* .$$

3.32. Show that the general ridge estimator $\widehat{\boldsymbol{\beta}}_{\boldsymbol{K}} = (\boldsymbol{X}'\boldsymbol{X} + \boldsymbol{K})^{-1}\boldsymbol{X}'\boldsymbol{y}$ is unbiased for $\boldsymbol{\beta}$ if and only if $\boldsymbol{K} = \boldsymbol{0}$.

3.33. Show that the Marquardt estimator $\widehat{\boldsymbol{\beta}}^{(r,c)}$, $c > 0$, is of the form of a general ridge estimator with $\boldsymbol{K} = ((1-c)/c)\,\boldsymbol{U}_2\boldsymbol{\Lambda}_2\boldsymbol{U}_2'$.

3.34. Show that the almost unbiased ridge estimator $\widehat{\boldsymbol{\beta}}_k^{\text{J}}$ is of the form of a general ridge estimator with $\boldsymbol{K} = k^2(\boldsymbol{X}'\boldsymbol{X} + 2k\boldsymbol{I}_p)^{-1}$.

3.35. Show that the iteration estimator $\widehat{\boldsymbol{\beta}}_{\delta,m}$ is of the form of a general ridge estimator with $\boldsymbol{K} = \boldsymbol{X}'\boldsymbol{X}(\boldsymbol{I}_p - (\boldsymbol{I}_p - \delta\boldsymbol{X}'\boldsymbol{X})^{m+1})^{-1} - \boldsymbol{X}'\boldsymbol{X}$. [Hint: $\boldsymbol{K} = \boldsymbol{U}(\boldsymbol{\Lambda}\boldsymbol{\Gamma}^{-1} - \boldsymbol{\Lambda})\boldsymbol{U}'$, where $\boldsymbol{X}'\boldsymbol{X} = \boldsymbol{U}\boldsymbol{\Lambda}\boldsymbol{U}'$ and $\boldsymbol{\Gamma}$ is matrix from Theorem 3.15.]

3.36. Sommers [106] considers the estimator

$$\widehat{\boldsymbol{\beta}}_k^m = ((\boldsymbol{X}'\boldsymbol{X})^m + k\boldsymbol{I}_p)^{-1}(\boldsymbol{X}'\boldsymbol{X})^m\widehat{\boldsymbol{\beta}}, \quad k \geq 0, \quad m = 1, 2, \ldots .$$

Show that $\widehat{\boldsymbol{\beta}}_k^m$ is of the form of a general ridge estimator with $\boldsymbol{K} = k[(\boldsymbol{X}'\boldsymbol{X})^{m-1}]^{-1}$.

3.37. Show that the ordinary least squares estimator $\widehat{\boldsymbol{\beta}}$ is the (uniquely determined) linear minimax estimator within the set $\mathcal{U}(\boldsymbol{\beta}) = \{\boldsymbol{A}\boldsymbol{y} : \boldsymbol{A}\boldsymbol{X} = \boldsymbol{I}_p\}$ when $\boldsymbol{\beta} \in \Theta_{\boldsymbol{\beta}} = \{\boldsymbol{\beta} \in \mathbb{R}^p : \boldsymbol{\beta}'\boldsymbol{T}\boldsymbol{\beta} \leq t^2\}$, i.e. the identity

$$\sup_{\boldsymbol{\beta} \in \Theta_{\boldsymbol{\beta}}} \rho(\boldsymbol{\beta}, \widehat{\boldsymbol{\beta}}) = \inf_{\widetilde{\boldsymbol{\beta}} \in \mathcal{U}(\boldsymbol{\beta})} \sup_{\boldsymbol{\beta} \in \Theta_{\boldsymbol{\beta}}} \rho(\boldsymbol{\beta}, \widetilde{\boldsymbol{\beta}}).$$

holds true.

3.38. Consider the linear regression model with $p > 1$ and assumptions (i) to (iv) and (vii), so that $\boldsymbol{X}'\boldsymbol{X} = \boldsymbol{I}_p$. Show that for $\rho(\boldsymbol{\beta}, \widetilde{\boldsymbol{\beta}}) = \mathrm{tr}[\mathrm{MSE}(\boldsymbol{\beta}, \widetilde{\boldsymbol{\beta}})]$ the condition

$$\rho(\boldsymbol{\beta}, \widehat{\boldsymbol{\beta}}_{p\boldsymbol{I}_p}) < \rho(\boldsymbol{\beta}, \widehat{\boldsymbol{\beta}}_{\boldsymbol{I}_p})$$

is satisfied for any vector $\boldsymbol{\beta} \in \mathbb{R}^p$ such that $\boldsymbol{\beta}'\boldsymbol{\beta} \leq \sigma^2$.

3.39. Consider the linear regression model with assumptions (i) to (iv), where $\boldsymbol{\beta} \in \mathbb{R}^p$ is the realization of a $p \times 1$ random vector $\boldsymbol{b}$ with $\mathrm{E}(\boldsymbol{b}\boldsymbol{b}') = \sigma^2\boldsymbol{M}$. The $p \times p$ matrix $\boldsymbol{M}$ is symmetric nonnegative definite.

(a) Show that the Bayes risk of a homogeneous linear estimator $\boldsymbol{A}\boldsymbol{y}$ for $\boldsymbol{\beta}$ can be written as

$$\rho^{\mathrm{B}}(\boldsymbol{\beta}, \boldsymbol{A}\boldsymbol{y}) = \sigma^2 \,\mathrm{tr}[\boldsymbol{A}(\boldsymbol{X}\boldsymbol{M}\boldsymbol{X}' + \boldsymbol{I}_n)\boldsymbol{A}' - \boldsymbol{A}\boldsymbol{X}\boldsymbol{M} - \boldsymbol{M}\boldsymbol{X}'\boldsymbol{A}' + \boldsymbol{M}] \,.$$

(b) Let $\boldsymbol{A}_* = \boldsymbol{M}\boldsymbol{X}'(\boldsymbol{X}\boldsymbol{M}\boldsymbol{X}' + \boldsymbol{I}_n)^{-1}$. Show that the Bayes risk of $\boldsymbol{A}_*\boldsymbol{y}$ is given by

$$\rho^{\mathrm{B}}(\boldsymbol{\beta}, \boldsymbol{A}_*\boldsymbol{y}) = \sigma^2 \,\mathrm{tr}[\boldsymbol{M} - \boldsymbol{M}\boldsymbol{X}'(\boldsymbol{X}\boldsymbol{M}\boldsymbol{X}' + \boldsymbol{I}_n)^{-1}\boldsymbol{X}\boldsymbol{M}] \,.$$

(c) Show that the Bayes risk of a homogeneous linear estimator $\boldsymbol{A}\boldsymbol{y}$ for $\boldsymbol{\beta}$ can be written as

$$\rho^{\mathrm{B}}(\boldsymbol{\beta}, \boldsymbol{A}\boldsymbol{y}) = \rho^{\mathrm{B}}(\boldsymbol{\beta}, \boldsymbol{A}_*\boldsymbol{y}) + \sigma^2 \,\mathrm{tr}[(\boldsymbol{A} - \boldsymbol{A}_*)(\boldsymbol{X}\boldsymbol{M}\boldsymbol{X}' + \boldsymbol{I}_n)(\boldsymbol{A} - \boldsymbol{A}_*)'] \,.$$

(d) Explain why we can call $\boldsymbol{A}_*\boldsymbol{y}$ a *homogeneous* linear Bayes estimator for $\boldsymbol{\beta}$.

3.40. Consider the linear regression model with assumptions (i) to (iv), where $\boldsymbol{\beta} \in \mathbb{R}^p$ is the realization of a $p \times 1$ random vector $\boldsymbol{b}$ with $\mathrm{E}(\boldsymbol{b}) = \boldsymbol{\mu}$ and $\mathrm{Cov}(\boldsymbol{b}) = \sigma^2\boldsymbol{T}$. The $p \times p$ matrix $\boldsymbol{T}$ is symmetric nonnegative definite.

(a) Show that the Bayes risk of a linear estimator $\boldsymbol{Ay}+\boldsymbol{a}$ for $\boldsymbol{\beta}$ can be written in the form $\rho^{\mathrm{B}}(\boldsymbol{\beta}, \boldsymbol{Ay}+\boldsymbol{a}) = f + g$, where

$$f = \sigma^2 \operatorname{tr}(\boldsymbol{AA}') + \sigma^2 \operatorname{tr}[(\boldsymbol{AX} - \boldsymbol{I}_p)\boldsymbol{T}(\boldsymbol{AX} - \boldsymbol{I}_p)']$$

and

$$g = \operatorname{tr}[((\boldsymbol{AX} - \boldsymbol{I}_p)\boldsymbol{\mu} + \boldsymbol{a})((\boldsymbol{AX} - \boldsymbol{I}_p)\boldsymbol{\mu} + \boldsymbol{a}))'] \,.$$

(b) Show that $\boldsymbol{A}_*\boldsymbol{y} + \boldsymbol{a}_*$ is the linear Bayes estimator for $\boldsymbol{\beta}$, if $\boldsymbol{A}_*$ minimizes the function f with respect to all $p \times n$ matrices $\boldsymbol{A}$, and moreover $\boldsymbol{a}_* = \boldsymbol{\mu} - \boldsymbol{A}_*\boldsymbol{X}\boldsymbol{\mu}$ holds true.
(c) Show that the function f is minimal with respect to all $p\times n$ matrices $\boldsymbol{A}$ for $\boldsymbol{A} = \boldsymbol{A}_*$ with $\boldsymbol{A}_* = \boldsymbol{TX}'(\boldsymbol{XTX}' + \boldsymbol{I}_n)^{-1}$.

3.41. Let $\boldsymbol{S}$ be a $p \times p$ symmetric positive definite matrix and let $\boldsymbol{T}$ be a $p \times p$ symmetric nonnegative definite matrix . Show that the matrix $\boldsymbol{T}(\boldsymbol{T} + \boldsymbol{S}^{-1})^{-1}\boldsymbol{S}^{-1}$ is symmetric, so that

$$\boldsymbol{T}(\boldsymbol{T} + \boldsymbol{S}^{-1})^{-1}\boldsymbol{S}^{-1} = \boldsymbol{S}^{-1}(\boldsymbol{T} + \boldsymbol{S}^{-1})^{-1}\boldsymbol{T}$$

holds true.

3.42. Show that under the linear regression model with assumptions (i) to (iv), the direction modified shrinkage estimator

$$\widehat{\boldsymbol{\beta}}_{\boldsymbol{\beta}_0}(\varrho) = (1-\alpha)\boldsymbol{\beta}_0 + \alpha\widehat{\boldsymbol{\beta}}, \quad \alpha = \frac{1}{1+\varrho}, \quad \varrho > 0\,,$$

is a linear Bayes estimator for $\boldsymbol{\beta}$, when $\boldsymbol{\beta} \in \mathbb{R}^p$ is supposed to be the realization of a $p\times 1$ random vector $\boldsymbol{b}$ with $\mathrm{E}(\boldsymbol{b}) = \boldsymbol{\beta}_0$ and $\mathrm{Cov}(\boldsymbol{b}) = \varrho\,\sigma^2(\boldsymbol{X}'\boldsymbol{X})^{-1}$.

3.43. Show that under the linear regression model with assumptions (i) to (iv), the estimator

$$\widehat{\boldsymbol{\beta}}_{k,\boldsymbol{\beta}_0} = (\boldsymbol{X}'\boldsymbol{X} + k\boldsymbol{I}_p)^{-1}(\boldsymbol{X}'\boldsymbol{y} + k\boldsymbol{\beta}_0), \quad k > 0,$$

is a linear Bayes estimator for $\boldsymbol{\beta}$, when $\boldsymbol{\beta} \in \mathbb{R}^p$ is supposed to be the realization of a $p \times 1$ random vector $\boldsymbol{b}$ with $\mathrm{E}(\boldsymbol{b}) = \boldsymbol{\beta}_0$ and $\mathrm{Cov}(\boldsymbol{b}) = k\,\sigma^2\boldsymbol{I}_p$.

3.44. Consider the situation from Example 3.29 and compute the ordinary least squares estimator $\widehat{\boldsymbol{\beta}}$ as well as its observed squared error loss when the data point (x_8, y_8) is deleted from Table 3.4.

3.45. Consider the situation from Example 3.29 and compute the weighted (non-iterated) least squares estimator for $\boldsymbol{\beta}$ from Andrews' $\sin(a)$-function with $a = 1.339$. Compare the result with the estimate from Problem 3.44.

4 Linear Admissibility

Chapters 2 and 3 deal with alternatives to the ordinary least squares estimator $\widehat{\boldsymbol{\beta}}$ for $\boldsymbol{\beta}$. Any of these can be viewed as being based on some linear estimator for $\boldsymbol{\beta}$, supposed to be admissible within the set of all linear estimators. This guarantees that the considered alternative linear estimator is better than the ordinary least squares estimator for at least one possible value of the unknown $\boldsymbol{\beta}$. The actual chapter investigates the structure of linearly admissible estimators.

4.1 Preliminary Considerations

We have already introduced a number of point estimators for $\boldsymbol{\beta}$ whose application can be beneficial under certain conditions. Nonetheless, it appears to be reasonable to consider only those estimators which cannot uniformly be outperformed by some other estimator. This is essentially the meaning of *admissibility*. However, admissibility is a property which can only be inspected within a specific context, depending on

- the *set of point estimators* we take into account and from which we take the possible competitors of a specific point estimator,
- the *loss function* and the corresponding *risk function* we apply for the assessment of a point estimator,
- the *parameter space* we assume the unknown parameter vector to lie in.

The Set of Point Estimators

We have already noted that the statistical properties of linear estimators are rather easy to determine, while, in general, the properties of corresponding non-linear variants, being used in practice, are rather difficult to derive. Linear estimators like $\widehat{\boldsymbol{\beta}}_k$ and $\widehat{\boldsymbol{\beta}}(\varrho)$ can also be seen as non-operational decision rules, which must be made practical for actual applications. This can be achieved by replacing the parameters k and ϱ by appropriate estimates.

Nonetheless, statistical properties of linear estimators are of significant importance. If a risk comparison of $\widehat{\boldsymbol{\beta}}_k$ (for non-stochastic k) and $\widehat{\boldsymbol{\beta}}$ had shown that no value $(\boldsymbol{\beta}, \sigma^2)$ would have existed such that $\widehat{\boldsymbol{\beta}}_k$ has a smaller risk than $\widehat{\beta}$, then we would not make an effort to find an appropriate method to determine k. Instead, we would have turned towards different linear estimators.

Therefore, it makes sense to investigate admissibility within the set of linear estimators. Only those linear estimators which appear to be reasonable choices under specific situations and *in addition* turn out to be admissible within the set of all linear estimators should be applied for practical purposes. Whether the non-linear variants of those estimators are also admissible within the set of all point estimators is a question of further investigation. In the following, we consider the set of linear estimators

$$\mathcal{L}(\beta) = \{Ay + a : \ A \in \mathbb{R}^{p\times n}, a \in \mathbb{R}^p\} ,$$

or the set of homogeneous linear estimators

$$\mathcal{L}^{\mathrm{h}}(\beta) = \{Ay : \ A \in \mathbb{R}^{p\times n}\}$$

as the set Δ of possible decision rules. A linear estimator is called *linearly admissible* for β if it is admissible within the set $\mathcal{L}(\beta)$ or the set $\mathcal{L}^{\mathrm{h}}(\beta)$.

The Loss Function

If the set Δ of possible decision rules is given by $\mathcal{L}(\beta)$ or $\mathcal{L}^{\mathrm{h}}(\beta)$, then two point estimators $\widetilde{\beta}_1, \widetilde{\beta}_2 \in \Delta$ satisfy the condition

$$\widetilde{\beta}_1 + G(\widetilde{\beta}_2 - \widetilde{\beta}_1) \in \Delta$$

for any $p \times p$ matrix G. Thus, in view of Theorem 1.1, the admissibility with respect to the unweighted squared error loss implies the admissibility with respect to any weighted squared error loss. Moreover, the former type of admissibility implies admissibility with respect to the matrix-valued squared error loss according to Theorem 1.3. For the purpose of guaranteeing linear admissibility of an estimator, it is meaningful to investigate the strongest form, so that in the following the considered loss function is the unweighted squared error loss

$$L(\beta, \widetilde{\beta}) = (\widetilde{\beta} - \beta)'(\widetilde{\beta} - \beta)$$

with corresponding risk $\rho(\beta, \widetilde{\beta}) = \mathrm{tr}[\mathrm{MSE}(\beta, \widetilde{\beta})]$.

The Parameter Space

For the estimation of the unknown parameter vector β three different parameter spaces come into focus, reflecting different states of prior knowledge about β.

Case 1. When we consider the linear regression model with assumptions (i) to (iv) and without any further demands on the parameter vector $\boldsymbol{\beta}$, then we assume that the true parameter vector $\boldsymbol{\beta}$ can be any value in $\mathbb{R}^p$. This means that the parameter space $\Theta_{\boldsymbol{\beta}}$ for $\boldsymbol{\beta}$ is given by $\Theta_{\boldsymbol{\beta}} = \mathbb{R}^p$. Then, an estimator is called uniformly better than another, if it has not greater risk for all values $\boldsymbol{\beta} \in \Theta_{\boldsymbol{\beta}}$ and all values $\sigma^2 \in (0, \infty)$ and a strictly smaller risk for at least one combination of values $\boldsymbol{\beta} \in \Theta_{\boldsymbol{\beta}}$ and $\sigma^2 \in (0, \infty)$. Hence, the whole parameter space we consider in this case is

$$\Theta = \Theta_{\boldsymbol{\beta}} \times (0, \infty) = \mathbb{R}^p \times (0, \infty) .$$

Case 2. When we assume that the parameter vector $\boldsymbol{\beta}$ satisfies linear restrictions $\boldsymbol{R}\boldsymbol{\beta} = \boldsymbol{r}$ for some known matrix $\boldsymbol{R}$ and some known vector $\boldsymbol{r}$, then of course $\boldsymbol{\beta}$ cannot be any element in $\mathbb{R}^p$. The required parameter space is $\Theta_{\boldsymbol{\beta}} = \mathcal{B}$ with $\mathcal{B} = \{\boldsymbol{\beta} \in \mathbb{R}^p : \boldsymbol{R}\boldsymbol{\beta} = \boldsymbol{r}\}$ and the whole parameter space is given by

$$\Theta = \Theta_{\boldsymbol{\beta}} \times (0, \infty) = \mathcal{B} \times (0, \infty) .$$

Case 3. When the unknown parameters $\boldsymbol{\beta}$ and σ^2 are supposed to satisfy the relationship $\boldsymbol{\beta}'\boldsymbol{T}\boldsymbol{\beta} \leq \sigma^2$ for some known symmetric positive definite matrix $\boldsymbol{T}$, then again the parameter space must be a restricted one. In contrast to the two cases above, we do not keep the two parameter spaces for $\boldsymbol{\beta}$ and σ^2 separate, but compare the risks of two estimators under variation of the possible values $\boldsymbol{\beta}$ satisfying the condition $\boldsymbol{\beta}'\boldsymbol{T}\boldsymbol{\beta} \leq \sigma^2$. Therefore, the whole parameter space is

$$\Theta = \Theta_{\boldsymbol{\beta}} = \{\boldsymbol{\beta} \in \mathbb{R}^p : \boldsymbol{\beta}'\boldsymbol{T}\boldsymbol{\beta} \leq \sigma^2\} .$$

The parameter σ^2 is thus considered to be unknown but fixed.

4.2 Linear Admissibility in the Non-Restricted Model

In this section we consider the question of linear admissibility for the case that the parameter space in the linear regression model is given by $\Theta = \mathbb{R}^p \times (0, \infty)$. The initial work on this field is given by Cohen [26]. For the characterization of linearly admissible estimators in linear models, Rao [90] derives basic results.

4.2.1 Linear Admissibility in the Simple Mean Shift Model

Before considering linear admissibility in the linear regression model with assumptions (i) to (iv), we may demonstrate the complexity of the problem under a simple linear model with only one parameter of interest. This model is described by the equation

$$\boldsymbol{y} = \mu \mathbf{1}_n + \boldsymbol{\varepsilon}, \quad \boldsymbol{\varepsilon} \sim (\mathbf{0}, \sigma^2 \boldsymbol{I}_n),$$

where $\mathbf{1}_n$ denotes the $n \times 1$ vector whose every element is equal to 1. This is a special case of the linear regression model with $p = 1$ and unknown regression parameter $\mu \in \mathbb{R}$. We will assume $n > 1$ in the following. The above described model will also be called the *simple mean shift model.*

A homogeneous linear estimator for μ is of the form $\boldsymbol{a}'\boldsymbol{y}$ for some $n \times 1$ vector $\boldsymbol{a}$. The unweighted squared error risk of $\boldsymbol{a}'\boldsymbol{y}$ is given as

$$\rho(\mu, \boldsymbol{a}'\boldsymbol{y}) = \mathrm{E}[(\boldsymbol{a}'\boldsymbol{y} - \mu)^2] = \sigma^2 \boldsymbol{a}'\boldsymbol{a} + \mu^2 (\mathbf{1}_n'\boldsymbol{a} - 1)^2 .$$

For simplicity we will consider the admissibility of an estimator $\boldsymbol{a}'\boldsymbol{y}$ within the set $\mathcal{L}^{\mathrm{h}}(\mu)$ of homogeneous linear estimators. The following theorem shows that an estimator $\boldsymbol{a}'\boldsymbol{y}$ must be of a specific form to guarantee than no other estimator from $\mathcal{L}^{\mathrm{h}}(\mu)$ is uniformly better.

Theorem 4.1. *Let $\boldsymbol{a}'\boldsymbol{y}$ be an estimator for μ in the simple mean shift model. If the vector $\boldsymbol{a}$ cannot be written in the form $\boldsymbol{a} = \alpha \mathbf{1}_n$ for some number α, then there exists a uniformly better estimator $\boldsymbol{b}'\boldsymbol{y}$ with $\boldsymbol{b} = \gamma \mathbf{1}_n$ for some number γ.*

Proof. We assume that there does not exist a number α such that $\boldsymbol{a} = \alpha \mathbf{1}_n$, and consider the estimator $\boldsymbol{b}'\boldsymbol{y}$ with

$$\boldsymbol{b} = \gamma \mathbf{1}_n, \quad \gamma = \frac{\boldsymbol{a}'\mathbf{1}_n}{n} .$$

Then, in view of $n = \mathbf{1}_n'\mathbf{1}_n$ the risk $\rho(\mu, \boldsymbol{b}'\boldsymbol{y})$ of $\boldsymbol{b}'\boldsymbol{y}$ is

$$\rho(\mu, \boldsymbol{b}'\boldsymbol{y}) = \sigma^2 \frac{1}{n} \boldsymbol{a}'\mathbf{1}_n\mathbf{1}_n'\boldsymbol{a} + \mu^2 (\mathbf{1}_n'\boldsymbol{a} - 1)^2 .$$

This gives

$$\sigma^{-2} n [\rho(\mu, \boldsymbol{a}'\boldsymbol{y}) - \rho(\mu, \boldsymbol{b}'\boldsymbol{y})] = \boldsymbol{a}'\boldsymbol{a}\mathbf{1}_n'\mathbf{1}_n - \boldsymbol{a}'\mathbf{1}_n\mathbf{1}_n'\boldsymbol{a} .$$

From the Cauchy-Schwarz inequality we conclude that $\boldsymbol{a}'\mathbf{1}_n\mathbf{1}_n'\boldsymbol{a} \leq \boldsymbol{a}'\boldsymbol{a}\mathbf{1}_n'\mathbf{1}_n$, where equality holds true if and only if $\boldsymbol{a} = \alpha \mathbf{1}_n$ for some number α. Since the latter is not satisfied by assumption, it follows that $\rho(\mu, \boldsymbol{b}'\boldsymbol{y}) < \rho(\mu, \boldsymbol{a}'\boldsymbol{y})$ for all possible values $0 < \sigma^2 < \infty$ and $\mu \in \mathbb{R}$. □

We may now derive derive necessary and sufficient conditions for a linear estimator $\boldsymbol{a}'\boldsymbol{y}$ to be linearly admissible within the set $\mathcal{L}^{\mathrm{h}}(\mu)$ of homogeneous linear estimators for μ.

- Any homogeneous linear estimator which is admissible within $L^{\mathrm{h}}(\mu)$ must be of the form $\boldsymbol{a}'\boldsymbol{y}$ such that $\boldsymbol{a} = \alpha \mathbf{1}_n$ for some α. Hence, for a characterization of the set of linear admissibility we can confine to such estimators and determine appropriate values for α.

- Initially, we consider the case $\alpha = 1/n$. Then $\rho(\mu, \alpha \mathbf{1}_n' \boldsymbol{y}) = \sigma^2/n$. If we suppose that there exists a uniformly better estimator $\boldsymbol{b}'\boldsymbol{y}$ than $\boldsymbol{a}'\boldsymbol{y}$, then without loss of generality we can assume that $\boldsymbol{b}'\boldsymbol{y} = \gamma \mathbf{1}_n' \boldsymbol{y}$ for some number $\gamma \neq 1/n$ in view of Theorem 4.1. But then we can always find some $0 < \sigma^2 < \infty$ and some $\mu \in \mathbb{R}$ such that the inequality

$$\rho(\mu, \gamma \mathbf{1}_n' \boldsymbol{y}) = \sigma^2 n \gamma^2 + \mu^2 \underbrace{(\gamma n - 1)^2}_{>0} > \sigma^2/n = \rho(\mu, \alpha \mathbf{1}_n' \boldsymbol{y})$$

  holds true. The latter is of course a contradiction to the assumption that $\gamma \mathbf{1}_n' \boldsymbol{y}$ is uniformly better than $\alpha \mathbf{1}_n' \boldsymbol{y}$. This shows that the estimator $\alpha \mathbf{1}_n' \boldsymbol{y}$ with $\alpha = 1/n$ must be admissible within the set $\mathcal{L}^{\mathrm{h}}(\mu)$.
- Let us now consider the estimator $\gamma \mathbf{1}_n' \boldsymbol{y}$ with

$$\gamma = \frac{1}{n}(1 + c(n\alpha - 1)), \quad 0 < c < 1 ,$$

  for the case that $\alpha \neq 1/n$. Then

$$\rho(\mu, \gamma \mathbf{1}_n' \boldsymbol{y}) = \sigma^2 n \gamma^2 + c^2 \mu^2 (n\alpha - 1)^2 .$$

  Thus, the estimator $\gamma \mathbf{1}_n' \boldsymbol{y}$ has a strictly smaller squared bias $c^2\mu^2(n\alpha - 1)^2$ for any μ than the estimator $\alpha \mathbf{1}_n' \boldsymbol{y}$, whose squared bias is $\mu^2(n\alpha - 1)^2$ and whose risk is

$$\rho(\mu, \gamma \mathbf{1}_n' \boldsymbol{y}) = \sigma^2 n \alpha^2 + \mu^2 (n\alpha - 1)^2 .$$

  Therefore a *necessary* condition for the admissibility of $\alpha \mathbf{1}_n' \boldsymbol{y}$ is that $n\alpha^2 < n\gamma^2$ holds true, since otherwise $\gamma \mathbf{1}_n' \boldsymbol{y}$ is uniformly better than $\alpha \mathbf{1}_n' \boldsymbol{y}$. The necessary condition can equivalently be rewritten as

$$(1 + c)\alpha^2 n < (1 - c)\frac{1}{n} + 2c\alpha .$$

  If we take limits for $c \to 1$, then the condition $\alpha^2 n < \alpha$ turns out to be necessary for the admissibility of the estimator $\alpha \mathbf{1}_n' \boldsymbol{y}$. The inequality $\alpha^2 n < \alpha$ can equivalently be expressed as $0 \leq \alpha < 1/n$.
- The above considerations have shown that a necessary condition for the admissibility of an estimator $\boldsymbol{a}'\boldsymbol{y}$ within the set $\mathcal{L}^{\mathrm{h}}(\mu)$ is that $\boldsymbol{a} = \alpha \mathbf{1}_n$ with α satisfying $\alpha = 1/n$ or $0 \leq \alpha < 1/n$. The latter two conditions can be summarized as $0 \leq \alpha \leq 1/n$. A more compact form of this necessary condition is given by the inequality

$$\mathbf{1}_n \boldsymbol{a}' \boldsymbol{a} \mathbf{1}_n' \leq_{\mathrm{L}} \mathbf{1}_n \boldsymbol{a}',$$

  see also Problem 4.1.
- When we consider the estimator $\boldsymbol{a}'\boldsymbol{y}$ for μ, where $\boldsymbol{a} = \alpha \mathbf{1}_n$ for some α satisfying $0 \leq \alpha \leq 1/n$, then we can ask whether any such estimator is already admissible within the set $\mathcal{L}^{\mathrm{h}}(\mu)$, which would mean that the

established necessary condition is also *sufficient.* As a matter of fact this is the case. For the derivation of the sufficiency we can exclude the case $\alpha = 1/n$, since we already know that in this case the estimator $\alpha \mathbf{1}_n' \boldsymbol{y}$ is linearly admissible for μ.

– Let us assume that there exists an estimator $\boldsymbol{b}'\boldsymbol{y}$ which is uniformly better than $\alpha \mathbf{1}_n' \boldsymbol{y}$ with $0 \leq \alpha < 1/n$. Again, in view of Theorem 4.1 we can assume without loss of generality that $\boldsymbol{b}'\boldsymbol{y}$ is of the form $\boldsymbol{b}'\boldsymbol{y} = \gamma \mathbf{1}_n' \boldsymbol{y}$ for some number γ. Then

$$\sigma^2 n\gamma^2 + \mu^2(n\gamma - 1)^2 \leq \sigma^2 n\alpha^2 + \mu^2(n\alpha - 1)^2$$

must hold for any $0 < \sigma^2 < \infty$ and any $0 \leq \mu^2$. Taking limits for $\sigma^2 \to 0$ yields

$$(n\gamma - 1)^2 \leq (n\alpha - 1)^2,$$

i.e. $n\gamma - 1 = c(n\alpha - 1)$ for $|c| \leq 1$, or, equivalently,

$$\gamma = \frac{1}{n}(1 + c(n\alpha - 1)), \quad -1 \leq c \leq 1\,.$$

Since $\gamma \mathbf{1}_n' \boldsymbol{y}$ is assumed to be uniformly better than $\alpha \mathbf{1}_n' \boldsymbol{y}$, it follows that $c \neq 1$, since otherwise the estimators $\gamma \mathbf{1}_n' \boldsymbol{y}$ and $\alpha \mathbf{1}_n' \boldsymbol{y}$ coincide. In addition, from setting $\mu = 0$ the inequality

$$n\gamma^2 \leq n\alpha^2 \quad \text{with } \gamma = \frac{1}{n}(1 + c(n\alpha - 1))$$

must hold true. On the other hand, it can easily be shown that in case $0 \leq \alpha < 1/n$ the inequality

$$n\alpha^2 < n\gamma^2 \quad \text{with } \gamma = \frac{1}{n}(1 + c(n\alpha - 1))$$

is valid for any c satisfying $-1 \leq c < 1$.

Hence, there cannot exist some uniformly better estimator in $\mathcal{L}^{\mathrm{h}}(\mu)$ than $\boldsymbol{a}'\boldsymbol{y}$ whenever $\boldsymbol{a} = \alpha \mathbf{1}_n$ for some $0 \leq \alpha < 1/n$.

Theorem 4.2. *Under the simple mean shift regression model, an estimator* $\boldsymbol{a}'\boldsymbol{y}$ *is admissible for* μ *within the set* $\mathcal{L}^{\mathrm{h}}(\mu)$ *and with respect to the unweighted squared error loss if and only if* $\boldsymbol{a}$ *satisfies the condition* $\mathbf{1}_n \boldsymbol{a}' \boldsymbol{a} \mathbf{1}_n' \leq_{\mathrm{L}} \mathbf{1}_n \boldsymbol{a}'$.

The above theorem gives a characterization of the set of admissible estimators within the set $\mathcal{L}^{\mathrm{h}}(\mu)$ under the simple mean shift model.

In addition to the above, it can be shown that the condition given in Theorem 4.2 is also necessary and sufficient for the admissibility of $\boldsymbol{a}'\boldsymbol{y}$ within the set $\mathcal{L}(\mu)$ of not necessarily homogeneous linear estimators. Moreover, a non-homogeneous linear estimator $\boldsymbol{a}'\boldsymbol{y} + d$ turns out to be admissible for μ within $\mathcal{L}(\mu)$ if and only if the homogeneous part $\boldsymbol{a}'\boldsymbol{y}$ of the estimator is admissible within the set $\mathcal{L}(\mu)$, and, in addition, d is an arbitrary multiple of the number $1 - \boldsymbol{a}'\mathbf{1}_n$.

4.2.2 Characterization of Linearly Admissible Estimators

When we consider a linear estimator $\boldsymbol{Ay}+\boldsymbol{a}$ for $\boldsymbol{\beta}$ under the linear regression model with assumptions (i) to (iv), then we can give necessary and sufficient conditions on the matrix $\boldsymbol{A}$ and the vector $\boldsymbol{a}$ such that $\boldsymbol{Ay}+\boldsymbol{a}$ is admissible for $\boldsymbol{\beta}$ within the set $\mathcal{L}(\boldsymbol{\beta})$ and with respect to the unweighted squared error loss.

Here, we can distinguish between more *implicit* conditions being related to algebraic properties, and more *explicit* conditions being related to the actual form of $\boldsymbol{A}$ and $\boldsymbol{a}$.

Implicit Characterization

The following characterization can be derived from the results in [8]. We omit the proof here, since a sound derivation of the result appears to be too involved. The reader might get an idea of the complexity of the problem by reconsidering the explanations given in Sect. 4.2.1 for the simple mean shift model.

Theorem 4.3. *Under the linear regression model with assumptions (i) to (iv), a linear estimator $\boldsymbol{Ay}+\boldsymbol{a}$ is admissible for $\boldsymbol{\beta}$ within the set $\mathcal{L}(\boldsymbol{\beta})$ and with respect to the unweighted squared error loss if and only if the inequality*

$$\boldsymbol{XAA'X'} \leq_{\mathrm{L}} \boldsymbol{XA}$$

is satisfied, and in addition $\boldsymbol{a} \in \mathcal{C}(\boldsymbol{I}_p - \boldsymbol{AX})$ holds true.

The condition $\boldsymbol{a} \in \mathcal{C}(\boldsymbol{I}_p - \boldsymbol{AX})$ means that the vector $\boldsymbol{a}$ lies in the same subspace as the bias vector $\mathrm{bias}(\boldsymbol{Ay}) = \mathrm{E}(\boldsymbol{Ay}) - \boldsymbol{\beta} = (\boldsymbol{AX} - \boldsymbol{I}_p)\boldsymbol{\beta}$ of the homogeneous part $\boldsymbol{Ay}$ of the estimator. This condition is of course satisfied when we consider the admissibility of homogeneous linear estimators, in which case $\boldsymbol{a} = \boldsymbol{0}$.

The condition $\boldsymbol{XAA'X'} \leq_{\mathrm{L}} \boldsymbol{XA}$ implies that the $n \times n$ matrix $\boldsymbol{XA}$ is symmetric. As shown by the following theorem, there exists an alternative characterization of this condition which is based on eigenvalues. For this, recall that $\sigma(\cdot)$ denotes the set of distinct eigenvalues of a square matrix.

Theorem 4.4. *Under the linear regression model with assumptions (i) to (iv), a linear estimator $\boldsymbol{Ay}+\boldsymbol{a}$ is admissible for $\boldsymbol{\beta}$ within the set $\mathcal{L}(\boldsymbol{\beta})$ and with respect to the unweighted squared error loss if and only if the three conditions*

$$\boldsymbol{XA} = \boldsymbol{A'X'}, \quad \sigma(\boldsymbol{XA}) \subset [0,1], \quad \boldsymbol{a} \in \mathcal{C}(\boldsymbol{I}_p - \boldsymbol{AX})$$

are satisfied.

Proof. If the matrix $\boldsymbol{XA} - \boldsymbol{XAA'X'}$ is symmetric nonnegative definite, then the matrix $\boldsymbol{XA}$ must be symmetric. Hence, the assertion follows from Theorem 4.3 when it can be shown that for symmetric $\boldsymbol{XA}$ the matrix $\boldsymbol{XA} - \boldsymbol{XAA'X'}$ is nonnegative definite if and only if all eigenvalues of $\boldsymbol{XA}$ lie in the closed interval $[0, 1]$. As a matter of fact, this equivalence holds true. To see this, note that when $\boldsymbol{XA}$ is symmetric, it can be written as $\boldsymbol{XA} = \boldsymbol{U\Lambda U'}$, where $\boldsymbol{U}$ is an orthogonal matrix and $\boldsymbol{\Lambda}$ is a diagonal matrix containing the eigenvalues $\lambda_1, \dots, \lambda_n$ of $\boldsymbol{XA}$ on its main diagonal. Then the matrix $\boldsymbol{XA} - \boldsymbol{XAA'X'}$ can be written as

$$\boldsymbol{XA} - \boldsymbol{XAA'X'} = \boldsymbol{U}(\boldsymbol{\Lambda} - \boldsymbol{\Lambda}^2)\boldsymbol{U}' .$$

This matrix is nonnegative definite if and only if $\boldsymbol{\Lambda} - \boldsymbol{\Lambda}^2$ is nonnegative definite, which in turn is equivalent to

$$\lambda_i - \lambda_i^2 \geq 0, \quad i = 1, \dots, n .$$

But the latter can hold if and only if $0 \leq \lambda_i \leq 1$ for $i = 1, \dots, n$, which concludes the proof. □

The symmetry of the matrix $\boldsymbol{XA}$ ensures that all eigenvalues of $\boldsymbol{XA}$ are real-valued (as opposed to complex-valued eigenvalues with non-vanishing imaginary parts). The condition $\sigma(\boldsymbol{XA}) \subset [0, 1]$ is equivalent to $\sigma(\boldsymbol{AX}) \subset [0, 1]$ since the non-zero eigenvalues of $\boldsymbol{XA}$ and $\boldsymbol{AX}$ coincide. However, it is clear that in contrast to $\boldsymbol{XA}$ the matrix $\boldsymbol{AX}$ is *not* necessarily symmetric.

Theorems 4.3 and 4.4 give necessary and sufficient conditions on $\boldsymbol{A}$ and $\boldsymbol{a}$ for the admissibility of $\boldsymbol{Ay} + \boldsymbol{a}$ within the set of linear estimators. One might ask whether there exist conditions under which a linear estimator can also be admissible within a broader set of estimators for $\boldsymbol{\beta}$. As a matter of fact, under normality only one further imposed condition is required to make a linear estimator admissible within the set $\{\widetilde{\boldsymbol{\beta}}(\boldsymbol{y})\}$ of all possible estimators.

Remark 4.1. Under the linear regression model with assumptions (i) to (iv) and (v), a linear estimator $\boldsymbol{Ay} + \boldsymbol{a}$ is admissible for $\boldsymbol{\beta}$ within the set $\{\widetilde{\boldsymbol{\beta}}(\boldsymbol{y})\}$ and with respect to the unweighted squared error loss if and only if $\boldsymbol{Ay} + \boldsymbol{a}$ is admissible for $\boldsymbol{\beta}$ within $\mathcal{L}(\boldsymbol{\beta})$, and in addition the inequality

$$p - 2 \leq \operatorname{rk}(\boldsymbol{I}_p - \boldsymbol{AX})$$

is satisfied; see [83, Sect. 7.2].

Example 4.1. Consider the simple mean shift model, described by

$$\boldsymbol{y} = \mu \boldsymbol{1}_n + \boldsymbol{\varepsilon}, \quad \boldsymbol{\varepsilon} \sim (\boldsymbol{0}, \sigma^2 \boldsymbol{I}_n) .$$

Let $\boldsymbol{a}'\boldsymbol{y}$ with

$$\boldsymbol{a}' = \frac{1}{n+k}\boldsymbol{1}_n'$$

be a homogeneous linear estimator for μ, where k is a fixed (non-stochastic) number. We can ask whether this estimator is admissible for μ within the set $\mathcal{L}(\mu)$ and with respect to the unweighted squared error loss. To find an answer, we evaluate the conditions given in Theorem 4.3. Since the estimator $\boldsymbol{a}'\boldsymbol{y}$ is homogeneous, it is admissible for μ if and only the condition $\mathbf{1}_n\boldsymbol{a}'\boldsymbol{a}\mathbf{1}_n' \leq_{\mathrm{L}} \mathbf{1}_n\boldsymbol{a}'$ is fulfilled. We can see that

$$\mathbf{1}_n\boldsymbol{a}' - \mathbf{1}_n\boldsymbol{a}'\boldsymbol{a}\mathbf{1}_n' = \frac{k}{(n+k)^2}\mathbf{1}_n\mathbf{1}_n' .$$

This matrix is clearly symmetric and it is nonnegative definite if and only if $k \geq 0$. Thus, for $k \geq 0$ the above estimator is admissible within the set $\mathcal{L}(\mu)$. It is not difficult to conclude that the considered estimator is in fact the ridge estimator for μ, see Problem 3.23.

Since we have $p = 1$, it follows from Remark 4.1 that $\boldsymbol{a}'\boldsymbol{y}$ is even admissible within the set of all possible estimators when normality is assumed. This admissibility, however, is not guaranteed when the parameter k is replaced by some estimator $\widehat{k}$ such that the resulting ridge estimator estimator becomes non-linear. □

Explicit Characterization

Theorems 4.3 and 4.4 provide implicit characterizations of linearly admissible estimators $\boldsymbol{A}\boldsymbol{y} + \boldsymbol{a}$ by giving necessary and sufficient algebraic conditions on $\boldsymbol{A}$ and $\boldsymbol{a}$. From the following theorem, the explicit form of $\boldsymbol{A}$ and $\boldsymbol{a}$ which is necessary and sufficient for the admissibility of $\boldsymbol{A}\boldsymbol{y} + \boldsymbol{a}$ for the parameter vector $\boldsymbol{\beta}$ can be deduced.

Theorem 4.5. *Under the linear regression model with assumptions (i) to (iv), a linear estimator $\boldsymbol{A}\boldsymbol{y} + \boldsymbol{a}$ is admissible for $\boldsymbol{\beta}$ within the set $\mathcal{L}(\boldsymbol{\beta})$ and with respect to the unweighted squared error loss if and only if $\boldsymbol{A}$ and $\boldsymbol{a}$ can be written as*

$$\boldsymbol{A} = (\boldsymbol{X}'\boldsymbol{X})^{-1/2}\boldsymbol{G}(\boldsymbol{X}'\boldsymbol{X})^{-1/2}\boldsymbol{X}'$$

and

$$\boldsymbol{a} = (\boldsymbol{X}'\boldsymbol{X})^{-1/2}(\boldsymbol{I}_p - \boldsymbol{G})(\boldsymbol{X}'\boldsymbol{X})^{1/2}\boldsymbol{g}$$

for some symmetric $p \times p$ matrix $\boldsymbol{G}$ such that $\sigma(\boldsymbol{G}) \subset [0,1]$, and some $p \times 1$ vector $\boldsymbol{g}$.

Proof. First, we demonstrate that the two conditions $\boldsymbol{X}\boldsymbol{A} = \boldsymbol{A}'\boldsymbol{X}'$ and $\sigma(\boldsymbol{X}\boldsymbol{A}) \subset [0,1]$ from Theorem 4.4 are satisfied if and only if $\boldsymbol{A}$ can be written in the asserted form.

Now, if $\boldsymbol{A}$ can be written as $\boldsymbol{A} = (\boldsymbol{X}'\boldsymbol{X})^{-1/2}\boldsymbol{G}(\boldsymbol{X}'\boldsymbol{X})^{-1/2}\boldsymbol{X}'$ for some symmetric $p \times p$ matrix $\boldsymbol{G}$ with $\sigma(\boldsymbol{G}) \subset [0,1]$, then the matrix $\boldsymbol{X}\boldsymbol{A} = \boldsymbol{X}(\boldsymbol{X}'\boldsymbol{X})^{-1/2}\boldsymbol{G}(\boldsymbol{X}'\boldsymbol{X})^{-1/2}\boldsymbol{X}'$ is clearly symmetric. The non-zero eigenvalues of $\boldsymbol{X}\boldsymbol{A}$ coincide with the non-zero eigenvalues of

$$G(X'X)^{-1/2}X'X(X'X)^{-1/2} = G\,,$$

implying that all eigenvalues of XA lie in the closed interval $[0,1]$.

Conversely, if $Ay + a$ is linearly admissible for β, then the symmetry of the matrix XA implies that the identity $AX = (X'X)^{-1}X'XAX$ is equivalent to the identity $AX = (X'X)^{-1}X'A'X'X$. Multiplication of the latter from the left with $(X'X)^{1/2}$ and from the right with $(X'X)^{-1/2}$ yields

$$(X'X)^{1/2}AX(X'X)^{-1/2} = (X'X)^{-1/2}X'A'(X'X)^{1/2}\,.$$

If we set

$$G = (X'X)^{-1/2}X'A'(X'X)^{1/2}\,,$$

then this matrix G must be symmetric. The non-zero eigenvalues of G coincide with the nonzero eigenvalues of $AX(X'X)^{-1/2}(X'X)^{1/2} = AX$, which in turn coincide with the non-zero eigenvalues of XA. This shows that (in view of Theorem 4.4) all eigenvalues of the matrix G lie in $[0,1]$. Eventually,

$$\begin{aligned}(X'X)^{-1/2}G(X'X)^{-1/2}X' &= (X'X)^{-1}X'A'X' \\ &= (X'X)^{-1}X'XA \\ &= A\,.\end{aligned}$$

This shows that the matrix A can be written in the asserted form, whenever the estimator $Ay + a$ is linearly admissible for β.

The proof can be concluded by noting that the condition $a \in \mathcal{C}(I_p - AX)$ from Theorem 4.4 is equivalent to the existence of some vector g with $a = (I_p - AX)g$. This identity ensures the asserted form of the vector a. □

4.2.3 Ordinary Least Squares and Linearly Admissible Estimator

From Theorem 4.5 it follows that any linearly admissible estimator for β can be written in the form

$$S^{-1/2}GS^{1/2}\widehat{\beta} + S^{-1/2}(I_p - G)S^{1/2}g\,,$$

where $S = X'X$. This shows that any linearly admissible estimator is an (affine) linear transformation of the ordinary least squares estimator $\widehat{\beta}$.

The following theorem demonstrates that the covariance matrix of a linearly admissible estimator for β is always comparable to the covariance matrix of the ordinary least squares estimator with respect to the Löwner partial ordering.

Theorem 4.6. *Under the linear regression model with assumptions (i) to (iv), let $Ay + a$ be admissible for β within the set $\mathcal{L}(\beta)$ and with respect to the unweighted squared error loss. Then the inequality*

$$\mathrm{Cov}(Ay + a) \leq_{\mathrm{L}} \mathrm{Cov}(\widehat{\beta}).$$

holds true.

Proof. Since the estimator $\boldsymbol{Ay}+\boldsymbol{a}$ is linearly admissible for $\boldsymbol{\beta}$, it follows that the matrix $\boldsymbol{XA}$ is symmetric. As in the proof of Theorem 4.4 we can write $\boldsymbol{XA} = \boldsymbol{U\Lambda U}'$, where $\boldsymbol{U}$ is an orthogonal matrix and $\boldsymbol{\Lambda}$ is a diagonal matrix containing the eigenvalues of $\boldsymbol{XA}$ on its main diagonal. Since all eigenvalues of $\boldsymbol{XA}$ lie in $[0,1]$, the matrix $\boldsymbol{I}_n - \boldsymbol{\Lambda}^2$ is nonnegative definite. But then also the matrix

$$\boldsymbol{U}(\boldsymbol{I}_n - \boldsymbol{\Lambda}^2)\boldsymbol{U}' = \boldsymbol{I}_n - \boldsymbol{XAA}'\boldsymbol{X}'$$

is nonnegative definite. Multiplying this matrix from the left by $(\boldsymbol{X}'\boldsymbol{X})^{-1}\boldsymbol{X}'$ and from the right by $\boldsymbol{X}(\boldsymbol{X}'\boldsymbol{X})^{-1}$ yields the nonnegative definite matrix

$$(\boldsymbol{X}'\boldsymbol{X})^{-1}\boldsymbol{X}'\boldsymbol{X}(\boldsymbol{X}'\boldsymbol{X})^{-1} - (\boldsymbol{X}'\boldsymbol{X})^{-1}\boldsymbol{X}'\boldsymbol{XAA}'\boldsymbol{X}'\boldsymbol{X}(\boldsymbol{X}'\boldsymbol{X})^{-1}\,,$$

being identical to

$$(\boldsymbol{X}'\boldsymbol{X})^{-1} - \boldsymbol{AA}'\,.$$

From this, the nonnegative definiteness of the matrix $\mathrm{Cov}(\widehat{\boldsymbol{\beta}}) - \mathrm{Cov}(\boldsymbol{Ay}+\boldsymbol{a})$ follows. □

The above result shows that any estimator $\boldsymbol{Ay}+\boldsymbol{a}$ which is linearly admissible for $\boldsymbol{\beta}$ must have a covariance matrix which cannot be greater than the covariance matrix of the ordinary least squares estimator $\widehat{\boldsymbol{\beta}}$ in the sense of the Löwner partial ordering. Then, of course, also the variance of each element of $\boldsymbol{Ay}+\boldsymbol{a}$ is smaller than or equal to the variance of the corresponding element of $\widehat{\boldsymbol{\beta}}$.

However, since the Gauss-Markov Theorem 2.9 (see p. 51) holds true, the above result also implies that linearly admissible estimators cannot be unbiased for $\boldsymbol{\beta}$. As matter of fact, the following holds true.

Theorem 4.7. *Under the linear regression model with assumptions (i) to (iv), let $\boldsymbol{Ay}+\boldsymbol{a}$ be admissible for $\boldsymbol{\beta}$ within the set $\mathcal{L}(\boldsymbol{\beta})$ and with respect to the unweighted squared error loss. Then $\boldsymbol{Ay}+\boldsymbol{a}$ is unbiased for $\boldsymbol{\beta}$ if and only if*

$$\boldsymbol{Ay}+\boldsymbol{a} = \widehat{\boldsymbol{\beta}}$$

hods true.

Proof. An estimator $\boldsymbol{Ay}+\boldsymbol{a}$ is unbiased for $\boldsymbol{\beta}$ if and only if $\boldsymbol{a} = \boldsymbol{0}$ and $\boldsymbol{AX} = \boldsymbol{I}_p$. The latter identity can equivalently be written as $\boldsymbol{XAX} = \boldsymbol{X}$. When $\boldsymbol{Ay}+\boldsymbol{a}$ is linearly admissible for $\boldsymbol{\beta}$, then $\boldsymbol{XA}$ is symmetric, implying that the identity $\boldsymbol{XAX} = \boldsymbol{X}$ is equivalent to $\boldsymbol{A}'\boldsymbol{X}'\boldsymbol{X} = \boldsymbol{X}$, i.e. $\boldsymbol{A}' = \boldsymbol{X}(\boldsymbol{X}'\boldsymbol{X})^{-1}$, so that $\boldsymbol{A} = (\boldsymbol{X}'\boldsymbol{X})^{-1}\boldsymbol{X}'$. □

Now, we have seen that a linearly admissible estimator $\boldsymbol{Ay}+\boldsymbol{a}$ for $\boldsymbol{\beta}$ which does not coincide with $\widehat{\boldsymbol{\beta}}$ is necessarily a biased estimator and has a covariance matrix which is not greater than the covariance matrix of $\widehat{\boldsymbol{\beta}}$ with respect to the Löwner partial ordering. The converse is however not true, meaning that there might exist biased estimators for $\boldsymbol{\beta}$ having a smaller covariance matrix than that of the ordinary least squares estimator, but nonetheless being *not* linearly admissible for $\boldsymbol{\beta}$. Such an estimator is considered in the following example.

Example 4.2. Consider the linear shrinkage estimator

$$\widehat{\boldsymbol{\beta}}^{-}(\varrho) = -\frac{1}{1+\varrho}\widehat{\boldsymbol{\beta}}, \quad \varrho \geq 0\,,$$

with a negative shrinkage parameter $-1/(1+\varrho)$. Then this estimator $\widehat{\boldsymbol{\beta}}^{-}(\varrho)$ has the same covariance matrix as the usual shrinkage estimator $\widehat{\boldsymbol{\beta}}(\varrho)$, i.e.

$$\mathrm{Cov}(\widehat{\boldsymbol{\beta}}^{-}(\varrho)) = \frac{\sigma^2}{(1+\varrho)^2}(\boldsymbol{X}'\boldsymbol{X})^{-1}\,.$$

Clearly, the difference

$$\mathrm{Cov}(\widehat{\boldsymbol{\beta}}) - \mathrm{Cov}(\widehat{\boldsymbol{\beta}}^{-}(\varrho)) = \frac{\varrho(2+\varrho)\,\sigma 2}{(1+\varrho)^2}(\boldsymbol{X}'\boldsymbol{X})^{-1}$$

is nonnegative definite whenever $\varrho \geq 0$. Moreover, $\mathrm{E}(\widehat{\boldsymbol{\beta}}^{-}(\varrho)) = -\frac{1}{1+\varrho}\boldsymbol{\beta} \neq \boldsymbol{\beta}$ for any vector $\boldsymbol{\beta} \neq \boldsymbol{0}$, so that $\widehat{\boldsymbol{\beta}}^{-}(\varrho)$ is not an unbiased estimator for $\boldsymbol{\beta}$.

To examine whether $\widehat{\boldsymbol{\beta}}^{-}(\varrho)$ is linearly admissible for $\boldsymbol{\beta}$ or not, we write

$$\widehat{\boldsymbol{\beta}}^{-}(\varrho) = \boldsymbol{A}\boldsymbol{y} + \boldsymbol{a}, \quad \boldsymbol{A} = -\frac{1}{1+\varrho}(\boldsymbol{X}'\boldsymbol{X})^{-1}\boldsymbol{X}', \quad \boldsymbol{a} = \boldsymbol{0}\,.$$

Then the eigenvalues of $\boldsymbol{A}\boldsymbol{X}$ are all equal to $-1/(1+\varrho)$ and thus dot not lie in the interval $[0, 1]$. Hence, from Theorem 4.4 it follows that the estimator $\widehat{\boldsymbol{\beta}}^{-}(\varrho)$ is not linearly admissible for $\boldsymbol{\beta}$.

An alternative method to prove the inadmissibility is, to demonstrate that the estimator $\widehat{\boldsymbol{\beta}}(\varrho)$ is uniformly better than $\widehat{\boldsymbol{\beta}}^{-}(\varrho)$ with respect to the unweighted squared error loss, see Problem 4.3. □

4.2.4 Linear Transforms of Ordinary Least Squares Estimator

Form Theorem 4.5 we can conclude that any *homogeneous* linear estimator being admissible for $\boldsymbol{\beta}$ within the set of linear (not necessarily homogeneous) estimators can be written as a linear transformation $\boldsymbol{L}\widehat{\boldsymbol{\beta}}$ of the ordinary least squares estimator for some specific matrix $\boldsymbol{L}$.

On the other hand, given some matrix $\boldsymbol{L}$, the linear transformation $\boldsymbol{L}\widehat{\boldsymbol{\beta}}$ is linearly admissible for $\boldsymbol{\beta}$ if and only if the conditions

$$\boldsymbol{X}'\boldsymbol{X}\boldsymbol{L} = \boldsymbol{L}'\boldsymbol{X}'\boldsymbol{X} \quad \text{and} \quad \sigma(\boldsymbol{L}) \subset [0, 1]$$

are satisfied, see Problem 4.4. If $\boldsymbol{L}$ is a symmetric matrix, then the condition $\boldsymbol{X}'\boldsymbol{X}\boldsymbol{L} = \boldsymbol{L}\boldsymbol{X}'\boldsymbol{X}$ is fulfilled if and only if there exists an orthogonal matrix $\boldsymbol{U}$ and diagonal matrices $\boldsymbol{\Lambda}$ and $\boldsymbol{\Gamma}$ such that

$$\boldsymbol{X}'\boldsymbol{X} = \boldsymbol{U}\boldsymbol{\Lambda}\boldsymbol{U}' \quad \text{and} \quad \boldsymbol{L} = \boldsymbol{U}\boldsymbol{\Gamma}\boldsymbol{U}' ,$$

see also Theorem A.47. Then the condition $\sigma(\boldsymbol{L}) \subset [0,1]$ means that every diagonal element γ_i, $i = 1, \ldots, p$, lies in the interval $[0,1]$.

This shows that such linear transformations $\boldsymbol{L}\widehat{\boldsymbol{\beta}}$ are related to the set of estimators introduced by Obenchain [81], which can be written as

$$\mathcal{L}^{\text{Ob}}(\boldsymbol{\beta}) = \boldsymbol{U}\boldsymbol{\Gamma}\boldsymbol{U}'\widehat{\boldsymbol{\beta}}, \quad \boldsymbol{\Gamma} = \text{diag}(\gamma_1, \ldots, \gamma_p), \quad 0 \le \gamma_i \le 1 ,$$

where the orthogonal matrix $\boldsymbol{U}$ stems from a spectral decomposition $\boldsymbol{U}\boldsymbol{\Lambda}\boldsymbol{U}'$ of $\boldsymbol{X}'\boldsymbol{X}$. In view of the above considerations we can also write

$$\mathcal{L}^{\text{Ob}}(\boldsymbol{\beta}) = \{\boldsymbol{L}\widehat{\boldsymbol{\beta}} : \boldsymbol{L} = \boldsymbol{L}', \, \boldsymbol{X}'\boldsymbol{X}\boldsymbol{L} = \boldsymbol{L}\boldsymbol{X}'\boldsymbol{X}, \, \sigma(\boldsymbol{L}) \subset [0,1]\} ,$$

which we will call the *Obenchain class* of estimators for $\boldsymbol{\beta}$.

Any element $\boldsymbol{L}\widehat{\boldsymbol{\beta}}$ from the Obenchain class is linearly admissible for $\boldsymbol{\beta}$ and in addition satisfies $||\boldsymbol{L}\widehat{\boldsymbol{\beta}}||^2 \le ||\widehat{\boldsymbol{\beta}}||^2$, see Problem 4.4.

In view of the latter, the Obenchain class is sometimes also called the *class of generalized ridge estimators* in the literature. We will not adopt this term here, since an estimator from the Obenchain class is not necessarily of the form of a general ridge estimator $\widehat{\boldsymbol{\beta}}_{\boldsymbol{K}} = (\boldsymbol{X}'\boldsymbol{X} + \boldsymbol{K})^{-1}\boldsymbol{X}'\boldsymbol{y}$, where $\boldsymbol{K}$ is a symmetric nonnegative definite matrix. It is, however, interesting to investigate conditions under which an element from $\mathcal{L}^{\text{Ob}}(\boldsymbol{\beta})$ can be written as $\widehat{\boldsymbol{\beta}}_{\boldsymbol{K}}$ for some $\boldsymbol{K}$.

Theorem 4.8. *Under the linear regression model with assumptions (i) to (iv), an estimator $\boldsymbol{L}\widehat{\boldsymbol{\beta}} \in \mathcal{L}^{\text{Ob}}(\boldsymbol{\beta})$ can be written as $\boldsymbol{L}\widehat{\boldsymbol{\beta}} = \widehat{\boldsymbol{\beta}}_{\boldsymbol{K}}$ for some symmetric nonnegative definite matrix $\boldsymbol{K}$ irrespective of the realization $\boldsymbol{y} \in \mathbb{R}^n$ if and only if $\sigma(\boldsymbol{L}) \subset (0,1]$.*

Proof. To start with, assume that $\boldsymbol{L}\widehat{\boldsymbol{\beta}} \in \mathcal{L}^{\text{Ob}}(\boldsymbol{\beta})$ can be written as $\boldsymbol{L}\widehat{\boldsymbol{\beta}} = \widehat{\boldsymbol{\beta}}_{\boldsymbol{K}}$ for some symmetric nonnegative definite matrix $\boldsymbol{K}$ and every $\boldsymbol{y} \in \mathbb{R}^n$. Then

$$\boldsymbol{L}(\boldsymbol{X}'\boldsymbol{X})^{-1}\boldsymbol{X}' = (\boldsymbol{X}'\boldsymbol{X} + \boldsymbol{K})^{-1}\boldsymbol{X}' ,$$

and therefore

$$\boldsymbol{L} = (\boldsymbol{X}'\boldsymbol{X} + \boldsymbol{K})^{-1}\boldsymbol{X}'\boldsymbol{X} .$$

The rank of the right-hand matrix equals p, while the rank of the left-hand matrix $\boldsymbol{L}$ coincides with the number of non-zero eigenvalues of $\boldsymbol{L}$, since $\boldsymbol{L}$ is symmetric. Hence, the above identity implies that all eigenvalues of $\boldsymbol{L}$ must be non-zero, so that $\sigma(\boldsymbol{L}) \subset (0,1]$.

Now, suppose that $\boldsymbol{L}\widehat{\boldsymbol{\beta}} \in \mathcal{L}^{\mathrm{Ob}}(\boldsymbol{\beta})$ with $\sigma(\boldsymbol{L}) \subset (0,1]$ and let $\boldsymbol{U}\boldsymbol{\Lambda}\boldsymbol{U}'$ be a spectral decomposition of $\boldsymbol{X}'\boldsymbol{X}$. Then the matrix $\boldsymbol{L}$ can be written as $\boldsymbol{L} = \boldsymbol{U}\boldsymbol{\Gamma}\boldsymbol{U}'$, where $\boldsymbol{\Gamma}$ is a diagonal matrix whose every diagonal element lies in the interval $(0,1]$. Therefore, the matrix

$$\boldsymbol{K} = \boldsymbol{U}(\boldsymbol{\Lambda}\boldsymbol{\Gamma}^{-1} - \boldsymbol{\Lambda})\boldsymbol{U}'$$

is symmetric nonnegative definite, and

$$\begin{aligned}(\boldsymbol{X}'\boldsymbol{X} + \boldsymbol{K})^{-1}\boldsymbol{X}'\boldsymbol{y} &= \boldsymbol{U}\boldsymbol{\Gamma}\boldsymbol{\Lambda}^{-1}\boldsymbol{U}'\boldsymbol{X}'\boldsymbol{y} \\ &= \boldsymbol{U}\boldsymbol{\Gamma}\boldsymbol{U}'\boldsymbol{U}\boldsymbol{\Lambda}^{-1}\boldsymbol{U}'\boldsymbol{X}'\boldsymbol{y} \\ &= \boldsymbol{L}\widehat{\boldsymbol{\beta}}\,,\end{aligned}$$

thus concluding the proof. □

Table 4.1 displays some estimators from the Obenchain class with respect to a given spectral decomposition $\boldsymbol{X}'\boldsymbol{X} = \boldsymbol{U}\boldsymbol{\Lambda}\boldsymbol{U}' = \boldsymbol{U}_1\boldsymbol{\Lambda}_1\boldsymbol{U}_1' + \boldsymbol{U}_2\boldsymbol{\Lambda}_2\boldsymbol{U}_2'$.

Table 4.1. Different estimators from the Obenchain class

estimator	symbol	$\boldsymbol{L}\widehat{\boldsymbol{\beta}} \in \mathcal{L}^{\mathrm{Ob}}(\boldsymbol{\beta})$
ordinary least squares	$\widehat{\boldsymbol{\beta}}$	$\boldsymbol{L} = \boldsymbol{I}_p$
principal components	$\widehat{\boldsymbol{\beta}}^{(r)}$	$\boldsymbol{L} = \boldsymbol{U}_1\boldsymbol{U}_1'$
Marquardt	$\widehat{\boldsymbol{\beta}}^{(r,c)}$	$\boldsymbol{L} = \boldsymbol{U}_1\boldsymbol{U}_1' + c\boldsymbol{U}_2\boldsymbol{U}_2'$
ridge	$\widehat{\boldsymbol{\beta}}_k$	$\boldsymbol{L} = \boldsymbol{U}(\boldsymbol{\Lambda} + k\boldsymbol{I}_p)^{-1}\boldsymbol{\Lambda}\boldsymbol{U}'$
Baye-Parker	$\widehat{\boldsymbol{\beta}}_k^{(r)}$	$\boldsymbol{L} = \boldsymbol{U}_1(\boldsymbol{\Lambda}_1 + k\boldsymbol{I}_r)^{-1}\boldsymbol{\Lambda}_1\boldsymbol{U}_1'$
almost unbiased ridge	$\widehat{\boldsymbol{\beta}}_k^{\mathrm{J}}$	$\boldsymbol{L} = \boldsymbol{U}(\boldsymbol{\Lambda}^2 + 2k\boldsymbol{\Lambda})(\boldsymbol{\Lambda} + k\boldsymbol{I}_p)^{-2}\boldsymbol{U}'$
iteration	$\widehat{\boldsymbol{\beta}}_{\delta,m}$	$\boldsymbol{L} = \boldsymbol{U}(\boldsymbol{I}_p - (\boldsymbol{I}_p - \delta\boldsymbol{\Lambda})^{m+1})\boldsymbol{U}'$
shrinkage	$\widehat{\boldsymbol{\beta}}(\varrho)$	$\boldsymbol{L} = (1+\varrho)^{-1}\boldsymbol{I}_p$

4.2.5 Linear Admissibility of Known Estimators

If some homogeneous linear estimator belongs to the Obenchain class, then it is also admissible for $\boldsymbol{\beta}$ within the set of all linear estimators. As shown in Table 4.1, quite a few of the known estimators in fact belong to $\mathcal{L}^{\mathrm{Ob}}(\boldsymbol{\beta})$. Nonetheless, there exist (possibly non-homogeneous) estimators of *general ridge* type as well as *restricted estimators*, which do not belong to this class and thus the admissibility of such estimators has to be checked separately.

General Ridge Estimator

A non-homogeneous general ridge estimator is of the form

$$\widehat{\boldsymbol{\beta}}_{\boldsymbol{K},\boldsymbol{\beta}_0} = \boldsymbol{\beta}_0 + (\boldsymbol{X}'\boldsymbol{X} + \boldsymbol{K})^{-1}\boldsymbol{X}'(\boldsymbol{y} - \boldsymbol{X}\boldsymbol{\beta}_0)\,.$$

When the $p \times 1$ vector $\boldsymbol{\beta}_0$ and the $p \times p$ symmetric nonnegative definite matrix $\boldsymbol{K}$ are chosen non-stochastically, then $\widehat{\boldsymbol{\beta}}_{\boldsymbol{K},\boldsymbol{\beta}_0}$ is also called a *linear estimator of ridge type.*

Theorem 4.9. *Under the linear regression model with assumptions (i) to (iv), let $\boldsymbol{K}$ be a given $p \times p$ symmetric nonnegative definite matrix and let $\boldsymbol{\beta}_0$ be a given $p \times 1$ vector. Then the general ridge estimator*

$$\widehat{\boldsymbol{\beta}}_{\boldsymbol{K},\boldsymbol{\beta}_0} = \boldsymbol{\beta}_0 + (\boldsymbol{X}'\boldsymbol{X} + \boldsymbol{K})^{-1}\boldsymbol{X}'(\boldsymbol{y} - \boldsymbol{X}\boldsymbol{\beta}_0)$$

is linearly admissible for $\boldsymbol{\beta}$.

Proof. We can write $\widehat{\boldsymbol{\beta}}_{\boldsymbol{K},\boldsymbol{\beta}_0} = \boldsymbol{A}\boldsymbol{y} + \boldsymbol{a}$ with

$$\boldsymbol{A} = (\boldsymbol{X}'\boldsymbol{X} + \boldsymbol{K})^{-1}\boldsymbol{X}'$$

and

$$\boldsymbol{a} = \boldsymbol{\beta}_0 - (\boldsymbol{X}'\boldsymbol{X} + \boldsymbol{K})^{-1}\boldsymbol{X}'\boldsymbol{X}\boldsymbol{\beta}_0 = (\boldsymbol{X}'\boldsymbol{X} + \boldsymbol{K})^{-1}\boldsymbol{K}\boldsymbol{\beta}_0\,.$$

In view of the identity $\boldsymbol{I}_p - \boldsymbol{A}\boldsymbol{X} = (\boldsymbol{X}'\boldsymbol{X} + \boldsymbol{K})^{-1}\boldsymbol{K}$, the condition $\boldsymbol{a} \in \mathcal{C}(\boldsymbol{I}_p - \boldsymbol{A}\boldsymbol{X})$ is satisfied. Moreover, it is obvious that the matrix $\boldsymbol{X}\boldsymbol{A}$ is symmetric. Thus, according to Theorem 4.4 it remains to show that the eigenvalues of $\boldsymbol{X}\boldsymbol{A}$ lie in the closed interval $[0,1]$.

Since $\boldsymbol{K}$ is symmetric nonnegative definite, the matrix $\boldsymbol{X}\boldsymbol{A}$ is symmetric nonnegative definite too, showing that all eigenvalues of $\boldsymbol{X}\boldsymbol{A}$ are real nonnegative. In addition,

$$\boldsymbol{X}'\boldsymbol{X} \leq_{\mathrm{L}} \boldsymbol{X}'\boldsymbol{X} + \boldsymbol{K}\,,$$

implying that

$$(\boldsymbol{X}'\boldsymbol{X} + \boldsymbol{K})^{-1} \leq_{\mathrm{L}} (\boldsymbol{X}'\boldsymbol{X})^{-1}\,,$$

see also Theorem A.67. Multiplication of this inequality from the left by $\boldsymbol{X}$ and from the right by $\boldsymbol{X}'$ shows

$$\boldsymbol{X}(\boldsymbol{X}'\boldsymbol{X} + \boldsymbol{K})^{-1}\boldsymbol{X}' \leq_{\mathrm{L}} \boldsymbol{X}(\boldsymbol{X}'\boldsymbol{X})^{-1}\boldsymbol{X}'\,.$$

But then, any eigenvalue of $\boldsymbol{X}(\boldsymbol{X}'\boldsymbol{X}+\boldsymbol{K})^{-1}\boldsymbol{X}'$ is not greater than the largest eigenvalue of $\boldsymbol{X}(\boldsymbol{X}'\boldsymbol{X})^{-1}\boldsymbol{X}'$. Since the eigenvalues of $\boldsymbol{X}(\boldsymbol{X}'\boldsymbol{X})^{-1}\boldsymbol{X}'$ are only 0 or 1 (with at least one eigenvalue 1 when $\boldsymbol{X}$ is non-zero), it is shown that the eigenvalues of $\boldsymbol{X}\boldsymbol{A}$ lie in $[0,1]$, and the assertion follows. □

Table 4.2 displays linear ridge type estimators, some of them also belonging to the Obenchain class. In view of the above result, any of these estimators is linearly admissible for $\boldsymbol{\beta}$.

Table 4.2. Different linear ridge type estimators

estimator	symbol	$\widehat{\boldsymbol{\beta}}_{\boldsymbol{K},\boldsymbol{\beta}_0}$
ordinary least squares	$\widehat{\boldsymbol{\beta}}$	$\boldsymbol{K}=\boldsymbol{0}$, $\boldsymbol{\beta}_0=\boldsymbol{0}$
Marquardt	$\widehat{\boldsymbol{\beta}}^{(r,c)}$, $c>0$	$\boldsymbol{K}=((1-c)/c)\,\boldsymbol{U}_2\boldsymbol{\Lambda}_2\boldsymbol{U}_2'$, $\boldsymbol{\beta}_0=\boldsymbol{0}$
ridge	$\widehat{\boldsymbol{\beta}}_k$	$\boldsymbol{K}=k\boldsymbol{I}_p$, $\boldsymbol{\beta}_0=\boldsymbol{0}$
ridge	$\widehat{\boldsymbol{\beta}}_{k_1,\ldots,k_p}$	$\boldsymbol{K}=\boldsymbol{U}\mathrm{diag}(k_1,\ldots,k_p)\boldsymbol{U}'$, $\boldsymbol{\beta}_0=\boldsymbol{0}$
Farebrother	$\widehat{\boldsymbol{\beta}}_{\mathrm{R}}(k)$, $k<\infty$	$\boldsymbol{K}=k\boldsymbol{R}'\boldsymbol{R}$, $\boldsymbol{\beta}_0=\boldsymbol{R}'(\boldsymbol{R}\boldsymbol{R}')^{-1}\boldsymbol{r}$
almost unbiased ridge	$\widehat{\boldsymbol{\beta}}_k^{\mathrm{J}}$	$\boldsymbol{K}=k^2(\boldsymbol{X}'\boldsymbol{X}+2k\boldsymbol{I}_p)^{-1}$, $\boldsymbol{\beta}_0=\boldsymbol{0}$
iteration	$\widehat{\boldsymbol{\beta}}_{\delta,m}$	$\boldsymbol{K}=\boldsymbol{X}'\boldsymbol{X}(\boldsymbol{T}^{-1}-\boldsymbol{I}_p)$, $\boldsymbol{\beta}_0=\boldsymbol{0}$, $\boldsymbol{T}=\boldsymbol{I}_p-(\boldsymbol{I}_p-\delta\boldsymbol{X}'\boldsymbol{X})^{m+1}$
shrinkage	$\widehat{\boldsymbol{\beta}}(\varrho)$	$\boldsymbol{K}=\varrho\boldsymbol{X}'\boldsymbol{X}$, $\boldsymbol{\beta}_0=\boldsymbol{0}$
direction modified shrinkage	$\widehat{\boldsymbol{\beta}}_{\boldsymbol{\beta}_0}(\varrho)$	$\boldsymbol{K}=\varrho\boldsymbol{X}'\boldsymbol{X}$

Restricted Estimator

A linear estimator $\widetilde{\boldsymbol{\beta}}\in\mathcal{L}(\boldsymbol{\beta})$ satisfying the identity

$$\boldsymbol{R}\widetilde{\boldsymbol{\beta}}=\boldsymbol{r}$$

for some given $\boldsymbol{R}\neq\boldsymbol{0}$, some given $\boldsymbol{r}$, and *any* possible realization $\boldsymbol{y}\in\mathrm{I\!R}^n$ can be called a *restricted* linear estimator.

A linear ridge type estimator $\widehat{\boldsymbol{\beta}}_{\boldsymbol{K},\boldsymbol{\beta}_0}$ cannot be a restricted estimator in this sense when the trivial case $(\boldsymbol{R},\boldsymbol{r})=(\boldsymbol{0},\boldsymbol{0})$ is excluded. On the other hand, the restricted least squares estimator $\widehat{\boldsymbol{\beta}}_{\mathrm{R}}$ is of course a linear restricted estimator. The admissibility of $\widehat{\boldsymbol{\beta}}_{\mathrm{R}}$ within the set of linear estimators is guaranteed by the following result.

Theorem 4.10. *Under the linear regression model with assumptions (i) to (iv), let $\boldsymbol{R}$ be a given $m\times p$ matrix of full row rank and let $\boldsymbol{r}$ be a given $m\times 1$ vector. Then the restricted least squares estimator*

$$\widehat{\boldsymbol{\beta}}_{\mathrm{R}}=\widehat{\boldsymbol{\beta}}-(\boldsymbol{X}'\boldsymbol{X})^{-1}\boldsymbol{R}'[\boldsymbol{R}(\boldsymbol{X}'\boldsymbol{X})^{-1}\boldsymbol{R}']^{-1}(\boldsymbol{R}\widehat{\boldsymbol{\beta}}-\boldsymbol{r})$$

is linearly admissible for $\boldsymbol{\beta}$.

Proof. We can write $\widehat{\boldsymbol{\beta}}_{\mathrm{R}}=\boldsymbol{A}\boldsymbol{y}+\boldsymbol{a}$ with

$$\begin{aligned}\boldsymbol{A}&=(\boldsymbol{X}'\boldsymbol{X})^{-1}\boldsymbol{X}'-(\boldsymbol{X}'\boldsymbol{X})^{-1}\boldsymbol{R}'[\boldsymbol{R}(\boldsymbol{X}'\boldsymbol{X})^{-1}\boldsymbol{R}']^{-1}\boldsymbol{R}(\boldsymbol{X}'\boldsymbol{X})^{-1}\boldsymbol{X}'\,,\\ \boldsymbol{a}&=(\boldsymbol{X}'\boldsymbol{X})^{-1}\boldsymbol{R}'[\boldsymbol{R}(\boldsymbol{X}'\boldsymbol{X})^{-1}\boldsymbol{R}']^{-1}\boldsymbol{r}\,.\end{aligned}$$

Then the matrix $\boldsymbol{X}\boldsymbol{A}$ is obviously symmetric. Moreover, it can easily be checked that the identity $\boldsymbol{X}\boldsymbol{A}\boldsymbol{X}\boldsymbol{A}=\boldsymbol{X}\boldsymbol{A}$ is satisfied. This implies that the matrix

$$XA - XAA'X' = 0$$

is symmetric nonnegative definite. Since $a = (I_p - AX)R'(RR')^{-1}r$, Theorem 4.3 shows that $\widehat{\beta}_{\mathrm{R}}$ is linearly admissible for β. □

From the above theorem it follows that the principal components estimator $\widehat{\beta}^{(r)}$ is linearly admissible for β, since for $r < p$ it is a specific restricted least squares estimator satisfying

$$U_2'\widehat{\beta}^{(r)} = 0\,,$$

where $U_1\Lambda_1U_1' + U_2\Lambda_2U_2'$ is a spectral decomposition of the matrix $X'X$. Of course, the linear admissibility of $\widehat{\beta}^{(r)}$ can also be concluded from $\widehat{\beta}^{(r)}$ belonging to the Obenchain class.

The estimator $\widehat{\beta}_k^{(r)}$ from Baye and Parker [9] is also a restricted estimator satisfying $U_2'\widehat{\beta}_k^{(r)} = 0$ for any realization $y \in \mathbb{R}^n$. Nonetheless, it is not a specific restricted least squares estimator. The linear admissibility of this estimator follows from the fact that $\widehat{\beta}_k^{(r)}$ belongs to the Obenchain class.

Corollary 4.1 shows that a linear restricted estimator being linearly admissible for β is also linearly admissible when it is assumed that β satisfies in fact the imposed restrictions, see Sect. 4.3.3 for a detailed discussion.

4.2.6 Shrinkage Property and Linear Admissibility

We have already noted that any estimator $L\widehat{\beta} \in \mathcal{L}^{\mathrm{Ob}}(\beta)$ satisfies the shrinkage property

$$\|L\widehat{\beta}\|^2 \leq \|\widehat{\beta}\|^2 \quad \text{for every realization } y \in \mathbb{R}^n.$$

Then, of course, also

$$\mathrm{E}(\|L\widehat{\beta}\|^2) \leq \mathrm{E}(\|\widehat{\beta}\|^2)$$

holds true. Now, it is rather natural to ask whether not only any element from the Obenchain class but also any element from the set of linearly admissible estimators has this shrinkage property. We can try to find some answer by investigating the set of all those homogeneous linear estimators which satisfy the shrinkage condition.

Theorem 4.11. *Under the linear regression model with assumptions (i) to (iv), a homogeneous linear estimator Ay of β satisfies the condition*

$$\|Ay\|^2 \leq \|\widehat{\beta}\|^2 \quad \textit{for every realization } y \in \mathbb{R}^n$$

if and only if the matrix A can be written as

$$A = L(X'X)^{-1}X'$$

for some $p \times p$ matrix L such that $\sigma(L'L) \subset [0,1]$.

Proof. Let us notice in advance that the condition $\sigma(\boldsymbol{L}'\boldsymbol{L}) \subset [0,1]$ is equivalent to $\boldsymbol{L}'\boldsymbol{L} \leq_{\mathrm{L}} \boldsymbol{I}_p$.

Now, the condition $||\boldsymbol{A}\boldsymbol{y}||^2 \leq ||\widehat{\boldsymbol{\beta}}||^2$ is satisfied for every realization $\boldsymbol{y} \in \mathbb{R}^n$ if and only if

$$\boldsymbol{A}'\boldsymbol{A} \leq_{\mathrm{L}} \boldsymbol{X}(\boldsymbol{X}'\boldsymbol{X})^{-2}\boldsymbol{X}' \,.$$

It is easy to see that this inequality is always fulfilled when $\boldsymbol{A} = \boldsymbol{L}(\boldsymbol{X}'\boldsymbol{X})^{-1}\boldsymbol{X}'$ for some $\boldsymbol{L}$ such that $\boldsymbol{L}'\boldsymbol{L} \leq_{\mathrm{L}} \boldsymbol{I}_p$.

On the other hand, this inequality implies

$$\mathcal{C}(\boldsymbol{A}'\boldsymbol{A}) \subseteq \mathcal{C}[\boldsymbol{X}(\boldsymbol{X}'\boldsymbol{X})^{-2}\boldsymbol{X}']$$

in view of Theorem A.65. Since $\mathcal{C}(\boldsymbol{A}'\boldsymbol{A}) = \mathcal{C}(\boldsymbol{A}')$ and $\mathcal{C}[\boldsymbol{X}(\boldsymbol{X}'\boldsymbol{X})^{-2}\boldsymbol{X}'] = \mathcal{C}(\boldsymbol{X})$ hold true, there exists some matrix $\boldsymbol{G}$ with $\boldsymbol{A}' = \boldsymbol{X}\boldsymbol{G}$, so that $\boldsymbol{A} = \boldsymbol{G}'\boldsymbol{X}'$. But then $\boldsymbol{A}$ can be written as $\boldsymbol{A} = \boldsymbol{L}(\boldsymbol{X}'\boldsymbol{X})^{-1}\boldsymbol{X}'$ for $\boldsymbol{L} = \boldsymbol{G}'\boldsymbol{X}'\boldsymbol{X}$. In addition, this matrix $\boldsymbol{L}$ must satisfy $\boldsymbol{L}'\boldsymbol{L} \leq_{\mathrm{L}} \boldsymbol{I}_p$, since otherwise the former inequality cannot be valid. □

The above theorem shows that any homogeneous linear estimator satisfying the shrinkage condition is of the form $\boldsymbol{L}\widehat{\boldsymbol{\beta}}$ with $\sigma(\boldsymbol{L}'\boldsymbol{L}) \subset [0,1]$.

According to our remarks in Sect. 4.2.4, an estimator of the form $\boldsymbol{L}\widehat{\boldsymbol{\beta}}$ can only be linearly admissible for $\boldsymbol{\beta}$ when $\boldsymbol{L}$ satisfies the conditions $\boldsymbol{X}'\boldsymbol{X}\boldsymbol{L} = \boldsymbol{L}'\boldsymbol{X}'\boldsymbol{X}$ and $\sigma(\boldsymbol{L}) \subset [0,1]$. From this, we can conclude that a homogeneous linear estimator satisfying the shrinkage condition is not necessarily linearly admissible for $\boldsymbol{\beta}$, since it is easy to find matrices $\boldsymbol{L}$ which satisfy the condition $\sigma(\boldsymbol{L}'\boldsymbol{L}) \subset [0,1]$ but which do not satisfy the condition $\boldsymbol{X}'\boldsymbol{X}\boldsymbol{L} = \boldsymbol{L}'\boldsymbol{X}'\boldsymbol{X}$.

The remaining question is whether any linearly admissible estimator satisfies the shrinkage condition. Since any linearly admissible estimator can be written as $\boldsymbol{L}\widehat{\boldsymbol{\beta}}$ for some matrix $\boldsymbol{L}$ which fulfills the conditions $\boldsymbol{X}'\boldsymbol{X}\boldsymbol{L} = \boldsymbol{L}'\boldsymbol{X}'\boldsymbol{X}$ and $\sigma(\boldsymbol{L}) \subset [0,1]$, this leads to the question whether the latter two conditions also imply the condition $\sigma(\boldsymbol{L}'\boldsymbol{L}) \subset [0,1]$. The following example shows that the answer is negative.

Example 4.3. Consider a linear regression model where the matrix $\boldsymbol{X}'\boldsymbol{X}$ is given as

$$\boldsymbol{X}'\boldsymbol{X} = \frac{1}{5}\begin{pmatrix} 23 & 51 \\ 51 & 117 \end{pmatrix} .$$

The matrix

$$\boldsymbol{L} = \frac{1}{10}\begin{pmatrix} 7 & -6 \\ 1 & 12 \end{pmatrix}$$

has eigenvalues $\lambda_1 = 9/10$ and $\lambda_2 = 1$, so that all eigenvalues of $\boldsymbol{L}$ lie in the closed interval $[0,1]$. Moreover, the matrix

$$\boldsymbol{X}'\boldsymbol{X}\boldsymbol{L} = \frac{1}{25}\begin{pmatrix} 106 & 237 \\ 237 & 549 \end{pmatrix}$$

is symmetric, showing that the estimator $\boldsymbol{L}\widehat{\boldsymbol{\beta}}$ is linearly admissible for $\boldsymbol{\beta}$. On the other hand the eigenvalues of $\boldsymbol{L}'\boldsymbol{L}$ are given as $\lambda_{1,2} = (23 \pm \sqrt{205})/20$, so that one eigenvalue of $\boldsymbol{L}'\boldsymbol{L}$ is strictly greater than 1 and the estimator $\boldsymbol{L}\widehat{\boldsymbol{\beta}}$ has not the above described shrinkage property. □

Now, from the above example we can conclude that there can exist linearly admissible estimators $\boldsymbol{L}\widehat{\boldsymbol{\beta}}$ for which the condition $\sigma(\boldsymbol{L}'\boldsymbol{L}) \subset [0,1]$ does not hold true.

However, if we confine to symmetric matrices $\boldsymbol{L}$ (implying that $\boldsymbol{L}\widehat{\boldsymbol{\beta}}$ is an element from the Obenchain class), then the condition $\sigma(\boldsymbol{L}) \subset [0,1]$ implies $\sigma(\boldsymbol{L}'\boldsymbol{L}) \subset [0,1]$. A natural generalization of the Obenchain class of estimators is therefore

$$\{\boldsymbol{L}\widehat{\boldsymbol{\beta}} : \boldsymbol{X}'\boldsymbol{X}\boldsymbol{L} = \boldsymbol{L}\boldsymbol{X}'\boldsymbol{X},\ \sigma(\boldsymbol{L}) \subset [0,1],\ \sigma(\boldsymbol{L}'\boldsymbol{L}) \subset [0,1]\}\ ,$$

which describes the set of all homogeneous linear estimators with the shrinkage property which are also linearly admissible for $\boldsymbol{\beta}$.

The matrix $\boldsymbol{L}$ from Example 4.3 can also be written in the form $\boldsymbol{L} = (\boldsymbol{X}'\boldsymbol{X} + \boldsymbol{K})^{-1}\boldsymbol{X}'\boldsymbol{X}$ for

$$\boldsymbol{K} = \frac{1}{5}\begin{pmatrix} 2 & 4 \\ 4 & 8 \end{pmatrix}\ ,$$

where the matrix $\boldsymbol{K}$ is symmetric nonnegative definite. From this, we can also conclude that not any estimator

$$\widehat{\boldsymbol{\beta}}_{\boldsymbol{K}} = (\boldsymbol{X}'\boldsymbol{X} + \boldsymbol{K})^{-1}\boldsymbol{X}'\boldsymbol{y} = (\boldsymbol{X}'\boldsymbol{X} + \boldsymbol{K})^{-1}\boldsymbol{X}'\boldsymbol{X}\widehat{\boldsymbol{\beta}}$$

has the shrinkage property.

Remark 4.2. If $\widehat{\boldsymbol{\beta}}_{\boldsymbol{K}} = (\boldsymbol{X}'\boldsymbol{X} + \boldsymbol{K})^{-1}\boldsymbol{X}'\boldsymbol{y}$ for some symmetric nonnegative definite matrix $\boldsymbol{K}$, then it does *not* necessarily follow that $\|\widehat{\boldsymbol{\beta}}_{\boldsymbol{K}}\|^2 \leq \|\widehat{\boldsymbol{\beta}}\|^2$.

4.2.7 Convex Combination of Estimators

If we consider two point estimators $\widetilde{\boldsymbol{\beta}}_1$ and $\widetilde{\boldsymbol{\beta}}_2$ for the parameter vector $\boldsymbol{\beta}$, then the observed weighted squared error loss of the convex combination

$$\widetilde{\boldsymbol{\beta}}_3 = \alpha\widetilde{\boldsymbol{\beta}}_1 + (1-\alpha)\widetilde{\boldsymbol{\beta}}_2, \quad 0 \leq \alpha \leq 1\ ,$$

cannot exceed the maximum of the individual observed squared error losses of $\widetilde{\boldsymbol{\beta}}_1$ and $\widetilde{\boldsymbol{\beta}}_2$.

Theorem 4.12. *Under the linear regression model with assumptions (i) to (iv), let*

$$L(\boldsymbol{\beta}, \widetilde{\boldsymbol{\beta}}) = (\boldsymbol{\beta} - \widetilde{\boldsymbol{\beta}})'\boldsymbol{W}(\boldsymbol{\beta} - \widetilde{\boldsymbol{\beta}})$$

be the loss function for an estimator $\widetilde{\boldsymbol{\beta}}$ of $\boldsymbol{\beta}$, where $\boldsymbol{W}$ is some $p \times p$ symmetric nonnegative definite matrix. If $\widetilde{\boldsymbol{\beta}}_1$ and $\widetilde{\boldsymbol{\beta}}_2$ are two estimators for $\boldsymbol{\beta}$, then

$$L(\boldsymbol{\beta},\widetilde{\boldsymbol{\beta}}_3) \leq \max\{L(\boldsymbol{\beta},\widetilde{\boldsymbol{\beta}}_1), L(\boldsymbol{\beta},\widetilde{\boldsymbol{\beta}}_2)\}$$

holds true for $\widetilde{\boldsymbol{\beta}}_3 = \alpha\widetilde{\boldsymbol{\beta}}_1 + (1-\alpha)\widetilde{\boldsymbol{\beta}}_2$, *where* α *is any number satisfying* $0 \leq \alpha \leq 1$.

Proof. The weighted squared error loss of $\widetilde{\boldsymbol{\beta}}_3$ can be written as

$$L(\boldsymbol{\beta},\widetilde{\boldsymbol{\beta}}_3) = \|\alpha\boldsymbol{x} + (1-\alpha)\boldsymbol{y}\|^2_{\boldsymbol{W}}, \quad \boldsymbol{x} = \boldsymbol{\beta} - \widetilde{\boldsymbol{\beta}}_1, \quad \boldsymbol{y} = \boldsymbol{\beta} - \widetilde{\boldsymbol{\beta}}_2\,,$$

where the mapping $\|\boldsymbol{z}\|_{\boldsymbol{W}} = \sqrt{\boldsymbol{z}'\boldsymbol{W}\boldsymbol{z}}$ specifies a vector(semi)norm. Hence, the assertion follows from the true inequalities

$$\|\alpha\boldsymbol{x} + (1-\alpha)\boldsymbol{y}\|_{\boldsymbol{W}} \leq \alpha\|\boldsymbol{x}\|_{\boldsymbol{W}} + (1-\alpha)\|\boldsymbol{y}\|_{\boldsymbol{W}} \leq \max\{\|\boldsymbol{x}\|_{\boldsymbol{W}}, \|\boldsymbol{y}\|_{\boldsymbol{W}}\}\,,$$

thus concluding the proof. □

Remark 4.3. A convex combination of two estimators can be useful when both estimators appear to be appropriate in a specific situation and an 'either or' choice between them is not desired. The observed loss of the convex combination is bounded by the maximum of the observed losses of the individual estimators. In addition, there is a possibility for the observed loss of the convex combination to be even smaller than both individual observed losses.

It is of course also possible to appropriately combine more than two point estimators as

$$\widetilde{\boldsymbol{\beta}} = \sum_{i=1}^{q} \alpha_i\widetilde{\boldsymbol{\beta}}_i \quad \sum_{i=1}^{q}\alpha_i = 1, \quad \alpha_i \geq 0\,.$$

In the non-trivial case $\alpha_q \neq 1$ this combination can also be seen as the convex combination

$$\widetilde{\boldsymbol{\beta}} = (1-\alpha_q)\sum_{i=1}^{q-1}\gamma_i\widetilde{\boldsymbol{\beta}}_i + \alpha_q\widetilde{\boldsymbol{\beta}}_q, \quad \gamma_i = \frac{\alpha_i}{1-\alpha_q}\,,$$

of the two estimators $\sum_{i=1}^{q-1}\gamma_i\widetilde{\boldsymbol{\beta}}_i$ and $\widetilde{\boldsymbol{\beta}}_q$. The estimator $\sum_{i=1}^{q-1}\gamma_i\widetilde{\boldsymbol{\beta}}_i$ can again be seen as the convex combination of two estimators. The simplest combination is obtained when all point estimators in question are equally weighted, so that the combination is nothing else but the mean

$$\widetilde{\boldsymbol{\beta}} = \frac{1}{q}\sum_{i=1}^{q}\widetilde{\boldsymbol{\beta}}_i$$

of q point estimators $\widetilde{\boldsymbol{\beta}}_1,\ldots,\widetilde{\boldsymbol{\beta}}_q$.

In Sect. 3.4.6 we have already mentioned that the convex combination of principal components and ridge estimator is a simple way to apply these two rather different methods at the same time to remedy collinearity effects. In Example 3.10 we have computed 30 principal components and ridge estimates to illustrate the different behavior of the two estimation procedures, see Fig. 3.13, p. 117. We can now complement this example by considering the respective convex combinations of the estimates.

Example 4.4. In Example 3.10 we consider the linear regression model $\boldsymbol{y} = \boldsymbol{X}\boldsymbol{\beta} + \boldsymbol{\varepsilon}$, where

$$\boldsymbol{X}' = \begin{pmatrix} 1 & 1 & 1 & 1 & 1 \\ 1.9 & 2.1 & 2 & 2 & 1.8 \end{pmatrix} .$$

We generate 30 realizations of the vector $\boldsymbol{y}$ and compute the corresponding principal components estimates $\widehat{\boldsymbol{\beta}}^{(1)}$ as well as the corresponding ridge estimates $\widehat{\boldsymbol{\beta}}_{1/5}$, see Fig. 3.13, p. 117. For these estimates the average observed unweighted squared error loss of $\widehat{\boldsymbol{\beta}}_{1/5}$ is 0.9533, while the average observed unweighted squared error loss of $\widehat{\boldsymbol{\beta}}^{(1)}$ is 0.8136.

Let us now consider the simple convex combination $0.5\widehat{\boldsymbol{\beta}}_{1/5} + 0.5\widehat{\boldsymbol{\beta}}^{(1)}$, i.e. the simple mean of $\widehat{\boldsymbol{\beta}}_{1/5}$ and $\widehat{\boldsymbol{\beta}}^{(1)}$. Figures 4.1 and 4.2 show a comparison between the ridge and the combined estimates.

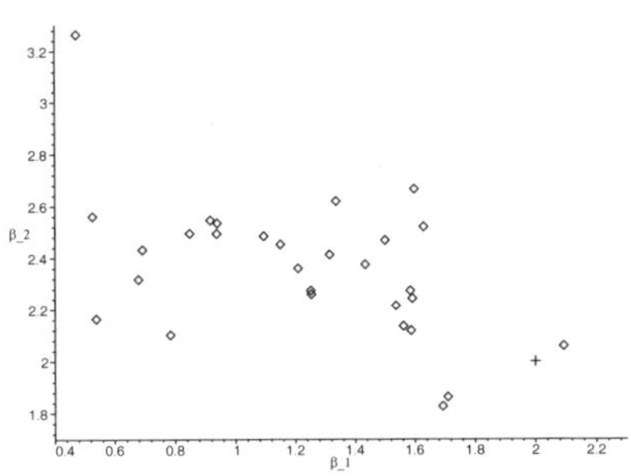

Fig. 4.1. 30 ridge estimates $\widehat{\boldsymbol{\beta}}_{1/5}$ in Example 4.4

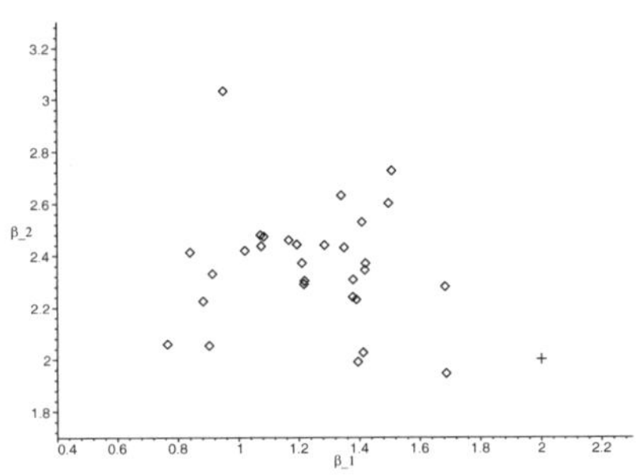

Fig. 4.2. 30 combined ridge and principal components estimates $0.5\widehat{\boldsymbol{\beta}}_{1/5} + 0.5\widehat{\boldsymbol{\beta}}^{(1)}$ in Example 4.4

The average observed unweighted squared error loss of $0.5\widehat{\boldsymbol{\beta}}_{1/5} + 0.5\widehat{\boldsymbol{\beta}}^{(1)}$ is 0.8231. □

If we wish to combine linearly admissible estimators, then we would like to know whether the convex combination is again linearly admissible. The following theorem considers the convex combination of two homogeneous linear estimators.

Theorem 4.13. *Under the linear regression model with assumptions (i) to (iv), let $\boldsymbol{A}_1\boldsymbol{y}$ and $\boldsymbol{A}_2\boldsymbol{y}$ be linearly admissible for $\boldsymbol{\beta}$ within the set $\mathcal{L}(\boldsymbol{\beta})$ and* *with respect to the unweighted squared error loss. Then for an arbitrary non-stochastic number* $0 \leq \alpha \leq 1$ *the estimator* $\alpha \boldsymbol{A}_1 \boldsymbol{y} + (1-\alpha)\boldsymbol{A}_2 \boldsymbol{y}$ *is linearly admissible for* $\boldsymbol{\beta}$.

Proof. Let $\boldsymbol{S} = \boldsymbol{X}'\boldsymbol{X}$. In view of Theorem 4.5, we can write

$$\boldsymbol{A}_1 = \boldsymbol{S}^{-1/2}\boldsymbol{G}_1\boldsymbol{S}^{-1/2}\boldsymbol{X}' \quad \text{and} \quad \boldsymbol{A}_2 = \boldsymbol{S}^{-1/2}\boldsymbol{G}_2\boldsymbol{S}^{-1/2}\boldsymbol{X}'$$

for symmetric matrices $\boldsymbol{G}_1$ and $\boldsymbol{G}_2$ whose eigenvalues all belong to $[0, 1]$. Then for the estimator $\boldsymbol{A}_3\boldsymbol{y} = \alpha\boldsymbol{A}_1\boldsymbol{y} + (1-\alpha)\boldsymbol{A}_2\boldsymbol{y}$ the matrix $\boldsymbol{A}_3$ is given as

$$\boldsymbol{A}_3 = \boldsymbol{S}^{-1/2}\boldsymbol{G}_3\boldsymbol{S}^{-1/2}\boldsymbol{X}', \quad \boldsymbol{G}_3 = \alpha\boldsymbol{G}_1 + (1-\alpha)\boldsymbol{G}_2 \, .$$

Since $\alpha\boldsymbol{G}_1$ and $(1-\alpha)\boldsymbol{G}_2$ are symmetric nonnegative definite, also the matrix $\boldsymbol{G}_3$ is symmetric nonnegative definite and all eigenvalues of $\boldsymbol{G}_3$ are thus real nonnegative numbers. Then from Theorem 4.5 the estimator $\boldsymbol{A}_3\boldsymbol{y}$ is linearly admissible for $\boldsymbol{\beta}$ when no eigenvalue of $\boldsymbol{G}_3$ is strictly greater than 1. If $||\cdot||$ denotes the spectral norm of a matrix, then $||\boldsymbol{G}_1|| \leq 1$ and $||\boldsymbol{G}_2|| \leq 1$, and therefore

$$||\boldsymbol{G}_3|| = ||\alpha\boldsymbol{G}_1 + (1-\alpha)\boldsymbol{G}_2|| \leq \alpha||\boldsymbol{G}_1|| + (1-\alpha)||\boldsymbol{G}_2|| \leq 1 \, .$$

But this shows that no eigenvalue of $\boldsymbol{G}_3$ can be strictly greater than 1, thus concluding the proof. □

Theorem 4.13 ensures that the convex combination of principal components and ridge estimator is linearly admissible for $\boldsymbol{\beta}$. Clearly, also for any other convex combination of homogeneous linear estimators being linearly admissible for $\boldsymbol{\beta}$, e.g. for elements from the Obenchain class, the admissibility property is maintained. The remaining question is whether this property is valid when non-homogeneous linear estimators for $\boldsymbol{\beta}$ are considered. The following theorem shows that the answer is affirmative.

Theorem 4.14. *Under the linear regression model with assumptions (i) to (iv), let* $\boldsymbol{A}_1\boldsymbol{y} + \boldsymbol{a}_1$ *and* $\boldsymbol{A}_2\boldsymbol{y} + \boldsymbol{a}_2$ *be linearly admissible for* $\boldsymbol{\beta}$ *within the set* $\mathcal{L}(\boldsymbol{\beta})$ *and with respect to the unweighted squared error loss. Then for an arbitrary non-stochastic number* $0 \leq \alpha \leq 1$ *the estimator*

$$\alpha(\boldsymbol{A}_1\boldsymbol{y} + \boldsymbol{a}_1) + (1-\alpha)(\boldsymbol{A}_2\boldsymbol{y} + \boldsymbol{a}_2)$$

is linearly admissible for $\boldsymbol{\beta}$.

Proof. Let $\boldsymbol{S} = \boldsymbol{X}'\boldsymbol{X}$. In view of Theorem 4.5 we can write

$$\boldsymbol{A}_1 = \boldsymbol{S}^{-1/2}\boldsymbol{G}_1\boldsymbol{S}^{-1/2}\boldsymbol{X}', \quad \boldsymbol{a}_1 = \boldsymbol{S}^{-1/2}(\boldsymbol{I}_p - \boldsymbol{G}_1)\boldsymbol{S}^{1/2}\boldsymbol{g}_1$$

and

$$\boldsymbol{A}_2 = \boldsymbol{S}^{-1/2}\boldsymbol{G}_2\boldsymbol{S}^{-1/2}\boldsymbol{X}', \quad \boldsymbol{a}_2 = \boldsymbol{S}^{-1/2}(\boldsymbol{I}_p - \boldsymbol{G}_2)\boldsymbol{S}^{1/2}\boldsymbol{g}_2 \, .$$

Here $\boldsymbol{G}_1$ and $\boldsymbol{G}_2$ are symmetric matrices, both having all their eigenvalues in the closed interval $[0,1]$, while $\boldsymbol{g}_1$ and $\boldsymbol{g}_2$ are $p \times 1$ vectors.

Now, from Theorems 4.5 and 4.13 it follows that the convex combination $\alpha(\boldsymbol{A}_1\boldsymbol{y} + \boldsymbol{a}_1) + (1-\alpha)(\boldsymbol{A}_2\boldsymbol{y} + \boldsymbol{a}_2)$ is linearly admissible for $\boldsymbol{\beta}$ if and only if there exists some $p \times 1$ vector $\boldsymbol{g}_3$ such that

$$\boldsymbol{a}_3 = \alpha\boldsymbol{a}_1 + (1-\alpha)\boldsymbol{a}_2$$

can be written as

$$\boldsymbol{a}_3 = (\boldsymbol{I}_p - \alpha\boldsymbol{A}_1\boldsymbol{X} - (1-\alpha)\boldsymbol{A}_2\boldsymbol{X})\boldsymbol{g}_3 = \boldsymbol{S}^{-1/2}(\boldsymbol{I}_p - \boldsymbol{G}_3)\boldsymbol{S}^{1/2}\boldsymbol{g}_3 \ ,$$

where $\boldsymbol{G}_3 = \alpha\boldsymbol{G}_1 + (1-\alpha)\boldsymbol{G}_2$. To see that such a vector $\boldsymbol{g}_3$ does in fact exist, let

$$\boldsymbol{H}_1 = \alpha(\boldsymbol{I}_p - \boldsymbol{G}_1) \quad \text{and} \quad \boldsymbol{H}_2 = (1-\alpha)(\boldsymbol{I}_p - \boldsymbol{G}_2) \ ,$$

where the matrices $\boldsymbol{H}_1$ and $\boldsymbol{H}_2$ are symmetric nonnegative definite. Then $\boldsymbol{I}_p - \boldsymbol{G}_3 = \boldsymbol{H}_1 + \boldsymbol{H}_2$ and the vector in question $\boldsymbol{g}_3$ exists if and only if $\boldsymbol{g}_3$ satisfies the equality

$$\boldsymbol{H}_1\boldsymbol{S}^{1/2}\boldsymbol{g}_1 + \boldsymbol{H}_2\boldsymbol{S}^{1/2}\boldsymbol{g}_2 = (\boldsymbol{H}_1 + \boldsymbol{H}_2)\boldsymbol{S}^{1/2}\boldsymbol{g}_3 \ .$$

But such a vector $\boldsymbol{g}_3$ can always be found, since for any two symmetric nonnegative definite matrices $\boldsymbol{A}$ and $\boldsymbol{B}$ the identity

$$\mathcal{C}(\boldsymbol{A}) + \mathcal{C}(\boldsymbol{B}) = \mathcal{C}(\boldsymbol{A} + \boldsymbol{B})$$

holds true. Now, since $\boldsymbol{H}_1\boldsymbol{S}^{1/2}\boldsymbol{g}_1 \in \mathcal{C}(\boldsymbol{H}_1)$ and $\boldsymbol{H}_2\boldsymbol{S}^{1/2}\boldsymbol{g}_2 \in \mathcal{C}(\boldsymbol{H}_2)$, there must exist some vector $\boldsymbol{z}$ with

$$\boldsymbol{H}_1\boldsymbol{S}^{1/2}\boldsymbol{g}_1 + \boldsymbol{H}_2\boldsymbol{S}^{1/2}\boldsymbol{g}_2 = (\boldsymbol{H}_1 + \boldsymbol{H}_2)\boldsymbol{z} \ .$$

The choice $\boldsymbol{g}_3 = \boldsymbol{S}^{-1/2}\boldsymbol{z}$ shows the required existence. □

We can reformulate the above theorem as follows, see also [83, p. 94].

Remark 4.4. Under the linear regression model with assumptions (i) to (iv), the set of linearly admissible estimators for $\boldsymbol{\beta}$ is convex.

As a further example for the application of a convex combination of a non-homogeneous and a homogeneous linear estimator consider

$$\widehat{\boldsymbol{\beta}}_{\mathrm{R},\alpha} = (1-\alpha)\widehat{\boldsymbol{\beta}}_{\mathrm{R}} + \alpha\widehat{\boldsymbol{\beta}} \ ,$$

which can be used when not only the ordinary least squares estimator but also some restricted least squares estimator $\widehat{\boldsymbol{\beta}}_{\mathrm{R}}$ should have a part in the estimation procedure.

The modified Bock estimator $\widehat{\boldsymbol{\beta}}^{+}_{\mathrm{Bk}}(d)$ is of this form but with stochastic $\alpha = [1 - d/F]_+$. Thus, the modified Bock estimator can be viewed as a possible non-linear variant of the linearly admissible estimator $\widehat{\boldsymbol{\beta}}_{\mathrm{R},\alpha}$.

4.2.8 Linear Bayes Estimator

Section 3.8 discusses the linear Bayes estimator

$$\widehat{\boldsymbol{\beta}}^{\mathrm{B}} = \boldsymbol{\mu} + \boldsymbol{T}[\boldsymbol{T} + (\boldsymbol{X}'\boldsymbol{X})^{-1}]^{-1}(\widehat{\boldsymbol{\beta}} - \boldsymbol{\mu}) ,$$

which can be applied under the assumption that $\boldsymbol{\beta}$ is the realization of a random vector $\boldsymbol{b}$ satisfying

$$\mathrm{E}(\boldsymbol{b}) = \boldsymbol{\mu} \quad \text{and} \quad \mathrm{Cov}(\boldsymbol{b}) = \sigma^2 \boldsymbol{T} .$$

By letting

$$\boldsymbol{G} = \boldsymbol{I}_p - \boldsymbol{S}^{-1/2}(\boldsymbol{T} + \boldsymbol{S}^{-1})^{-1}\boldsymbol{S}^{-1/2}, \quad \boldsymbol{S} = \boldsymbol{X}'\boldsymbol{X} ,$$

and noting

$$\boldsymbol{S}^{-1/2}\boldsymbol{G}\boldsymbol{S}^{1/2} = \boldsymbol{I}_p - \boldsymbol{S}^{-1}(\boldsymbol{T} + \boldsymbol{S}^{-1})^{-1} = \boldsymbol{T}(\boldsymbol{T} + \boldsymbol{S}^{-1})^{-1} ,$$

the linear Bayes estimator can also be written as

$$\widehat{\boldsymbol{\beta}}^{\mathrm{B}} = \boldsymbol{\mu} + \boldsymbol{S}^{-1/2}\boldsymbol{G}\boldsymbol{S}^{1/2}(\widehat{\boldsymbol{\beta}} - \boldsymbol{\mu}) .$$

The above matrix $\boldsymbol{G}$ is symmetric and has all eigenvalues in the interval $[0, 1)$, see Problem 4.9.

Hence, with $\boldsymbol{g} := \boldsymbol{\mu}$ it follows that a linear Bayes estimator can always be written as

$$\widehat{\boldsymbol{\beta}}^{\mathrm{B}} = \boldsymbol{S}^{-1/2}\boldsymbol{G}\boldsymbol{S}^{1/2}\widehat{\boldsymbol{\beta}} + \boldsymbol{S}^{-1/2}(\boldsymbol{I}_p - \boldsymbol{G})\boldsymbol{S}^{1/2}\boldsymbol{g}$$

for some symmetric matrix $\boldsymbol{G}$ such that $\sigma(\boldsymbol{G}) \subset [0, 1)$. Then, Theorem 4.5 shows that any linear Bayes estimator is linearly admissible for $\boldsymbol{\beta}$. On the other hand, as noted in the following theorem, any linear estimator of the form

$$\boldsymbol{S}^{-1/2}\boldsymbol{G}\boldsymbol{S}^{1/2}\widehat{\boldsymbol{\beta}} + \boldsymbol{S}^{-1/2}(\boldsymbol{I}_p - \boldsymbol{G})\boldsymbol{S}^{1/2}\boldsymbol{g}$$

is also a linear Bayes estimator.

Theorem 4.15. *Under the linear regression model with assumptions (i) to (iv), the estimator $\boldsymbol{A}\boldsymbol{y} + \boldsymbol{a}$ is a linear Bayes estimator for $\boldsymbol{\beta}$ if and only if the matrix $\boldsymbol{A}$ and the vector $\boldsymbol{a}$ can be written as*

$$\boldsymbol{A} = (\boldsymbol{X}'\boldsymbol{X})^{-1/2}\boldsymbol{G}(\boldsymbol{X}'\boldsymbol{X})^{-1/2}\boldsymbol{X}'$$

and

$$\boldsymbol{a} = (\boldsymbol{X}'\boldsymbol{X})^{-1/2}(\boldsymbol{I}_p - \boldsymbol{G})(\boldsymbol{X}'\boldsymbol{X})^{1/2}\boldsymbol{g}$$

for some symmetric $p \times p$ matrix $\boldsymbol{G}$ such that $\sigma(\boldsymbol{G}) \subset [0, 1)$ and some $p \times 1$ vector $\boldsymbol{g}$.

Proof. We have already shown above that any linear Bayes estimator can be written in the form

$$\widehat{\boldsymbol{\beta}}^{\mathrm{B}} = \boldsymbol{S}^{-1/2}\boldsymbol{G}\boldsymbol{S}^{1/2}\widehat{\boldsymbol{\beta}} + \boldsymbol{S}^{-1/2}(\boldsymbol{I}_p - \boldsymbol{G})\boldsymbol{S}^{1/2}\boldsymbol{g}$$

for some symmetric matrix $\boldsymbol{G}$ such that $\sigma(\boldsymbol{G}) \subset [0,1)$.

On the other hand, let $\boldsymbol{Ay}+\boldsymbol{a}$ be a linear estimator for $\boldsymbol{\beta}$, where $\boldsymbol{A}$ and $\boldsymbol{a}$ can be written in the asserted form. Let $\boldsymbol{S} = \boldsymbol{X}'\boldsymbol{X}$. Since $\boldsymbol{G}$ is a symmetric matrix with $\sigma(\boldsymbol{G}) \subset [0,1)$, the matrix $\boldsymbol{S}^{1/2}(\boldsymbol{I}_p-\boldsymbol{G})\boldsymbol{S}^{1/2}$ is symmetric positive definite. Moreover, the matrix

$$\boldsymbol{T} := [\boldsymbol{S}^{1/2}(\boldsymbol{I}_p - \boldsymbol{G})\boldsymbol{S}^{1/2}]^{-1} - \boldsymbol{S}^{-1}$$

is symmetric nonnegative definite. For this matrix $\boldsymbol{T}$ the identity

$$\boldsymbol{G} = \boldsymbol{I}_p - \boldsymbol{S}^{-1/2}(\boldsymbol{T} + \boldsymbol{S}^{-1})^{-1}\boldsymbol{S}^{-1/2}$$

holds true, showing that the linear estimator $\boldsymbol{Ay} + \boldsymbol{a}$ can be written as

$$\boldsymbol{Ay} + \boldsymbol{a} = \boldsymbol{g} + \boldsymbol{T}[\boldsymbol{T} + (\boldsymbol{X}'\boldsymbol{X})^{-1}]^{-1}(\widehat{\boldsymbol{\beta}} - \boldsymbol{g}) .$$

But this is in fact a linear Bayes estimator for $\boldsymbol{\beta}$ when $\boldsymbol{\beta}$ is viewed as the realization of a random vector $\boldsymbol{b}$ satisfying

$$\mathrm{E}(\boldsymbol{b}) = \boldsymbol{g} \quad \text{and} \quad \mathrm{Cov}(\boldsymbol{b}) = \sigma^2\left([\boldsymbol{S}^{1/2}(\boldsymbol{I}_p - \boldsymbol{G})\boldsymbol{S}^{1/2}]^{-1} - \boldsymbol{S}^{-1}\right) .$$

This concludes the proof. □

If we compare the above theorem with Theorem 4.5, then we can ascertain a strong similarity between the set of linear Bayes estimators and the set of linearly admissible estimators for $\boldsymbol{\beta}$. The only difference is that the symmetric matrix $\boldsymbol{G}$ can have all eigenvalues in $[0,1]$ when the estimator is linearly admissible, while it can only have all eigenvalues in $[0,1)$ when the estimator is linear Bayes.

If, for example, we choose $\boldsymbol{G} = \boldsymbol{I}_p$ and $\boldsymbol{g} = \boldsymbol{0}$, then the corresponding ordinary least squares estimator $\widehat{\boldsymbol{\beta}}$ is linearly admissible for $\boldsymbol{\beta}$ but cannot be a linear Bayes estimator.

On the other hand, Sect. 3.8.4 shows that the ordinary least squares estimator $\widehat{\boldsymbol{\beta}}$ can also be seen as a linear Bayes estimator when we admit a matrix $\boldsymbol{T} = \lim_{t\to\infty} t\boldsymbol{I}_p$, so that all variances of the elements of the corresponding random vector $\boldsymbol{b}$ are not finite. Hence, if such matrices $\boldsymbol{T}$ are allowed, then also possible eigenvalues 1 of the matrix $\boldsymbol{G}$ corresponding to the linear Bayes estimator must be allowed. When the latter is true, then the estimator is also called *generalized linear Bayes estimator* for $\boldsymbol{\beta}$ and the following can obviously be stated:

Remark 4.5. The set of generalized linear Bayes estimators coincides with the set of linearly admissible estimators for $\boldsymbol{\beta}$.

4.3 Linear Admissibility Under Linear Restrictions

We now turn to investigating the set of linearly admissible estimators for $\boldsymbol{\beta}$, when the parameter vector $\boldsymbol{\beta}$ satisfies additional linear restrictions $\boldsymbol{R\beta} = \boldsymbol{r}$, where it is assumed that the matrix $\boldsymbol{R}$ is of full row rank. Then the parameter space for $\boldsymbol{\beta}$ is given by $\Theta_{\boldsymbol{\beta}} = \mathcal{B}$, where

$$\mathcal{B} = \{\boldsymbol{\beta} \in \mathrm{I\!R}^p : \boldsymbol{R\beta} = \boldsymbol{r}\} ,$$

and thus the overall parameter space for $\boldsymbol{\beta}$ and σ^2 is $\Theta = \Theta_{\boldsymbol{\beta}} \times (0, \infty) = \mathcal{B} \times (0, \infty)$.

4.3.1 The Assumption of a Full Rank Restriction Matrix

In Sect. 3.1 the restricted least squares estimator $\widehat{\boldsymbol{\beta}}_{\mathrm{R}}$ is derived under the assumption of linear restrictions $\boldsymbol{R\beta} = \boldsymbol{r}$, where the $m \times p$ matrix $\boldsymbol{R}$ is of full row rank m. The full row rank assumption ensures that the linear system $\boldsymbol{R\beta} = \boldsymbol{r}$ does not contain superfluous restrictions explained by linear combination of others. However, sometimes one is confronted with a given solvable linear system

$$\boldsymbol{R}_*\boldsymbol{\beta} = \boldsymbol{r}_* ,$$

where the $\overline{m} \times p$ matrix $\boldsymbol{R}_*$ has rank $m < \overline{m}$, and thus is not of full row rank. When this linear system is solvable, then there exists a vector $\boldsymbol{g}_*$ such that $\boldsymbol{r}_* = \boldsymbol{R}_*\boldsymbol{g}_*$ holds true. One may, for example, choose $\boldsymbol{g}_* = \boldsymbol{R}_*^+\boldsymbol{r}$, where $\boldsymbol{R}_*^+$ denotes the Moore-Penrose inverse of $\boldsymbol{R}_*$.

Now, any $\overline{m} \times p$ matrix $\boldsymbol{R}_*$ with $\mathrm{rk}(\boldsymbol{R}_*) = m$ can be written as

$$\boldsymbol{R}_* = \boldsymbol{CR} ,$$

where $\boldsymbol{C}$ is a $\overline{m} \times m$ matrix with $\mathrm{rk}(\boldsymbol{C}) = m$, and $\boldsymbol{R}$ is a $m \times p$ matrix with $\mathrm{rk}(\boldsymbol{R}) = m$. This is a so-called *full rank decomposition* of $\boldsymbol{R}_*$. It can be derived from the *singular value decomposition* $\boldsymbol{R}_* = \boldsymbol{UDV}'$ of the $\overline{m} \times p$ matrix $\boldsymbol{R}_*$, which does always exist. Here, $\boldsymbol{U}$ is a $\overline{m} \times m$ matrix satisfying $\boldsymbol{U}'\boldsymbol{U} = \boldsymbol{I}_m$, $\boldsymbol{V}$ is a $p \times m$ matrix satisfying $\boldsymbol{V}'\boldsymbol{V} = \boldsymbol{I}_m$, and $\boldsymbol{D}$ is a $m \times m$ diagonal matrix with strictly positive main diagonal elements. Then, the choices $\boldsymbol{C} = \boldsymbol{UD}^{1/2}$ and $\boldsymbol{R} = \boldsymbol{D}^{1/2}\boldsymbol{V}'$ give the desired multiplicative decomposition of $\boldsymbol{R}_*$. (Notice that the Moore-Penrose inverse $\boldsymbol{R}_*^+$ is given as $\boldsymbol{R}_*^+ = \boldsymbol{VD}^{-1}\boldsymbol{U}'$.) Since the matrix $\boldsymbol{C}$ is of full column rank, it follows

$$\boldsymbol{R}_*\boldsymbol{\beta} = \boldsymbol{r}_* \quad \Leftrightarrow \quad \boldsymbol{CR\beta} = \boldsymbol{CRg}_* \quad \Leftrightarrow \quad \boldsymbol{R\beta} = \boldsymbol{r} ,$$

where $\boldsymbol{r} = \boldsymbol{Rg}_*$. We can thus state the following remark.

Remark 4.6. A solvable linear system $\boldsymbol{R}_*\boldsymbol{\beta} = \boldsymbol{r}_*$ can always be transformed into an equivalent linear system $\boldsymbol{R\beta} = \boldsymbol{r}$ with a full row rank coefficient matrix $\boldsymbol{R}$.

Since the two linear systems $\boldsymbol{R}_*\boldsymbol{\beta} = \boldsymbol{r}_*$ and $\boldsymbol{R}\boldsymbol{\beta} = \boldsymbol{r}$ specify the same restrictions on the parameter vector $\boldsymbol{\beta}$ (the set of possible solutions is identical), we can without loss of generality assume that the restrictions are given by $\boldsymbol{R}\boldsymbol{\beta} = \boldsymbol{r}$, where the matrix $\boldsymbol{R}$ is of full row rank.

4.3.2 Restricted Estimator

If restrictions of the form $\boldsymbol{R}\boldsymbol{\beta} = \boldsymbol{r}$ for some $m \times p$ matrix $\boldsymbol{R}$ are given, then the linear regression model with assumptions (i) to (iv) and (viii) can also be described by

$$\boldsymbol{y} = \boldsymbol{X}\boldsymbol{\beta}|_{\boldsymbol{R}\boldsymbol{\beta}=\boldsymbol{r}} + \boldsymbol{\varepsilon}, \quad \boldsymbol{\varepsilon} \sim (\boldsymbol{0}, \sigma^2 \boldsymbol{I}_n) \,.$$

Since in this situation we assume that the restrictions are in fact satisfied, it is desired to appropriately incorporate them when estimators for $\boldsymbol{\beta}$ are constructed. Sect. 3.1 shows that the application of the principle of least squares with respect to $\boldsymbol{R}\boldsymbol{\beta} = \boldsymbol{r}$ yields

$$\widehat{\boldsymbol{\beta}}_{\mathrm{R}} = \widehat{\boldsymbol{\beta}} - (\boldsymbol{X}'\boldsymbol{X})^{-1}\boldsymbol{R}'[\boldsymbol{R}(\boldsymbol{X}'\boldsymbol{X})^{-1}\boldsymbol{R}']^{-1}(\boldsymbol{R}\widehat{\boldsymbol{\beta}} - \boldsymbol{r}) \,,$$

as opposed to the ordinary least squares estimator $\widehat{\boldsymbol{\beta}} = (\boldsymbol{X}'\boldsymbol{X})^{-1}\boldsymbol{X}'\boldsymbol{y}$, obtained from the unrestricted principle of least squares.

Similarly, other estimation approaches presented in Chap. 3 should take the given linear system $\boldsymbol{R}\boldsymbol{\beta} = \boldsymbol{r}$ into account and then yield modified estimators.

In the following we exemplarily consider the so-called *restricted ridge estimator* for $\boldsymbol{\beta}$. For this, let

$$\boldsymbol{\beta}_0 = \boldsymbol{R}'(\boldsymbol{R}\boldsymbol{R}')^{-1}\boldsymbol{r} \,.$$

Then the vector $\boldsymbol{\beta}_0$ satisfies the identity $\boldsymbol{R}\boldsymbol{\beta}_0 = \boldsymbol{r}$, and in addition the inequality

$$||\boldsymbol{\beta}_0||^2 \leq ||\boldsymbol{\beta}||^2$$

for any vector $\boldsymbol{\beta}$ fulfilling the identity $\boldsymbol{R}\boldsymbol{\beta} = \boldsymbol{r}$. Thus, the vector $\boldsymbol{\beta}_0 \in \mathcal{B}$ is shorter (or at least not longer) than any other vector from the set $\mathcal{B}$, see also Problem 4.10. Now, let

$$\widehat{\boldsymbol{\beta}}_{k,\boldsymbol{\beta}_0} = (\boldsymbol{X}'\boldsymbol{X} + k\boldsymbol{I}_p)^{-1}(\boldsymbol{X}'\boldsymbol{y} + k\boldsymbol{\beta}_0), \quad k \geq 0 \,,$$

the ridge estimator for $\boldsymbol{\beta}$, which shrinks the ordinary least squares estimator $\widehat{\boldsymbol{\beta}}$ towards $\boldsymbol{\beta}_0$ (direction modified ridge estimator from Swindel [112], see page 136). Then the estimator

$$\widehat{\boldsymbol{\beta}}_{\mathrm{R},k} = \widehat{\boldsymbol{\beta}}_{k,\boldsymbol{\beta}_0} - (\boldsymbol{X}'\boldsymbol{X} + k\boldsymbol{I}_p)^{-1}\boldsymbol{R}'[\boldsymbol{R}(\boldsymbol{X}'\boldsymbol{X} + k\boldsymbol{I}_p)^{-1}\boldsymbol{R}']^{-1}(\boldsymbol{R}\widehat{\boldsymbol{\beta}}_{k,\boldsymbol{\beta}_0} - \boldsymbol{r})$$

is called the *restricted ridge estimator* for $\boldsymbol{\beta}$. Note that this estimator is not the ridge-shrinkage restricted least squares estimator introduced by Sarkar

[94], see page 122, although Sarkar also uses the name 'restricted ridge estimator'.

It is easy to see that

$$\boldsymbol{R}\widehat{\boldsymbol{\beta}}_{\mathrm{R},k} = \boldsymbol{r} \quad \forall \boldsymbol{y} \in \mathbb{R}^n \ ,$$

so that $\widehat{\boldsymbol{\beta}}_{\mathrm{R},k}$ is in fact a restricted estimator. Moreover, $\widehat{\boldsymbol{\beta}}_{\mathrm{R},0} = \widehat{\boldsymbol{\beta}}_{\mathrm{R}}$ holds true and the following theorem shows that $\widehat{\boldsymbol{\beta}}_{\mathrm{R},\infty} = \boldsymbol{\beta}_0$. Thus, that $\widehat{\boldsymbol{\beta}}_{\mathrm{R},k}$ can be interpreted as shrinking the restricted least squares estimator $\widehat{\boldsymbol{\beta}}_{\mathrm{R}}$ towards the shortest vector $\boldsymbol{\beta}_0$ satisfying the restrictions.

Theorem 4.16. *Under the linear regression model with assumptions (i) to (iv), the identity*

$$\lim_{k\to\infty} \widehat{\boldsymbol{\beta}}_{\mathrm{R},k} = \boldsymbol{\beta}_0$$

holds true, where $\boldsymbol{\beta}_0 = \boldsymbol{R}'(\boldsymbol{R}\boldsymbol{R}')^{-1}\boldsymbol{r}$.

Proof. Let $\boldsymbol{T}_k = \boldsymbol{X}'\boldsymbol{X} + k\boldsymbol{I}_p$ and let

$$\boldsymbol{M}_k = \boldsymbol{T}_k^{-1} - \boldsymbol{T}_k^{-1}\boldsymbol{R}'[\boldsymbol{R}\boldsymbol{T}_k^{-1}\boldsymbol{R}']^{-1}\boldsymbol{R}\boldsymbol{T}_k^{-1} \ .$$

Then the estimator $\widehat{\boldsymbol{\beta}}_{\mathrm{R},k}$ can be written in the form

$$\widehat{\boldsymbol{\beta}}_{\mathrm{R},k} = \boldsymbol{M}_k(\boldsymbol{X}'\boldsymbol{y} + k\boldsymbol{\beta}_0) + \boldsymbol{T}_k^{-1}\boldsymbol{R}'[\boldsymbol{R}\boldsymbol{T}_k^{-1}\boldsymbol{R}']^{-1}\boldsymbol{r} \ .$$

In view of $\boldsymbol{r} = \boldsymbol{R}\boldsymbol{\beta}_0$ and

$$\boldsymbol{T}_k^{-1}\boldsymbol{R}'[\boldsymbol{R}\boldsymbol{T}_k^{-1}\boldsymbol{R}']^{-1}\boldsymbol{R} = \boldsymbol{I}_p - \boldsymbol{M}_k\boldsymbol{T}_k \ ,$$

it follows that

$$\begin{aligned}\widehat{\boldsymbol{\beta}}_{\mathrm{R},k} &= \boldsymbol{M}_k\boldsymbol{X}'\boldsymbol{y} + k\boldsymbol{M}_k\boldsymbol{\beta}_0 + \boldsymbol{\beta}_0 - \boldsymbol{M}_k\boldsymbol{T}_k\boldsymbol{\beta}_0 \\ &= \boldsymbol{M}_k\boldsymbol{X}'\boldsymbol{y} + \boldsymbol{\beta}_0 - \boldsymbol{M}_k\boldsymbol{X}'\boldsymbol{X}\boldsymbol{\beta}_0 \ .\end{aligned}$$

Then the assertion is shown when $\lim_{k\to\infty} \boldsymbol{M}_k = \boldsymbol{0}_{p\times p}$ is satisfied. To see that the latter is in fact the case, write

$$\boldsymbol{M}_k = (\boldsymbol{Q}\boldsymbol{T}_k\boldsymbol{Q})^+, \quad \boldsymbol{Q} = \boldsymbol{I}_p - \boldsymbol{R}'(\boldsymbol{R}\boldsymbol{R}')^{-1}\boldsymbol{R} \ ,$$

where $(\boldsymbol{Q}\boldsymbol{T}_k\boldsymbol{Q})^+$ is the Moore-Penrose inverse of

$$\boldsymbol{Q}\boldsymbol{T}_k\boldsymbol{Q} = \boldsymbol{Q}\boldsymbol{X}'\boldsymbol{X}\boldsymbol{Q} + k\boldsymbol{Q} \ .$$

(That $\boldsymbol{M}_k$ is in fact the Moore-Penrose inverse of the matrix $\boldsymbol{Q}\boldsymbol{T}_k\boldsymbol{Q}$ can be verified by examining the four Moore-Penrose conditions, see Theorem A.17.) The matrix $\boldsymbol{Q}$ is symmetric and idempotent and has rank $p-m$. Thus $\boldsymbol{Q}$ has $p-m$ eigenvalues equal to 1, while all other eigenvalues are equal to 0. The matrix $\boldsymbol{QX'XQ}$ is symmetric nonnegative definite and has rank not greater than $p-m$, so that at least m eigenvalues are equal to 0. Since the matrices $\boldsymbol{QX'XQ}$ and $\boldsymbol{Q}$ commute, there exists a common spectral decomposition of the form

$$\boldsymbol{QX'XQ} = \boldsymbol{U\Lambda_1U'} \quad \text{and} \quad \boldsymbol{Q} = \boldsymbol{U\Lambda_2U'} ,$$

where $\boldsymbol{U}$ is a $p \times p$ orthogonal matrix, and $\boldsymbol{\Lambda}_1$ and $\boldsymbol{\Lambda}_2$ are $p \times p$ diagonal matrices whose main diagonal elements are the eigenvalues of $\boldsymbol{QX'XQ}$ and $\boldsymbol{Q}$, respectively.

As noted above, the matrix $\boldsymbol{Q}$ is idempotent, showing that the main diagonal elements of $\boldsymbol{\Lambda}_2$ are either 0 or 1. Moreover, the column space of $\boldsymbol{\Lambda}_1$ is contained in the column space of $\boldsymbol{\Lambda}_2$, see Problem 4.11, showing that the j-th main diagonal element of $\boldsymbol{\Lambda}_1$ can only differ from 0 when the j-th main diagonal element of $\boldsymbol{\Lambda}_2$ is equal to 1. This means that there must exist some orthogonal matrix $\boldsymbol{V}$ such that

$$\boldsymbol{QX'XQ} = \boldsymbol{V}\begin{pmatrix} \boldsymbol{\Lambda} & \boldsymbol{0} \\ \boldsymbol{0} & \boldsymbol{0} \end{pmatrix}\boldsymbol{V'} \quad \text{and} \quad \boldsymbol{Q} = \boldsymbol{V}\begin{pmatrix} \boldsymbol{I}_{p-m} & \boldsymbol{0} \\ \boldsymbol{0} & \boldsymbol{0} \end{pmatrix}\boldsymbol{V'} ,$$

where $\boldsymbol{\Lambda}$ is a $(p-m) \times (p-m)$ diagonal matrix with nonnegative (not necessarily positive) main diagonal elements. From these representations it follows that

$$\lim_{k\to\infty} \boldsymbol{M}_k = \lim_{k\to\infty} (\boldsymbol{QT}_k\boldsymbol{Q})^+ = \lim_{k\to\infty} \boldsymbol{V}\begin{pmatrix} (\boldsymbol{\Lambda} + k\boldsymbol{I}_{p-m})^{-1} & \boldsymbol{0} \\ \boldsymbol{0} & \boldsymbol{0} \end{pmatrix}\boldsymbol{V'} = \boldsymbol{0}_{p\times p} ,$$

thus concluding the proof. □

To illustrate how the restricted ridge estimator $\widehat{\boldsymbol{\beta}}_{\mathrm{R},k}$ works, we consider the following example.

Example 4.5. Consider the linear regression model $\boldsymbol{y} = \boldsymbol{X\beta} + \boldsymbol{\varepsilon}$ with

$$\boldsymbol{y} = \begin{pmatrix} 6.0521 \\ 7.0280 \\ 7.1230 \\ 4.4441 \\ 5.0813 \end{pmatrix}, \quad \boldsymbol{X} = \begin{pmatrix} 1 & 1.9 \\ 1 & 2.1 \\ 1 & 2 \\ 1 & 2 \\ 1 & 1.8 \end{pmatrix}, \quad \boldsymbol{\beta} = \begin{pmatrix} \beta_1 \\ \beta_2 \end{pmatrix} .$$

It is assumed that the restriction

$$\boldsymbol{R\beta} = \boldsymbol{r}, \qquad \boldsymbol{R} = (17, 38), \quad \boldsymbol{r} = 110 ,$$

holds true. The restricted least squares estimator is computed as $\widehat{\boldsymbol{\beta}}_{\mathrm{R}} = (1.3200, 2.3042)'$.

To apply the restricted ridge estimator we consider the ridge trace of $\widehat{\boldsymbol{\beta}}_{\mathrm{R},k} - \boldsymbol{\beta}_0$, where $\boldsymbol{\beta}_0 = \boldsymbol{R}'(\boldsymbol{RR}')^{-1}\boldsymbol{r} = (1.0791, 2.4120)'$, as displayed in Fig. 4.3.

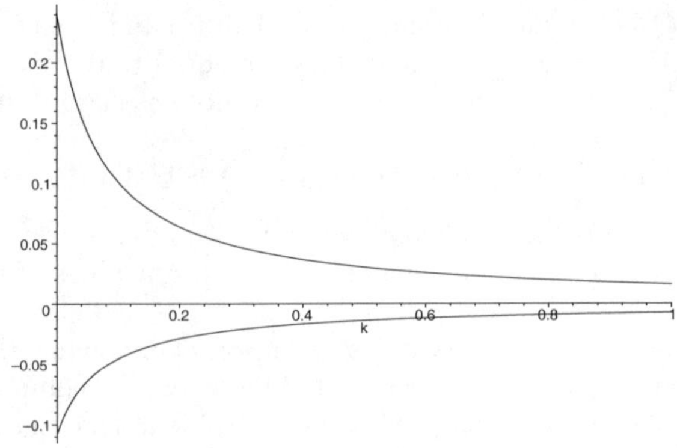

Fig. 4.3. Ridge trace of $\widehat{\boldsymbol{\beta}}_{\mathrm{R},k} - \boldsymbol{\beta}_0$ in Example 4.5

On the basis of this graphic we choose $k = 1/5$. This yields the restricted ridge estimate

$$\widehat{\boldsymbol{\beta}}_{\mathrm{R},k} = (1.1427, 2.3835)',$$

which is a little farther away from the true parameter vector $\boldsymbol{\beta} = (2,2)'$ than the restricted least squares estimate $\widehat{\boldsymbol{\beta}}_{\mathrm{R}}$. This is not surprising, since here no choice of k can yield a better estimate than the restricted least squares estimate. The reason for this is that any shrinking towards $\boldsymbol{\beta}_0$ yields an estimate which is farther away from the true parameter vector than the restricted least squares estimate, as illustrated by Fig. 4.4, showing the true parameter vector (cross), the restricted least squares estimate (box), and the restricted ridge estimate (circle). The arrow represents the vector $\boldsymbol{\beta}_0$.

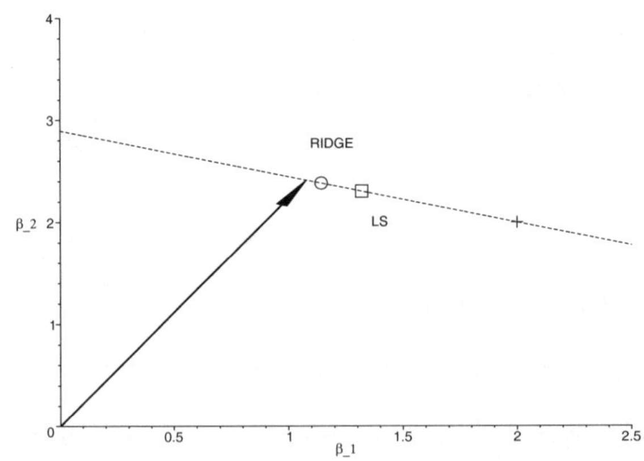

Fig. 4.4. Estimates $\widehat{\boldsymbol{\beta}}_{\mathrm{R}}$ and $\widehat{\boldsymbol{\beta}}_{\mathrm{R},k=1/5}$ in Example 4.5

But what can be said when more than only one estimate is available and the respective average losses are considered? To obtain an impression of the behavior of the estimators, we generate 30 realizations of $\boldsymbol{y}$ and compute the corresponding estimates $\widehat{\boldsymbol{\beta}}_{\mathrm{R}}$ and $\widehat{\boldsymbol{\beta}}_{\mathrm{R},k}$ for $k = 1/5$. Figures 4.5 and 4.6 show the result.

The average observed unweighted squared error loss of $\widehat{\boldsymbol{\beta}}_{\mathrm{R}}$ is 10.3551, while the average observed unweighted squared error loss of $\widehat{\boldsymbol{\beta}}_{\mathrm{R},k=1/5}$ is 1.5416. (The average observed loss of the ordinary least squares estimator $\widehat{\boldsymbol{\beta}}$ is 86.0342.) The better performance of the restricted ridge estimator can mainly be accounted for the closeness of the vector $\boldsymbol{\beta}_0$ to the true parameter vector $\boldsymbol{\beta} = (2, 2)'$. (The average observed unweighted squared error loss of the 'estimator' $\boldsymbol{\beta}_0$ is 1.0179.) □

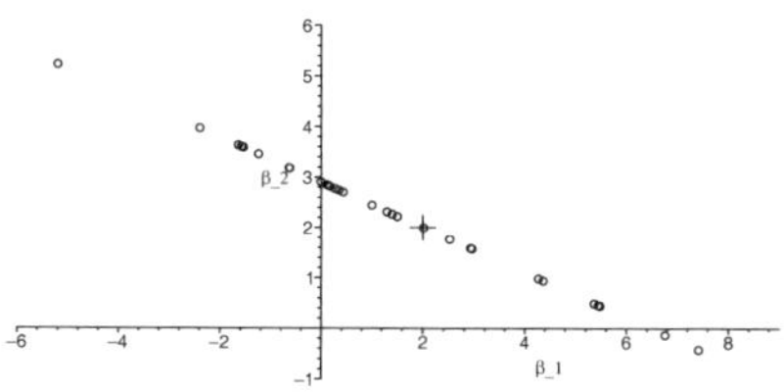

Fig. 4.5. 30 estimates $\widehat{\boldsymbol{\beta}}_{\mathrm{R}}$ in Example 4.5

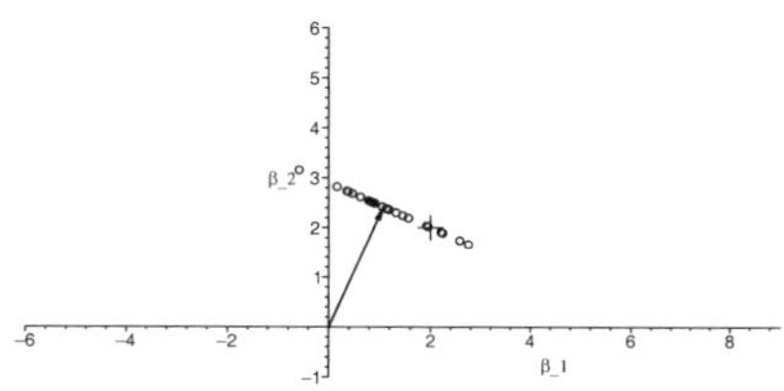

Fig. 4.6. 30 estimates $\widehat{\boldsymbol{\beta}}_{\mathrm{R},k}$ for $k = 1/5$ in Example 4.5

The above example shows that on principle it is *possible* to achieve better restricted ridge estimates than restricted least squares estimates with respect to the average unweighted squared error loss. It is, however, rather difficult to identify situations under which this effect can be expected. If collinearity is present and if one of the restrictions is almost parallel to one of the main directions of dispersion of the ordinary least squares estimator, as it is the case in the above example, then also the restricted least squares estimator will have a rather large variance with respect to this direction. Then one may expect some improvement from shrinking the estimates, achieved by the application of the restricted ridge estimator. If, on the other hand, the restrictions are orthogonal to the main directions of dispersion of the ordinary least squares estimator (as it is e.g. the case for the artificial restrictions corresponding to the principal components estimator), then the variances of the restricted least squares estimator can be expected to be rather small and a possible improvement from application of the restricted ridge estimator is doubtful.

The restricted ridge estimator $\widehat{\boldsymbol{\beta}}_{\mathrm{R},k}$ can serve as an example for an alternative to the restricted least squares estimator when linear restrictions $\boldsymbol{R\beta} = \boldsymbol{r}$ are present. One may also construct a more general class of alternatives by replacing the matrix $k\boldsymbol{I}_p$ in $\widehat{\boldsymbol{\beta}}_{\mathrm{R},k}$ by some $p \times p$ symmetric nonnegative definite matrix $\boldsymbol{K}$. Then

$$\widehat{\beta}_{\mathrm{R},\boldsymbol{K}}$$
$$= \widehat{\beta}_{\boldsymbol{K},\beta_0} - (\boldsymbol{X}'\boldsymbol{X} + \boldsymbol{K})^{-1}\boldsymbol{R}'[\boldsymbol{R}(\boldsymbol{X}'\boldsymbol{X} + \boldsymbol{K})^{-1}\boldsymbol{R}']^{-1}(\boldsymbol{R}\widehat{\beta}_{\boldsymbol{K},\beta_0} - \boldsymbol{r})$$

can be called *general restricted ridge estimator* for the parameter vector β. Here, the estimator

$$\widehat{\beta}_{\boldsymbol{K},\beta_0} = (\boldsymbol{X}'\boldsymbol{X} + \boldsymbol{K})^{-1}(\boldsymbol{X}'\boldsymbol{y} + \boldsymbol{K}\beta_0)$$

is a non-homogeneous general ridge estimator for β, which can also be written in the form

$$\widehat{\beta}_{\boldsymbol{K},\beta_0} = \beta_0 + (\boldsymbol{X}'\boldsymbol{X} + \boldsymbol{K})^{-1}\boldsymbol{X}'(\boldsymbol{y} - \boldsymbol{X}\beta_0) ,$$

compare also Sect. 4.2.5.

This shows that in a situation when linear restrictions on the parameter vector are given, there do in fact exist alternatives to the restricted least squares estimator, possibly obtainable from different choices of the matrix $\boldsymbol{K}$. However, it is reasonable to take only those into account, which turn out to be admissible for β under the restricted linear regression model.

4.3.3 Characterization of Linearly Admissible Estimators

When we consider a linear estimator $\boldsymbol{Ay}+\boldsymbol{a}$ under the linear regression model with assumptions (i) to (iv) and (viii), then we can give conditions on $\boldsymbol{A}$ and $\boldsymbol{a}$ such the linear estimator is admissible for $\beta \in \{\beta : \boldsymbol{R}\beta = \boldsymbol{r}\}$ within the set $\mathcal{L}(\beta)$ of linear estimators and with respect to the unweighted squared error loss.

Theorem 4.17. *Under the linear regression model with assumptions (i) to (iv) and (viii), a linear estimator $\boldsymbol{Ay} + \boldsymbol{a}$ is admissible for β within the set $\mathcal{L}(\beta)$ and with respect to the unweighted squared error loss if and only if the three conditions*

$$\boldsymbol{XA} = \boldsymbol{A}'\boldsymbol{X}', \quad \sigma(\boldsymbol{XA}) \subset [0,1], \quad \boldsymbol{a} \in \mathcal{C}(\boldsymbol{I}_p - \boldsymbol{AX}) ,$$

hold together with

$$\boldsymbol{RA} = \boldsymbol{0} \quad \textit{and} \quad \boldsymbol{Ra} = \boldsymbol{r} .$$

The above theorem follows from the results in [6, 7] and provides a more implicit characterization of the set of linearly admissible estimators for β. Similarly to Sect. 4.2.2 it is also possible to derive a more explicit characterization of this set of estimators, see Problem 4.12.

It is easy to see that a linear estimator $\widetilde{\beta} = \boldsymbol{Ay} + \boldsymbol{a}$ satisfies

$$\boldsymbol{R}\widetilde{\beta} = \boldsymbol{r} \quad \forall \boldsymbol{y} \in \mathbb{R}^n$$

if and only if $\boldsymbol{RA} = \boldsymbol{0}$ and $\boldsymbol{Ra} = \boldsymbol{r}$ hold true. Any such estimator $\widetilde{\beta}$ can be called a *restricted linear estimator*. By combining Theorems 4.4 and 4.17, we can obtain the following characterization.

Corollary 4.1. *An estimator* $\boldsymbol{Ay}+\boldsymbol{a}$ *is linearly admissible for* $\boldsymbol{\beta}$ *under the linear regression model with assumptions (i) to (iv) and parameter space* $\Theta_{\boldsymbol{\beta}}=\{\boldsymbol{\beta}:\boldsymbol{R\beta}=\boldsymbol{r}\}$ *if and only if it is a restricted linear estimator (with respect to* $\boldsymbol{R\beta}=\boldsymbol{r}$*) and, in addition, it is linearly admissible for* $\boldsymbol{\beta}$ *when the parameter space is* $\Theta_{\boldsymbol{\beta}}=\mathbb{R}^p$.

If under the linear regression model with assumption (i) to (iv) we denote by $\mathcal{L}_r(\boldsymbol{\beta})$ the set of restricted linear estimators with respect to $\boldsymbol{R\beta}=\boldsymbol{r}$ (whether these restrictions are satisfied or not) and by $\mathcal{A}(\boldsymbol{\beta})$ the set of linearly admissible estimators for $\boldsymbol{\beta}$, then the set

$$\mathcal{A}_r(\boldsymbol{\beta}):=\mathcal{L}_r(\boldsymbol{\beta})\cap\mathcal{A}(\boldsymbol{\beta})$$

is exactly the set of linearly admissible estimators for $\boldsymbol{\beta}$ when the restrictions $\boldsymbol{R\beta}=\boldsymbol{r}$ are assumed to hold true.

Unbiased Estimator. Consider the set

$$\mathcal{U}(\boldsymbol{\beta})=\{\boldsymbol{Ay}+\boldsymbol{a}:\ \boldsymbol{AX}=\boldsymbol{I}_p,\boldsymbol{a}=\boldsymbol{0}\}\ ,$$

of unbiased linear estimators for $\boldsymbol{\beta}$ when the parameter space is $\Theta_{\boldsymbol{\beta}}=\mathbb{R}^p$, being a subset of the set of unbiased estimators for $\boldsymbol{\beta}$ when $\Theta_{\boldsymbol{\beta}}=\{\boldsymbol{\beta}:\boldsymbol{R\beta}=\boldsymbol{r}\}$ is the considered parameter space for $\boldsymbol{\beta}$, see also Remark 3.1. Then we can see from the above corollary that an element from $\mathcal{U}(\boldsymbol{\beta})$ can only be admissible for $\boldsymbol{\beta}$ under the linear regression model with assumptions (i) to (iv) and (viii) if $\boldsymbol{r}=\boldsymbol{0}$ and $\boldsymbol{R}=\boldsymbol{0}$ hold true, in which case the restrictions are meaningless. This is not a surprising fact, since from Corollary 3.2 the restricted least squares estimator $\widehat{\boldsymbol{\beta}}_{\mathrm{R}}$ with respect to $\boldsymbol{R\beta}=\boldsymbol{r}$ is uniformly better than $\widehat{\boldsymbol{\beta}}$, and thus than any estimator from $\mathcal{U}(\boldsymbol{\beta})$, whenever the restrictions $\boldsymbol{R\beta}=\boldsymbol{r}$ are assumed to hold true.

Restricted Least Squares Estimator. Section 4.2.5 shows that the restricted least squares estimator $\widehat{\boldsymbol{\beta}}_{\mathrm{R}}$ with respect to arbitrary restrictions $\boldsymbol{R\beta}=\boldsymbol{r}$ is always linearly admissible for $\boldsymbol{\beta}$ when the parameter space is $\Theta_{\boldsymbol{\beta}}=\mathbb{R}^p$. Since $\widehat{\boldsymbol{\beta}}_{\mathrm{R}}$ satisfies the condition $\boldsymbol{R}\widehat{\boldsymbol{\beta}}_{\mathrm{R}}=\boldsymbol{r}$ for any realization $\boldsymbol{y}\in\mathbb{R}^n$, it immediately follows that the restricted least squares estimator is also linearly admissible for $\boldsymbol{\beta}$ when the parameter space is $\Theta_{\boldsymbol{\beta}}=\{\boldsymbol{\beta}:\boldsymbol{R\beta}=\boldsymbol{r}\}$.

General Restricted Ridge Estimator. The restricted least squares estimator $\widehat{\boldsymbol{\beta}}_{\mathrm{R}}$ can be viewed as a special case of the general restricted ridge estimator $\widehat{\boldsymbol{\beta}}_{\mathrm{R},\boldsymbol{K}}$ with $\boldsymbol{K}=\boldsymbol{0}$. The following theorem shows that any estimator $\widehat{\boldsymbol{\beta}}_{\mathrm{R},\boldsymbol{K}}$ is also linearly admissible for $\boldsymbol{\beta}$.

Theorem 4.18. *Under the linear regression model with assumptions (i) to (iv) and (viii), the general restricted ridge estimator* $\widehat{\boldsymbol{\beta}}_{\mathrm{R},\boldsymbol{K}}$ *is linearly admissible for* $\boldsymbol{\beta}$.

Proof. The general restricted ridge estimator $\widehat{\beta}_{\mathrm{R},\boldsymbol{K}}$ can be written in the form $\widehat{\beta}_{\mathrm{R},\boldsymbol{K}} = \boldsymbol{A}\boldsymbol{y} + \boldsymbol{a}$, where

$$\boldsymbol{A} = \boldsymbol{M}_{\boldsymbol{K}}\boldsymbol{X}' \quad \text{and} \quad \boldsymbol{a} = (\boldsymbol{I}_p - \boldsymbol{M}_{\boldsymbol{K}}\boldsymbol{X}'\boldsymbol{X})\boldsymbol{\beta}_0 \, .$$

Here,

$$\boldsymbol{M}_{\boldsymbol{K}} = \boldsymbol{T}_{\boldsymbol{K}}^{-1} - \boldsymbol{T}_{\boldsymbol{K}}^{-1}\boldsymbol{R}'[\boldsymbol{R}\boldsymbol{T}_{\boldsymbol{K}}^{-1}\boldsymbol{R}']^{-1}\boldsymbol{R}\boldsymbol{T}_{\boldsymbol{K}}^{-1} \, ,$$

where $\boldsymbol{T}_{\boldsymbol{K}} = \boldsymbol{X}'\boldsymbol{X} + \boldsymbol{K}$. In view of $\boldsymbol{R}\boldsymbol{M}_{\boldsymbol{K}} = \boldsymbol{0}$ and $\boldsymbol{R}\boldsymbol{\beta}_0 = \boldsymbol{r}$, it follows that the conditions $\boldsymbol{R}\boldsymbol{A} = \boldsymbol{0}$ and $\boldsymbol{R}\boldsymbol{a} = \boldsymbol{r}$ are satisfied. Moreover, $\boldsymbol{X}\boldsymbol{A} = \boldsymbol{A}'\boldsymbol{X}'$ and $\boldsymbol{a} \in \mathcal{C}(\boldsymbol{I}_p - \boldsymbol{A}\boldsymbol{X})$ hold true. The proof of Theorem 4.9 shows that the inequality $\boldsymbol{X}\boldsymbol{T}_K^{-1}\boldsymbol{X}' \leq_{\mathrm{L}} \boldsymbol{X}(\boldsymbol{X}'\boldsymbol{X})^{-1}\boldsymbol{X}'$ is fulfilled. Then $\boldsymbol{X}\boldsymbol{T}_K^{-1}\boldsymbol{X}' \leq_{\mathrm{L}} \boldsymbol{I}_n$ in view of $\boldsymbol{X}(\boldsymbol{X}'\boldsymbol{X})^{-1}\boldsymbol{X}' \leq_{\mathrm{L}} \boldsymbol{I}_n$, showing that $\boldsymbol{X}\boldsymbol{M}_{\boldsymbol{K}}\boldsymbol{X}' \leq_{\mathrm{L}} \boldsymbol{I}_n$ holds true. The latter implies $\sigma(\boldsymbol{X}\boldsymbol{A}) \subset [0,1]$, thus concluding the proof. □

4.4 Linear Admissibility Under Elliptical Restrictions

In this section we identify the set of linearly admissible estimators for $\boldsymbol{\beta}$ when the unknown parameters $\boldsymbol{\beta}$ and σ^2 satisfy the relationship $\boldsymbol{\beta}'\boldsymbol{T}\boldsymbol{\beta} \leq \sigma^2$ for some known positive definite matrix $\boldsymbol{T}$. The considered parameter space is given by

$$\Theta = \Theta_{\boldsymbol{\beta}} = \{\boldsymbol{\beta} \in \mathbb{R}^p : \boldsymbol{\beta}'\boldsymbol{T}\boldsymbol{\beta} \leq \sigma^2\} \, ,$$

since we compare risks of estimators under variation of $\boldsymbol{\beta}$ solely. The parameter $\sigma^2 \in (0, \infty)$ is thus considered as arbitrary but fixed.

In the following we refer to the condition $\boldsymbol{\beta}'\boldsymbol{T}\boldsymbol{\beta} \leq \sigma^2$ as assumption

$$\text{(ix)} \quad \boldsymbol{\beta}'\boldsymbol{T}\boldsymbol{\beta} \leq \sigma^2$$

of the linear regression model. If the given prior information cannot be specified as in (ix) but is instead of the form

$$(\boldsymbol{\beta} - \boldsymbol{\beta}_0)'\boldsymbol{T}(\boldsymbol{\beta} - \boldsymbol{\beta}_0) \leq \sigma^2$$

for some known vector $\boldsymbol{\beta}_0 \in \mathbb{R}^p$, then we may consider the linear regression model

$$\overline{\boldsymbol{y}} = \boldsymbol{X}\overline{\boldsymbol{\beta}} + \boldsymbol{\varepsilon}, \quad \overline{\boldsymbol{y}} = \boldsymbol{y} - \boldsymbol{X}\boldsymbol{\beta}_0, \quad \overline{\boldsymbol{\beta}} = \boldsymbol{\beta} - \boldsymbol{\beta}_0, \quad \overline{\boldsymbol{\beta}}'\boldsymbol{T}\overline{\boldsymbol{\beta}} \leq \sigma^2 \, .$$

In such a case the results in this section carry over to linear estimators $\boldsymbol{A}\overline{\boldsymbol{y}}$ of the vector $\overline{\boldsymbol{\beta}}$. If some estimator $\boldsymbol{A}\overline{\boldsymbol{y}}$ is chosen, then the estimator $\boldsymbol{\beta}_0 + \boldsymbol{A}\overline{\boldsymbol{y}}$ is an obvious corresponding estimator for the original parameter vector $\boldsymbol{\beta}$.

4.4.1 Characterization of Linearly Admissible Estimators

We now consider the admissibility of a homogeneous linear estimator for $\boldsymbol{\beta}$ within the set

$$\mathcal{L}^{\mathrm{h}}(\boldsymbol{\beta}) = \{\boldsymbol{Ay} : \boldsymbol{A} \in \mathbb{R}^{p \times n}\}$$

under the linear regression model with assumptions (i) to (iv) and (ix). The following theorems provide different characterizations of this set of estimators for $\boldsymbol{\beta}$.

Theorem 4.19. *Under the linear regression model with assumptions (i) to (iv) and (ix), the estimator $\boldsymbol{Ay}$ is admissible for $\boldsymbol{\beta}$ within the set $\mathcal{L}^{\mathrm{h}}(\boldsymbol{\beta})$ and with respect to the unweighted squared error loss if and only if $\boldsymbol{A}$ can be written as*

$$\boldsymbol{A} = \boldsymbol{S}^{-1/2}\boldsymbol{G}\boldsymbol{S}^{-1/2}\boldsymbol{X}'$$

for some symmetric $p \times p$ matrix $\boldsymbol{G}$ satisfying the two conditions

$$\mathrm{tr}[(\boldsymbol{I}_p - \boldsymbol{G})^{-1}\boldsymbol{S}^{-1/2}\boldsymbol{T}\boldsymbol{S}^{-1/2}] \leq 1 + \mathrm{tr}[\boldsymbol{S}^{-1/2}\boldsymbol{T}\boldsymbol{S}^{-1/2}]$$

and

$$\sigma(\boldsymbol{G}) \subset [0, 1)\,,$$

where $\boldsymbol{S} = \boldsymbol{X}'\boldsymbol{X}$.

We do not give a proof for this result but refer the reader to [54]. A more implicit characterization of the set of linearly admissible estimators for $\boldsymbol{\beta}$ is provided by the following theorem.

Theorem 4.20. *Under the linear regression model with assumptions (i) to (iv) and (ix), the estimator $\boldsymbol{Ay}$ is admissible for $\boldsymbol{\beta}$ within the set $\mathcal{L}^{\mathrm{h}}(\boldsymbol{\beta})$ and with respect to the unweighted squared error loss if and only if the conditions*

$$\boldsymbol{XA} = \boldsymbol{A}'\boldsymbol{X}', \quad \sigma(\boldsymbol{XA}) \subset [0, 1)\,,$$

and

$$\mathrm{tr}[(\boldsymbol{I}_p - \boldsymbol{AX})^{-1}(\boldsymbol{X}'\boldsymbol{X})^{-1}\boldsymbol{T}] \leq 1 + \mathrm{tr}[(\boldsymbol{X}'\boldsymbol{X})^{-1}\boldsymbol{T}]$$

are satisfied.

Proof. Similarly to the proof of Theorem 4.5 it can be shown that the conditions $\boldsymbol{XA} = \boldsymbol{A}'\boldsymbol{X}'$ and $\sigma(\boldsymbol{XA}) \subset [0, 1)$ are equivalent to $\boldsymbol{A} = (\boldsymbol{X}'\boldsymbol{X})^{-1/2}\boldsymbol{G}(\boldsymbol{X}'\boldsymbol{X})^{-1/2}\boldsymbol{X}'$ for some matrix $\boldsymbol{G}$ with $\boldsymbol{G} = \boldsymbol{G}'$ and $\sigma(\boldsymbol{G}) \subset [0, 1)$. For this matrix $\boldsymbol{A}$, the identity

$$\boldsymbol{I}_p - \boldsymbol{AX} = \boldsymbol{S}^{-1/2}(\boldsymbol{I}_p - \boldsymbol{G})\boldsymbol{S}^{1/2}$$

holds true, where $\boldsymbol{S} = \boldsymbol{X}'\boldsymbol{X}$. Since all eigenvalues of $\boldsymbol{G}$ lie in $[0, 1)$, the matrix $\boldsymbol{I}_p - \boldsymbol{G}$ is nonsingular, and

$$(\boldsymbol{I}_p - \boldsymbol{AX})^{-1} = \boldsymbol{S}^{-1/2}(\boldsymbol{I}_p - \boldsymbol{G})^{-1}\boldsymbol{S}^{1/2}\,.$$

Then the assertion follows in view of Theorem 4.19. □

The matrix $\boldsymbol{I}_p - \boldsymbol{A}\boldsymbol{X}$ in the above theorem is nonsingular, since all eigenvalues of $\boldsymbol{X}\boldsymbol{A}$ and thus of $\boldsymbol{A}\boldsymbol{X}$ lie in the interval $[0,1)$.

If $\mathcal{A}(\boldsymbol{\beta})$ denotes the set of linearly admissible estimators under the linear regression model with parameter space $\Theta_{\boldsymbol{\beta}} = \mathbb{R}^p$, then Theorems 4.4 and Theorem 4.20 show that the set of linearly admissible estimators with considered parameter space $\Theta_{\boldsymbol{\beta}} = \{\boldsymbol{\beta} \in \mathbb{R}^p : \boldsymbol{\beta}'\boldsymbol{T}\boldsymbol{\beta} \leq \sigma^2\}$ is a subset of $\mathcal{A}(\boldsymbol{\beta})$.

According to Theorem 4.15 an estimator $\widetilde{\boldsymbol{\beta}} = \boldsymbol{S}^{-1/2}\boldsymbol{G}\boldsymbol{S}^{-1/2}\boldsymbol{X}'\boldsymbol{y}$ with $\boldsymbol{G} = \boldsymbol{G}'$ and $\sigma(\boldsymbol{G}) \subset [0,1)$ is a homogeneous linear Bayes estimator for $\boldsymbol{\beta}$. Then from Corollary 3.6 the estimator $\widetilde{\boldsymbol{\beta}}$ can also be written as $\widetilde{\boldsymbol{\beta}} = \boldsymbol{M}(\boldsymbol{M}+\boldsymbol{S}^{-1})^{-1}\widehat{\boldsymbol{\beta}}$ for some symmetric nonnegative definite matrix $\boldsymbol{M}$. Thus, the characterization of the set of linearly admissible estimators can also be based on this representation of linear Bayes estimators.

Theorem 4.21. *Under the linear regression model with assumptions (i) to (iv) and (ix), consider the following three conditions:*

(i) $\boldsymbol{A}\boldsymbol{y}$ *is linearly admissible for* $\boldsymbol{\beta}$*;*
(ii) $\boldsymbol{A} = \boldsymbol{M}(\boldsymbol{M} + (\boldsymbol{X}'\boldsymbol{X})^{-1})^{-1}(\boldsymbol{X}'\boldsymbol{X})^{-1}\boldsymbol{X}'$ *for some* $p \times p$ *symmetric nonnegative definite matrix* $\boldsymbol{M}$*;*
(iii) $\mathrm{tr}(\boldsymbol{M}\boldsymbol{T}) \leq 1$.

Then the following two statements are true:

(a) *Condition (i) is satisfied if and only if conditions (ii) and (iii) are both valid.*
(b) *If condition (i) and (ii) are satisfied, then condition (iii) is valid.*

Proof. As noted above, the conditions $\boldsymbol{X}\boldsymbol{A} = \boldsymbol{A}'\boldsymbol{X}'$ and $\sigma(\boldsymbol{X}\boldsymbol{A}) \subset [0,1)$ are satisfied if and only if $\boldsymbol{A}$ can be written as $\boldsymbol{A} = \boldsymbol{M}(\boldsymbol{M} + \boldsymbol{S}^{-1})^{-1}\boldsymbol{S}^{-1}\boldsymbol{X}'$, $\boldsymbol{S} = \boldsymbol{X}'\boldsymbol{X}$, for some $p \times p$ symmetric nonnegative definite matrix $\boldsymbol{M}$. Then

$$\boldsymbol{I}_p - \boldsymbol{A}\boldsymbol{X} = \boldsymbol{S}^{-1}(\boldsymbol{M} + \boldsymbol{S}^{-1})^{-1} ,$$

and thus $(\boldsymbol{I}_p - \boldsymbol{A}\boldsymbol{X})^{-1} = \boldsymbol{M}\boldsymbol{S} + \boldsymbol{I}_p$. Then, statements (a) and (b) follow from Theorem 4.20. □

The above theorem shows that any homogeneous linearly admissible estimator for $\boldsymbol{\beta}$ can be written as

$$\widetilde{\boldsymbol{\beta}} = \boldsymbol{M}(\boldsymbol{M} + (\boldsymbol{X}'\boldsymbol{X})^{-1})^{-1}\widehat{\boldsymbol{\beta}} .$$

Problem 3.39 shows that this estimator can be seen as a homogeneous linear Bayes estimator when $\boldsymbol{\beta} \in \mathbb{R}^p$ is the realization of a $p \times 1$ random vector $\boldsymbol{b}$ with $\mathrm{E}(\boldsymbol{b}\boldsymbol{b}') = \sigma^2\boldsymbol{M}$. However, the above theorem also shows that not any symmetric nonnegative definite matrix $\boldsymbol{M}$ is suitable for ensuring admissibility. The same conclusion can be drawn from the Bayesian point of view, since under the additional model assumption (ix) the parameter vector $\boldsymbol{\beta}$ cannot be any value from $\mathbb{R}^p$. Thus the set of possible matrices $\boldsymbol{M}$ must be somehow restricted. As noted in [83, p. 232], appropriate matrices $\boldsymbol{M}$ under assumption (ix) are just those, satisfying the additional condition $\mathrm{tr}(\boldsymbol{MT}) \leq 1$. Hence, the above theorem truly involves the set of homogeneous linear Bayes estimators for $\boldsymbol{\beta}$ under the linear regression model with assumptions (i) to (iv) and (ix).

Remark 4.7. Under the linear regression model with assumptions (i) to (iv) and (ix), a homogeneous linear estimator is admissible for $\boldsymbol{\beta}$ within the set of homogeneous linear estimators if and only if it a homogeneous linear Bayes estimator for $\boldsymbol{\beta}$.

The coincidence of the set of linearly admissible estimators and the set of linear Bayes estimators remains true when we take arbitrary compact parameter spaces $\Theta_{\boldsymbol{\beta}} \subset \mathbb{R}^p$ into account and consider $\Delta = \mathcal{L}(\boldsymbol{\beta})$ as the set of possible decision rules, see Theorem 16.2 in [83]. Notice that $\Theta_{\boldsymbol{\beta}} = \{\boldsymbol{\beta} : \boldsymbol{\beta}'\boldsymbol{T}\boldsymbol{\beta} \leq \sigma^2\}$ is a compact set when the parameter $\sigma^2 \in (0, \infty)$ is assumed to be arbitrary but fixed.

4.4.2 Linear Admissibility of Certain Linear Estimators

We have already noted that under given restrictions $\boldsymbol{\beta}'\boldsymbol{T}\boldsymbol{\beta} \leq \sigma^2$, a homogeneous linear estimator $\boldsymbol{Ay}$ does not necessarily satisfy these restrictions, simply because the inequality $\boldsymbol{y}'\boldsymbol{A}'\boldsymbol{T}\boldsymbol{A}\boldsymbol{y} \leq \sigma^2$ will not hold true for any realization $\boldsymbol{y} \in \mathbb{R}^n$.

This is also true when the estimator is linearly admissible for $\boldsymbol{\beta}$. Hence, the situation of elliptical restrictions is different from the situation of given linear restrictions $\boldsymbol{R\beta} = \boldsymbol{r}$, since in the latter case linearly admissible estimators for $\boldsymbol{\beta}$ must necessarily satisfy these restrictions. In fact, Corollary 4.1 claims that the set of linearly admissible estimators is exactly the subset of linearly admissible estimators with respect to the parameter space $\Theta_{\boldsymbol{\beta}} = \mathbb{R}^p$ which satisfy the linear restrictions for every realization $\boldsymbol{y}$.

Nonetheless, the two situations have something in common, namely in both cases the set of linearly admissible estimators is some subset of the set $\mathcal{A}(\boldsymbol{\beta})$ of linearly admissible estimators when the considered parameter space is $\Theta_{\boldsymbol{\beta}} = \mathbb{R}^p$.

Hence, under the additional model assumption (ix) any considered homogeneous linear estimator from Chap. 3 (being linearly admissible for $\boldsymbol{\beta}$ with respect to $\Theta_{\boldsymbol{\beta}} = \mathbb{R}^p$) is a possible candidate as an estimator for the parameter vector $\boldsymbol{\beta}$. In the following we consider some of these in more detail.

Unbiased Estimator. If we consider an estimator $\boldsymbol{Ay}$ from the set

$$\mathcal{U}(\boldsymbol{\beta}) = \{\boldsymbol{Ay} : \boldsymbol{AX} = \boldsymbol{I}_p\} ,$$

then the matrix $\boldsymbol{XA}$ has exactly p eigenvalues equal to 1. Thus an element from this set cannot be linearly admissible for $\boldsymbol{\beta}$ under the linear regression model with assumptions (i) to (iv) and (ix). In particular, the ordinary least squares estimator $\widehat{\beta}$, being the only linearly admissible from $\mathcal{U}(\beta)$ when $\Theta_\beta = \mathbb{R}^p$, is *not* linearly admissible for β when $\Theta_\beta = \{\beta : \beta' T \beta \leq \sigma^2\}$.

Restricted Least Squares Estimator. If R is a $m \times p$ matrix of full row rank, then the restricted least squares estimator $\widehat{\beta}_{\mathrm{R}}$ corresponding to the restrictions $R\beta = \mathbf{0}$ is given as

$$\widehat{\beta}_{\mathrm{R}} = Ay, \quad A = S^{-1}X' - S^{-1}R'(RS^{-1}R')^{-1}RS^{-1}X' ,$$

where $S = X'X$. The matrix

$$AX = I_p - S^{-1}R'(RS^{-1}R')^{-1}R$$

is idempotent and has $p - m$ eigenvalues equal to 1 and m eigenvalues equal to 0. But then, also the matrix XA has $p-m$ eigenvalues equal to 1, showing that in case $m < p$ the restricted least squares estimator $\widehat{\beta}_{\mathrm{R}}$ is *not* linearly admissible for β under the linear regression model with assumptions (i) to (iv) and (ix). In particular, the principal components estimator $\widehat{\beta}^{(r)}$ is not linearly admissible for β under elliptical restrictions.

General Ridge Estimator. Let us now consider the general ridge estimator

$$\widehat{\beta}_K = (X'X + K)^{-1}X'y$$

for some $p \times p$ symmetric nonnegative definite matrix K. When we write $A = (X'X + K)^{-1}X'$, then

$$I_p - AX = (X'X + K)^{-1}K .$$

In view of Theorem 4.20 an estimator Ay can only be linearly admissible for β when $I_p - AX$ is non-singular. Thus, the non-singularity of the matrix K is necessary for the linear admissibility of $\widehat{\beta}_K$ under the linear regression model with assumptions (i) to (iv) and (ix). On the other hand, if K and thus $I_p - AX$ is non-singular, then all eigenvalues of $AX = I_p - (X'X + K)^{-1}K$ lie in the interval $[0, 1)$. In addition, if the matrix K is non-singular, then $\widehat{\beta}_K$ can be written as

$$\widehat{\beta}_K = K^{-1}(K^{-1} + (X'X)^{-1})^{-1}\widehat{\beta} .$$

In view of the above we can state the following result.

Theorem 4.22. *Under the linear regression model with assumptions (i) to (iv) and (ix), let K be a $p \times p$ symmetric nonnegative definite matrix. Then the general ridge estimator $\widehat{\beta}_K = (X'X + K)^{-1}X'y$ is linearly admissible for β if and only if K is nonsingular and the inequality $\mathrm{tr}(K^{-1}T) \leq 1$ holds true.*

Section 3.7 introduces the upper quasi minimax estimator $\widehat{\beta}_T$ and the lower quasi minimax estimator $\widehat{\beta}_{pT}$ as possible estimators for β under given restrictions $\beta' T \beta \leq \sigma^2$. In view of the above theorem, the estimator $\widehat{\beta}_T$ turns out to be *not* linearly admissible for β when $p \geq 2$, while this is not true for $\widehat{\beta}_{pT}$, i.e $\widehat{\beta}_{pT}$ is in fact linearly admissible for β. Moreover, any estimator

$$\widehat{\beta}_{kT} = (X'X + kT)^{-1} X'y, \quad k \geq p\,,$$

is linearly admissible for β.

4.4.3 Admissible Improvements Over Ordinary Least Squares

Section 3.7.5 shows that the inequality

$$\mathrm{MSE}(\beta, \widehat{\beta}_T) \leq_{\mathrm{L}} \mathrm{MSE}(\beta, \widehat{\beta})$$

holds true for any vector $\beta \in \mathbb{R}^p$ satisfying the condition $\beta' T \beta \leq \sigma^2$. Moreover, since $\mathrm{MSE}(\beta, \widehat{\beta}_T) \neq \mathrm{MSE}(\beta, \widehat{\beta})$ for $p > 1$, it follows that

$$\rho(\beta, \widehat{\beta}_T) < \rho(\beta, \widehat{\beta}) \quad \text{for all } \beta \in \mathbb{R}^p \text{ such that } \beta' T \beta \leq \sigma^2\,,$$

where $\rho(\beta, \widetilde{\beta}) = \mathrm{tr}[\mathrm{MSE}(\beta, \widetilde{\beta})]$. Hence, the upper quasi minimax estimator $\widehat{\beta}_T$ is strictly better than the ordinary least squares estimator $\widehat{\beta}$ with respect to the unweighted squared error risk for any vector β satisfying the condition $\beta' T \beta \leq \sigma^2$.

On the other hand, although $\widehat{\beta}_T$ is uniformly better than $\widehat{\beta}$ under the linear regression model with assumptions (i) to (iv) and (ix), it is not at the same time an admissible estimator for β within the set of homogeneous linear estimators.

The situation is reversed when we consider the estimator $\widehat{\beta}_{pT}$ instead of $\widehat{\beta}_T$. Then, $\widehat{\beta}_{pT}$ is linearly admissible for β when $\beta' T \beta \leq \sigma^2$ holds true, but it is not necessarily better than $\widehat{\beta}$ for every vector β satisfying $\beta' T \beta \leq \sigma^2$.

The respective limitations of upper and lower quasi minimax estimator may lead our interest to the question of the existence of linear estimators satisfying both properties at the same time, namely being linearly admissible for β, and, in addition, being uniformly better than the ordinary least squares estimator for all those vectors β satisfying the condition $\beta' T \beta \leq \sigma^2$.

As the suitable candidate, we consider the homogeneous linear Bayes estimator

$$\widetilde{\beta} = M(M + (X'X)^{-1})^{-1} \widehat{\beta}$$

for some $p \times p$ symmetric nonnegative definite matrix M, known to be the necessary form of a linearly admissible estimator for β from Theorem 4.21. It is thus of interest to find a necessary and sufficient condition for the improvement of $\widetilde{\beta}$ upon the ordinary least squares estimator $\widehat{\beta}$.

Theorem 4.23. *Under the linear regression model with assumptions (i) to (iv), let* $\widetilde{\boldsymbol{\beta}} = \boldsymbol{M}(\boldsymbol{M} + (\boldsymbol{X}'\boldsymbol{X})^{-1})^{-1}\widehat{\boldsymbol{\beta}}$ *for some* $p \times p$ *symmetric nonnegative definite matrix* $\boldsymbol{M}$*. Moreover, let* $\boldsymbol{T}$ *be a* $p \times p$ *symmetric positive definite matrix. Then*

$$\mathrm{MSE}(\boldsymbol{\beta}, \widetilde{\boldsymbol{\beta}}) \leq_{\mathrm{L}} \mathrm{MSE}(\boldsymbol{\beta}, \widehat{\boldsymbol{\beta}}) \quad \forall \boldsymbol{\beta} \in \{\boldsymbol{\beta} : \boldsymbol{\beta}'\boldsymbol{T}\boldsymbol{\beta} \leq \sigma^2\}$$

holds true if and only if $\boldsymbol{T}^{-1} \leq_{\mathrm{L}} 2\boldsymbol{M} + (\boldsymbol{X}'\boldsymbol{X})^{-1}$ *is satisfied.*

Proof. Let $\boldsymbol{S} = \boldsymbol{X}'\boldsymbol{X}$ and $\boldsymbol{L} = \boldsymbol{M}(\boldsymbol{M} + \boldsymbol{S}^{-1})^{-1}$. Then the difference

$$\boldsymbol{\Delta} = \mathrm{MSE}(\boldsymbol{\beta}, \widehat{\boldsymbol{\beta}}) - \mathrm{MSE}(\boldsymbol{\beta}, \widetilde{\boldsymbol{\beta}})$$

can be written as

$$\boldsymbol{\Delta} = \sigma^2\boldsymbol{S}^{-1} - \sigma^2\boldsymbol{L}\boldsymbol{S}^{-1}\boldsymbol{L}' - (\boldsymbol{I}_p - \boldsymbol{L})\boldsymbol{\beta}\boldsymbol{\beta}'(\boldsymbol{I}_p - \boldsymbol{L})' .$$

Since $\boldsymbol{L}\boldsymbol{S}^{-1}$ is symmetric, it follows that the identity $\boldsymbol{L}\boldsymbol{S}^{-1}\boldsymbol{S}\boldsymbol{S}^{-1}\boldsymbol{L}' = \boldsymbol{S}^{-1}\boldsymbol{L}'\boldsymbol{S}\boldsymbol{L}\boldsymbol{S}^{-1}$ holds true. Moreover, $\boldsymbol{S}^{-1}(\boldsymbol{M} + \boldsymbol{S}^{-1})^{-1} = \boldsymbol{I}_p - \boldsymbol{L}$, so that

$$\boldsymbol{\Delta} = \sigma^2\boldsymbol{S}^{-1} - \sigma^2(\boldsymbol{I}_p - \boldsymbol{L})\boldsymbol{M}\boldsymbol{S}\boldsymbol{M}(\boldsymbol{I}_p - \boldsymbol{L})' - (\boldsymbol{I}_p - \boldsymbol{L})\boldsymbol{\beta}\boldsymbol{\beta}'(\boldsymbol{I}_p - \boldsymbol{L})' .$$

The matrix $\boldsymbol{\Delta}$ is nonnegative definite if and only if $(\boldsymbol{I}_p - \boldsymbol{L})^{-1}\boldsymbol{\Delta}[(\boldsymbol{I}_p - \boldsymbol{L})']^{-1}$ is nonnegative definite, where $(\boldsymbol{I}_p - \boldsymbol{L})^{-1} = (\boldsymbol{M} + \boldsymbol{S}^{-1})\boldsymbol{S} = \boldsymbol{M}\boldsymbol{S} + \boldsymbol{I}_p$. In view of

$$(\boldsymbol{I}_p - \boldsymbol{L})^{-1}\boldsymbol{\Delta}[(\boldsymbol{I}_p - \boldsymbol{L})']^{-1} = \sigma^2(2\boldsymbol{M} + \boldsymbol{S}^{-1}) - \boldsymbol{\beta}\boldsymbol{\beta}'$$

it follows that $\boldsymbol{\Delta}$ is nonnegative definite if and only if

$$\boldsymbol{\beta}'(2\boldsymbol{M} + \boldsymbol{S}^{-1})^{-1}\boldsymbol{\beta} \leq \sigma^2$$

is satisfied. Hence, it remains to show that this inequality is satisfied for all $\boldsymbol{\beta}$ with $\boldsymbol{\beta}'\boldsymbol{T}\boldsymbol{\beta} \leq \sigma^2$ if and only if $(2\boldsymbol{M} + \boldsymbol{S}^{-1})^{-1} \leq_{\mathrm{L}} \boldsymbol{T}$, or, equivalently, $\boldsymbol{T}^{-1} \leq_{\mathrm{L}} 2\boldsymbol{M} + \boldsymbol{S}^{-1}$ holds true.

If the inequality $(2\boldsymbol{M}+\boldsymbol{S}^{-1})^{-1} \leq_{\mathrm{L}} \boldsymbol{T}$ is satisfied, then $\boldsymbol{\beta}'(2\boldsymbol{M}+\boldsymbol{S}^{-1})^{-1}\boldsymbol{\beta} \leq \boldsymbol{\beta}'\boldsymbol{T}\boldsymbol{\beta}$ for every vector $\boldsymbol{\beta} \in \mathbb{R}^p$. Thus $\boldsymbol{\beta}'(2\boldsymbol{M}+\boldsymbol{S}^{-1})^{-1}\boldsymbol{\beta} \leq \sigma^2$ for every vector $\boldsymbol{\beta}$ satisfying $\boldsymbol{\beta}'\boldsymbol{T}\boldsymbol{\beta} \leq \sigma^2$. On the other hand, let

$$\boldsymbol{\beta}'(2\boldsymbol{M} + \boldsymbol{S}^{-1})^{-1}\boldsymbol{\beta} \leq \sigma^2 \quad \forall \boldsymbol{\beta} \in \{\boldsymbol{\beta} : \boldsymbol{\beta}'\boldsymbol{T}\boldsymbol{\beta} \leq \sigma^2\}$$

be true and assume that then the inequality $(2\boldsymbol{M} + \boldsymbol{S}^{-1})^{-1} \leq_{\mathrm{L}} \boldsymbol{T}$ does not hold. Then there exists some vector $\boldsymbol{\beta}_1 \in \mathbb{R}^p$ with

$$\boldsymbol{\beta}_1'(2\boldsymbol{M} + \boldsymbol{S}^{-1})^{-1}\boldsymbol{\beta}_1 > \boldsymbol{\beta}_1'\boldsymbol{T}\boldsymbol{\beta}_1 .$$

Moreover, we can always find some $\varepsilon > 0$ with $\boldsymbol{\beta}_2 = \varepsilon\boldsymbol{\beta}_1$ and $\boldsymbol{\beta}_2'\boldsymbol{T}\boldsymbol{\beta}_2 = \sigma^2$. Then

$$\boldsymbol{\beta}_2'(2\boldsymbol{M} + \boldsymbol{S}^{-1})^{-1}\boldsymbol{\beta}_2 > \underbrace{\boldsymbol{\beta}_2'\boldsymbol{T}\boldsymbol{\beta}_2}_{=\sigma^2} .$$

But then $\beta_2'(2M + S^{-1})^{-1}\beta_2 > \sigma^2$ for some vector β_2 with $\beta_2' T \beta_2 \leq \sigma^2$. This contradicts the above assumption, so that the inequality $(2M + S^{-1})^{-1} \leq_L T$ must be true if Δ is nonnegative definite for all β satisfying $\beta' T \beta \leq \sigma^2$. This concludes the proof. □

The above theorem gives a necessary and sufficient condition for a homogeneous linear Bayes estimator

$$\widetilde{\beta} = M(M + (X'X)^{-1})^{-1}\widehat{\beta}$$

to be uniformly not worse than $\widehat{\beta}$ with respect to MSE under the linear regression model with assumptions (i) to (iv) and (ix). Moreover, Problem 4.15 shows that in case $p > 1$ the difference of the MSE matrices of these estimators cannot be zero, justifying the term 'improvement' of $\widetilde{\beta}$ upon $\widehat{\beta}$.

From the choice $M = T^{-1}$ it can be seen that $\widetilde{\beta} = \widehat{\beta}_T$ is uniformly not worse than $\widehat{\beta}$ in this sense (as already noted before), but since the condition $\mathrm{tr}(MT) \leq 1$ is not fulfilled, the estimator $\widehat{\beta}_T$ is not linearly admissible for β.

From Theorems 4.21 and 4.23 we can easily conclude that under the linear regression model with assumptions (i) to (iv) and (ix), a homogeneous linear estimator is uniformly not worse than $\widehat{\beta}$ with respect to MSE, and, in addition, linearly admissible for β if and only if it can be written as

$$\widetilde{\beta} = M(M + (X'X)^{-1})^{-1}\widehat{\beta}$$

for some $p \times p$ symmetric nonnegative definite matrix M satisfying the two conditions

$$T^{-1} \leq_L 2M + (X'X)^{-1} \quad \text{and} \quad \mathrm{tr}(MT) \leq 1\,.$$

But does there always exist such an estimator? Unfortunately, the answer is negative. To see this, note that the condition $T^{-1} \leq_L 2M + (X'X)^{-1}$ is equivalent to

$$T^{1/2}T^{-1}T^{1/2} \leq_L 2T^{1/2}MT^{1/2} + T^{1/2}(X'X)^{-1}T^{1/2}\,.$$

This inequality yields

$$\mathrm{tr}[T^{1/2}T^{-1}T^{1/2}] \leq 2\,\mathrm{tr}[T^{1/2}MT^{1/2}] + \mathrm{tr}[T^{1/2}(X'X)^{-1}T^{1/2}]\,.$$

Hence, the inequalities

$$\frac{1}{2}(p - \mathrm{tr}[(X'X)^{-1}T]) \leq \mathrm{tr}(MT) \leq 1$$

turn out to be *necessary* for $\widetilde{\beta}$ to be uniformly not worse than $\widehat{\beta}$ with respect to MSE, and, in addition, to be linearly admissible for β. But it is easy to find appropriate matrices X and T such that $\mathrm{tr}[(X'X)^{-1}T] < p - 2$ holds true, in which case there cannot exist some matrix M satisfying the required inequalities.

Remark 4.8. Under the linear regression model with assumptions (i) to (iv) and (ix), a homogeneous linear estimator for $\boldsymbol{\beta}$ is a linearly admissible improvement upon the ordinary least squares estimator with respect to the matrix-valued mean squared error risk if and only if it is a homogeneous linear Bayes estimator when $\boldsymbol{\beta}$ is the realization of a $p \times 1$ random vector $\boldsymbol{b}$ with $\mathrm{E}(\boldsymbol{b}\boldsymbol{b}') = \sigma^2 \boldsymbol{M}$ satisfying

$$\boldsymbol{T}^{-1} \leq_{\mathrm{L}} 2\boldsymbol{M} + (\boldsymbol{X}'\boldsymbol{X})^{-1} \quad \text{and} \quad \mathrm{tr}(\boldsymbol{M}\boldsymbol{T}) \leq 1 \,.$$

Such an estimator does not necessarily exist.

4.5 Problems

4.1. Show that the inequality $\mathbf{1}_n \boldsymbol{a}'\boldsymbol{a}\mathbf{1}_n' \leq_{\mathrm{L}} \mathbf{1}_n \boldsymbol{a}'$ holds true if and only if the $n \times 1$ vector $\boldsymbol{a}$ can be written as $\boldsymbol{a} = \alpha \mathbf{1}_n$ for some number $0 \leq \alpha \leq 1/n$. [Hint: Note that $\mathbf{1}_n \boldsymbol{a}'\boldsymbol{a}\mathbf{1}_n' \leq_{\mathrm{L}} \mathbf{1}_n \boldsymbol{a}'$ implies the symmetry of the $n \times n$ matrix $\mathbf{1}_n \boldsymbol{a}'$. What can then be said about $\boldsymbol{a}$?]

4.2. Consider the linear regression model with

$$\boldsymbol{y} = \begin{pmatrix} y_1 \\ y_2 \\ y_3 \\ y_4 \end{pmatrix}, \quad \boldsymbol{X} = \begin{pmatrix} 1 & 0 \\ 1 & 0 \\ 0 & 1 \\ 0 & 1 \end{pmatrix}, \quad \boldsymbol{\beta} = \begin{pmatrix} \beta_1 \\ \beta_2 \end{pmatrix}.$$

(a) Check whether the two estimators

$$\widetilde{\boldsymbol{\beta}}_1 = \frac{1}{2} \begin{pmatrix} 2y_1 \\ -y_1 + y_2 - y_3 + 3y_4 \end{pmatrix}, \quad \widetilde{\boldsymbol{\beta}}_2 = \begin{pmatrix} 2y_1 + y_2 + y_3 + y_4 \\ y_1 + y_2 + y_3 + 2y_4 \end{pmatrix}$$

are linearly admissible for $\boldsymbol{\beta}$.

(b) Consider the estimator $\boldsymbol{A}\boldsymbol{y}$ for $\boldsymbol{\beta}$, where

$$\boldsymbol{A} = \begin{pmatrix} a & a & 0 & 0 \\ 0 & 0 & a & a \end{pmatrix},$$

and a is an arbitrary number with $0 \leq a \leq 1/2$. Check whether $\boldsymbol{A}\boldsymbol{y}$ is linearly admissible for $\boldsymbol{\beta}$.

4.3. Show that the estimator $\widehat{\boldsymbol{\beta}}(\varrho)$ is uniformly better than the estimator $\widehat{\boldsymbol{\beta}}^{-}(\varrho)$ from Example 4.2 with respect to $\rho(\boldsymbol{\beta}, \widetilde{\boldsymbol{\beta}}) = \mathrm{E}[(\widetilde{\boldsymbol{\beta}} - \boldsymbol{\beta})'(\widetilde{\boldsymbol{\beta}} - \boldsymbol{\beta})]$.

4.4. Consider the estimator $\boldsymbol{L}\widehat{\boldsymbol{\beta}}$ for some $p \times p$ matrix $\boldsymbol{L}$.

(a) Show that $\boldsymbol{L}\widehat{\boldsymbol{\beta}}$ is linearly admissible for $\boldsymbol{\beta}$ if and only if $\boldsymbol{X}'\boldsymbol{X}\boldsymbol{L} = \boldsymbol{L}'\boldsymbol{X}'\boldsymbol{X}$ and in addition all eigenvalues of $\boldsymbol{L}$ lie in $[0, 1]$.

(b) Show that if $\boldsymbol{L}\widehat{\boldsymbol{\beta}}$ is linearly admissible for $\boldsymbol{\beta}$ and in addition $\boldsymbol{L} = \boldsymbol{L}'$ holds true, then it follows $\|\boldsymbol{L}\widehat{\boldsymbol{\beta}}\|^2 \leq \|\widehat{\boldsymbol{\beta}}\|^2$.

4.5. Under the linear regression model with assumptions (i) to (iv) consider the estimator

$$\widehat{\boldsymbol{\beta}}_{k,l} = (\boldsymbol{X}'\boldsymbol{X} + k\boldsymbol{I}_p + l(\boldsymbol{X}'\boldsymbol{X} + k\boldsymbol{I}_p)^{-1})\boldsymbol{X}'\boldsymbol{y} ,$$

where $k, l \geq 0$ are non-stochastic, see [14]. Check whether $\widehat{\boldsymbol{\beta}}_{k,l}$ is linearly admissible for $\boldsymbol{\beta}$.

4.6. Check whether the estimator $\boldsymbol{0}$ is linearly admissible for $\boldsymbol{\beta}$ under the linear regression model with assumptions (i) to (iv).

4.7. Show that the Obenchain class $\mathcal{L}^{\mathrm{Ob}}(\boldsymbol{\beta})$ is a convex set.

4.8. Consider the linear regression model with assumptions (i) to (iv) and (v). Use Remark 4.1 to demonstrate that the general ridge estimator $\widehat{\boldsymbol{\beta}}_{\boldsymbol{K}} = (\boldsymbol{X}'\boldsymbol{X} + \boldsymbol{K})^{-1}\boldsymbol{X}'\boldsymbol{y}$ is admissible for $\boldsymbol{\beta}$ within the set of all estimators (and with respect to the unweighted squared error loss) if the condition

$$p - 2 \leq \mathrm{rk}(\boldsymbol{K})$$

holds true. What can then be said about $\widehat{\boldsymbol{\beta}}$, $\widehat{\boldsymbol{\beta}}_k$ and $\widehat{\boldsymbol{\beta}}(\varrho)$?

4.9. Let $\boldsymbol{T}$ be a $p \times p$ symmetric nonnegative definite matrix and let $\boldsymbol{S}$ be a $p \times p$ symmetric positive definite matrix. Moreover, let $\boldsymbol{G} = \boldsymbol{I}_p - \boldsymbol{S}^{-1/2}(\boldsymbol{T} + \boldsymbol{S}^{-1})^{-1}\boldsymbol{S}^{-1/2}$.

(a) Show that the matrix $\boldsymbol{G}$ is symmetric nonnegative definite.
(b) Show that all eigenvalues of $\boldsymbol{G}$ are strictly smaller than 1.

4.10. Show that $\boldsymbol{\beta}_0 = \boldsymbol{R}'(\boldsymbol{R}\boldsymbol{R}')^{-1}\boldsymbol{r}$ satisfies the inequality

$$\|\boldsymbol{\beta}_0\|^2 \leq \|\boldsymbol{\beta}\|^2 \quad \forall \boldsymbol{\beta}|_{\boldsymbol{R}\boldsymbol{\beta}=\boldsymbol{r}} .$$

[Hint: Consider the set of all possible solutions of $\boldsymbol{R}\boldsymbol{\beta} = \boldsymbol{r}$ with regard to the specific solution $\boldsymbol{\beta}_0$.]

4.11. Consider two $p \times p$ symmetric nonnegative definite matrices $\boldsymbol{A}$ and $\boldsymbol{B}$ satisfying $\mathcal{C}(\boldsymbol{A}) \subseteq \mathcal{C}(\boldsymbol{B})$ and $\boldsymbol{A}\boldsymbol{B} = \boldsymbol{B}\boldsymbol{A}$. Let $\boldsymbol{A} = \boldsymbol{U}\boldsymbol{\Lambda}_1\boldsymbol{U}'$ and $\boldsymbol{B} = \boldsymbol{U}\boldsymbol{\Lambda}_2\boldsymbol{U}'$ spectral decompositions of $\boldsymbol{A}$ and $\boldsymbol{B}$ via *the same* orthogonal matrix $\boldsymbol{U}$. (Such a *simultaneous* spectral decomposition does exist in view of the assumptions.) Show that $\mathcal{C}(\boldsymbol{\Lambda}_1) \subseteq \mathcal{C}(\boldsymbol{\Lambda}_2)$ is satisfied.

4.12. Consider the linear regression model with assumptions (i) to (iv) and (viii), $\boldsymbol{R}\boldsymbol{\beta} = \boldsymbol{r}$. Show that a linear estimator $\widetilde{\boldsymbol{\beta}}$ is linearly admissible for $\boldsymbol{\beta}$ if and only if

$$\widetilde{\boldsymbol{\beta}} = \boldsymbol{S}^{-1/2}\boldsymbol{G}\boldsymbol{S}^{1/2}\widehat{\boldsymbol{\beta}} + \boldsymbol{S}^{-1/2}(\boldsymbol{I}_p - \boldsymbol{G})\boldsymbol{S}^{1/2}\boldsymbol{g} ,$$

where $\boldsymbol{S} = \boldsymbol{X}'\boldsymbol{X}$, and $\boldsymbol{G}$ and $\boldsymbol{g}$ satisfy the conditions $\boldsymbol{G} = \boldsymbol{G}'$, $\sigma(\boldsymbol{G}) \subset [0, 1]$, $\boldsymbol{R}\boldsymbol{S}^{-1/2}\boldsymbol{G} = \boldsymbol{0}$, and $\boldsymbol{R}\boldsymbol{g} = r$.

4.13. Under the linear regression model with assumptions (i) to (iv), a linear estimator $\boldsymbol{Ay}+\boldsymbol{a}$ is called *weakly restricted* if it satisfies

$$\mathrm{E}[\boldsymbol{R}(\boldsymbol{Ay}+\boldsymbol{a})]=\boldsymbol{r} \quad \forall \boldsymbol{\beta} \in \mathrm{IR}^p$$

for some $m \times p$ matrix $\boldsymbol{R}$ of full row rank and some $m \times 1$ vector $\boldsymbol{r}$. Show that $\boldsymbol{Ay}+\boldsymbol{a}$ is linearly admissible for $\boldsymbol{\beta}$ under the linear regression model with assumptions (i) to (iv) and (viii), $\boldsymbol{R\beta}=\boldsymbol{r}$, if and only if under the unrestricted linear regression model with assumptions (i) to (iv) it is (a) linearly admissible for $\boldsymbol{\beta}$ and (b) weakly restricted with respect to $\boldsymbol{R\beta}=\boldsymbol{r}$.

4.14. Check whether $\boldsymbol{\beta}_0=\boldsymbol{R}'(\boldsymbol{RR}')^{-1}\boldsymbol{r}$ is linearly admissible for $\boldsymbol{\beta}$ under the linear regression model with assumptions (i) to (iv) and (viii), $\boldsymbol{R\beta}=\boldsymbol{r}$.

4.15. Under the linear regression model with assumptions (i) to (iv), let $\boldsymbol{M}$ be a $p \times p$ symmetric nonnegative definite matrix. Show that for the estimator $\widetilde{\boldsymbol{\beta}}=\boldsymbol{M}(\boldsymbol{M}+(\boldsymbol{X}'\boldsymbol{X})^{-1})^{-1}\widehat{\boldsymbol{\beta}}$ the difference

$$\boldsymbol{\Delta}=\mathrm{MSE}(\boldsymbol{\beta},\widehat{\boldsymbol{\beta}})-\mathrm{MSE}(\boldsymbol{\beta},\widetilde{\boldsymbol{\beta}})$$

is nonzero whenever $p>1$.

4.16. Under the linear regression model with assumptions (i) to (iv), let $\boldsymbol{\beta}'\boldsymbol{\beta} \le \sigma^2$. Show that the ridge estimator $\widehat{\boldsymbol{\beta}}_k=(\boldsymbol{X}'\boldsymbol{X}+k\boldsymbol{I}_p)^{-1}\boldsymbol{X}'\boldsymbol{y}$ is uniformly not worse than $\widehat{\boldsymbol{\beta}}$ with respect to MSE, and, in addition, linearly admissible for $\boldsymbol{\beta}$ if and only if the inequalities

$$p \le k \le \frac{2}{1-\lambda_{\max}}$$

are satisfied, where $\lambda_{\max}$ denotes the largest eigenvalue of $\boldsymbol{X}'\boldsymbol{X}$.

4.17. Let $\boldsymbol{T}$ be a $p \times p$ symmetric positive definite matrix. Use Theorem 4.23 to show that under the linear regression model with assumptions (i) to (iv), the inequality

$$\mathrm{MSE}(\boldsymbol{\beta},\widehat{\boldsymbol{\beta}}_{p\boldsymbol{T}}) \le_{\mathrm{L}} \mathrm{MSE}(\boldsymbol{\beta},\widehat{\boldsymbol{\beta}}) \quad \forall \boldsymbol{\beta} \in \{\boldsymbol{\beta} : \boldsymbol{\beta}'\boldsymbol{T}\boldsymbol{\beta} \le \sigma^2\}$$

holds true if and only if

$$\boldsymbol{0} \le_{\mathrm{L}} \frac{2-p}{p}\boldsymbol{T}^{-1}+(\boldsymbol{X}'\boldsymbol{X})^{-1}$$

is satisfied. [Hint: $\widehat{\boldsymbol{\beta}}_{p\boldsymbol{T}}$ can be written as $\widehat{\boldsymbol{\beta}}_{p\boldsymbol{T}}=\boldsymbol{M}(\boldsymbol{M}+(\boldsymbol{X}'\boldsymbol{X})^{-1})^{-1}\widehat{\boldsymbol{\beta}}$ for $\boldsymbol{M}=(1/p)\boldsymbol{T}^{-1}$.]

Part III

Miscellaneous Topics

5 The Covariance Matrix of the Error Vector

Assumption (iv) of the linear regression model claims the covariance matrix of the error vector ε to be $\text{Cov}(\varepsilon) = \sigma^2 \boldsymbol{I}_n$ with an unknown parameter $\sigma^2 \in (0, \infty)$. This chapter discusses the estimation of σ^2 in detail, and introduces situations under which it appears to be reasonable to extend assumption (iv) to $\text{Cov}(\varepsilon) = \sigma^2 \boldsymbol{V}$ for some symmetric positive/nonnegative definite matrix $\boldsymbol{V} \neq \boldsymbol{I}_n$.

5.1 Estimation of the Error Variance

Section 2.2.4 introduces the least squares variance estimator

$$\widehat{\sigma}^2 = \frac{1}{n-p}(\boldsymbol{y} - \boldsymbol{X}\widehat{\boldsymbol{\beta}})'(\boldsymbol{y} - \boldsymbol{X}\widehat{\boldsymbol{\beta}})$$

as an unbiased estimator for σ^2 and mentions the existence of other possible estimators like $\widehat{\sigma}^2_{\text{T}}$ and $\widehat{\sigma}^2_{\text{ML}}$. This section deals with the estimation of σ^2 in more detail.

5.1.1 The Sample Variance

In the linear regression model $\boldsymbol{y} = \boldsymbol{X}\boldsymbol{\beta} + \varepsilon$, the random variables $y_1, \ldots, y_n$ can also be seen as a sample of size n, which, however, does not meet the requirements of a *simple random sample* [79, Definition 2.2.2], since the individual y_i are not identically distributed as long as not all independent variables are constant. The appropriate measure for the dispersion in a simple random sample is the *sample variance*

$$\widehat{\sigma}^2_{\boldsymbol{y}} = \frac{1}{n-1}\sum_{i=1}^{n}(y_i - \overline{y})^2 .$$

It coincides with the least squares variance estimator for σ^2 in the special case that the linear regression model is the simple mean shift model, described by the equation

$$\boldsymbol{y} = \mu \boldsymbol{1}_n + \varepsilon, \quad \varepsilon \sim (\boldsymbol{0}, \sigma^2 \boldsymbol{I}_n) ,$$

where $\mathbf{1}_n$ denotes the $n \times 1$ vector whose every element is equal to 1. In that case the ordinary least squares estimator for μ is $\widehat{\mu} = (1/n)\mathbf{1}_n'\boldsymbol{y}$, and the vector of ordinary least squares residuals is given by $\widehat{\boldsymbol{\varepsilon}} = \boldsymbol{C}\boldsymbol{y}$, where

$$\boldsymbol{C} = \boldsymbol{I}_n - \frac{1}{n}\mathbf{1}_n\mathbf{1}_n'$$

is the centering matrix. Then

$$\widehat{\sigma}^2 = \frac{1}{n-1}\boldsymbol{y}'\boldsymbol{C}\boldsymbol{y}$$

is identical to $\widehat{\sigma}^2_{\boldsymbol{y}}$. Hence, under the simple mean shift model, the estimator $\widehat{\sigma}^2_{\boldsymbol{y}}$ is unbiased for σ^2. If we consider $\widehat{\sigma}^2_{\boldsymbol{y}}$ as an estimator for σ^2 under the linear regression model $\boldsymbol{y} = \boldsymbol{X\beta} + \boldsymbol{\varepsilon}$ with assumptions (i) to (iv), then

$$\mathrm{E}(\widehat{\sigma}^2_{\boldsymbol{y}}) = \sigma^2\frac{\mathrm{tr}(\boldsymbol{C})}{n-1} + \frac{\boldsymbol{\beta}'\boldsymbol{X}'\boldsymbol{C}\boldsymbol{X}\boldsymbol{\beta}}{n-1} ,$$

where $\mathrm{tr}(\boldsymbol{C}) = n - 1$. Hence, we have $\mathrm{E}(\widehat{\sigma}^2_{\boldsymbol{y}}) = \sigma^2$ for every $\sigma^2 \in (0,\infty)$ and every $\boldsymbol{\beta} \in \mathbb{R}^p$ if and only if $\boldsymbol{CX} = \mathbf{0}$. The latter means that the column space of $\boldsymbol{X}$ is contained in the column space of $\mathbf{1}_n$ which can only happen if each column of $\boldsymbol{X}$ consists of identical elements.

Hence, the sample variance as an estimator for σ^2 is usually biased upwards in the linear regression model, meaning that the actual estimate must be expected to be greater than the true σ^2. Therefore, with respect to the bias, the least squares estimator $\widehat{\sigma}^2$ is more favorable than the sample variance $\widehat{\sigma}^2_{\boldsymbol{y}}$.

This is not surprising result, since under the linear regression model the random variables $y_1, \ldots, y_n$ do not have identical expectations (unless all independent variables are constant), thus contradicting the usual assumptions for a reasonable application of the sample variance.

5.1.2 Nonnegative Unbiased Estimation

For the estimation of the positive parameter σ^2, it is quite natural to consider quadratic estimators of the form

$$s^2 = \boldsymbol{y}'\boldsymbol{A}\boldsymbol{y} ,$$

supposed to yield nonnegative estimates only. This can be guaranteed by assuming the $n \times n$ matrix $\boldsymbol{A}$ to be symmetric nonnegative definite. A quadratic estimator $s^2 = \boldsymbol{y}'\boldsymbol{A}\boldsymbol{y}$ with $\mathbf{0} \leq_{\mathrm{L}} \boldsymbol{A}$ will be called *nonnegative*. Moreover, a quadratic estimator $s^2 = \boldsymbol{y}'\boldsymbol{A}\boldsymbol{y}$ will be called *unbiased* for σ^2, if

$$\mathrm{E}(s^2) = \sigma^2 \quad \forall(\boldsymbol{\beta}, \sigma^2) \in \mathbb{R}^p \times (0,\infty)$$

is satisfied. The following theorem gives a characterization of the set of nonnegative quadratic estimators, being unbiased for σ^2.

Theorem 5.1. *Under the linear regression model with assumptions (i) to (iv), a nonnegative quadratic estimator* $s^2 = \boldsymbol{y}'\boldsymbol{A}\boldsymbol{y}$ *is unbiased for* σ^2 *if and only if* $\operatorname{tr}(\boldsymbol{A}) = 1$ *and* $\boldsymbol{A}\boldsymbol{X} = \boldsymbol{0}$.

Proof. We have

$$\mathrm{E}(\boldsymbol{y}'\boldsymbol{A}\boldsymbol{y}) = \sigma^2 \operatorname{tr}(\boldsymbol{A}) + \boldsymbol{\beta}'\boldsymbol{X}'\boldsymbol{A}\boldsymbol{X}\boldsymbol{\beta} .$$

If the conditions $\operatorname{tr}(\boldsymbol{A}) = 1$ and $\boldsymbol{A}\boldsymbol{X} = \boldsymbol{0}$ are both satisfied, then clearly $\mathrm{E}(\boldsymbol{y}'\boldsymbol{A}\boldsymbol{y}) = \sigma^2$ for every $(\boldsymbol{\beta}, \sigma^2) \in \mathbb{R}^p \times (0, \infty)$. Conversely, if

$$\mathrm{E}(\boldsymbol{y}'\boldsymbol{A}\boldsymbol{y}) = \sigma^2 \operatorname{tr}(\boldsymbol{A}) + \boldsymbol{\beta}'\boldsymbol{X}'\boldsymbol{A}\boldsymbol{X}\boldsymbol{\beta} = \sigma^2$$

holds true for all $(\boldsymbol{\beta}, \sigma^2) \in \mathbb{R}^p \times (0, \infty)$, then it does also hold for $\boldsymbol{\beta} = \boldsymbol{0}$. This implies that $\sigma^2 \operatorname{tr}(\boldsymbol{A}) = \sigma^2$ must be valid every $\sigma^2 > 0$, so that $\operatorname{tr}(\boldsymbol{A}) = 1$. Now, if $\operatorname{tr}(\boldsymbol{A}) = 1$ is satisfied, then $\boldsymbol{\beta}'\boldsymbol{X}'\boldsymbol{A}\boldsymbol{X}\boldsymbol{\beta} = 0$ must also be fulfilled for every $\boldsymbol{\beta} \in \mathbb{R}^p$. Since $\boldsymbol{A}$ is symmetric nonnegative definite, the latter is equivalent to $\boldsymbol{A}\boldsymbol{X} = \boldsymbol{0}$. □

In the following, the symbol $\mathcal{Q}^+(\sigma^2)$ denotes the set of nonnegative quadratic estimators for σ^2, i.e

$$\mathcal{Q}^+(\sigma^2) = \{\boldsymbol{y}'\boldsymbol{A}\boldsymbol{y} : \boldsymbol{0} \leq_{\mathrm{L}} \boldsymbol{A}\} ,$$

while the symbol $\mathcal{Q}^+_u(\sigma^2)$ denotes the subset of $\mathcal{Q}^+(\sigma^2)$ consisting of unbiased estimators, i.e.

$$\mathcal{Q}^+(\sigma^2) = \{\boldsymbol{y}'\boldsymbol{A}\boldsymbol{y} : \boldsymbol{0} \leq_{\mathrm{L}} \boldsymbol{A}, \operatorname{tr}(\boldsymbol{A}) = 1, \boldsymbol{A}\boldsymbol{X} = \boldsymbol{0}\} .$$

Obviously, the least squares variance estimator

$$\widehat{\sigma}^2 = \frac{1}{n-p}\boldsymbol{y}'\boldsymbol{M}\boldsymbol{y}, \quad \boldsymbol{M} = \boldsymbol{I}_p - \boldsymbol{X}(\boldsymbol{X}'\boldsymbol{X})^{-1}\boldsymbol{X}',$$

belongs to the set $\mathcal{Q}^+_u(\sigma^2)$.

5.1.3 Optimality of the Least Squares Variance Estimator

If under the linear regression model the vector $\boldsymbol{y}$ is n-variate normally distributed, i.e. model assumption (v) is satisfied, then

$$\operatorname{Var}(\boldsymbol{y}'\boldsymbol{A}\boldsymbol{y}) = 2\sigma^4 \operatorname{tr}(\boldsymbol{A}^2) + 4\sigma^2\boldsymbol{\beta}'\boldsymbol{X}'\boldsymbol{A}^2\boldsymbol{X}\boldsymbol{\beta}$$

for any symmetric $n \times n$ matrix $\boldsymbol{A}$, see also Theorem B.15. This formula can be applied to show that under normality the estimator $\widehat{\sigma}^2$ is not only nonnegative unbiased but also has smallest variance within the set $\mathcal{Q}^+_u(\sigma^2)$.

Theorem 5.2. *Under the linear regression model with assumptions (i) to (iv) and (v), the inequality* $\operatorname{Var}(\widehat{\sigma}^2) \leq \operatorname{Var}(s^2)$ *holds true for any* $s^2 \in \mathcal{Q}^+_u(\sigma^2)$.

Proof. For $(n-p)\widehat{\sigma}^2 = \boldsymbol{y}'\boldsymbol{M}\boldsymbol{y}$ we have

$$\mathrm{Var}(\widehat{\sigma}^2) = 2\sigma^4 \,\mathrm{tr}[\frac{1}{(n-p)^2}\boldsymbol{M}^2] = 2\sigma^4 \frac{1}{(n-p)^2} \underbrace{\mathrm{tr}(\boldsymbol{M})}_{n-p} = \frac{2\sigma^4}{n-p}\,.$$

Let us now write an estimator $s^2 = \boldsymbol{y}'\boldsymbol{A}\boldsymbol{y} \in \mathcal{Q}_u^+(\sigma^2)$ as

$$s^2 = \frac{1}{n-p}\boldsymbol{y}'\boldsymbol{M}\boldsymbol{y} + \boldsymbol{y}'\boldsymbol{B}\boldsymbol{y}\,,$$

where $\boldsymbol{B} = \boldsymbol{A} - \frac{1}{n-p}\boldsymbol{M}$. Since $s^2 \in \mathcal{Q}_u^+(\sigma^2)$, the matrix $\boldsymbol{A}$ satisfies $\boldsymbol{A} = \boldsymbol{A}'$, $\mathrm{tr}(\boldsymbol{A}) = 1$, and $\boldsymbol{A}\boldsymbol{X} = \boldsymbol{0}$. Thus,

$$\boldsymbol{B} = \boldsymbol{B}', \quad \mathrm{tr}(\boldsymbol{B}) = 0, \quad \text{and} \quad \boldsymbol{B}\boldsymbol{X} = \boldsymbol{0}\,.$$

Now, the variance of $s^2 \in \mathcal{Q}_u^+(\sigma^2)$ is easily seen to be

$$\mathrm{Var}(s^2) = 2\sigma^4\,\mathrm{tr}(\boldsymbol{A}^2)\,,$$

where

$$\begin{aligned}
\mathrm{tr}(\boldsymbol{A}^2) &= \mathrm{tr}[(\boldsymbol{B} + \frac{1}{n-p}\boldsymbol{M})(\boldsymbol{B} + \frac{1}{n-p}\boldsymbol{M})] \\
&= \mathrm{tr}[\boldsymbol{B}\boldsymbol{B} + \frac{1}{n-p}\underbrace{\boldsymbol{M}\boldsymbol{B}}_{\boldsymbol{B}} + \frac{1}{n-p}\underbrace{\boldsymbol{B}\boldsymbol{M}}_{\boldsymbol{B}} + \frac{1}{(n-p)^2}\underbrace{\boldsymbol{M}\boldsymbol{M}}_{\boldsymbol{M}}] \\
&= \mathrm{tr}(\boldsymbol{B}^2) + \frac{1}{n-p}\,.
\end{aligned}$$

This shows that

$$\mathrm{Var}(s^2) = \mathrm{Var}(\widehat{\sigma}^2) + 2\sigma^4\,\mathrm{tr}(\boldsymbol{B}^2)\,.$$

Since the matrix $\boldsymbol{B}$ is symmetric, the matrix $\boldsymbol{B}^2$ is symmetric nonnegative definite, implying $\mathrm{tr}(\boldsymbol{B}^2) \geq 0$. Hence, the assertion follows from the latter identity. □

5.1.4 Non-Admissibility of the Least Squares Variance Estimator

The unweighted squared error loss of a quadratic estimator s^2 for σ^2 is

$$L(\sigma^2, s^2) = (s^2 - \sigma^2)^2$$

with corresponding risk

$$\rho(\sigma^2, s^2) = \mathrm{E}[(s^2 - \sigma^2)^2] = \mathrm{Var}(s^2) + [\mathrm{E}(s^2) - \sigma^2]^2.$$

Hence, Theorem 5.2 states that under normality

$$\rho(\sigma^2, \widehat{\sigma}^2) \leq \rho(\sigma^2, s^2) \quad \forall (\boldsymbol{\beta}, \sigma^2) \in \mathbb{R}^p \times (0, \infty)$$

for any $s^2 \in \mathcal{Q}_u^+(\sigma^2)$. Nonetheless, one may ask whether there could exist nonnegative quadratic estimators for σ^2 being uniformly better than $\widehat{\sigma}^2$ with respect to the unweighted squared error risk. As it turns out, this is in fact the case. Before giving the corresponding result, we derive the risk for a specific subset of quadratic estimators for σ^2.

Theorem 5.3. *Under the linear regression model with assumptions (i) to (iv) and (v), let $s^2 = \boldsymbol{y}'\boldsymbol{A}\boldsymbol{y}$ be a quadratic estimator for σ^2 such that $\boldsymbol{A} = \boldsymbol{A}'$ and $\boldsymbol{A}\boldsymbol{X} = \boldsymbol{0}$. Then, the unweighted squared error risk of s^2 is given by*

$$\rho(\sigma^2, s^2) = \sigma^4[2\,\mathrm{tr}(\boldsymbol{A}^2) + (1 - \mathrm{tr}(\boldsymbol{A}))^2] \, .$$

Proof. The identity

$$\mathrm{E}(\boldsymbol{y}'\boldsymbol{A}\boldsymbol{y}) = \sigma^2\,\mathrm{tr}(\boldsymbol{A}) + \boldsymbol{\beta}'\boldsymbol{X}'\boldsymbol{A}\boldsymbol{X}\boldsymbol{\beta}$$

holds true even if $\boldsymbol{A}$ is not symmetric and $\boldsymbol{y}$ is not n-variate normally distributed, while the identity

$$\mathrm{Var}(\boldsymbol{y}'\boldsymbol{A}\boldsymbol{y}) = 2\sigma^4\,\mathrm{tr}(\boldsymbol{A}^2) + 4\sigma^2\boldsymbol{\beta}'\boldsymbol{X}'\boldsymbol{A}^2\boldsymbol{X}\boldsymbol{\beta}$$

is valid for the symmetric matrix $\boldsymbol{A}$ and the n-variate normally distributed vector $\boldsymbol{y}$. Since the unweighted squared error risk $\rho(\sigma^2, \boldsymbol{y}'\boldsymbol{A}\boldsymbol{y})$ is given by

$$\rho(\sigma^2, \boldsymbol{y}'\boldsymbol{A}\boldsymbol{y}) = \mathrm{Var}(\boldsymbol{y}'\boldsymbol{A}\boldsymbol{y}) + [\mathrm{E}(\boldsymbol{y}'\boldsymbol{A}\boldsymbol{y}) - \sigma^2]^2 \, ,$$

under the assumption $\boldsymbol{A}\boldsymbol{X} = \boldsymbol{0}$ it follows that

$$\rho(\sigma^2, \boldsymbol{y}'\boldsymbol{A}\boldsymbol{y}) = 2\sigma^4\,\mathrm{tr}(\boldsymbol{A}^2) + [\sigma^2\,\mathrm{tr}(\boldsymbol{A}) - \sigma^2]^2 \, ,$$

thus showing the assertion. □

Now we can show that, although under normality $\widehat{\sigma}^2$ cannot uniformly be outperformed by some nonnegative unbiased estimator for σ^2, it can nonetheless uniformly be outperformed by some estimator from the set of of nonnegative but not necessarily unbiased estimators for σ^2.

Theorem 5.4. *Under the linear regression model with assumptions (i) to (iv) and (v), the estimator*

$$\widehat{\sigma}_{\mathrm{T}}^2 = \frac{1}{n-p+2}\boldsymbol{y}'\boldsymbol{M}\boldsymbol{y}, \quad \boldsymbol{M} = \boldsymbol{I}_n - \boldsymbol{X}(\boldsymbol{X}'\boldsymbol{X})^{-1}\boldsymbol{X}' \, ,$$

is uniformly better than $\widehat{\sigma}^2$ with respect to the unweighted squared error risk.

Proof. In view of Theorem 5.3, the unweighted squared error risk of the estimator $\widehat{\sigma}_{\mathrm{T}}^2 = \boldsymbol{y}'\boldsymbol{A}\boldsymbol{y}$, where $\boldsymbol{A} = (n-p+2)^{-1}\boldsymbol{M}$, is given by

$$\rho(\sigma^2, \widehat{\sigma}_{\mathrm{T}}^2) = \sigma^4[2\frac{\mathrm{tr}(\boldsymbol{M}^2)}{(n-p+2)^2} + (1 - \frac{\mathrm{tr}(\boldsymbol{M})}{n-p+2})^2] \,.$$

Using $\mathrm{tr}(\boldsymbol{M}) = \mathrm{tr}(\boldsymbol{M}^2) = n - p$, it follows that

$$\rho(\sigma^2, \widehat{\sigma}_{\mathrm{T}}^2) = \frac{2\sigma^4}{n-p+2}$$

is satisfied. Then in view of

$$\rho(\sigma^2, \widehat{\sigma}^2) = \frac{2\sigma^4}{n-p} \,,$$

the inequality

$$\rho(\sigma^2, \widehat{\sigma}_{\mathrm{T}}^2) < \rho(\sigma^2, \widehat{\sigma}^2)$$

is satisfied for every $(\boldsymbol{\beta}, \sigma^2) \in \mathrm{I\!R}^p \times (0, \infty)$. □

Since $\widehat{\sigma}^2$ and $\widehat{\sigma}_{\mathrm{T}}^2$ are both estimators from the set $\Delta = \mathcal{Q}^+(\sigma^2)$, the above theorem shows that $\widehat{\sigma}^2$ is inadmissible for σ^2 within Δ and with respect to the unweighted squared error risk under the linear regression model with assumptions (i) to (iv) and (v).

Example 5.1. Consider the linear regression model $\boldsymbol{y} = \boldsymbol{X\beta} + \boldsymbol{\varepsilon}$, where

$$\boldsymbol{X}' = \begin{pmatrix} 1 & 1 & 1 & 1 & 1 \\ 1 & 2 & 3 & 4 & 5 \end{pmatrix} \,.$$

Moreover, the elements ε_i of the error vector $\boldsymbol{\varepsilon}$ are known to be independently $N(0, \sigma^2)$ distributed for $i = 1, \ldots, 5$. The true parameters are given as $\boldsymbol{\beta} = (0.2, 0.9)'$ and $\sigma^2 = 1$. The unweighted squared error risk of $\widehat{\sigma}^2$ is $2/3$, while it is $2/5$ for the estimator $\widehat{\sigma}_{\mathrm{T}}^2$. The unweighted squared error risk of the sample variance $\widehat{\sigma}_{\boldsymbol{y}}^2$ is $10601/1600 = 6.625625$. Fig. 5.1 shows the boxplots of 100 estimates $\widehat{\sigma}_{\boldsymbol{y}}^2$, $\widehat{\sigma}^2$ and $\widehat{\sigma}_{\mathrm{T}}^2$ with average observed losses of 7.4977, 0.6235, and 0.3550, respectively. □

The above example illustrates that under normality the estimator $\widehat{\sigma}_{\mathrm{T}}^2$ has smaller unweighted squared risk than $\widehat{\sigma}^2$. The achieved improvement will, however, not be relevant when the sample size n is rather large, since in this case the difference between the two estimates will only be marginal. In addition, an expected improvement cannot be guaranteed when the error vector is not normally distributed. These two facts can be seen as the reason for the preference of the unbiased estimator $\widehat{\sigma}^2$ over the biased estimator $\widehat{\sigma}_{\mathrm{T}}^2$ in practical applications.

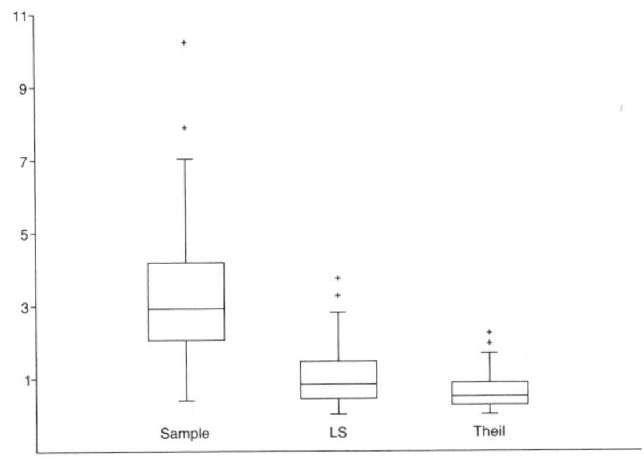

Fig. 5.1. Boxplots of 100 estimates $\widehat{\sigma}_y^2$ (Sample), $\widehat{\sigma}^2$ (LS), and $\widehat{\sigma}_\mathrm{T}^2$ (Theil)

5.2 Non-Scalar Covariance Matrix

Assumption (iv) of the linear regression model $\boldsymbol{y} = \boldsymbol{X\beta} + \boldsymbol{\varepsilon}$ claims that $\mathrm{Cov}(\boldsymbol{\varepsilon}) = \sigma^2 \boldsymbol{I}_n$. Such a covariance matrix is also called a positive *scalar matrix*, being the identity matrix multiplied by a positive scalar. The assumption $\mathrm{Cov}(\boldsymbol{\varepsilon}) = \sigma^2 \boldsymbol{I}_n$ is equivalent to demanding that the individual elements ε_i of $\boldsymbol{\varepsilon}$ are

- *homoscedastic*, i.e. $\mathrm{Var}(\varepsilon_i) = \sigma^2$ for $i = 1, \ldots, n$, and
- *uncorrelated*, i.e. $\mathrm{Cov}(\varepsilon_i, \varepsilon_j) = 0$ for $i \neq j$.

If the errors ε_i satisfy these two properties, then they are also called *spherical* errors. In practical applications there occur situations under which it appears to be reasonable to acknowledge that homoscedasticity and/or uncorrelatedness cannot be satisfied. Two popular non-spherical covariance structures are known as *heteroscedasticity*, in which case

$$\mathrm{Cov}(\boldsymbol{\varepsilon}) = \begin{pmatrix} \sigma_1^2 & 0 & 0 & \cdots & 0 \\ 0 & \sigma_2^2 & 0 & \cdots & 0 \\ 0 & 0 & \sigma_3^2 & \cdots & 0 \\ \vdots & \vdots & \vdots & & \vdots \\ 0 & 0 & 0 & \cdots & \sigma_n^2 \end{pmatrix}$$

(non-homogeneous but uncorrelated errors), and *serial correlation* (also called *autocorrelation*), in which case

$$\mathrm{Cov}(\boldsymbol{\varepsilon}) = \sigma^2 \begin{pmatrix} 1 & \varrho_1 & \varrho_2 & \cdots & \varrho_{n-1} \\ \varrho_1 & 1 & \varrho_1 & \cdots & \varrho_{n-2} \\ \varrho_2 & \varrho_1 & 1 & \cdots & \varrho_{n-3} \\ \vdots & \vdots & \vdots & & \vdots \\ \varrho_{n-1} & \varrho_{n-2} & \varrho_{n-3} & \cdots & 1 \end{pmatrix}$$

(homogeneous but correlated errors). Sect. 5.3 gives examples for such situations. Then it is of interest to investigate the effect of $\text{Cov}(\varepsilon) = \sigma^2 \boldsymbol{V}$ for some matrix $\boldsymbol{V} \neq \boldsymbol{I}_n$ on the performance of estimators.

5.2.1 Preliminary Considerations

Consider the linear regression model, described by

$$\boldsymbol{y} = \boldsymbol{X}\boldsymbol{\beta} + \boldsymbol{\varepsilon}$$

with assumptions (i) to (iii) and in addition (iv∗), where

(iv∗) the elements of the $n \times 1$ vector $\boldsymbol{\varepsilon}$ are non-observable random variables such that $\text{E}(\boldsymbol{\varepsilon}) = \boldsymbol{0}$ and $\text{Cov}(\boldsymbol{\varepsilon}) = \sigma^2 \boldsymbol{V}$, where $\boldsymbol{V}$ is a known $n \times n$ symmetric positive definite matrix and $\sigma^2 > 0$ is an unknown parameter.

Now, what are the consequences of considering the linear regression model with assumptions (i) to (iii) and (iv∗) instead of (iv)? Our main concern is with linear estimation of $\boldsymbol{\beta}$ here.

Linear Estimation

It is clear that when $\text{Cov}(\boldsymbol{\varepsilon}) = \sigma^2 \boldsymbol{V}$, also $\text{Cov}(\boldsymbol{y}) = \sigma^2 \boldsymbol{V}$. Thus, the covariance matrix of a linear estimator $\boldsymbol{A}\boldsymbol{y} + \boldsymbol{a}$ of $\boldsymbol{\beta}$ is given as

$$\text{Cov}(\boldsymbol{A}\boldsymbol{y} + \boldsymbol{a}) = \sigma^2 \boldsymbol{A}\boldsymbol{V}\boldsymbol{A}' \,.$$

The bias of $\boldsymbol{A}\boldsymbol{y} + \boldsymbol{a}$ is

$$\text{bias}(\boldsymbol{A}\boldsymbol{y} + \boldsymbol{a}) = \text{E}(\boldsymbol{A}\boldsymbol{y} + \boldsymbol{a}) - \boldsymbol{\beta} = (\boldsymbol{A}\boldsymbol{X} - \boldsymbol{I}_p)\boldsymbol{\beta} + \boldsymbol{a} \,,$$

and is therefore not effected by the matrix $\boldsymbol{V}$ for given $\boldsymbol{A}$ and $\boldsymbol{a}$.

Remark 5.1. Under the linear regression model with assumption (i) to (iii), the bias of a given linear estimator $\boldsymbol{A}\boldsymbol{y} + \boldsymbol{a}$ of $\boldsymbol{\beta}$ does not depend on the actual covariance matrix $\text{Cov}(\boldsymbol{y}) = \sigma^2 \boldsymbol{V}$.

If we consider the matrix-valued squared error risk

$$\text{MSE}(\boldsymbol{\beta}, \boldsymbol{A}\boldsymbol{y} + \boldsymbol{a}) = \text{Cov}(\boldsymbol{A}\boldsymbol{y} + \boldsymbol{a}) + \text{bias}(\boldsymbol{A}\boldsymbol{y} + \boldsymbol{a})\, \text{bias}(\boldsymbol{A}\boldsymbol{y} + \boldsymbol{a})'$$

of a given linear estimator $\boldsymbol{A}\boldsymbol{y} + \boldsymbol{a}$, then it is obvious that this risk does in fact depend on the matrix $\boldsymbol{V}$. In other words, the risk of a given linear estimator (for fixed $\boldsymbol{A}$ and $\boldsymbol{a}$) will change for different choices of $\boldsymbol{V}$.

The linear estimators for $\boldsymbol{\beta}$ considered in Chap. 3 clearly do not depend on some symmetric positive definite $\boldsymbol{V}$, and, naturally, their risk performance has been studied under the assumption (iv∗) with $\boldsymbol{V} = \boldsymbol{I}_n$. If, however, $\boldsymbol{V} \neq \boldsymbol{I}_n$, then it is obvious that the principles behind specific estimators should allow for the situation of a non-scalar covariance matrix of $\boldsymbol{y}$. Thus, it appears to be necessary to appropriately adjust estimators.

5.2.2 The Transformed Model

How can known estimation principles carry over to produce appropriate estimators under the linear regression model with assumptions (i) to (iii) and (iv∗)? One possibility is offered by the transformation

$$\boldsymbol{V}^{-1/2}\boldsymbol{y} = \boldsymbol{V}^{-1/2}\boldsymbol{X}\boldsymbol{\beta} + \boldsymbol{V}^{-1/2}\boldsymbol{\varepsilon}$$

of the model equation $\boldsymbol{y} = \boldsymbol{X}\boldsymbol{\beta} + \boldsymbol{\varepsilon}$. Setting $\boldsymbol{y}_* = \boldsymbol{V}^{-1/2}\boldsymbol{y}$, $\boldsymbol{X}_* = \boldsymbol{V}^{-1/2}\boldsymbol{X}$, and $\boldsymbol{\varepsilon}_* = \boldsymbol{V}^{-1/2}\boldsymbol{\varepsilon}$, the linear regression model described by

$$\boldsymbol{y}_* = \boldsymbol{X}_*\boldsymbol{\beta} + \boldsymbol{\varepsilon}_*$$

satisfies the usual assumptions (i) to (iv). Hence estimators for the parameters $\boldsymbol{\beta}$ and σ^2 may be obtained as before when $\boldsymbol{y}$ is replaced by $\boldsymbol{y}_*$ and $\boldsymbol{X}$ is replaced by $\boldsymbol{X}_*$. Such estimators clearly depend on $\boldsymbol{V}^{-1/2}$ and thus on $\boldsymbol{V}$.

Remark 5.2. Instead of the uniquely determined matrix $\boldsymbol{V}^{-1/2}$, one may also apply any other nonsingular $n \times n$ matrix $\boldsymbol{F}$ with $\boldsymbol{F}'\boldsymbol{F} = \boldsymbol{V}^{-1}$ to obtain a transformed model

$$\underbrace{\boldsymbol{F}\boldsymbol{y}}_{\boldsymbol{y}_*} = \underbrace{\boldsymbol{F}\boldsymbol{X}}_{\boldsymbol{X}_*}\boldsymbol{\beta} + \underbrace{\boldsymbol{F}\boldsymbol{\varepsilon}}_{\boldsymbol{\varepsilon}_*},$$

where $\boldsymbol{\varepsilon}_* \sim (\boldsymbol{0}, \sigma^2\boldsymbol{I}_n)$, see also Problem 5.3. This can be advantageous, when an easy matrix $\boldsymbol{F}$ with $\boldsymbol{F}'\boldsymbol{F} = \boldsymbol{V}^{-1}$ exists. The observation matrices $(\boldsymbol{y}_*, \boldsymbol{X}_*)$ are not the same for different transformations, but the transformed models may be considered as being equivalent for estimation of $\boldsymbol{\beta}$ and σ^2.

Ordinary and Generalized Least Squares Estimator

The ordinary least squares estimator in the transformed regression model is given by

$$\widehat{\boldsymbol{\beta}}_* = (\boldsymbol{X}_*'\boldsymbol{X}_*)^{-1}\boldsymbol{X}_*'\boldsymbol{y}_* = (\boldsymbol{X}'\boldsymbol{V}^{-1}\boldsymbol{X})^{-1}\boldsymbol{X}'\boldsymbol{V}^{-1}\boldsymbol{y}\,.$$

If we consider the estimator $\widehat{\boldsymbol{\beta}}_* = (\boldsymbol{X}_*'\boldsymbol{X}_*)^{-1}\boldsymbol{X}_*'\boldsymbol{y}_*$ under the transformed model, or the same estimator under the original model with assumptions (i) to (iii) and (iv∗), then in both cases

$$\mathrm{MSE}(\boldsymbol{\beta}, \widehat{\boldsymbol{\beta}}_*) = \mathrm{Cov}(\widehat{\boldsymbol{\beta}}_*) = \sigma^2(\boldsymbol{X}'\boldsymbol{V}^{-1}\boldsymbol{X})^{-1}.$$

If $\boldsymbol{A}\boldsymbol{y} + \boldsymbol{a}$ is any linear unbiased estimator for $\boldsymbol{\beta}$ satisfying $\boldsymbol{A}\boldsymbol{X} = \boldsymbol{I}_n$ and $\boldsymbol{a} = \boldsymbol{0}$, then under the original model the identity

$$\mathrm{MSE}(\boldsymbol{\beta}, \boldsymbol{A}\boldsymbol{y} + \boldsymbol{a}) = \mathrm{Cov}(\boldsymbol{A}\boldsymbol{y} + \boldsymbol{a}) = \sigma^2\boldsymbol{B}\boldsymbol{B}' + \sigma^2(\boldsymbol{X}'\boldsymbol{V}^{-1}\boldsymbol{X})^{-1}$$

can be shown to hold true for some matrix $\boldsymbol{B}$ analogously to Lemma 2.1.

Theorem 5.5. *Under the linear regression model with assumptions (i) to (iii) and (iv∗), the estimator $\widehat{\boldsymbol{\beta}}_*$ is uniformly not worse than any other linear unbiased estimator with respect to* $\mathrm{MSE}(\boldsymbol{\beta}, \widetilde{\boldsymbol{\beta}})$.

Comparing this result with Theorem 2.9 (Gauss-Markov Theorem), it can be seen that the estimator $\widehat{\boldsymbol{\beta}}_*$ takes over the role of $\widehat{\boldsymbol{\beta}} = (\boldsymbol{X}'\boldsymbol{X})^{-1}\boldsymbol{X}'\boldsymbol{y}$ when model assumption (iv) is replaced by assumption (iv∗). Theorem 5.5 may also be called *generalized Gauss-Markov Theorem.*

On the other hand, the estimator $\widehat{\boldsymbol{\beta}}$ is still available under the regression model with assumptions (i) to (iii) and (iv∗), and it still minimizes the function

$$f(\widetilde{\boldsymbol{\beta}}) = \|\boldsymbol{y} - \boldsymbol{X}\widetilde{\boldsymbol{\beta}}\|^2 = (\boldsymbol{y} - \boldsymbol{X}\widetilde{\boldsymbol{\beta}})'(\boldsymbol{y} - \boldsymbol{X}\widetilde{\boldsymbol{\beta}})$$

with respect to $\widetilde{\boldsymbol{\beta}}$. Hence $\widehat{\boldsymbol{\beta}}$ may still be called *ordinary least squares estimator* under the linear regression model with assumptions (i) to (iii) and (iv∗). It is also still an unbiased estimator, but, as noted above, it is no longer the optimal choice with respect to the matrix-valued mean squared error risk.

This can happen, since the principle of least squares is not effected by the randomness and distribution of the vector $\boldsymbol{y}$, it is in fact not a statistical but a purely mathematical principle. On the other hand, the statistical properties of $\widehat{\boldsymbol{\beta}}$ considered as an estimator for $\boldsymbol{\beta}$ obviously depend on the distribution of $\boldsymbol{y}$. If $\mathrm{Cov}(\boldsymbol{y}) = \sigma^2\boldsymbol{V}$, then

$$\mathrm{Cov}(\widehat{\boldsymbol{\beta}}) = \sigma^2(\boldsymbol{X}'\boldsymbol{X})^{-1}\boldsymbol{X}'\boldsymbol{V}\boldsymbol{X}(\boldsymbol{X}'\boldsymbol{X})^{-1} .$$

As noted above, this is no longer the optimal covariance matrix with respect to the Löwner partial ordering generated by linear unbiased estimators. As a matter of fact, the inequality

$$(\boldsymbol{X}'\boldsymbol{V}^{-1}\boldsymbol{X})^{-1} \leq_{\mathrm{L}} (\boldsymbol{X}'\boldsymbol{X})^{-1}\boldsymbol{X}'\boldsymbol{V}\boldsymbol{X}(\boldsymbol{X}'\boldsymbol{X})^{-1}$$

holds true, see Problem 5.4, where usually the two matrices on the left-hand and the right-hand side do not coincide.

We have already noted that $\widehat{\boldsymbol{\beta}}_*$ is the ordinary least squares estimator under the transformed model and thus minimizes the function $g(\widetilde{\boldsymbol{\beta}}) = \|\boldsymbol{y}_* - \boldsymbol{X}_*\widetilde{\boldsymbol{\beta}}\|^2$ with respect to $\widetilde{\boldsymbol{\beta}}$. Since the function

$$g(\widetilde{\boldsymbol{\beta}}) = (\boldsymbol{y} - \boldsymbol{X}\widetilde{\boldsymbol{\beta}})'\boldsymbol{V}^{-1}(\boldsymbol{y} - \boldsymbol{X}\widetilde{\boldsymbol{\beta}}) ,$$

coincides with the above function $f(\widetilde{\boldsymbol{\beta}})$ in case $\boldsymbol{V} = \boldsymbol{I}_n$, $g(\cdot)$ may be seen as a generalization of $f(\cdot)$, and it is convenient to call the estimator $\widehat{\boldsymbol{\beta}}_*$ the *generalized least squares estimator* under the linear regression model with assumptions (i) to (iii) and (iv∗). Since Theorem 5.5 is due to Aitken [1], the estimator $\widehat{\boldsymbol{\beta}}_*$ is also called *Aitken estimator.*

Estimation of the Error Variance

The least squares variance estimator for σ^2 under the transformed model is

$$\begin{aligned}\widehat{\sigma}_*^2 &= \frac{1}{n-p}(\boldsymbol{y}_* - \boldsymbol{X}_*\widehat{\boldsymbol{\beta}}_*)'(\boldsymbol{y}_* - \boldsymbol{X}_*\widehat{\boldsymbol{\beta}}_*) \\ &= \frac{1}{n-p}(\boldsymbol{y} - \boldsymbol{X}\widehat{\boldsymbol{\beta}}_*)'\boldsymbol{V}^{-1}(\boldsymbol{y} - \boldsymbol{X}\widehat{\boldsymbol{\beta}}_*) \,.\end{aligned}$$

It can be be shown that $\widehat{\sigma}_*^2$ is unbiased for σ^2 under the linear regression model with assumptions (i) to (iii) and (iv∗), see Problem 5.5, while such a statement is usually not true for the previous estimator

$$\widehat{\sigma}^2 = \frac{1}{n-p}(\boldsymbol{y} - \boldsymbol{X}\widehat{\boldsymbol{\beta}})'(\boldsymbol{y} - \widehat{\boldsymbol{X}\boldsymbol{\beta}}) \,.$$

Whether this estimator is biased upwards or downwards depends on the ratio

$$\frac{\operatorname{tr}(\boldsymbol{M}\boldsymbol{V})}{n-p}, \qquad \boldsymbol{M} = \boldsymbol{I}_n - \boldsymbol{X}(\boldsymbol{X}'\boldsymbol{X})^{-1}\boldsymbol{X}' \,,$$

being greater or smaller than 1. In view of this, the estimator $\widehat{\sigma}_*^2$, allowing for $\operatorname{Cov}(\boldsymbol{\varepsilon}) = \sigma^2\boldsymbol{V}$, appears to be more appropriate than $\widehat{\sigma}^2$ for estimating σ^2.

Alternative Estimators

The application of the principle of least squares yields the generalized least squares estimator $\widehat{\boldsymbol{\beta}}_*$, performing better under the linear regression model with assumptions (i) to (iii) and (iv∗) than the previous estimator $\widehat{\boldsymbol{\beta}}$. In contrast to $\widehat{\boldsymbol{\beta}}$, the estimator $\widehat{\boldsymbol{\beta}}_*$ allows for $\operatorname{Cov}(\boldsymbol{\varepsilon}) = \sigma^2\boldsymbol{V}$. Hence, it is natural to derive alternatives to $\widehat{\boldsymbol{\beta}}_*$ in a similar fashion, thus obtaining estimators suiting the model assumptions. For example, one may specify the ridge estimator for $\boldsymbol{\beta}$ as

$$\begin{aligned}\widehat{\boldsymbol{\beta}}_{*k} &= (\boldsymbol{X}_*'\boldsymbol{X}_* + k\boldsymbol{I}_p)^{-1}\boldsymbol{X}_*'\boldsymbol{y}_* \\ &= (\boldsymbol{X}'\boldsymbol{V}^{-1}\boldsymbol{X} + k\boldsymbol{I}_p)^{-1}\boldsymbol{X}'\boldsymbol{V}^{-1}\boldsymbol{y} \,.\end{aligned}$$

Then, from Theorem 3.8 it follows that under the transformed model

$$\operatorname{MSE}(\boldsymbol{\beta}, \widehat{\boldsymbol{\beta}}_*) \leq_{\mathrm{L}} \operatorname{MSE}(\boldsymbol{\beta}, \widehat{\boldsymbol{\beta}}_{*k})$$

if and only if

$$\boldsymbol{\beta}'\left[\frac{2}{k}\boldsymbol{I}_p + (\boldsymbol{X}_*'\boldsymbol{X}_*)^{-1}\right]^{-1}\boldsymbol{\beta} \leq \sigma^2$$

is satisfied. It is easy to show that this is also a necessary and sufficient condition for $\operatorname{MSE}(\boldsymbol{\beta}, \widehat{\boldsymbol{\beta}}_*) \leq_{\mathrm{L}} \operatorname{MSE}(\boldsymbol{\beta}, \widehat{\boldsymbol{\beta}}_{*k})$ to be true under the original model with assumptions (i) to (iii) and (iv∗).

For an application of this estimator one might inspect the ridge trace of $\widehat{\boldsymbol{\beta}}_{*k}$ and/or estimate the ridge parameter k, e.g. by

$$\widehat{k}_{\mathrm{HKB}} = \frac{p\widehat{\sigma}_*^2}{\widehat{\boldsymbol{\beta}}_*'\widehat{\boldsymbol{\beta}}_*} ,$$

being the estimator from Hoerl, Kennard and Baldwin, see page 126, adjusted to suit model assumption (iv∗).

Remark 5.3. Properties of the matrix $\boldsymbol{X}$ other than the full column rank are not necessarily preserved by the transformation $\boldsymbol{X}_* = \boldsymbol{X}$. For example, if one of the independent variables is constant (all elements of some column of $\boldsymbol{X}$ are identical), then the matrix $\boldsymbol{X}_*$ does not necessarily has a constant column too. Moreover, if $\boldsymbol{X}$ is known to admit collinearity, then this it not necessarily the case for the matrix $\boldsymbol{X}_*$.

5.2.3 Two-Stage Estimation

It is a major drawback of the linear regression model with a non-scalar covariance matrix of the error vector that often in practice the matrix $\boldsymbol{V}$ is not completely known. Although there exist situations where it is reasonable to allow for $\boldsymbol{V} \neq \boldsymbol{I}_n$, the matrix $\boldsymbol{V}$ may depend on one or more unknown parameters $\omega_1, \ldots, \omega_s$, so that $\boldsymbol{V} \equiv \boldsymbol{V}(\boldsymbol{\omega})$, where $\boldsymbol{\omega} = (\omega_1, \ldots, \omega_s)'$. Then the above procedure, obtaining estimators via the transformed model

$$\boldsymbol{V}^{-1/2}\boldsymbol{y} = \boldsymbol{V}^{-1/2}\boldsymbol{X}\boldsymbol{\beta} + \boldsymbol{V}^{-1/2}\boldsymbol{\varepsilon} ,$$

will not yield practical estimators.

One may, however, at first estimate the parameter vector $\boldsymbol{\omega}$ and thus the matrix $\boldsymbol{V}$, and then apply the transformation with $\boldsymbol{V}$ replaced by $\widehat{\boldsymbol{V}} := \boldsymbol{V}(\widehat{\boldsymbol{\omega}})$. The resulting estimators are often called *two-stage estimators*. One example is the two-stage generalized least squares estimator

$$\widehat{\boldsymbol{\beta}}_*(\widehat{\boldsymbol{V}}) = (\boldsymbol{X}'\widehat{\boldsymbol{V}}^{-1}\boldsymbol{X})^{-1}\boldsymbol{X}'\widehat{\boldsymbol{V}}^{-1}\boldsymbol{y} ,$$

which is, of course a non-linear estimator and thus has not the same statistical properties as the generalized least squares estimator $\widehat{\boldsymbol{\beta}}_*$. Under certain assumptions on $\boldsymbol{V}$ and $\widehat{\boldsymbol{V}}$ it is possible to demonstrate that the two-stage generalized least squares estimator asymptotically behaves like the one-stage estimator, see e.g. [46, Sect. 10.5.2], but in general the derivation of exact small sample properties is rather difficult.

The situation is even worse when an alternative like the two-stage nonlinear ridge estimator of the form

$$\widehat{\boldsymbol{\beta}}_{*\widehat{k}}(\widehat{\boldsymbol{V}}) = (\boldsymbol{X}'\widehat{\boldsymbol{V}}^{-1}\boldsymbol{X} + \widehat{k}\boldsymbol{I}_p)^{-1}\boldsymbol{X}'\widehat{\boldsymbol{V}}^{-1}\boldsymbol{y}$$

is considered, since here not only $\boldsymbol{V}$ but also k is replaced by an estimate. Nonetheless, the application of such a two-stage ridge estimator may be beneficial over the two-stage generalized least squares estimator when the matrix $\boldsymbol{X}'\widehat{\boldsymbol{V}}^{-1}\boldsymbol{X}$ is badly conditioned.

5.3 Occurrence of Non-Scalar Covariance Matrices

We now consider some examples for different types of covariance matrices for the error vector.

5.3.1 Seemingly Unrelated Regression

A seemingly unrelated regression model, introduced by Zellner [130], can be seen as an example for a linear regression model with a non-scalar covariance matrix of the error vector.

Suppose that we have observation matrices $(\boldsymbol{y}_1, \boldsymbol{X}_1)$ and $(\boldsymbol{y}_2, \boldsymbol{X}_2)$, where $\boldsymbol{y}_1$ contains n observations of gross investment of firm A in the years denoted by $1, \ldots, n$ and $\boldsymbol{y}_2$ contains n observations of gross investment of firm B in the same years $1, \ldots, n$, where firms A and B are also operating in the same industry. Then we may apply two linear models $\boldsymbol{y}_1 = \boldsymbol{X}_1\boldsymbol{\beta}_1 + \boldsymbol{\varepsilon}_1$ and $\boldsymbol{y}_2 = \boldsymbol{X}_2\boldsymbol{\beta}_2 + \boldsymbol{\varepsilon}_2$ separately, and obtain estimators $(\widehat{\boldsymbol{\beta}}_1, \widehat{\sigma}_1)$ and $(\widehat{\boldsymbol{\beta}}_2, \widehat{\sigma}_2)$ which may prove useful to give insight into the gross investment of firms A and B, respectively.

Both linear models are *seemingly unrelated*, since they are related to two different firms and the parameter vectors $\boldsymbol{\beta}_1$ and $\boldsymbol{\beta}_2$ might be quite different, even if the involved variables in the two models are the same (each one related to a specific firm, however). On the other hand, since the observations are taken in the same years and the firms operate in the same industry, it is not farfetched to assume that there exist effects (such as government policy, international events, etc.) which are not explicitly incorporated within the model equations and thus influence the i-th element ε_{1i} of $\boldsymbol{\varepsilon}_1$ as well as the i-th element ε_{2i} of $\boldsymbol{\varepsilon}_2$. We can make allowance for this fact by assuming that the random variables ε_{1i} and ε_{2i} are correlated.

If we assume that the degree of correlation is the same in each year $1, \ldots, n$ and if in addition the variances of ε_{1i} are the same for $i = 1, \ldots, n$ and the variances of ε_{2i} are the same for $i = 1, \ldots, n$, then also the covariances between ε_{1i} and ε_{2i} are the same for $i = 1, \ldots, n$. Thus, we can assume

$$\operatorname{Var}(\varepsilon_{1i}) = \sigma_{11}, \quad \operatorname{Var}(\varepsilon_{2i}) = \sigma_{22}, \quad \text{and} \quad \operatorname{Cov}(\varepsilon_{1i}, \varepsilon_{2i}) = \sigma_{12}$$

for $i = 1, \ldots, n$, where $\sigma_{11} > 0$, $\sigma_{22} > 0$ and σ_{12} with $|\sigma_{12}| < \sqrt{\sigma_{11}\sigma_{22}}$ are unknown parameters. If in addition there is no nonzero correlation between the errors of different years (neither between two errors of different years for the same firm, nor between two errors of different years for the two firms), then

$$\mathrm{Cov}\begin{pmatrix}\varepsilon_1\\ \varepsilon_2\end{pmatrix} = \begin{pmatrix}\sigma_{11}\boldsymbol{I}_n & \sigma_{12}\boldsymbol{I}_n\\ \sigma_{12}\boldsymbol{I}_n & \sigma_{22}\boldsymbol{I}_n\end{pmatrix} =: \boldsymbol{V} ,$$

which can alternatively be expressed as

$$\boldsymbol{V} = \boldsymbol{\Sigma} \otimes \boldsymbol{I}_n, \quad \boldsymbol{\Sigma} = \begin{pmatrix}\sigma_{11} & \sigma_{12}\\ \sigma_{12} & \sigma_{22}\end{pmatrix} ,$$

where the symbol $\otimes$ denotes the Kronecker product of matrices. Note that the conditions $\sigma_{11} > 0$, $\sigma_{22} > 0$ and $|\sigma_{12}| < \sqrt{\sigma_{11}\sigma_{22}}$ ensure the positive definiteness of $\boldsymbol{\Sigma}$ and hence the positive definiteness of $\boldsymbol{V}$, see also Theorem A.59. The inequality $|\sigma_{12}| < \sqrt{\sigma_{11}\sigma_{22}}$ is equivalent to the fact that the correlation $\varrho = \sigma_{12}/\sqrt{\sigma_{11}\sigma_{22}}$ between ε_{1i} and ε_{2i} is strictly between -1 and 1.

Now, the two linear models $\boldsymbol{y}_1 = \boldsymbol{X}_1\boldsymbol{\beta}_1 + \varepsilon_1$ and $\boldsymbol{y}_2 = \boldsymbol{X}_2\boldsymbol{\beta}_2 + \varepsilon_2$ are in fact related via the distribution of their errors, and it seems appropriate to consider a combined linear model of the form

$$\boldsymbol{y} = \boldsymbol{X}\boldsymbol{\beta} + \varepsilon, \quad \varepsilon \sim (\boldsymbol{0}, \boldsymbol{V}) ,$$

where

$$\boldsymbol{y} = \begin{pmatrix}\boldsymbol{y}_1\\ \boldsymbol{y}_2\end{pmatrix}, \boldsymbol{X} = \begin{pmatrix}\boldsymbol{X}_1 & \boldsymbol{0}\\ \boldsymbol{0} & \boldsymbol{X}_2\end{pmatrix}, \boldsymbol{\beta} = \begin{pmatrix}\boldsymbol{\beta}_1\\ \boldsymbol{\beta}_2\end{pmatrix}, \varepsilon = \begin{pmatrix}\varepsilon_1\\ \varepsilon_2\end{pmatrix} .$$

Since it is unrealistic to assume that σ_{11}, σ_{22} and σ_{12} are known, the generalized least squares estimator

$$\widehat{\boldsymbol{\beta}}_* = (\boldsymbol{X}'\boldsymbol{V}^{-1}\boldsymbol{X})^{-1}\boldsymbol{X}'\boldsymbol{V}^{-1}\boldsymbol{y}$$

for $\boldsymbol{\beta} = (\boldsymbol{\beta}_1', \boldsymbol{\beta}_2')'$ is not practical. (Formally, model assumption (iv$*$) is not exactly satisfied here, since we have written $\mathrm{Cov}(\varepsilon) = \boldsymbol{V}$ and not $\mathrm{Cov}(\varepsilon) = \sigma^2\boldsymbol{V}$, but this is irrelevant to the estimation of $\boldsymbol{\beta}$.) As proposed in [130], one may estimate σ_{st} by

$$\widehat{\sigma}_{st} = \frac{1}{n}(\boldsymbol{y}_s - \boldsymbol{X}_s\widehat{\boldsymbol{\beta}}_s)'(\boldsymbol{y}_t - \boldsymbol{X}_t\widehat{\boldsymbol{\beta}}_t), \quad s,t \in \{1,2\} ,$$

where $\widehat{\boldsymbol{\beta}}_s = (\boldsymbol{X}_s'\boldsymbol{X}_s)^{-1}\boldsymbol{X}_s'\boldsymbol{y}_s$, $s \in \{1,2\}$, and then estimate $\boldsymbol{\beta}$ by

$$\widehat{\boldsymbol{\beta}}_*(\widehat{\boldsymbol{V}}) = (\boldsymbol{X}'\widehat{\boldsymbol{V}}^{-1}\boldsymbol{X})^{-1}\boldsymbol{X}'\widehat{\boldsymbol{V}}^{-1}\boldsymbol{y} ,$$

where

$$\widehat{\boldsymbol{V}} = \begin{pmatrix}\widehat{\sigma}_{11}\boldsymbol{I}_n & \widehat{\sigma}_{12}\boldsymbol{I}_n\\ \widehat{\sigma}_{12}\boldsymbol{I}_n & \widehat{\sigma}_{22}\boldsymbol{I}_n\end{pmatrix} .$$

The Cauchy-Schwarz inequality ensures that always $|\widehat{\sigma}_{12}| \leq \sqrt{\widehat{\sigma}_{11}\widehat{\sigma}_{22}}$, see Problem 5.6, where equality (yielding a singular nonnegative definite matrix $\widehat{\boldsymbol{V}}$) will presumably not occur in practice.

One may expect some benefit from this two-stage combination procedure over the uncombined separate estimation of the parameter vectors when the true correlation $\varrho = \sigma_{12}/\sqrt{\sigma_{11}\sigma_{22}}$ is fairly high.

Example 5.2. Consider the two observations matrices

$$(\boldsymbol{y}_1, \boldsymbol{X}_1) = \begin{pmatrix} 2.6844 & 1 & 1.5 \\ 1.7478 & 1 & 1.9 \\ 1.9040 & 1 & 2.0 \\ 1.4318 & 1 & 1.3 \\ 3.5169 & 1 & 2.7 \end{pmatrix} \quad \text{and} \quad (\boldsymbol{y}_2, \boldsymbol{X}_2) = \begin{pmatrix} 4.5964 & 1 & 3.2 \\ 3.3617 & 1 & 2.7 \\ 1.9453 & 1 & 1.4 \\ 2.8440 & 1 & 2.1 \\ 3.0656 & 1 & 1.9 \end{pmatrix},$$

where $\boldsymbol{y}_1$ and $\boldsymbol{y}_2$ are generated from a seemingly unrelated regression model with true parameters

$$\boldsymbol{\beta}_1 = \begin{pmatrix} 0.5 \\ 1 \end{pmatrix}, \boldsymbol{\beta}_2 = \begin{pmatrix} 1 \\ 1 \end{pmatrix}, \sigma_{11} = 0.65, \sigma_{22} = 0.25, \sigma_{12} = 0.4 .$$

Then $\varrho = \sigma_{12}/\sqrt{\sigma_{11}\sigma_{22}} = 0.9923$. Now, the estimates

$$\widehat{\sigma}_{11} = 0.2790, \quad \widehat{\sigma}_{22} = 0.0687, \quad \widehat{\sigma}_{12} = 0.1148$$

are not very close to the true values. Nonetheless, the two-stage generalized least squares estimate

$$\widehat{\boldsymbol{\beta}}_*(\widehat{\boldsymbol{V}}) = \begin{pmatrix} 0.4508 \\ 0.9608 \\ 0.7753 \\ 1.0563 \end{pmatrix}$$

obviously provides better estimates for $\boldsymbol{\beta}_1$ and $\boldsymbol{\beta}_2$ than the ordinary least squares estimates

$$\widehat{\boldsymbol{\beta}}_1 = \begin{pmatrix} 0.1697 \\ 1.1103 \end{pmatrix} \quad \text{and} \quad \widehat{\boldsymbol{\beta}}_2 = \begin{pmatrix} 0.2194 \\ 1.3023 \end{pmatrix} .$$

This is also confirmed when 30 vectors $\boldsymbol{y}_1$ and $\boldsymbol{y}_2$ are generated and two-stage generalized least squares and ordinary least squares estimates are compared. See Figs. 5.2 and 5.3, showing 30 ordinary least squares and 30 two-stage generalized least squares estimates of $\boldsymbol{\beta}_2$. □

See also [108] for an extensive discussion of seemingly unrelated regression equations (SURE) models and note that the described procedure can be generalized to more than two observation matrices

$$(\boldsymbol{y}_1, \boldsymbol{X}_1), (\boldsymbol{y}_2, \boldsymbol{X}_2), \ldots, (\boldsymbol{y}_m, \boldsymbol{X}_m)$$

by assuming

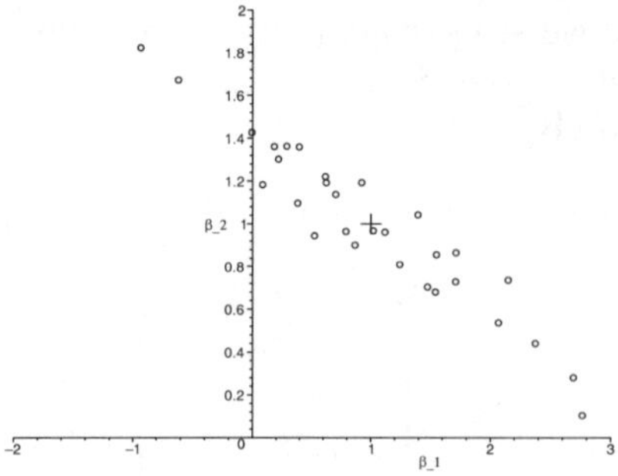

Fig. 5.2. 30 ordinary least squares estimates of $\boldsymbol{\beta}_2$ in Example 3.6

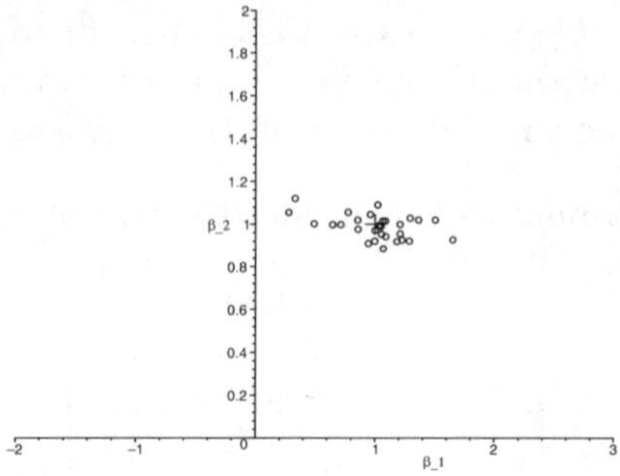

Fig. 5.3. 30 two-stage generalized least squares estimates of $\boldsymbol{\beta}_2$ in Example 3.6

$$\operatorname{Cov}\begin{pmatrix}\varepsilon_1\\ \varepsilon_2\\ \vdots\\ \varepsilon_m\end{pmatrix} = \begin{pmatrix}\sigma_{11}\boldsymbol{I}_n & \sigma_{12}\boldsymbol{I}_n & \cdots & \sigma_{1m}\boldsymbol{I}_n\\ \sigma_{12}\boldsymbol{I}_n & \sigma_{22}\boldsymbol{I}_n & \cdots & \sigma_{2m}\boldsymbol{I}_n\\ \vdots & \vdots & & \vdots\\ \sigma_{1m}\boldsymbol{I}_n & \sigma_{2m}\boldsymbol{I}_n & \cdots & \sigma_{mm}\boldsymbol{I}_n\end{pmatrix}$$

for the error vector of the combined linear regression model.

5.3.2 Heteroscedastic Errors

Under the linear regression model

$$y_i = \beta_1 x_{i,1} + \cdots + \beta_p x_{i,p} + \varepsilon_i, \quad i = 1, \ldots, n ,$$

with assumptions (i) to (iv), the errors ε_i have identical variance σ^2 for $i = 1, \ldots, n$, which is often referred as the case of *homoscedastic* errors. Sometimes, it is however more appropriate to assume that the errors are *heteroscedastic* in the sense that the variances of at least some of the individual error variables ε_i and ε_j, $i \neq j$ are different. Usually, heteroscedasticity is understood in the pure sense, meaning that ε_i and ε_j remain uncorrelated for any $i, j \in \{1, \ldots, n\}$ with $i \neq j$, being equivalent to $\operatorname{Cov}(\varepsilon_i, \varepsilon_j) = 0$ for any $i \neq j$. In the following, we shortly discuss three situations where pure heteroscedasticity occurs.

Aggregated Data

Consider the linear regression model $\boldsymbol{y} = \boldsymbol{X}\boldsymbol{\beta} + \boldsymbol{\varepsilon}$ with assumptions (i) to (iv). Then the i-th equation may be written as

$$y_i = \boldsymbol{x}'_{[i]}\boldsymbol{\beta} + \varepsilon_i, \quad i = 1, \ldots, n ,$$

where $\boldsymbol{x}'_{[i]}$ denotes the i-th row of the matrix $\boldsymbol{X}$ and hence

$$(y_i, \boldsymbol{x}'_{[i]}) = (y_i, x_{i,1}, \ldots, x_{i,p})$$

is the i-th observation of the variables y and $x_1, \ldots, x_p$. Now, suppose that the data $(y_i, \boldsymbol{x}'_{[i]})$ is divided into m disjoint groups $G_1, \ldots, G_m$, and the individual $(y_i, \boldsymbol{x}'_{[i]})$ cannot be observed, but only the mean values

$$(\overline{y}_1, \overline{x}_{1,1}, \ldots, \overline{x}_{1,p}), \ldots, (\overline{y}_m, \overline{x}_{m,1}, \ldots, \overline{x}_{m,p})$$

in each group, where

$$\overline{y}_k = \frac{1}{n_k} \sum_{i \in G_k} y_i \quad \text{and} \quad \overline{x}_{k,j} = \frac{1}{n_k} \sum_{i \in G_k} x_{i,j}$$

for $k = 1, \ldots, m$ and $j = 1, \ldots, p$, and n_k is the number of observations in group G_k with $\sum_{k=1}^{m} n_k = n$. How can we apply a linear regression model to this data?

Model for Aggregated Data. We can write

$$\overline{y}_k = \frac{1}{n_k} \sum_{i \in G_k} (\boldsymbol{x}'_{[i]}\boldsymbol{\beta} + \varepsilon_i) = \frac{1}{n_k} \sum_{i \in G_k} \boldsymbol{x}'_{[i]}\boldsymbol{\beta} + \frac{1}{n_k} \sum_{i \in G_k} \varepsilon_i ,$$

where

$$\frac{1}{n_k} \sum_{i \in G_k} \boldsymbol{x}'_{[i]}\boldsymbol{\beta} = \frac{1}{n_k} \sum_{i \in G_k} \sum_{j=1}^{p} x_{i,j}\beta_j = \sum_{j=1}^{p} \beta_j \frac{1}{n_k} \sum_{i \in G_k} x_{i,j} = \overline{\boldsymbol{x}}'_{[k]}\boldsymbol{\beta}$$

with $\overline{\boldsymbol{x}}'_{[k]} = (\overline{x}_{k,1}, \ldots, \overline{x}_{k,p})$ for $k = 1, \ldots, m$. Thus we have

$$\overline{y}_k = \overline{\boldsymbol{x}}'_{[k]}\boldsymbol{\beta} + \overline{\varepsilon}_k, \quad k = 1, \ldots, m ,$$

with $\overline{\varepsilon}_k = (1/n_k) \sum_{i \in G_k} \varepsilon_i$, or, in corresponding matrix notation,

$$\overline{\boldsymbol{y}} = \overline{\boldsymbol{X}}\boldsymbol{\beta} + \overline{\boldsymbol{\varepsilon}} ,$$

being the appropriate linear model for analysis of the aggregated data $(\overline{y}_k, \overline{\boldsymbol{x}}'_{[k]})$, $k = 1, \ldots, m$, when the original observations $(y_i, \boldsymbol{x}'_{[i]})$, $i = 1, \ldots, n$, are assumed to follow a linear regression model with assumptions (i) to (iv).

But what is the expectation vector and the covariance matrix of the error vector $\overline{\boldsymbol{\varepsilon}}$ in this model?

Distribution of the Error Vector. Since $\mathrm{E}(\varepsilon_i) = 0$ for $i = 1, \ldots, n$, it follows

$$\mathrm{E}(\overline{\varepsilon}_k) = \mathrm{E}(\frac{1}{n_k} \sum_{i \in G_k} \varepsilon_i) = \frac{1}{n_k} \sum_{i \in G_k} \mathrm{E}(\varepsilon_i) = 0 ,$$

showing $\mathrm{E}(\overline{\boldsymbol{\varepsilon}}) = \mathbf{0}$. Moreover, $\mathrm{Var}(\varepsilon_i) = \sigma^2$ and $\mathrm{Cov}(\varepsilon_i, \varepsilon_j) = 0$ for $i, j = 1, \ldots, n$, $i \neq j$, yield

$$\mathrm{Var}(\overline{\varepsilon}_k) = \mathrm{Var}(\frac{1}{n_k}\sum_{i\in G_k}\varepsilon_i) = \frac{1}{n_k^2}\sum_{i\in G_k}\mathrm{Var}(\varepsilon_i) = \frac{1}{n_k^2}n_k\sigma^2 = \frac{\sigma^2}{n_k}$$

and

$$\mathrm{Var}(\overline{\varepsilon}_k, \overline{\varepsilon}_l) = \frac{1}{n_k}\frac{1}{n_l}\mathrm{Cov}(\sum_{i\in G_k}\varepsilon_i, \sum_{j\in G_l}\varepsilon_j) = \frac{1}{n_k}\frac{1}{n_l}\sum_{i\in G_k}\sum_{j\in G_l}\mathrm{Cov}(\varepsilon_i, \varepsilon_j) = 0$$

for $k, l = 1, \ldots, m$, $k \neq l$. (Note that $i \in G_k$ and $j \in G_l$ cannot coincide for $k \neq l$ since the groups are disjoint.) Thus,

$$\mathrm{E}(\overline{\varepsilon}) = \mathbf{0} \quad \text{and} \quad \mathrm{Cov}(\overline{\varepsilon}) = \sigma^2 \boldsymbol{V}, \quad \boldsymbol{V} = \begin{pmatrix} \frac{1}{n_1} & 0 & \cdots & 0 \\ 0 & \frac{1}{n_2} & \cdots & 0 \\ & & \vdots & \\ 0 & 0 & \cdots & \frac{1}{n_m} \end{pmatrix},$$

and $\overline{\boldsymbol{y}} = \overline{\boldsymbol{X}}\boldsymbol{\beta} + \overline{\varepsilon}$ is a linear regression model satisfying assumptions (i) to (iii) and (iv$*$) if the model matrix $\overline{\boldsymbol{X}}$ is of full column rank and the number m of groups is strictly greater than the number p of independent variables.

Transformation of the Original Model. When the data under the original model $\boldsymbol{y} = \boldsymbol{X}\boldsymbol{\beta} + \varepsilon$ is given as

$$\boldsymbol{y} = \begin{pmatrix} \boldsymbol{y}_1 \\ \vdots \\ \boldsymbol{y}_m \end{pmatrix} \quad \text{and} \quad \boldsymbol{X} = \begin{pmatrix} \boldsymbol{X}_1 \\ \vdots \\ \boldsymbol{X}_m \end{pmatrix},$$

where $(\boldsymbol{y}_k, \boldsymbol{X}_k)$ is the observation matrix in group G_k, $k = 1, \ldots, m$, then the model $\overline{\boldsymbol{y}} = \overline{\boldsymbol{X}}\boldsymbol{\beta} + \overline{\varepsilon}$ can also be obtained as the linear transformation

$$\underbrace{\boldsymbol{F}\boldsymbol{y}}_{\overline{\boldsymbol{y}}} = \underbrace{\boldsymbol{F}\boldsymbol{X}}_{\overline{\boldsymbol{X}}}\boldsymbol{\beta} + \underbrace{\boldsymbol{F}\varepsilon}_{\overline{\varepsilon}}, \quad \boldsymbol{F} = \begin{pmatrix} \frac{1}{n_1}\mathbf{1}'_{n_1} & \mathbf{0} & \cdots & \mathbf{0} \\ \mathbf{0} & \frac{1}{n_2}\mathbf{1}'_{n_2} & \cdots & \mathbf{0} \\ & & \vdots & \\ \mathbf{0} & \mathbf{0} & \cdots & \frac{1}{n_m}\mathbf{1}'_{n_m} \end{pmatrix},$$

where $\mathrm{E}(\boldsymbol{F}\varepsilon) = \boldsymbol{F}\,\mathrm{E}(\varepsilon) = \mathbf{0}$ and $\mathrm{Cov}(\boldsymbol{F}\varepsilon) = \sigma^2\boldsymbol{F}\boldsymbol{F}' = \sigma^2\boldsymbol{V}$ with $\boldsymbol{V}$ as above.

Estimation of $\boldsymbol{\beta}$. For estimating $\boldsymbol{\beta}$ the generalized least squares estimator

$$\widehat{\boldsymbol{\beta}}_* = (\overline{\boldsymbol{X}}'\boldsymbol{V}^{-1}\overline{\boldsymbol{X}})^{-1}\overline{\boldsymbol{X}}'\boldsymbol{V}^{-1}\boldsymbol{y}$$

can be applied when the number n_k of observations in each group is known.

Grouped Data

Even if the data is not aggregated as assumed before, it might occur that the observations are divided into m groups such that $\boldsymbol{y} = (\boldsymbol{y}_1', \ldots, \boldsymbol{y}_m')'$ and $\boldsymbol{y}_k$ contains the n_k observations from group G_k. For example, the dependent variable y may be the expenditure of households being observed within different income classes, so that $\boldsymbol{y}_k$ contains the observations of n_k individual household expenditures in income class G_k.

In this case, it appears to be reasonable that the variance within each class is nearly the same, while it will be quite different for different classes. One may allow for this by considering a linear regression model $\boldsymbol{y} = \boldsymbol{X\beta} + \boldsymbol{\varepsilon}$ with $\mathrm{E}(\boldsymbol{\varepsilon}) = \boldsymbol{0}$ and

$$\mathrm{Cov}(\boldsymbol{\varepsilon}) = \boldsymbol{V}, \quad \boldsymbol{V} = \begin{pmatrix} \sigma_1^2 \boldsymbol{I}_{n_1} & \boldsymbol{0} & \cdots & \boldsymbol{0} \\ \boldsymbol{0} & \sigma_2^2 \boldsymbol{I}_{n_2} & \cdots & \boldsymbol{0} \\ & & \vdots & \\ \boldsymbol{0} & \boldsymbol{0} & \cdots & \sigma_m^2 \boldsymbol{I}_{n_m} \end{pmatrix},$$

where $\sigma_1^2, \ldots, \sigma_m^2$ are unknown positive parameters. This is often referred as a situation of *grouped heteroscedasticity*. Clearly, the generalized least squares estimator

$$\widehat{\boldsymbol{\beta}}_* = (\boldsymbol{X}'\boldsymbol{V}^{-1}\boldsymbol{X})^{-1}\boldsymbol{X}'\boldsymbol{V}^{-1}\boldsymbol{y}$$

is not a practical estimator for $\boldsymbol{\beta}$, and a two-stage procedure involves estimators for the unknowns $\sigma_1^2, \ldots, \sigma_m^2$. If $n_k > p$ for every $k = 1, \ldots, m$, then one may estimate $\widehat{\sigma}_k^2$ by

$$\widehat{\sigma}_k^2 = \frac{1}{n_k - p}(\boldsymbol{y}_k - \boldsymbol{X}_k\widehat{\boldsymbol{\beta}}^{(k)})'(\boldsymbol{y}_k - \boldsymbol{X}_k\widehat{\boldsymbol{\beta}}^{(k)}),$$

where $\widehat{\boldsymbol{\beta}}^{(k)}$ is the ordinary least squares estimator for $\boldsymbol{\beta}$ obtained under the model

$$\boldsymbol{y}_k = \boldsymbol{X}_k\boldsymbol{\beta} + \boldsymbol{\varepsilon}_k, \quad \boldsymbol{\varepsilon}_k \sim (\boldsymbol{0}, \sigma_k^2 \boldsymbol{I}_{n_k}).$$

In a second step, one will estimate $\boldsymbol{\beta}$ by

$$\widehat{\boldsymbol{\beta}}_*(\widehat{\boldsymbol{V}}) = (\boldsymbol{X}'\widehat{\boldsymbol{V}}^{-1}\boldsymbol{X})^{-1}\boldsymbol{X}'\widehat{\boldsymbol{V}}^{-1}\boldsymbol{y},$$

where $\widehat{\boldsymbol{V}}$ is obtained from $\boldsymbol{V}$ by replacing σ_k^2 by $\widehat{\sigma}_k^2$ for $k = 1, \ldots, m$.

When we assume that the expectation of the elements of $\boldsymbol{y}_k$ is nearly the same within each group G_k, then $\boldsymbol{X}_k\boldsymbol{\beta} \approx \boldsymbol{1}_{n_k}\mu_k$ for some μ_k and the least squares variance estimator for σ_k^2 under the model

$$\boldsymbol{y}_k = \boldsymbol{1}_{n_k}\mu_k + \boldsymbol{\varepsilon}_k, \quad \boldsymbol{\varepsilon}_k \sim (\boldsymbol{0}, \sigma_k^2 \boldsymbol{I}_{n_k}).$$

coincides with the sample variance

$$\widehat{\sigma}^2_{\boldsymbol{y}_k} = \frac{1}{n_k - 1} \sum_{i \in G_k} (y_i - \overline{y})^2 .$$

Hence, an alternative two-stage generalized least squares estimator for $\boldsymbol{\beta}$ is derived when $\widehat{\boldsymbol{V}}$ is obtained from $\boldsymbol{V}$ by replacing σ_k^2 by the sample variances $\widehat{\sigma}^2_{\boldsymbol{y}_k}$ for $k = 1, \ldots, m$.

Variance as Functions of Variables

Consider the linear regression model

$$y_i = \beta_1 x_{i,1} + \cdots + \beta_p x_{i,p} + \varepsilon_i, \quad i = 1, \ldots, n ,$$

where $\mathrm{Var}(y_i) = \sigma_i^2 = \mathrm{Var}(\varepsilon_i)$ depends on $x_{i,j}$ for some $j \in \{1, \ldots, p\}$. For example, the variable y_i may be the profit of firm i, while $x_{i,j}$ is the research expenditure of firm i. Then the variance of y_i might become greater for rising values of $x_{i,j}$, since large research expenditures might imply greater deviations from the expected profit of the firm. One may allow for this by assuming that σ_i^2 is proportional to the square of $x_{i,j}$, i.e.

$$\sigma_i^2 = \sigma^2 x_{i,j}^2 ,$$

leading to the consideration of a linear regression model $\boldsymbol{y} = \boldsymbol{X}\boldsymbol{\beta} + \boldsymbol{\varepsilon}$ with $\mathrm{E}(\boldsymbol{\varepsilon}) = \boldsymbol{0}$ and

$$\mathrm{Cov}(\boldsymbol{\varepsilon}) = \sigma^2 \boldsymbol{V}, \quad \boldsymbol{V} = \begin{pmatrix} x_{1,j}^2 & 0 & \cdots & 0 \\ 0 & x_{2,j}^2 & \cdots & 0 \\ & & \vdots & \\ 0 & 0 & \cdots & x_{n,j}^2 \end{pmatrix} .$$

This is a further instance of a heteroscedastic model, where the generalized least squares estimator

$$\widehat{\boldsymbol{\beta}}_* = (\boldsymbol{X}'\boldsymbol{V}^{-1}\boldsymbol{X})^{-1}\boldsymbol{X}'\boldsymbol{V}^{-1}\boldsymbol{y}$$

for $\boldsymbol{\beta}$ is practical, since the matrix $\boldsymbol{V}$ is known. One may find indication for such a model by plotting the data points $(x_{i,j}, y_i)$, $i = 1, \ldots, n$, where x_j is the variable being suspected to cause heteroscedasticity. If a pattern as in Fig. 5.4 is revealed, then this suggests the application of the above model.

It is of course possible that $\mathrm{Var}(y_i)$ is a more complicated function of one or even more variables. For example, as also noted in [113, p. 244/245], specific data (expenditure on tea of households) may indicate that the variance is proportional to the square of the mean, i.e.

$$\mathrm{Var}(y_i) = \sigma^2 (\mathrm{E}(y_i))^2 = \sigma^2 (\boldsymbol{x}'_{[i]}\boldsymbol{\beta})^2 ,$$

where $\boldsymbol{x}'_{[i]}$ is the i-th row of $\boldsymbol{X}$. In that case

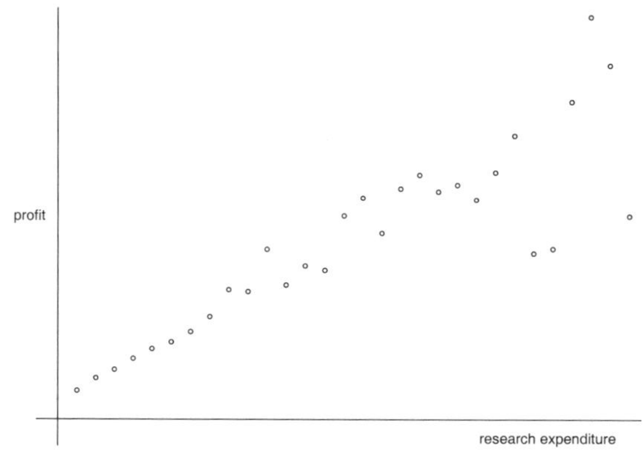

Fig. 5.4. Data points indicating $\sigma_i^2 = \sigma^2 x_{ij}^2$, where y stands for profit and x_j for research expenditure

$$\boldsymbol{V} = \begin{pmatrix} (\boldsymbol{x}'_{[1]}\boldsymbol{\beta})^2 & 0 & \cdots & 0 \\ 0 & (\boldsymbol{x}'_{[2]}\boldsymbol{\beta})^2 & \cdots & 0 \\ & & \vdots & \\ 0 & 0 & \cdots & (\boldsymbol{x}'_{[n]}\boldsymbol{\beta})^2 \end{pmatrix}$$

is not known. A possible estimator for $\boldsymbol{\beta}$ is the two-stage generalized least squares estimator, where $\widehat{\boldsymbol{V}}$ is obtained form $\boldsymbol{V}$ by replacing $\boldsymbol{\beta}$ by the ordinary least squares estimator $\widehat{\boldsymbol{\beta}} = (\boldsymbol{X}'\boldsymbol{X})^{-1}\boldsymbol{X}'\boldsymbol{y}$.

Ordinary Least Squares Estimation

When heteroscedasticity is present, then the above approaches suggest the application of the generalized least squares or the two-stage generalized least squares estimator for estimating $\boldsymbol{\beta}$. Nonetheless, when there is no information about the possible reason for the heteroscedasticity, the estimation of

$$\mathrm{Cov}(\boldsymbol{\varepsilon}) = \begin{pmatrix} \mathrm{E}(\varepsilon_1^2) & 0 & \cdots & 0 \\ 0 & \mathrm{E}(\varepsilon_2^2) & \cdots & 0 \\ & & \vdots & \\ 0 & 0 & \cdots & \mathrm{E}(\varepsilon_n^2) \end{pmatrix}$$

is rather uncertain. A very rough estimator might be

$$\widehat{\mathrm{Cov}(\boldsymbol{\varepsilon})} = \begin{pmatrix} \widehat{\varepsilon}_1^2 & 0 & \cdots & 0 \\ 0 & \widehat{\varepsilon}_2^2 & \cdots & 0 \\ & & \vdots & \\ 0 & 0 & \cdots & \widehat{\varepsilon}_n^2 \end{pmatrix},$$

where $\widehat{\varepsilon}_i$ are the elements of $\widehat{\varepsilon} = \boldsymbol{y} - \boldsymbol{X}\widehat{\boldsymbol{\beta}}$. Now, instead of using $\widehat{\mathrm{Cov}(\varepsilon)}$ for two-stage generalized least squares estimation, one may confine to $\widehat{\boldsymbol{\beta}}$ as an unbiased estimator for $\boldsymbol{\beta}$, and apply $\widehat{\mathrm{Cov}(\varepsilon)}$ for obtaining some estimator for $\mathrm{Cov}(\widehat{\boldsymbol{\beta}})$. As a matter of fact $\mathrm{Cov}(\widehat{\boldsymbol{\beta}}) = (\boldsymbol{X}'\boldsymbol{X})^{-1}\boldsymbol{X}'\,\mathrm{Cov}(\varepsilon)\boldsymbol{X}(\boldsymbol{X}'\boldsymbol{X})^{-1}$, and hence

$$\widehat{\mathrm{Cov}(\widehat{\boldsymbol{\beta}})} = (\boldsymbol{X}'\boldsymbol{X})^{-1}\boldsymbol{X}'\widehat{\mathrm{Cov}(\varepsilon)}\boldsymbol{X}(\boldsymbol{X}'\boldsymbol{X})^{-1}\,.$$

This estimator is due to White [126], see also [72] for a review on certain modifications. Davidson and MacKinnon [28, p. 554] strongly recommend to replace $\widehat{\varepsilon}_i$ by $\widehat{\varepsilon}_i/(1-p_{ii})^2$, where p_{ii} is the i-th main diagonal element of $\boldsymbol{P} = \boldsymbol{X}(\boldsymbol{X}'\boldsymbol{X})^{-1}\boldsymbol{X}'$.

The squares roots of the main diagonal elements of $\widehat{\mathrm{Cov}(\widehat{\boldsymbol{\beta}})}$ may be used as standard errors of the individual elements of $\widehat{\boldsymbol{\beta}}$.

5.3.3 Equicorrelated Errors

Suppose that in the linear regression model

$$y_i = \beta_1 x_{i,1} + \cdots + \beta_p x_{i,p} + \varepsilon_i, \quad i = 1, \ldots, n$$

the errors are homoscedastic, i.e. $\mathrm{Var}(\varepsilon_i) = \sigma^2$ for $i = 1, \ldots, n$, but the correlation $\varrho = \mathrm{Cov}(\varepsilon_i, \varepsilon_j)/\sigma^2$ between different elements ε_i and ε_j, $i \neq j$, is not necessarily identical to zero. Then

$$\mathrm{Cov}(\varepsilon) = \begin{pmatrix} \sigma^2 & \sigma^2\varrho & \cdots & \sigma^2\varrho \\ \sigma^2\varrho & \sigma^2 & \cdots & \sigma^2\varrho \\ & & \vdots & \\ \sigma^2\varrho & \cdots & \sigma^2\varrho & \sigma^2 \end{pmatrix} = \sigma^2\boldsymbol{V}, \quad \boldsymbol{V} = (1-\varrho)\boldsymbol{I}_n + \varrho\boldsymbol{1}_n\boldsymbol{1}_n'\,.$$

It can be shown that the symmetric matrix $\boldsymbol{V}$ is positive definite if and only if

$$\frac{-1}{n-1} < \varrho < 1\,,$$

see Problem 5.7. This is referred in the literature as the case of *equicorrelated* errors, generalizing the usual assumption $\varrho = 0$. Equicorrelation is a special case of serial correlation with $\varrho_1 = \cdots = \varrho_{n-1} = \varrho$.

A specific instance of equicorrelation occurs when the error vector ε can be decomposed as

$$\varepsilon = \boldsymbol{1}_n\xi_0 + \boldsymbol{\xi}, \quad \boldsymbol{\xi} \sim (\boldsymbol{0}, \sigma_1^2\boldsymbol{I}_n), \quad \xi_0 \sim (0, \sigma_0^2), \quad \mathrm{Cov}(\xi_0, \boldsymbol{\xi}) = \boldsymbol{0}\,.$$

Then

$$\mathrm{Cov}(\varepsilon) = \begin{pmatrix} \sigma_0^2+\sigma_1^2 & \sigma_0^2 & \cdots & \sigma_0^2 \\ \sigma_0^2 & \sigma_0^2+\sigma_1^2 & \cdots & \sigma_0^2 \\ & & \vdots & \\ \sigma_0^2 & \cdots & \sigma_0^2 & \sigma_0^2+\sigma_1^2 \end{pmatrix},$$

and $0 < \varrho = \sigma_0^2/(\sigma_0^2+\sigma_1^2) < 1$ whenever $\sigma_0^2, \sigma_1^2 > 0$.

Estimation of $\boldsymbol{\beta}$. Now, one might think that an appropriate estimator for $\boldsymbol{\beta}$ is the two-stage generalized least squares estimator $\widehat{\boldsymbol{\beta}}_*(\widehat{\boldsymbol{V}})$ when $\widehat{\boldsymbol{V}}$ is obtained from $\boldsymbol{V}$ by replacing ϱ by some appropriate estimator $\widehat{\varrho}$. However, as noted in Sect. 5.5, it follows that

$$\widehat{\boldsymbol{\beta}}_* = (\boldsymbol{X}'\boldsymbol{V}^{-1}\boldsymbol{X})^{-1}\boldsymbol{X}'\boldsymbol{V}^{-1}\boldsymbol{y}$$

coincides with $\widehat{\boldsymbol{\beta}} = (\boldsymbol{X}'\boldsymbol{X})^{-1}\boldsymbol{X}'\boldsymbol{y}$ whenever the vector $\mathbf{1}_n$ is contained in the column space of $\boldsymbol{X}$. Then, the actual value of ϱ is irrelevant to estimation of $\boldsymbol{\beta}$ when we confine to least squares estimation, showing that ordinary least squares can also be appropriate in cases of non-spherical covariance matrices of the error vector.

5.3.4 Autocorrelated Errors

Especially when the y_i from the equation

$$y_i = \beta_1 x_{i,1} + \cdots + \beta_p x_{i,p} + \varepsilon_i, \quad i = 1, \ldots, n ,$$

are observed at consecutive points of time denoted by $i = 1, \ldots, n$, then it is often reasonable to assume that the possible correlation between ε_i and ε_j becomes smaller when the distance between i and j increases, so that each ε_i may represent a persistent shock dying away over time.

The Error Process. A simple error process allowing for such a situation is the so-called *first-order autoregressive* or *AR(1)* process, described by

$$\varepsilon_i = \varrho\varepsilon_{i-1} + u_i ,$$

where $u_i \sim (0, \sigma_u^2)$ for all possible i and $\mathrm{Cov}(u_i, u_j) = 0$ whenever $i \neq j$. The parameter ϱ is assumed to satisfy $|\varrho| < 1$. It that case the errors are called *AR(1) errors* .

Note that the process behaves quite differently, depending on the sign of ϱ. If ϱ is positive, then successive values of the ε's will tend to be close together, while for negative ϱ successive ε's will be quite far away from each other and on opposite sides of the horizontal zero line.

Figures 5.5 and 5.6 illustrate this behavior by showing AR(1) errors with positive and negative sign of ϱ at 100 points of time (x-axis), respectively. The ε's are connected with lines in order to bring out the development over time.

When we assume that the process began in the infinite past, then we can write

$$\begin{aligned}
\varepsilon_i &= \varrho\varepsilon_{i-1} + u_i \\
&= \varrho^2\varepsilon_{i-2} + \varrho u_{i-1} + u_i \\
&\;\vdots \\
&= \sum_{\tau=0}^{\infty} \varrho^\tau u_{i-\tau} .
\end{aligned}$$

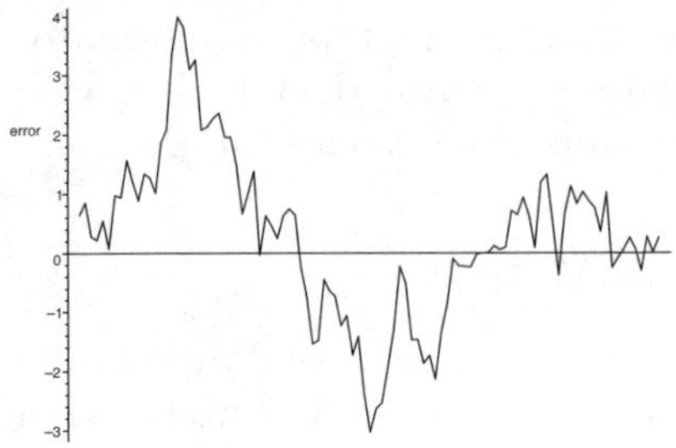

Fig. 5.5. AR(1) errors with positive ϱ at 100 points of time

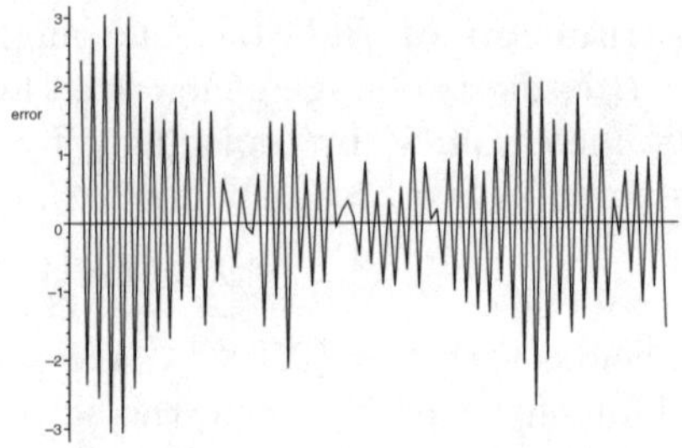

Fig. 5.6. AR(1) errors with negative ϱ at 100 points of time

This is the *moving average* representation of the process, showing that all past $u_{i-\tau}$ influence the error ε_i but with decreasing weight ϱ^τ for increasing values τ.

What are expectation vector and covariance matrix of the error vector $\boldsymbol{\varepsilon}$?

Distribution of the Error Vector. The expectation of the i-th element ε_i of $\boldsymbol{\varepsilon}$ satisfies

$$\mathrm{E}(\varepsilon_i) = \sum_{\tau=0}^{\infty} \varrho^\tau\, \mathrm{E}(u_{i-\tau}) = 0\,.$$

Hence, $\mathrm{Var}(\varepsilon_i) = \mathrm{E}(\varepsilon_i^2)$. The square ε_i^2 can be written as

$$\varepsilon_i^2 = (u_i + \varrho u_{i-1} + \varrho^2 u_{i-2} + \cdots)(u_i + \varrho u_{i-1} + \varrho^2 u_{i-2} + \cdots)\,.$$

In view of $\mathrm{E}(u_i u_j) = 0$ for $i \neq j$ it follows that

$$\begin{aligned}\mathrm{E}(\varepsilon_i^2) &= \underbrace{\mathrm{E}(u_i^2)}_{=\sigma_u^2} + \varrho^2 \underbrace{\mathrm{E}(u_{i-1}^2)}_{=\sigma_u^2} + \varrho^4 \underbrace{\mathrm{E}(u_{i-2}^2)}_{=\sigma_u^2} + \cdots \\ &= \sigma_u^2 \sum_{\tau=0}^{\infty} \varrho^{2\tau} \\ &= \sigma_u^2 \frac{1}{1-\varrho^2}\,.\end{aligned}$$

Therefore

$$\mathrm{Var}(\varepsilon_i) = \sigma^2 := \sigma_u^2/(1-\varrho^2)$$

for every i, showing that the errors are homoscedastic.

The covariance between ε_i and ε_{i-k} is given by $\mathrm{Cov}(\varepsilon_i, \varepsilon_{i-k}) = \mathrm{E}(\varepsilon_i \varepsilon_{i-k})$, where for $k > 0$ the identity

$$\begin{aligned}\varepsilon_i \varepsilon_{i-k} = {} & (u_i + \varrho u_{i-1} + \cdots + \varrho^k u_{i-k} + \varrho^{k+1} u_{i-k-1} + \cdots) \\ & \times (u_{i-k} + \varrho u_{i-k-1} + \varrho^2 u_{i-k-2} + \cdots)\end{aligned}$$

holds true. In view of $\mathrm{E}(u_i u_j) = 0$, $i \neq j$, it follows

$$\begin{aligned}\mathrm{Cov}(\varepsilon_i, \varepsilon_{i-k}) &= \varrho^k \,\mathrm{E}(u_{i-k}^2) + \varrho^{k+2}\,\mathrm{E}(u_{i-k-1}^2) + \varrho^{k+4}\,\mathrm{E}(u_{i-k-2}^2) + \cdots \\ &= \varrho^k \sigma_u^2 \sum_{\tau=0}^{\infty} \varrho^{2\tau} \\ &= \varrho^k \sigma^2 \,.\end{aligned}$$

This result may also be applied for the case that $k < 0$ by writing $g = -k$. Then

$$\mathrm{Cov}(\varepsilon_i, \varepsilon_{i-k}) = \mathrm{Cov}(\varepsilon_i, \varepsilon_{i+g}) = \mathrm{Cov}(\varepsilon_{i+g}, \varepsilon_{i+g-g}) \,.$$

Since $g = |k|$ is positive, similarly to the above it follows

$$\mathrm{Cov}(\varepsilon_i, \varepsilon_{i-k}) = \varrho^{|k|}\sigma^2 \,.$$

This means that

$$\mathrm{Cov}(\varepsilon_i, \varepsilon_j) = \sigma^2 \varrho^{|i-j|}$$

for $i, j = 1, \ldots, n$, and, in summary,

$$\mathrm{E}(\boldsymbol{\varepsilon}) = \mathbf{0} \quad \text{and} \quad \mathrm{Cov}(\boldsymbol{\varepsilon}) = \sigma^2 \boldsymbol{V}, \quad \boldsymbol{V} = \begin{pmatrix} 1 & \varrho & \varrho^2 & \cdots & \varrho^{n-1} \\ \varrho & 1 & \varrho & \cdots & \varrho^{n-2} \\ \varrho^2 & \varrho & 1 & \cdots & \varrho^{n-3} \\ \vdots & \vdots & \vdots & & \vdots \\ \varrho^{n-1} & \varrho^{n-2} & \varrho^{n-3} & \cdots & 1 \end{pmatrix} .$$

Autocorrelation generated by AR(1) errors is thus a special case of serial correlation with $\varrho_i = \varrho^i$ for $i = 1, \ldots, n-1$.

Note that we obtain the same covariance matrix when we do not assume that the process started in the infinite past but that the initial ε_1 satisfies

$$\mathrm{E}(\varepsilon_1) = 0, \qquad \mathrm{Var}(\varepsilon_1) = \frac{\sigma_u^2}{1-\varrho^2}, \qquad \mathrm{Cov}(u_i, \varepsilon_1) = 0$$

for $i = 2, \ldots, n$. Then from the finite representation

$$\varepsilon_i = \sum_{\tau=0}^{i-2} \varrho^\tau u_{i-\tau} + \varrho^{i-1}\varepsilon_1, \qquad i = 2, \ldots, n$$

one may as well deduce $\mathrm{E}(\boldsymbol{\varepsilon}) = \mathbf{0}$ and $\mathrm{Cov}(\boldsymbol{\varepsilon}) = \sigma^2 \boldsymbol{V}$ with $\boldsymbol{V}$ as above.

Estimation of $\boldsymbol{\beta}$. The inverse of the matrix $\boldsymbol{V}$ is given by

$$\boldsymbol{V}^{-1} = \frac{1}{1-\varrho^2} \begin{pmatrix} 1 & -\varrho & 0 & \cdots & 0 & 0 \\ -\varrho & 1+\varrho^2 & -\varrho & \cdots & 0 & 0 \\ 0 & -\varrho & 1+\varrho^2 & \cdots & 0 & 0 \\ 0 & 0 & 0 & \cdots & 1+\varrho^2 & -\varrho \\ 0 & 0 & 0 & \cdots & -\varrho & 1 \end{pmatrix},$$

but usually, the generalized least squares estimator

$$\widehat{\boldsymbol{\beta}}_* = (\boldsymbol{X}'\boldsymbol{V}^{-1}\boldsymbol{X})^{-1}\boldsymbol{X}'\boldsymbol{V}^{-1}\boldsymbol{y}$$

is not practical for $\boldsymbol{\beta}$, since the parameter ϱ is not known. A possible estimator for ϱ can be obtained from the equation $\varepsilon_i = \varrho\varepsilon_{i-1}+u_i$, where ε_i is replaced by $\widehat{\varepsilon}_i$ being the i-th element of the vector $\widehat{\boldsymbol{\varepsilon}} = \boldsymbol{y} - \boldsymbol{X}\widehat{\boldsymbol{\beta}}$ with $\widehat{\boldsymbol{\beta}} = (\boldsymbol{X}'\boldsymbol{X})^{-1}\boldsymbol{X}'\boldsymbol{y}$. Then the equation

$$\widehat{\varepsilon}_i = \varrho\widehat{\varepsilon}_{i-1} + u_i, \quad i = 2, \ldots, n ,$$

can be understood as a linear regression model, although the $\widehat{\varepsilon}_{i-1}$ on the right-hand side are realizations of random variables and thus the classical model assumptions do not hold here. Nonetheless, by ignoring the latter objection, the ordinary least squares estimator for ϱ can be computed as

$$\widehat{\varrho} = \frac{\sum_{i=2}^{n} \widehat{\varepsilon}_{i-1}\widehat{\varepsilon}_i}{\sum_{i=2}^{n} \widehat{\varepsilon}_{i-1}^2} ,$$

being a possible estimator for ϱ. As an alternative, Theil [113] proposes

$$\widehat{\varrho}_{\mathrm{T}} = \frac{\frac{1}{n-1}\sum_{i=2}^{n} \widehat{\varepsilon}_{i-1}\widehat{\varepsilon}_i}{\frac{1}{n-p}\sum_{i=1}^{n} \widehat{\varepsilon}_{i-1}^2} .$$

as an estimator for ϱ.

5.4 Singular Covariance Matrices

Model assumption (iv$*$) implies that $\sigma^{-2}\,\mathrm{Cov}(\boldsymbol{\varepsilon}) = \boldsymbol{V}$ is symmetric and positive definite under the linear regression model $\boldsymbol{y} = \boldsymbol{X}\boldsymbol{\beta}+\boldsymbol{\varepsilon}$. On the other hand, from Theorem B.3, the covariance matrix of a random vector is necessarily symmetric and nonnegative definite but not necessarily nonsingular and thus not necessarily positive definite. Hence, the assumption (iv$*$) excludes nonsingular matrices as possible covariance matrices for the error vector, although such matrices do in fact exist, as shown by the following example.

Example 5.3. Let $y_1 \sim (0, \sigma_1^2)$ and $y_2 \sim (0, \sigma_2^2)$ be two random variables with $\mathrm{Cov}(y_1, y_2) = 0$. The covariance matrix of the vector $\boldsymbol{y} = (y_1, y_2, y_3)'$ with $y_3 = y_1 + y_2$ is given by

$$\mathrm{Cov}(\boldsymbol{y}) = \begin{pmatrix} \sigma_1^2 & 0 & \sigma_1^2 \\ 0 & \sigma_2^2 & \sigma_2^2 \\ \sigma_1^2 & \sigma_2^2 & \sigma_1^2 + \sigma_2^2 \end{pmatrix} ,$$

which is clearly singular and thus not positive definite. □

The above example gives already some indication for situations under which singular covariance matrices of ε and hence of $\boldsymbol{y}$ may occur under the model $\boldsymbol{y} = \boldsymbol{X\beta} + \boldsymbol{\varepsilon}$. As a matter of fact, for the above vector $\boldsymbol{y}$ and for $\boldsymbol{c} = (1, 1, -1)'$ the identity $\boldsymbol{c}'\boldsymbol{y} = 0$ is true, irrespective of the possible outcome of $\boldsymbol{y}$. Thus, the random vector $\boldsymbol{y}$ is not free to realize in $\mathbb{R}^3$, but is restricted in the sense that the linear equation $\boldsymbol{c}'\boldsymbol{y} = 0$ must be satisfied for any possible outcome of $\boldsymbol{y}$. This describes the situation under which singularity of $\mathrm{Cov}(\boldsymbol{y})$ will occur.

Theorem 5.6. *Under the linear regression model with assumptions (i) to (iii),* $\mathrm{Cov}(\boldsymbol{y})$ *is singular if and only if there exists some nonzero non-stochastic vector* $\boldsymbol{c}$ *and some non-stochastic number* d *such that the identity*

$$\boldsymbol{c}'\boldsymbol{y} = d$$

is satisfied with probability 1.

Proof. Note in advance that $\mathrm{Prob}(\boldsymbol{c}'\boldsymbol{y} = d) = 1$ if and only if

$$\mathrm{E}(\boldsymbol{c}'\boldsymbol{y}) = d \quad \text{and} \quad \mathrm{Var}(\boldsymbol{c}'\boldsymbol{y}) = 0$$

are satisfied.

Now, suppose that $\mathrm{E}(\boldsymbol{c}'\boldsymbol{y}) = d$ and $\mathrm{Var}(\boldsymbol{c}'\boldsymbol{y}) = 0$ are fulfilled for some nonzero $\boldsymbol{c}$ and some number d. Then the latter identity is equivalent to $\boldsymbol{c}'\,\mathrm{Cov}(\boldsymbol{y})\boldsymbol{c} = 0$, which in turn is satisfied if and only if $\mathrm{Cov}(\boldsymbol{y})\boldsymbol{c} = 0$, since $\mathrm{Cov}(\boldsymbol{y})$ is symmetric nonnegative definite. But since $\boldsymbol{c}$ is nonzero, the identity $\mathrm{Cov}(\boldsymbol{y})\boldsymbol{c} = 0$ implies that $\mathrm{Cov}(\boldsymbol{y})$ must be singular.

Conversely, assume that $\mathrm{Cov}(\boldsymbol{y})$ is singular. Then there must exist some nonzero vector $\boldsymbol{c}$ with $\mathrm{Cov}(\boldsymbol{y})\boldsymbol{c} = 0$. But then also $\mathrm{Var}(\boldsymbol{c}'\boldsymbol{y}) = 0$ and $\mathrm{E}(\boldsymbol{c}'\boldsymbol{y}) = d$ with $d = \boldsymbol{c}'\boldsymbol{X\beta}$, showing that $\mathrm{Prob}(\boldsymbol{c}'\boldsymbol{y} = d) = 1$. □

Hence, whenever the observable random vector $\boldsymbol{y}$ under the linear regression model $\boldsymbol{y} = \boldsymbol{X\beta} + \boldsymbol{\varepsilon}$ satisfies linear restrictions irrespective of its actual outcome, then $\mathrm{Cov}(\boldsymbol{y})$ must be singular. This may occur for example in econometrics, see e.g. [46, Sect. 14.3]. We reconsider the example given in [113, Sect. 6.7] here.

Consider the three model equations

$$\begin{aligned} y_i^{(1)} &= \beta_1 + \beta_2 x_i + \varepsilon_i^{(1)}, \\ y_i^{(2)} &= \gamma_1 + \gamma_2 x_i + \varepsilon_i^{(2)}, \\ y_i^{(3)} &= \delta_1 + \delta_2 x_i + \varepsilon_i^{(3)} \end{aligned}$$

for $i = 1, \ldots, n$, where $y_i^{(1)}$, $y_i^{(2)}$, and $y_i^{(3)}$ stand for expenditure on food, durables and the rest, respectively, of household i, while the non-random x_i stands for the total expenditure of household i. Then, clearly the realizations of $y_i^{(1)}$, $y_i^{(2)}$, and $y_i^{(3)}$ must sum up to the non-random x_i for any $i = 1, \ldots, n$.

Hence, if we combine the three equations into one linear regression model $\boldsymbol{y} = \boldsymbol{X}\boldsymbol{\beta} + \boldsymbol{\varepsilon}$ such that

$$\boldsymbol{y} = (y_1^{(1)}, y_1^{(2)}, y_1^{(3)}, \ldots, y_n^{(1)}, y_n^{(2)}, y_n^{(3)})' ,$$

then the covariance matrix of $\boldsymbol{y}$ must be singular, since e.g. the identity $\boldsymbol{c}'\boldsymbol{y} = x_1$ is satisfied for $\boldsymbol{c} = (1, 1, 1, 0, \ldots, 0)'$. To avoid this, one may simply leave out the third equation, modeling the expenditure on the rest, since it provides some kind of superfluous information causing $\mathrm{Cov}(\boldsymbol{y})$ to become singular.

A different example for the occurrence of a singular covariance matrix of the error vector is the incorporation of linear restrictions $\boldsymbol{R}\boldsymbol{\beta} = \boldsymbol{r}$ on the parameter vector $\boldsymbol{\beta}$. As an alternative approach to restricted least squares estimation as considered in Sect. 3.1, one may combine the two equations $\boldsymbol{y} = \boldsymbol{X}\boldsymbol{\beta} + \boldsymbol{\varepsilon}$ and $\boldsymbol{R}\boldsymbol{\beta} = \boldsymbol{r}$ into one model equation

$$\begin{pmatrix} \boldsymbol{y} \\ \boldsymbol{r} \end{pmatrix} = \begin{pmatrix} \boldsymbol{X} \\ \boldsymbol{R} \end{pmatrix} \boldsymbol{\beta} + \begin{pmatrix} \boldsymbol{\varepsilon} \\ \boldsymbol{0} \end{pmatrix} ,$$

where

$$\mathrm{E}\begin{pmatrix} \boldsymbol{\varepsilon} \\ \boldsymbol{0} \end{pmatrix} = \begin{pmatrix} \boldsymbol{0} \\ \boldsymbol{0} \end{pmatrix} \quad \text{and} \quad \mathrm{Cov}\begin{pmatrix} \boldsymbol{\varepsilon} \\ \boldsymbol{0} \end{pmatrix} = \begin{pmatrix} \sigma^2\boldsymbol{V} & \boldsymbol{0} \\ \boldsymbol{0} & \boldsymbol{0} \end{pmatrix} .$$

Hence, even if $\boldsymbol{V}$ is symmetric positive definite, the covariance matrix of the error vector in the combined model is singular. How can we estimate $\boldsymbol{\beta}$ under such a model with singular covariance matrix of the error vector?

Clearly, under the linear regression model with assumption (i) to (iii) and singular covariance matrix $\mathrm{Cov}(\boldsymbol{\varepsilon}) = \sigma^2\boldsymbol{V}$, the generalized least squares estimator

$$\widehat{\boldsymbol{\beta}}_* = (\boldsymbol{X}'\boldsymbol{V}^{-1}\boldsymbol{X})^{-1}\boldsymbol{X}'\boldsymbol{V}^{-1}\boldsymbol{y}$$

does not exist. Nonetheless, there does in fact exist an estimator $\widehat{\widehat{\boldsymbol{\beta}}}$ satisfying

$$\mathrm{Cov}(\widehat{\widehat{\boldsymbol{\beta}}}) \leq_{\mathrm{L}} \mathrm{Cov}(\boldsymbol{A}\boldsymbol{y} + \boldsymbol{a})$$

for every linear estimator $\boldsymbol{A}\boldsymbol{y} + \boldsymbol{a}$ with $\mathrm{E}(\boldsymbol{A}\boldsymbol{y} + \boldsymbol{a}) = \boldsymbol{\beta}$ for all $\boldsymbol{\beta} \in \mathrm{I\!R}^p$. This estimator has a number of different representations, which do all coincide with probability one. As a matter of fact, one possible representation is

$$\widehat{\widehat{\boldsymbol{\beta}}} = (\boldsymbol{X}'\boldsymbol{T}^{+}\boldsymbol{X})^{-1}\boldsymbol{X}'\boldsymbol{T}^{+}\boldsymbol{y}, \quad \boldsymbol{T} = \boldsymbol{V} + k\boldsymbol{X}\boldsymbol{X}' ,$$

where $\boldsymbol{T}^{+}$ is the Moore-Penrose inverse of $\boldsymbol{T}$ and k is an arbitrary nonnegative number which must be chosen strictly positive when $\mathcal{C}(\boldsymbol{X}) \not\subseteq \mathcal{C}(\boldsymbol{V})$, but can be chosen identical to zero when $\mathcal{C}(\boldsymbol{X}) \subseteq \mathcal{C}(\boldsymbol{V})$. The latter is always satisfied when $\boldsymbol{V}$ is nonsingular, showing that the estimator $\widehat{\widehat{\boldsymbol{\beta}}}$ coincides with the generalized least squares estimator for $\boldsymbol{\beta}$ then. On the other hand, the number k can always be chosen identical to 1, even when $\boldsymbol{V} = \boldsymbol{I}_n$, in which case

$$\widehat{\widehat{\boldsymbol{\beta}}} = (\boldsymbol{X}'(\boldsymbol{I}_n + \boldsymbol{X}\boldsymbol{X}')^{-1}\boldsymbol{X})^{-1}\boldsymbol{X}'(\boldsymbol{I}_n + \boldsymbol{X}\boldsymbol{X}')^{-1}\boldsymbol{y}$$

is simply a more involved representation of the ordinary least squares estimator $\widehat{\boldsymbol{\beta}} = (\boldsymbol{X}'\boldsymbol{X})^{-1}\boldsymbol{X}'\boldsymbol{y}$, see also Problem 5.8.

Theorem 5.7. *Under the linear regression model with assumptions (i) to (iii) and possibly singular covariance matrix* $\mathrm{Cov}(\boldsymbol{\varepsilon}) = \sigma^2\boldsymbol{V}$, *the estimator*

$$\widehat{\widehat{\boldsymbol{\beta}}} = (\boldsymbol{X}'\boldsymbol{T}^+\boldsymbol{X})^{-1}\boldsymbol{X}'\boldsymbol{T}^+\boldsymbol{y},\ \boldsymbol{T} = \boldsymbol{V} + k\boldsymbol{X}\boldsymbol{X}',\ k\begin{cases} > 0 & \text{if } \mathcal{C}(\boldsymbol{X}) \not\subseteq \mathcal{C}(\boldsymbol{V}) \\ \geq 0 & \text{otherwise} \end{cases},$$

is uniformly not worse than any estimator $\widetilde{\boldsymbol{\beta}}$ *from* $\mathcal{U}(\boldsymbol{\beta})$ *with respect to* $\mathrm{MSE}(\boldsymbol{\beta}, \widetilde{\boldsymbol{\beta}})$.

Proof. For any estimator $\widetilde{\boldsymbol{\beta}}$ in

$$\mathcal{U}(\boldsymbol{\beta}) = \{\boldsymbol{A}\boldsymbol{y} + \boldsymbol{a} : \boldsymbol{A}\boldsymbol{X} = \boldsymbol{I}_p,\ \boldsymbol{a} = \boldsymbol{0}\},$$

it follows that $\mathrm{MSE}(\boldsymbol{\beta}, \widetilde{\boldsymbol{\beta}}) = \mathrm{Cov}(\widetilde{\boldsymbol{\beta}})$. Since $\widehat{\widehat{\boldsymbol{\beta}}} \in \mathcal{U}(\boldsymbol{\beta})$, we can conclude analogously to Sect. 2.3.2, that the assertion holds true if $\widehat{\widehat{\boldsymbol{\beta}}}$ is uncorrelated with every linear zero function, being any linear transformation $\boldsymbol{A}\boldsymbol{y}$ with $\boldsymbol{A}\boldsymbol{X} = \boldsymbol{0}$. Now,

$$\mathrm{Cov}(\widehat{\widehat{\boldsymbol{\beta}}}, \boldsymbol{A}\boldsymbol{y}) = \sigma^2(\boldsymbol{X}'\boldsymbol{T}^+\boldsymbol{X})^{-1}\boldsymbol{X}'\boldsymbol{T}^+\boldsymbol{V}\boldsymbol{A}'.$$

By writing $\boldsymbol{V} = \boldsymbol{T} - k\boldsymbol{X}\boldsymbol{X}'$ and using $\boldsymbol{X}'\boldsymbol{T}^+\boldsymbol{T} = \boldsymbol{X}'$, it follows

$$\mathrm{Cov}(\widehat{\widehat{\boldsymbol{\beta}}}, \boldsymbol{A}\boldsymbol{y}) = \sigma^2(\boldsymbol{X}'\boldsymbol{T}^+\boldsymbol{X})^{-1}(\boldsymbol{I}_p - k\boldsymbol{X}'\boldsymbol{T}^+\boldsymbol{X})\boldsymbol{X}'\boldsymbol{A}',$$

which in view of $\boldsymbol{X}'\boldsymbol{A}' = \boldsymbol{0}$ equals the zero matrix. This shows the assertion. □

The above theorem can be seen as a further generalization of Theorem 5.5, and may also be called *generalized Gauss-Markov Theorem.*

5.5 Equality of Ordinary and Generalized Least Squares

When in the linear regression model with assumptions (i) to (iv∗), the ordinary least squares estimator $\widehat{\boldsymbol{\beta}} = (\boldsymbol{X}'\boldsymbol{X})^{-1}\boldsymbol{X}'\boldsymbol{y}$ coincides with the generalized least squares estimator $\widehat{\boldsymbol{\beta}}_* = (\boldsymbol{X}'\boldsymbol{V}^{-1}\boldsymbol{X})^{-1}\boldsymbol{X}'\boldsymbol{V}^{-1}\boldsymbol{y}$, then we neither have to compute $\boldsymbol{V}^{-1}$, nor do we have to apply some two-stage procedure when $\boldsymbol{V}$ is not completely known. We can simply use the ordinary least squares estimator $\widehat{\boldsymbol{\beta}}$ not depending on $\boldsymbol{V}$.

Although this coincidence rarely occurs, it is of interest to have an easy as possible characterization.

Theorem 5.8. *Under the linear regression model with assumptions (i) to (iv*), the identity*

$$(\boldsymbol{X}'\boldsymbol{V}^{-1}\boldsymbol{X})^{-1}\boldsymbol{X}'\boldsymbol{V}^{-1}\boldsymbol{y} = (\boldsymbol{X}'\boldsymbol{X})^{-1}\boldsymbol{X}'\boldsymbol{y}$$

holds true for every $\boldsymbol{y} \in \mathbb{R}^n$ *if and only if*

$$\mathcal{C}(\boldsymbol{V}\boldsymbol{X}) \subseteq \mathcal{C}(\boldsymbol{X}),$$

where $\mathcal{C}(\cdot)$ *denote the column space of a matrix.*

Proof. Let $(\boldsymbol{X}'\boldsymbol{V}^{-1}\boldsymbol{X})^{-1}\boldsymbol{X}'\boldsymbol{V}^{-1}\boldsymbol{y} = (\boldsymbol{X}'\boldsymbol{X})^{-1}\boldsymbol{X}'\boldsymbol{y}$ be true for every $\boldsymbol{y} \in \mathbb{R}^n$. Then

$$(\boldsymbol{X}'\boldsymbol{V}^{-1}\boldsymbol{X})^{-1}\boldsymbol{X}'\boldsymbol{V}^{-1} = (\boldsymbol{X}'\boldsymbol{X})^{-1}\boldsymbol{X}',$$

implying that

$$\boldsymbol{X}(\boldsymbol{X}'\boldsymbol{V}^{-1}\boldsymbol{X})^{-1}\boldsymbol{X}'\boldsymbol{V}^{-1} = \boldsymbol{X}(\boldsymbol{X}'\boldsymbol{X})^{-1}\boldsymbol{X}'.$$

Transposing both sides and the premultiplying by $\boldsymbol{V}$ yields

$$\boldsymbol{X}(\boldsymbol{X}'\boldsymbol{V}^{-1}\boldsymbol{X})^{-1}\boldsymbol{X}' = \boldsymbol{V}\boldsymbol{X}(\boldsymbol{X}'\boldsymbol{X})^{-1}\boldsymbol{X}'.$$

Now, since $\mathcal{C}(\boldsymbol{V}\boldsymbol{X}(\boldsymbol{X}'\boldsymbol{X})^{-1}\boldsymbol{X}') = \mathcal{C}(\boldsymbol{V}\boldsymbol{X})$ it follows that

$$\mathcal{C}(\boldsymbol{V}\boldsymbol{X}) = \mathcal{C}(\boldsymbol{X}(\boldsymbol{X}'\boldsymbol{V}^{-1}\boldsymbol{X})^{-1}\boldsymbol{X}') \subseteq \mathcal{C}(\boldsymbol{X}).$$

Conversely, let $\mathcal{C}(\boldsymbol{V}\boldsymbol{X}) \subseteq \mathcal{C}(\boldsymbol{X})$ be satisfied. Since $\boldsymbol{V}$ is nonsingular, the identity $\mathrm{rk}(\boldsymbol{V}\boldsymbol{X}) = \mathrm{rk}(\boldsymbol{X})$ is true, showing that $\mathcal{C}(\boldsymbol{V}\boldsymbol{X}) \subseteq \mathcal{C}(\boldsymbol{X})$ is satisfied if and only if $\mathcal{C}(\boldsymbol{V}\boldsymbol{X}) = \mathcal{C}(\boldsymbol{X})$ is valid. This identity is equivalent to $\mathcal{C}(\boldsymbol{X}) = \mathcal{C}(\boldsymbol{V}^{-1}\boldsymbol{X})$, which in turn is satisfied if and only $\mathcal{C}(\boldsymbol{V}^{-1}\boldsymbol{X}) \subseteq \mathcal{C}(\boldsymbol{X})$. Then $\boldsymbol{V}^{-1}\boldsymbol{X} = \boldsymbol{X}\boldsymbol{G}$ for some matrix $\boldsymbol{G}$, implying that $\boldsymbol{X}'\boldsymbol{V}^{-1} = \boldsymbol{G}'\boldsymbol{X}'$. Then

$$\boldsymbol{X}'\boldsymbol{V}^{-1} = \boldsymbol{G}'\boldsymbol{X}' = \boldsymbol{G}'\boldsymbol{X}'\boldsymbol{X}(\boldsymbol{X}'\boldsymbol{X})^{-1}\boldsymbol{X}' = \boldsymbol{X}'\boldsymbol{V}^{-1}\boldsymbol{X}(\boldsymbol{X}'\boldsymbol{X})^{-1}\boldsymbol{X}'.$$

This implies $(\boldsymbol{X}'\boldsymbol{V}^{-1}\boldsymbol{X})^{-1}\boldsymbol{X}'\boldsymbol{V}^{-1} = (\boldsymbol{X}'\boldsymbol{X})^{-1}\boldsymbol{X}'$, thus showing the assertion. □

The condition $\mathcal{C}(\boldsymbol{V}\boldsymbol{X}) \subseteq \mathcal{C}(\boldsymbol{X})$ is often referred in the literature as *the invariance of* $\mathcal{C}(\boldsymbol{X})$ *under* $\boldsymbol{V}$. As noted in the proof of Theorem 5.8, when $\boldsymbol{V}$ is nonsingular, then the four conditions

$$\mathcal{C}(\boldsymbol{V}\boldsymbol{X}) \subseteq \mathcal{C}(\boldsymbol{X}), \mathcal{C}(\boldsymbol{V}\boldsymbol{X}) = \mathcal{C}(\boldsymbol{X}), \mathcal{C}(\boldsymbol{V}^{-1}\boldsymbol{X}) = \mathcal{C}(\boldsymbol{X}), \mathcal{C}(\boldsymbol{V}^{-1}\boldsymbol{X}) \subseteq \mathcal{C}(\boldsymbol{X})$$

are equivalent to each other. In the general case, when $\boldsymbol{V}$ can be singular, then $\mathcal{C}(\boldsymbol{V}\boldsymbol{X}) \subseteq \mathcal{C}(\boldsymbol{X})$ is not necessarily equivalent to $\mathcal{C}(\boldsymbol{V}\boldsymbol{X}) = \mathcal{C}(\boldsymbol{X})$. Nonetheless, the condition $\mathcal{C}(\boldsymbol{V}\boldsymbol{X}) \subseteq \mathcal{C}(\boldsymbol{X})$ is necessary and sufficient for

$$\widehat{\boldsymbol{X}\boldsymbol{\beta}} = \boldsymbol{X}(\boldsymbol{X}'\boldsymbol{X})^{+}\boldsymbol{X}'\boldsymbol{y}$$

to be the best linear unbiased estimator for $\boldsymbol{X\beta}$ even when $\boldsymbol{V}$ is singular and $\boldsymbol{X}$ is not of full column rank, see also [84] for a discussion of further equivalent conditions.

When the matrix $\boldsymbol{V}$ is completely unknown, then it is not possible to check the condition $\mathcal{C}(\boldsymbol{VX}) \subseteq \mathcal{C}(\boldsymbol{X})$ and one does not know whether ordinary and generalized least squares estimator coincide. However, usually the matrix $\boldsymbol{V}$ is at least partially known and there exist instances when the condition $\mathcal{C}(\boldsymbol{VX}) \subseteq \mathcal{C}(\boldsymbol{X})$ can be verified. Consider for example the matrix $\boldsymbol{V} = (1 - \varrho)\boldsymbol{I}_n + \varrho \boldsymbol{1}_n \boldsymbol{1}_n'$, occurring in the equicorrelated case. Then

$$\boldsymbol{VX} = (1 - \varrho)\boldsymbol{X} + \varrho \boldsymbol{1}_n \boldsymbol{1}_n' \boldsymbol{X} .$$

If in addition $\boldsymbol{1}_n \in \mathcal{C}(\boldsymbol{X})$, then there exists some vector $\boldsymbol{a}$ such that $\boldsymbol{1}_n = \boldsymbol{Xa}$, giving

$$\boldsymbol{VX} = \boldsymbol{Xb}, \quad \boldsymbol{b} = (1 - \varrho)\boldsymbol{I}_n + \varrho \boldsymbol{a} \boldsymbol{1}_n' \boldsymbol{X} .$$

This shows that the condition $\mathcal{C}(\boldsymbol{VX}) \subseteq \mathcal{C}(\boldsymbol{X})$ is in fact satisfied and we may apply the ordinary least squares estimator for estimating $\boldsymbol{\beta}$.

5.6 Problems

5.1. Under the linear regression model with assumptions (i) to (iv) and (v), consider the maximum likelihood estimator

$$\widehat{\sigma}^2_{\mathrm{ML}} = \frac{1}{n} \boldsymbol{y}'\boldsymbol{M}\boldsymbol{y}, \quad \boldsymbol{M} = \boldsymbol{I}_n - \boldsymbol{X}(\boldsymbol{X}'\boldsymbol{X})^{-1}\boldsymbol{X}' ,$$

for σ^2.

(a) Determine the unweighted squared error risk $\rho(\sigma^2, \widehat{\sigma}^2_{\mathrm{ML}}) = \mathrm{E}[(\widehat{\sigma}^2_{\mathrm{ML}} - \sigma^2)^2]$ of the maximum likelihood estimator for σ^2.
(b) Show that for $p \neq 2$ the maximum likelihood estimator $\widehat{\sigma}^2_{\mathrm{ML}}$ is inadmissible by verifying that

$$\rho(\sigma^2, \widehat{\sigma}^2_{\mathrm{T}}) < \rho(\sigma^2, \widehat{\sigma}^2_{\mathrm{ML}})$$

for every $(\boldsymbol{\beta}, \sigma^2) \in \mathrm{I\!R}^p \times (0, \infty)$.

5.2. Show that under the linear regression model with assumptions (i) to (iv) and (v) the unweighted squared error risk of the sample variance $\widehat{\sigma}^2_{\boldsymbol{y}}$ is given by

$$\rho(\sigma^2, \widehat{\sigma}^2_{\boldsymbol{y}}) = \frac{2\sigma^2}{n-1}(\sigma^2 + 2b) + b^2 ,$$

where $b = \mathrm{E}(\widehat{\sigma}^2_{\boldsymbol{y}}) - \sigma^2 = \frac{1}{n-1}\boldsymbol{\beta}'\boldsymbol{X}'\boldsymbol{C}\boldsymbol{X}\boldsymbol{\beta}$ is the bias of $\widehat{\sigma}^2_{\boldsymbol{y}}$ and $\boldsymbol{C} = \boldsymbol{I}_n - \frac{1}{n}\boldsymbol{1}_n\boldsymbol{1}_n'$ is the centering matrix.

5.3. Let $\boldsymbol{V}$ be an $n \times n$ symmetric positive definite matrix, and let $\boldsymbol{F}$ be a nonsingular $n \times n$ matrix with $\boldsymbol{F}'\boldsymbol{F} = \boldsymbol{V}^{-1}$. Show that $\boldsymbol{F}\boldsymbol{V}\boldsymbol{F}' = \boldsymbol{I}_n$.

5.4. Consider the linear regression model with assumptions (i) to (iii) and (iv∗).

(a) Show that the inequality

$$\boldsymbol{X}(\boldsymbol{X}'\boldsymbol{V}^{-1}\boldsymbol{X})^{-1}\boldsymbol{X}' \leq_{\mathrm{L}} \boldsymbol{V}$$

holds true. [Hint: Apply Theorem A.65.]

(b) Use the inequality from (a) to demonstrate that

$$\mathrm{Cov}(\widehat{\boldsymbol{\beta}}_*) \leq_{\mathrm{L}} \mathrm{Cov}(\widehat{\boldsymbol{\beta}})$$

holds true, where $\widehat{\boldsymbol{\beta}}_*$ and $\widehat{\boldsymbol{\beta}}$ are generalized and ordinary least squares estimator for $\boldsymbol{\beta}$, respectively.

5.5. Consider the linear regression model with assumptions (i) to (iii) and (iv∗).

(a) Show that $\widehat{\sigma}_*^2$ can be written as

$$\widehat{\sigma}_*^2 = \frac{1}{n-p}\boldsymbol{y}'(\boldsymbol{I}_n - \boldsymbol{X}(\boldsymbol{X}'\boldsymbol{V}^{-1}\boldsymbol{X})^{-1}\boldsymbol{X}'\boldsymbol{V}^{-1})\boldsymbol{y}\,.$$

(b) Show that $\mathrm{E}(\widehat{\sigma}_*^2) = \sigma^2$ for every $(\boldsymbol{\beta}, \sigma^2) \in \mathrm{I\!R}^p \times (0, \infty)$.

5.6. For the seemingly unrelated regression model consider the estimators $\widehat{\sigma}_{st} = (1/n)\boldsymbol{z}_s'\boldsymbol{z}_t$ with $\boldsymbol{z}_s = \boldsymbol{y}_s - \boldsymbol{X}_s(\boldsymbol{X}_s'\boldsymbol{X}_s)^{-1}\boldsymbol{X}_s\boldsymbol{y}_s$ for $s, t \in \{1, 2\}$. Show that $|\widehat{\sigma}_{12}| \leq \sqrt{\widehat{\sigma}_{11}\widehat{\sigma}_{22}}$.

5.7. (a) Consider the ratio

$$r(\boldsymbol{x}) := \frac{\boldsymbol{x}'\frac{1}{n}\boldsymbol{1}_n\boldsymbol{1}_n'\boldsymbol{x}}{\boldsymbol{x}'\boldsymbol{x}}$$

and use Theorem A.48 to show that $r(\boldsymbol{x})$ takes its values in the closed interval $[0, 1]$ as $\boldsymbol{x}$ varies over $\mathrm{I\!R}^n \setminus \{\boldsymbol{0}\}$.

(b) Consider the symmetric matrix $\boldsymbol{V} = (1-\varrho)\boldsymbol{I}_n + \varrho\boldsymbol{1}_n\boldsymbol{1}_n'$ (occurring in the case of equicorrelation) and show that $\boldsymbol{x}'\boldsymbol{V}\boldsymbol{x} > 0$ for every $\boldsymbol{x} \in \mathrm{I\!R}^n \setminus \{\boldsymbol{0}\}$ if and only if $-1/(n-1) < \varrho < 1$.

5.8. Let $\boldsymbol{X}$ be an $n \times p$ matrix of full column rank.

(a) Use Theorem A.23 to show that

$$(\boldsymbol{I}_n + \boldsymbol{X}\boldsymbol{X}')^{-1} = \boldsymbol{I}_n - \boldsymbol{X}(\boldsymbol{I}_p + \boldsymbol{X}'\boldsymbol{X})^{-1}\boldsymbol{X}'\,.$$

(b) Show that

$$(\boldsymbol{X}'(\boldsymbol{I}_n + \boldsymbol{X}\boldsymbol{X}')^{-1}\boldsymbol{X})^{-1} = (\boldsymbol{X}'\boldsymbol{X})^{-1}(\boldsymbol{I}_p + \boldsymbol{X}'\boldsymbol{X}) .$$

(c) Show that

$$\boldsymbol{X}'(\boldsymbol{I}_n + \boldsymbol{X}\boldsymbol{X}')^{-1} = (\boldsymbol{I}_p + \boldsymbol{X}'\boldsymbol{X})^{-1}\boldsymbol{X}' .$$

(d) Show that the ordinary least squares estimator $\widehat{\beta} = (\boldsymbol{X}'\boldsymbol{X})^{-1}\boldsymbol{X}'\boldsymbol{y}$ can be written as

$$\widehat{\beta} = (\boldsymbol{X}'(\boldsymbol{I}_n + \boldsymbol{X}\boldsymbol{X}')^{-1}\boldsymbol{X})^{-1}\boldsymbol{X}'(\boldsymbol{I}_n + \boldsymbol{X}\boldsymbol{X}')^{-1}\boldsymbol{y} .$$

5.9. Let $\boldsymbol{V}$ be an $n \times n$ symmetric nonnegative definite matrix, and let $\boldsymbol{X}$ be an $n \times p$ matrix with $n > p$. Show that the condition $\mathcal{C}(\boldsymbol{V}\boldsymbol{X}) \subseteq \mathcal{C}(\boldsymbol{X})$ is equivalent to $\boldsymbol{P}\boldsymbol{V} = \boldsymbol{V}\boldsymbol{P}$, where $\boldsymbol{P} = \boldsymbol{X}\boldsymbol{X}^+$ is the orthogonal projector onto $\mathcal{C}(\boldsymbol{X})$.

5.10. Consider the linear regression model with assumptions (i) to (iii) and (iv$*$) with

$$\boldsymbol{V} = k^2\boldsymbol{I}_n + \boldsymbol{V}_1 + \boldsymbol{V}_2 ,$$

where $k \in \mathbb{R}$, $\boldsymbol{V}_1$ is symmetric nonnegative definite with $\mathcal{C}(\boldsymbol{V}_1) \subseteq \mathcal{C}(\boldsymbol{X})$, and $\boldsymbol{V}_2$ is symmetric nonnegative definite with $\mathcal{C}(\boldsymbol{V}_2) \subseteq \mathcal{C}(\boldsymbol{X})^\perp$.

(a) Explain why $\boldsymbol{V} = (1-\varrho)\boldsymbol{I}_n + \varrho\boldsymbol{1}_n\boldsymbol{1}_n'$ is a special case of the above matrix $\boldsymbol{V}$.
(b) Show that ordinary and generalized least squares estimator coincide.

6 Regression Diagnostics

When linear regression methods are applied to given data, useful results can be expected when the chosen model is considerably plausible, meaning that no substantial indications for inconsistencies and violation of model assumptions can be found. This concerns e.g. the choice and number of incorporated variables, the goodness of fit, the degree of collinearity, the distribution of the errors (normality, homogeneity, uncorrelatedness), the linearity of the assumed relationship, and the influence of individual observations.

In order to find indication for the plausibility of a linear regression model, several diagnostic procedures are known. In this chapter a short review of some of these is presented.

6.1 Selecting Independent Variables

Suppose that we have identified a dependent variable y which should be explained by independent variables from a set $\{x_1, \ldots, x_s\}$ via a linear regression model. Then, which variables out of this set should be employed? Quite naturally, we would like to include only important independent variables, so that from a 'rather small' number of independent variables the dependent variable can predicted with 'sufficient preciseness'.

There may, of course, be formal reasons for choosing a variable, e.g. economic theory might suggest a relationship between y and a specific x_j, but, in general, one also wishes to measure how 'good' a specific model is in the above sense.

6.1.1 Mallows' C_p

Consider a linear regression model, described by $\boldsymbol{y} = \boldsymbol{X}\boldsymbol{\beta} + \boldsymbol{\varepsilon}$ with $\boldsymbol{X} = (\boldsymbol{X}_1, \boldsymbol{X}_2)$, where $\boldsymbol{X}_1$ is an $n \times p$ matrix and $\boldsymbol{X}_2$ is an $n \times q$ matrix, so that

$$\boldsymbol{y} = \boldsymbol{X}_1\boldsymbol{\beta}_1 + \boldsymbol{X}_2\boldsymbol{\beta}_2 + \boldsymbol{\varepsilon}, \quad \boldsymbol{\varepsilon} \sim (\boldsymbol{0}, \sigma^2\boldsymbol{I}_n) ,$$

where $\boldsymbol{\beta} = (\boldsymbol{\beta}_1', \boldsymbol{\beta}_2')'$. Suppose we predict $\boldsymbol{y}$ from the reduced model

$$\boldsymbol{y} = \boldsymbol{X}_1\boldsymbol{\beta}_1 + \boldsymbol{u}, \quad \boldsymbol{u} \sim (\boldsymbol{0}, \sigma^2\boldsymbol{I}_n) ,$$

as $\widehat{\boldsymbol{y}} = \boldsymbol{X}_1\widehat{\boldsymbol{\beta}}_1$ with $\widehat{\boldsymbol{\beta}}_1 = (\boldsymbol{X}_1'\boldsymbol{X}_1)^{-1}\boldsymbol{X}_1'\boldsymbol{y}$. Since prediction of $\boldsymbol{y}$ can also be seen as the estimation of $\mathrm{E}(\boldsymbol{y}) = \boldsymbol{X}\boldsymbol{\beta}$, the expected unweighted squared error loss of $\widehat{\boldsymbol{y}}$ is given by

$$\Gamma_p := \mathrm{E}\left[(\boldsymbol{X}_1\widehat{\boldsymbol{\beta}}_1 - \boldsymbol{X}\boldsymbol{\beta})'(\boldsymbol{X}_1\widehat{\boldsymbol{\beta}}_1 - \boldsymbol{X}\boldsymbol{\beta})\right] .$$

Now, the reduced model $\boldsymbol{y} = \boldsymbol{X}_1\boldsymbol{\beta}_1 + \boldsymbol{u}$ may be considered as an appropriate model if Γ_p is rather small. Writing $\boldsymbol{P}_1 = \boldsymbol{X}_1(\boldsymbol{X}_1'\boldsymbol{X}_1)^{-1}\boldsymbol{X}_1'$, we compute

$$\Gamma_p = \mathrm{E}(\boldsymbol{y}'\boldsymbol{P}_1\boldsymbol{y}) - 2\boldsymbol{\beta}'\boldsymbol{X}'\boldsymbol{P}_1\boldsymbol{X}\boldsymbol{\beta} + \boldsymbol{\beta}'\boldsymbol{X}'\boldsymbol{X}\boldsymbol{\beta}$$

with

$$\mathrm{E}(\boldsymbol{y}'\boldsymbol{P}_1\boldsymbol{y}) = \sigma^2 \underbrace{\mathrm{tr}(\boldsymbol{P}_1)}_{=p} + \boldsymbol{\beta}'\boldsymbol{X}'\boldsymbol{P}_1\boldsymbol{X}\boldsymbol{\beta} .$$

Then

$$\Gamma_p = \sigma^2 p + \boldsymbol{\beta}'\boldsymbol{X}'\boldsymbol{M}_1\boldsymbol{X}\boldsymbol{\beta}, \quad \boldsymbol{M}_1 = \boldsymbol{I}_n - \boldsymbol{P}_1 .$$

Since we have $\mathrm{E}(\boldsymbol{y}'\boldsymbol{M}_1\boldsymbol{y}) = \sigma^2\,\mathrm{tr}(\boldsymbol{M}_1) + \boldsymbol{\beta}'\boldsymbol{X}'\boldsymbol{M}_1\boldsymbol{X}\boldsymbol{\beta}$ with $\mathrm{tr}(\boldsymbol{M}_1) = n - p$, it follows that

$$\Gamma_p = \sigma^2(2p - n) + \mathrm{E}(\boldsymbol{y}'\boldsymbol{M}_1\boldsymbol{y}) .$$

Clearly, Γ_p depends on the unknowns $\boldsymbol{\beta}$ and σ^2 and thus we cannot say whether Γ_p is small or not. However, a natural estimator is given by

$$\widehat{\Gamma}_p = \widehat{\sigma}^2(2p - n) + \mathrm{RSS}_p ,$$

where $\widehat{\sigma}^2$ is an estimator for σ^2 and $\mathrm{RSS}_p = \boldsymbol{y}'\boldsymbol{M}_1\boldsymbol{y}$ is the residual sum of squares in the reduced model. The rescaled quantity

$$C_p = (2p - n) + \frac{\mathrm{RSS}_p}{\widehat{\sigma}^2}$$

is known as Mallows' C_p for the model $\boldsymbol{y} = \boldsymbol{X}_1\boldsymbol{\beta}_1 + \boldsymbol{u}$, see [74]. Usually, the estimator

$$\widehat{\sigma}^2 = \frac{1}{n-p-q}(\boldsymbol{y} - \boldsymbol{X}\widehat{\boldsymbol{\beta}})'(\boldsymbol{y} - \boldsymbol{X}\widehat{\boldsymbol{\beta}}), \quad \widehat{\boldsymbol{\beta}} = (\boldsymbol{X}'\boldsymbol{X})^{-1}\boldsymbol{X}'\boldsymbol{y} ,$$

from the original model is used for estimating σ^2.

When we consider different reduced models (including the non-reduced original model), then the models with the smallest C_p appear to be appropriate. When the above estimator for σ^2 is used, then we have $C_p = p$ in the non-reduced model.

6.1.2 Stepwise Regression

Stepwise regression procedures delete or add one variable at a time and evaluate the resulting model. The model which performs best is then chosen. An algorithm based on Mallows' C_p can be outlined as follows.

Stepwise Regression (Backward Selection).

Stage 1. Consider the linear regression model with all independent variables available. Compute the least squares variance estimate $\widehat{\sigma}^2$ for σ^2 under this model and compute C_p (being identical to p).
Stage 2. Delete one variable and compute C_p in the reduced model. Repeat this procedure for every independent variable in the model. Choose the model with the smallest C_p (including the starting model).
Stage 3. The chosen model from stage 2 is the new starting model but with $\widehat{\sigma}^2$ from above *not* being recomputed. Goto stage 2 until the new starting model does not change any more.

The outlined procedure can also be done in a forward manner. The starting model is then the linear regression model containing only the intercept. However, in this case the estimator $\widehat{\sigma}^2$ coincides with the sample variance, being a rather weak estimator for σ^2 when the true regression involves several (non-constant) regressors, as it is usually the case. Backward and forward selection do not necessarily yield the same model.

The above algorithm essentially describes the effect of the function `step` in S-PLUS. The only difference is that `step` does not compute C_p but

$$\widetilde{C}_p = \widehat{\sigma}^2(C_p + n) ,$$

being obviously equivalent to C_p for comparison purposes and thus yielding the same results. (In addition to backward and forward selection, `step` also allows a mixture between both.)

Remark 6.1. Stepwise regression methods can be criticized for various reasons. They handle the independent variables one at a time and thus may fail to yield the best model with respect to a given criterion. In addition, the resulting model is chosen on the basis of the observation $\boldsymbol{y}$ and could be quite different for other observations, see also the discussion in [21, Sect. 14.3]. Nonetheless, they may be seen as an easy aid when the user is confronted with the problem of fixing a model equation.

6.1.3 Alternative Criteria

Instead of Mallows' C_p, alternative criteria for selecting independent variables may be applied.

AIC Criterion. Although S-PLUS and R are both based on the S language, the function `step` is different under R. As a matter of fact, the criterion for choosing a model is not C_p but AIC_p (An Information Criterion), given as

$$\mathrm{AIC}_p = n \ln(\mathrm{RSS}_p / n) + 2p ,$$

where $\text{RSS}_p = \boldsymbol{y}'\boldsymbol{M}_1\boldsymbol{y}$ is the residual sum of squares from the model $\boldsymbol{y} = \boldsymbol{X}_1\boldsymbol{\beta}_1 + \boldsymbol{u}$ as before. AIC has been introduced by Akaike [2] (and is thus also called Akaike's Information Criterion) as

$$\text{AIC} = -2\,\text{maximized log-likelihood} + 2(\#\text{parameters}) .$$

Under the linear regression model with assumptions (i) to (iv) and (v), the likelihood, i.e. the density of $\boldsymbol{y}$ considered as a function of $\boldsymbol{\beta}$ and σ^2, is given as

$$L(\boldsymbol{\beta}, \sigma^2; \boldsymbol{y}) = \frac{1}{(2\pi\sigma^2)^{n/2}} \exp\left\{-\frac{1}{2}\frac{(\boldsymbol{y}-\boldsymbol{X\beta})'(\boldsymbol{y}-\boldsymbol{X\beta})}{\sigma^2}\right\} ,$$

and hence the log-likelihood is

$$L^*(\boldsymbol{\beta}, \sigma^2; \boldsymbol{y}) = -\frac{n}{2}\ln(2\pi) - \frac{n}{2}\ln(\sigma^2) - \frac{1}{2}\frac{(\boldsymbol{y}-\boldsymbol{X\beta})'(\boldsymbol{y}-\boldsymbol{X\beta})}{\sigma^2} .$$

This function is maximized with respect to $\boldsymbol{\beta}$ and σ^2 for $\boldsymbol{\beta} \doteq \widehat{\boldsymbol{\beta}}$ and $\sigma^2 \doteq \frac{n-p}{n}\widehat{\sigma}^2$, where $\widehat{\boldsymbol{\beta}}$ and $\widehat{\sigma}^2$ are ordinary least squares estimator and least squares variance estimator, respectively. Then

$$\begin{aligned} \text{AIC} &= -2L^*(\widehat{\boldsymbol{\beta}}, \tfrac{n-p}{n}\widehat{\sigma}^2; \boldsymbol{y}) + 2p \\ &= n\ln(\text{RSS}\,/n) + 2p + n(\ln(2\pi) + 1) , \end{aligned}$$

where $\text{RSS} = \boldsymbol{y}'\boldsymbol{M}\boldsymbol{y}$ with $\boldsymbol{M} = \boldsymbol{I}_n - \boldsymbol{X}(\boldsymbol{X}'\boldsymbol{X})^{-1}\boldsymbol{X}'$. Since the additive term $n(\ln(2\pi)+1)$ only contains the number of observations n as information from the model, and n remains the same when models with different parameters but the same observation $\boldsymbol{y}$ are compared, the term $n(\ln(2\pi)+1)$ is irrelevant to comparison of different AICs. Thus we may consider AIC_p from above as a criterion for model selection.

One may argue that AIC_p is advantageous over C_p since the former is a widely accepted criterion and does only depend on n, p and RSS_p, whereas C_p additionally depends on the scale factor $\widehat{\sigma}^2$. The latter may lead to misjudgments when $\widehat{\sigma}^2$ is far away from the true σ^2.

The above outlined algorithm for stepwise regression can of course easily be adapted to work with AIC_p instead of C_p.

Example 6.1. Consider the cement data given in Table 1.1, p. 4. We have a dependent variable y (heat), four independent variables $x_1, \ldots, x_4$ (ingredients) and the intercept variable $\imath$ at our disposal. Suppose we wish to decide which of the independent variables (including the intercept) to incorporate into a linear regression model on the basis of stepwise regression. Applying the S-PLUS function `step` for backward selection, starting from a model with variables $\{\imath, x_1, x_2, x_3, x_4\}$, gives an estimate $\widehat{\sigma}^2 = 5.9830$ and results in the model with variables $\{\imath, x_1, x_2\}$. (Note that if $\imath$ should be treated as any other independent variable, then a linear model in S-PLUS and R should be defined without the prescribed intercept but with an extra variable having only ones as values.)

Applying the same function to forward selection, starting from a model with variable $\imath$, yields $\widehat{\sigma}^2 = 226.3136$ and results in the model with variables $\{\imath, x_1, x_4\}$. This shows that backward and forward selection do not necessarily yield the same models. The values $\widetilde{C}_p = 93.802$ and $\widetilde{C}_p = 1432.644$ from the two resulting models are not comparable to each other since they are related to the different estimates of σ^2.

Applying the R function `step` for backward selection results in the model with variables $\{\imath, x_1, x_2, x_4\}$ and $\text{AIC}_p = 24.97$ for this model. The forward selection produces the same model with the same AIC_p. This shows that the two criteria C_p and AIC_p do not necessarily yield the same models.

When we compute the AIC_p for the model with variables $\{\imath, x_1, x_2\}$, then $\text{AIC}_p = 25.42$, while for the model with variables $\{\imath, x_1, x_4\}$ we obtain $\text{AIC}_p = 28.74$. Different AIC_ps are comparable to each other and 25.42 is somewhat close to 24.97. Hence, if we prefer a simple model, then AIC_p does not strongly act against a model with $\{\imath, x_1, x_2\}$, and we may end up with this choice. The same model is also recommended in [99, p. 367]. □

Least Squares Variance Estimation. A further decision rule, which can also be performed in a stepwise manner, is simply 'choose the model with the smallest least squares variance estimate'. This rule is based on the observation that the least squares variance estimator in biased upwardly whenever the linear model is misspecified. To see this, assume that

$$y = X\beta + \varepsilon, \quad \varepsilon \sim (\mathbf{0}, \sigma^2 I_n) ,$$

is the true model, but σ^2 is estimated from the model

$$y = Z\gamma + u, \quad u \sim (\mathbf{0}, \sigma^2 I_n) ,$$

as

$$\widehat{\sigma}^2 = \frac{1}{n-q} y' M_Z y, \quad M_Z = I_n - Z(Z'Z)^{-1}Z' ,$$

where $q < n$ is the number of columns of Z. Then

$$\begin{aligned} \text{E}(\widehat{\sigma}^2) &= \frac{1}{n-q}(\sigma^2 \underbrace{\text{tr}(M_Z)}_{=n-q} + \beta' X' M_Z X \beta) \\ &= \sigma^2 + \frac{1}{n-q} \beta' X' M_Z X \beta , \end{aligned}$$

where $\beta' X' M_Z X \beta / (n-q) \geq 0$. Hence, the estimate of σ^2 from an incorrect model is expected to be greater than the true σ^2.

Choosing a model on the basis of the smallest least squares variance estimate is identical to choosing a model on the basis of the greatest *adjusted coefficient of determination*, see Remark 6.3 in the following section.

6.2 Assessing Goodness of Fit

The coefficient of determination can be seen as a measure on how good the independent variables determine the dependent variable. It may thus also be seen as a measure for the quality of the considered model. Nonetheless, the comparison of different regression models with the help of the coefficient of determination needs some caution.

Classical Coefficient of Determination. Under the linear regression model $\boldsymbol{y} = \boldsymbol{X}\boldsymbol{\beta} + \boldsymbol{\varepsilon}$, $\boldsymbol{\varepsilon} \sim (\boldsymbol{0}, \sigma^2 \boldsymbol{I}_n)$, we have already introduced the classical coefficient of determination

$$R^2 = \frac{||\boldsymbol{X}\widehat{\boldsymbol{\beta}}||^2}{||\boldsymbol{y}||^2}, \quad \boldsymbol{y} \neq \boldsymbol{0},$$

putting the prediction $\widehat{\boldsymbol{y}} = \boldsymbol{X}\widehat{\boldsymbol{\beta}}$ of $\boldsymbol{y}$ in relation to the actual observation $\boldsymbol{y}$. The closer $\widehat{\boldsymbol{y}}$ is to $\boldsymbol{y}$ the greater is R^2. It is easy to see that

$$0 \leq R^2 \leq 1 .$$

Nonnegativity follows, since nominator and denominator are nonnegative and positive (for $\boldsymbol{y} \neq \boldsymbol{0}$), respectively. Moreover, the inequality $R^2 \leq 1$, i.e. $||\boldsymbol{X}\widehat{\boldsymbol{\beta}}||^2 \leq ||\boldsymbol{y}||^2$ can equivalently be expressed as $\boldsymbol{y}'\boldsymbol{P}\boldsymbol{y} \leq \boldsymbol{y}'\boldsymbol{y}$, where $\boldsymbol{P} = \boldsymbol{X}(\boldsymbol{X}'\boldsymbol{X})^{-1}\boldsymbol{X}'$. But since $\boldsymbol{I}_n - \boldsymbol{P}$ is symmetric nonnegative definite, it follows $0 \leq \boldsymbol{y}'(\boldsymbol{I}_n - \boldsymbol{P})\boldsymbol{y}$ for any $\boldsymbol{y}$, showing $R^2 \leq 1$.

Centered Coefficient of Determination. When the linear regression model $\boldsymbol{y} = \boldsymbol{X}\boldsymbol{\beta} + \boldsymbol{\varepsilon}$ contains an intercept, then often the centered coefficient of determination

$$R_{\mathrm{c}}^2 = \frac{||\boldsymbol{X}\widehat{\boldsymbol{\beta}}||^2 - ||\boldsymbol{1}_n\widehat{\mu}||^2}{||\boldsymbol{y}||^2 - ||\boldsymbol{1}_n\widehat{\mu}||^2}$$

is considered. The vector $\boldsymbol{1}_n\widehat{\mu}$ is the prediction of $\boldsymbol{y}$ from the model

$$\boldsymbol{y} = \mu\boldsymbol{1}_n + \boldsymbol{u}, \quad \boldsymbol{u} \sim (\boldsymbol{0}, \sigma^2\boldsymbol{I}_n) ,$$

where $\widehat{\mu} = (1/n)\boldsymbol{1}_n'\boldsymbol{y} = \overline{y}$ is the ordinary least squares estimator for μ and thus $\boldsymbol{1}_n\widehat{\mu} = \boldsymbol{1}_n\overline{y}$. The centered coefficient of determination can also be written as

$$R_{\mathrm{c}}^2 = \frac{\boldsymbol{y}'\boldsymbol{P}\boldsymbol{y} - \boldsymbol{y}'\boldsymbol{P}_1\boldsymbol{y}}{\boldsymbol{y}'\boldsymbol{y} - \boldsymbol{y}'\boldsymbol{P}_1\boldsymbol{y}} ,$$

where $\boldsymbol{P} = \boldsymbol{X}(\boldsymbol{X}'\boldsymbol{X})^{-1}\boldsymbol{X}'$ and $\boldsymbol{P}_1 = (1/n)\boldsymbol{1}_n\boldsymbol{1}_n'$. Since the matrix $\boldsymbol{I}_n - \boldsymbol{P}_1$ (the centering matrix) is symmetric nonnegative definite, the denominator is zero if and only if $(\boldsymbol{I}_n - \boldsymbol{P}_1)\boldsymbol{y} = \boldsymbol{0}$, which can only occur if all elements of $\boldsymbol{y}$ are identical. The latter will of course not be the case for real data and can thus be excluded from the analysis.

Theorem 6.1. *Under the linear regression model with assumptions (i) to (iv), the following two statements hold true:*

(a) *The inequality $R_c^2 \le R^2$ is valid, with equality holding if and only if $\overline{y} = 0$ or $R^2 = 1$.*
(b) *If $\mathbf{1}_n$ belongs to the column space of $\boldsymbol{X}$, then $0 \le R_c^2$.*

Proof. To see statement (a), note that the inequality $R_c^2 \le R^2$ is equivalent to

$$(||\boldsymbol{X}\widehat{\boldsymbol{\beta}}||^2 - ||\mathbf{1}_n\overline{y}||^2)||\boldsymbol{y}||^2 \le (||\boldsymbol{y}||^2 - ||\mathbf{1}_n\overline{y}||^2)||\boldsymbol{X}\widehat{\boldsymbol{\beta}}||^2 ,$$

which can be simplified to

$$-||\mathbf{1}_n\overline{y}||^2||\boldsymbol{y}||^2 \le -||\mathbf{1}_n\overline{y}||^2||\boldsymbol{X}\widehat{\boldsymbol{\beta}}||^2 .$$

If $||\mathbf{1}_n\overline{y}||^2 = 0$, i.e. $\overline{y} = 0$, then this inequality is trivially satisfied, while otherwise it is equivalent to $||\boldsymbol{X}\widehat{\boldsymbol{\beta}}||^2 \le ||\boldsymbol{y}||^2$, which is always satisfied. Equality can occur only if $\overline{y} = 0$ or $||\boldsymbol{X}\widehat{\boldsymbol{\beta}}||^2 = ||\boldsymbol{y}||^2$, the latter being equivalent to $R^2 = 1$.

For statement (b) note that we have excluded the case of a zero denominator $\boldsymbol{y}'(\boldsymbol{I}_n - \boldsymbol{P}_1)\boldsymbol{y} \ge 0$. The nominator $\boldsymbol{y}'(\boldsymbol{P} - \boldsymbol{P}_1)\boldsymbol{y}$ is nonnegative, since the matrix $\boldsymbol{P} - \boldsymbol{P}_1$ is symmetric nonnegative definite. This follows from the fact that $\boldsymbol{P}_1 \le_{\mathrm{L}} \boldsymbol{P}$ if and only if $\mathcal{C}(\boldsymbol{P}_1) \subseteq \mathcal{C}(\boldsymbol{P})$, see Theorem A.69, the latter inclusion being satisfied if and only if $\mathbf{1}_n$ belongs to the column space of $\boldsymbol{X}$. □

Usually a linear regression model contains an intercept, and in that case the centered coefficient of determination is applied, satisfying

$$0 \le R_c^2 \le 1 .$$

The classical coefficient of determination can of course also be computed (but is in fact usually only applied for models without intercept) and will yield a greater value.

A fairly large R_c^2 can be seen as an indication for a 'good' model and hence it may also serve as a criterion for the comparison of different models with the same $\boldsymbol{y}$ as the vector of observations of the dependent variable. This is, however, only reasonable when the models in question contain *the same number of independent variables.*

The reason for this is, that adding independent variables to a linear regression model will usually increase (but never decrease) the value R_c^2. Thus, adding variables will always produce a better model with respect to R_c^2. To see this, consider the centered coefficients of determination $R_c^2(1)$ under the model $\boldsymbol{y} = \boldsymbol{X}_1\boldsymbol{\beta}_1 + \boldsymbol{u}$ and $R_c^2(2)$ under the model $\boldsymbol{y} = \boldsymbol{X}_1\boldsymbol{\beta}_1 + \boldsymbol{X}_2\boldsymbol{\beta}_2 + \boldsymbol{\varepsilon}$, respectively. Then

$$R_c^2(1) = \frac{\boldsymbol{y}'\boldsymbol{P}_{\boldsymbol{X}_1}\boldsymbol{y} - \boldsymbol{y}'\boldsymbol{P}_1\boldsymbol{y}}{\boldsymbol{y}'\boldsymbol{y} - \boldsymbol{y}'\boldsymbol{P}_1\boldsymbol{y}}, \quad \boldsymbol{P}_{\boldsymbol{X}_1} = \boldsymbol{X}_1(\boldsymbol{X}_1'\boldsymbol{X}_1)^{-1}\boldsymbol{X}_1'$$

and

$$R_c^2(2) = \frac{\boldsymbol{y}'\boldsymbol{P}\boldsymbol{y} - \boldsymbol{y}'\boldsymbol{P}_1\boldsymbol{y}}{\boldsymbol{y}'\boldsymbol{y} - \boldsymbol{y}'\boldsymbol{P}_1\boldsymbol{y}}, \quad \boldsymbol{P} = \boldsymbol{X}(\boldsymbol{X}'\boldsymbol{X})^{-1}\boldsymbol{X}'$$

for $\boldsymbol{X} = (\boldsymbol{X}_1, \boldsymbol{X}_2)$. Since the values $\boldsymbol{y}'\boldsymbol{y}$ and $\boldsymbol{y}'\boldsymbol{P}_1\boldsymbol{y}$ are the same in both models, the inequality

$$R_c^2(1) \leq R_c^2(2)$$

is equivalent to $\boldsymbol{y}'\boldsymbol{P}_{\boldsymbol{X}_1}\boldsymbol{y} \leq \boldsymbol{y}'\boldsymbol{P}\boldsymbol{y}$. The latter is always satisfied since the column space of $\boldsymbol{X}_1$ is clearly contained in the column space of $\boldsymbol{X} = (\boldsymbol{X}_1, \boldsymbol{X}_2)$, giving $\boldsymbol{P}_{\boldsymbol{X}_1} \leq_{\mathrm{L}} \boldsymbol{P}$, see Theorem A.69. A similar reasoning holds true for the classical coefficient of determination R^2.

Remark 6.2. Consider the comparison of two linear regression models with different independent variables but the same observation $\boldsymbol{y}$. When the number of independent variables in both models is identical, then one may choose the model with greater R_c^2 (or R^2). Otherwise neither R_c^2 nor R^2 provide an adequate criterion for selection.

Adjusted Coefficient of Determination. In the linear model $\boldsymbol{y} = \boldsymbol{X}\boldsymbol{\beta} + \boldsymbol{\varepsilon}$, it is easy to see that centered coefficient of determination can also be written as

$$R_c^2 = 1 - \frac{\boldsymbol{y}'\boldsymbol{M}\boldsymbol{y}}{\boldsymbol{y}'\boldsymbol{C}\boldsymbol{y}},$$

where $\boldsymbol{M} = \boldsymbol{I}_n - \boldsymbol{P}$ and $\boldsymbol{C} = \boldsymbol{I}_n - \boldsymbol{P}_1$. Then,

$$R_c^2 = 1 - \frac{(n-p)\widehat{\sigma}^2}{(n-1)\widehat{\sigma}_{\boldsymbol{y}}^2},$$

where $\widehat{\sigma}^2$ is the least squares variance estimator for σ^2 and $\widehat{\sigma}_{\boldsymbol{y}}^2$ is the sample variance. Adjusting for the respective degrees of freedom yields the *adjusted coefficient of determination*

$$\overline{R}^2 = 1 - \frac{\widehat{\sigma}^2}{\widehat{\sigma}_{\boldsymbol{y}}^2}.$$

It can also be written as

$$\overline{R}^2 = 1 - \frac{n-1}{n-p}(1 - R_c^2).$$

The adjusted coefficient of determination $\overline{R}^2$ will be smaller than R_c^2, but can also become negative even if the model contains an intercept.

Theorem 6.2. *Under the linear regression model with assumptions (i) to (iv), the following two statements hold true:*

(a) *If $p \geq 2$, then the inequality $\overline{R}^2 < R_c^2$ is satisfied if and only if $R_c^2 < 1$.*

(b) *The inequality* $\overline{R}^2 < 0$ *is satisfied if and only if* $R_c^2 < (p-1)/(n-1)$.

Proof. For statement (a) note that the inequality $\overline{R}^2 < R_c^2$ is equivalent to

$$1 - \frac{n-1}{n-p}(1 - R_c^2) < R_c^2 \,,$$

which is satisfied if and only if

$$\frac{1-p}{n-1} < \frac{1-p}{n-1} R_c^2 \,.$$

Since $p \geq 2$, the latter is equivalent to $R_c^2 < 1$. Statement (b) follows by noting that

$$1 - \frac{n-1}{n-p}(1 - R_c^2) < 0$$

is equivalent to $R_c^2 < (p-1)(n-1)$. □

The main difference between R_c^2 and $\overline{R}^2$ is, that $\overline{R}^2$ can also decrease when independent variables are added to the model. To see this, consider the adjusted coefficient of determination $\overline{R}^2(1)$ and the least squares variance estimate $\widehat{\sigma}^2(1)$ under the model $\boldsymbol{y} = \boldsymbol{X}_1\boldsymbol{\beta}_1 + \boldsymbol{u}$ as well as $\overline{R}^2(2)$ and $\widehat{\sigma}^2(2)$ under the model $\boldsymbol{y} = \boldsymbol{X}_1\boldsymbol{\beta}_1 + \boldsymbol{X}_2\boldsymbol{\beta}_2 + \varepsilon$, where the observation $\boldsymbol{y}$ is the same in both models. Then from the representation $\overline{R}^2 = 1 - \widehat{\sigma}^2/\widehat{\sigma}_{\boldsymbol{y}}^2$ it follows that

$$\overline{R}^2(1) \leq \overline{R}^2(2)$$

if and only if $\widehat{\sigma}^2(2) \leq \widehat{\sigma}^2(1)$. But the latter inequality must not necessarily be true. Especially when the model $\boldsymbol{y} = \boldsymbol{X}_1\boldsymbol{\beta}_1 + \boldsymbol{u}$ is close to the true model and $\boldsymbol{y} = \boldsymbol{X}_1\boldsymbol{\beta}_1 + \boldsymbol{X}_2\boldsymbol{\beta}_2 + \varepsilon$ is misspecified, then one may expect that the estimate $\widehat{\sigma}^2(1)$ is smaller than $\widehat{\sigma}^2(2)$ as noted in the previous section.

Remark 6.3. Consider the comparison of two linear regression models with different independent variables but the same observation $\boldsymbol{y}$. Then choosing the model with greater $\overline{R}^2$ is identical to choosing the model with smaller least squares variance estimate.

Table 6.1. R_c^2, $\overline{R}^2$, and AIC_p for different models in Example 6.2

	$\{\imath, x_1, x_2\}$	$\{\imath, x_1, x_4\}$	$\{\imath, x_1, x_2, x_4\}$	$\{\imath, x_1, x_2, x_3, x_4\}$
R_c^2	0.9787	0.9725	0.9823	0.9824
$\overline{R}^2$	0.9744	0.9670	0.9764	0.9736
AIC_p	25.42	28.74	24.97	26.94

Example 6.2. Consider again the cement data investigated in Example 6.1. The stepwise regression procedures yield models with variables $\{\imath, x_1, x_2\}$, $\{\imath, x_1, x_4\}$ and $\{\imath, x_1, x_2, x_4\}$ as possible results. We may also compare these models with respect to the adjusted coefficient of determination, see Table 6.1. When we consider the models with variables $\{\imath, x_1, x_2\}$ and $\{\imath, x_1, x_4\}$, then all three criteria R_c^2, $\overline{R}^2$, and AIC$_p$ act in favor of the model with $\{\imath, x_1, x_2\}$. It is not surprising that the criteria yield the same result here, since for models with the same number of parameters they are all equivalent, see Problem 6.1.

When we compare $\{\imath, x_1, x_2\}$ with $\{\imath, x_1, x_2, x_4\}$, then $\overline{R}^2$ and AIC$_p$ both act in favor of the model with variables $\{\imath, x_1, x_2, x_4\}$, but the difference to $\{\imath, x_1, x_2\}$ is not very large in both cases.

Table 6.1 also contains the respective values for the model with all variables $\{\imath, x_1, x_2, x_3, x_4\}$. When regarding R_c^2 for the different models, then it is confirmed that the values increase with increasing number of independent variables. This is however not necessarily the case for $\overline{R}^2$. As a matter of fact, the value of $\overline{R}^2$ is smaller for the model with variables $\{\imath, x_1, x_2, x_3, x_4\}$ than for the models with $\{\imath, x_1, x_2, x_4\}$ and with $\{\imath, x_1, x_2\}$. □

6.3 Diagnosing Collinearity

For the diagnose of collinearity a number of methods are available in the literature, see [10, Chap. 3] and [11] for a comprehensive treatment of this topic. In the following, we review two rather different approaches.

6.3.1 Variance Inflation Factors

When collinearity is present in the linear regression model with assumptions (i) to (iv), then some of the elements of the ordinary least squares estimator $\widehat{\boldsymbol{\beta}} = (\boldsymbol{X}'\boldsymbol{X})^{-1}\boldsymbol{X}'\boldsymbol{y}$ have a rather large variance. But how can this be measured?

Variance Inflation. To consider the variance of the j-th element of the ordinary least squares estimator $\mathrm{Var}(\widehat{\beta}_j)$ in more detail, we write

$$\boldsymbol{X} = (\boldsymbol{x}_j, \boldsymbol{X}_j) ,$$

where the elements of the $n \times 1$ vector $\boldsymbol{x}_j$ are the observations of the variable x_j and $\boldsymbol{X}_j$ is the observation matrix of the remaining variables. (Note that the order of the independent variables in the linear regression model is irrelevant to the ordinary least squares estimates of the corresponding parameters.) Then $\mathrm{Cov}(\widehat{\boldsymbol{\beta}}) = \sigma^2(\boldsymbol{X}'\boldsymbol{X})^{-1}$, where

$$(\boldsymbol{X}'\boldsymbol{X})^{-1} = \begin{pmatrix} \boldsymbol{x}_j'\boldsymbol{x}_j & \boldsymbol{x}_j'\boldsymbol{X}_j \\ \boldsymbol{X}_j'\boldsymbol{x}_j & \boldsymbol{X}_j'\boldsymbol{X}_j \end{pmatrix}^{-1} .$$

From Theorem A.24, the first main diagonal element of $(\boldsymbol{X}'\boldsymbol{X})^{-1}$ is given as

$$(\boldsymbol{x}_j'\boldsymbol{M}_j\boldsymbol{x}_j)^{-1}, \quad \boldsymbol{M}_j = \boldsymbol{I}_n - \boldsymbol{X}_j(\boldsymbol{X}_j'\boldsymbol{X}_j)^{-1}\boldsymbol{X}_j ,$$

implying

$$\mathrm{Var}(\widehat{\beta}_j) = \frac{\sigma^2}{\boldsymbol{x}_j'\boldsymbol{M}_j\boldsymbol{x}_j} .$$

Two extreme cases may be considered.

Case 1. The vector $\boldsymbol{x}_j$ is contained in the column space of $\boldsymbol{X}_j$, i.e. it is completely determined as a linear function of the columns of $\boldsymbol{X}_j$. This is the case of complete collinearity associated with $\boldsymbol{x}_j$. Then $\boldsymbol{M}_j\boldsymbol{x}_j = \boldsymbol{0}$ and the above formula for $\mathrm{Var}(\widehat{\beta}_j)$ makes no sense, which is not surprising, since the inverse $(\boldsymbol{X}'\boldsymbol{X})^{-1}$ does not exist.

Case 2. The vector $\boldsymbol{x}_j$ is contained in the orthogonal complement of the column space of $\boldsymbol{X}_j$, i.e. it is orthogonal to each of the columns in $\boldsymbol{X}_j$. This is the case of complete absence of collinearity associated with $\boldsymbol{x}_j$. Then $\boldsymbol{M}_j\boldsymbol{x}_j = \boldsymbol{x}_j$ and $\mathrm{Var}(\widehat{\beta}_j) = \sigma^2/\boldsymbol{x}_j'\boldsymbol{x}_j$.

For real data, none of the two cases will occur. One may, however, put the true variance $\sigma^2/\boldsymbol{x}_j'\boldsymbol{M}_j\boldsymbol{x}_j$ in relation to the 'optimal' variance $\sigma^2/\boldsymbol{x}_j'\boldsymbol{x}_j$, in order to see how much $\mathrm{Var}(\widehat{\beta}_j)$ is 'inflated' by collinearity. The *variance inflation factor* (VIF) associated with the variable x_j is therefore defined as

$$\mathrm{VIF}_j = \frac{\frac{\sigma^2}{\boldsymbol{x}_j'\boldsymbol{M}_j\boldsymbol{x}_j}}{\frac{\sigma^2}{\boldsymbol{x}_j'\boldsymbol{x}_j}} = \frac{\boldsymbol{x}_j'\boldsymbol{x}_j}{\boldsymbol{x}_j'\boldsymbol{M}_j\boldsymbol{x}_j} .$$

It may also be written as

$$\mathrm{VIF}_j = \frac{1}{1-R_j^2}, \quad R_j^2 = 1 - \frac{\boldsymbol{x}_j'\boldsymbol{M}_j\boldsymbol{x}_j}{\boldsymbol{x}_j'\boldsymbol{x}_j} .$$

The quantity R_j^2 is the classical coefficient of determination in the model

$$\boldsymbol{x}_j = \boldsymbol{X}_j\boldsymbol{\gamma} + \boldsymbol{u}, \quad \boldsymbol{u} \sim (\boldsymbol{0}, \tau^2\boldsymbol{I}_n) .$$

Clearly, the closer R_j^2 is to 1, the better the vector $\boldsymbol{x}_j$ can be explained as a linear function of $\boldsymbol{X}_j$, i.e. the more is $\boldsymbol{x}_j$ related to collinearity. It is, however, very difficult to say what is close enough to 1 to indicate collinearity.

Uncentered and Centered VIFs. Instead of using the classical coefficient of determination in the definition of VIF_j, one may also apply the centered coefficient of determination. As a matter of fact, the latter definition is more common. We may call VIF_j uncentered or centered, depending on whether the classical or the centered coefficient of determination is used. From the above considerations, a centered VIF_j only makes sense when the matrix $\boldsymbol{X}_j$ contains $\mathbf{1}_n$ as a column and the variable x_j is not constant. But then, as pointed out in [11, p.29], the centered VIF has no ability to discover collinearity involving the intercept. As a possible reason for the popularity of the centered VIF, one may see that often centering is recommended as a tool for removing unnecessary collinearity effects, which, however, is criticized in [11, Chap. 6]. A further reason could be that the centered VIFs turn out to be the main diagonal elements of the inverse $\boldsymbol{R}^{-1}$ of the correlation matrix $\boldsymbol{R}$ of the independent variables (apart from the intercept). The latter is often provided by computer packages, so that computation of centered VIFs becomes quite easy.

Critical Values. It is often recommended in the literature that centered VIFs greater than 10 indicate harmful collinearity. This corresponds to centered coefficient of determinations greater than 0.9. Since the classical coefficient of determination is always greater than the centered one (apart from special cases, see Theorem 6.1), the critical value for the uncentered VIF should be higher.

Example 6.3. Consider the cement data investigated in Examples 6.1 and 6.2. When all variables $\{\imath, x_1, x_2, x_3, x_4\}$ are included in the linear regression model, then centered as well as uncentered VIFs indicate strong collinearity among them, see Table 6.2.

Table 6.2. VIFs for the model with variables $\{\imath, x_1, x_2, x_3, x_4\}$ in Example 6.3

	Variable				
VIF	$\imath$	x_1	x_2	x_3	x_4
centered	–	38.4962	254.4232	46.8684	282.5129
uncentered	10668.5095	105.5971	2893.8665	218.2968	1265.6778

The uncentered VIFs reveal that collinearity is strongly related to the intercept, which is not surprising, since the values of the four independent variables are percentages, nearly summing up to the constant 100 for each observation.

This may suggest a model with variables $\{x_1, x_2, x_3, x_4\}$, excluding the intercept $\imath$. As matter of fact, the uncentered VIFs given in Table 6.3 do not give strong indication for harmful collinearity. □

Table 6.3. VIFs for the model with variables $\{x_1, x_2, x_3, x_4\}$ in Example 6.3

VIF	Variable			
	x_1	x_2	x_3	x_4
uncentered	6.688643	12.995434	9.981159	4.418264

Note that in Example 3.16 the correlation matrix $\boldsymbol{R}$ of the four independent variables (apart from the intercept) being included in the linear model in Example 6.3 is already computed. Its inverse is

$$\boldsymbol{R}^{-1} = \begin{pmatrix} 38.4962 & 94.1197 & 41.8841 & 99.7858 \\ 94.1197 & 254.4232 & 105.0914 & 267.5395 \\ 41.8841 & 105.0914 & 46.8684 & 111.1451 \\ 99.7858 & 267.5395 & 111.1451 & 282.5129 \end{pmatrix},$$

thus confirming the fact that centered VIFs can be obtained as the diagonal elements of $\boldsymbol{R}^{-1}$.

Remark 6.4. The number of near linear dependencies cannot be concluded from VIFs. If, for example, there is a near dependency involving variables x_1, x_2, x_3, and x_4, the VIFs for all four variables will be large. Then a further possible near dependency involving x_2 and x_4 cannot be discovered from inspecting the VIFs.

6.3.2 Scaled Condition Indexes

As an alternative procedure to diagnose collinearity, we present the method described in [11, p. 67], which can be carried out in three stages.

Stage 1. Rescale the columns of the matrix $\boldsymbol{X}$ such that each column has length 1. This is simply achieved by dividing each element of the j-th column $\boldsymbol{x}_j$ of $\boldsymbol{X}$ by $\sqrt{\boldsymbol{x}_j'\boldsymbol{x}_j}$, $j = 1, \ldots, p$. For simplicity, the so-obtained matrix is still denoted by $\boldsymbol{X}$ in the following.

Stage 2. Compute the p (scaled) *condition indexes*

$$\eta_j = \frac{\sqrt{\lambda_{\max}}}{\sqrt{\lambda_j}}, \quad j = 1, \ldots, p,$$

of the $n \times p$ matrix $\boldsymbol{X}$, where λ_j is the j-th eigenvalue of $\boldsymbol{X}'\boldsymbol{X}$ and $\lambda_{\max}$ is the largest eigenvalue of $\boldsymbol{X}'\boldsymbol{X}$. (Note that the positive square roots of the eigenvalues of $\boldsymbol{X}'\boldsymbol{X}$ are also called its *singular values.*)
Identify 'harmful' condition indexes (greater than 30 say). The number of harmful condition indexes gives the number of near linear dependencies.

Stage 3. For each condition index η_j compute the row vector

$$\boldsymbol{\pi}'_{[j]} = (\pi_{j1}, \ldots, \pi_{jp})$$

of *variance-decomposition proportions.* These are obtained as follows. Let $\boldsymbol{U}\boldsymbol{\Lambda}\boldsymbol{U}'$ be a spectral decomposition of $\boldsymbol{X}'\boldsymbol{X}$. Then from $\mathrm{Cov}(\widehat{\boldsymbol{\beta}}) = \sigma^2(\boldsymbol{X}'\boldsymbol{X})^{-1} = \sigma^2\boldsymbol{U}\boldsymbol{\Lambda}^{-1}\boldsymbol{U}'$ it follows that

$$\mathrm{Var}(\widehat{\beta}_k) = \sigma^2 \sum_{j=1}^{p} \frac{u_{kj}^2}{\lambda_j}.$$

Then the variance-decomposition proportion π_{jk} is the proportion of $\mathrm{Var}(\widehat{\beta}_k)$ associated with $\frac{u_{kj}^2}{\lambda_j}$ being related to λ_j. Let

$$\phi_{kj} = \frac{u_{kj}^2}{\lambda_j} \quad \text{and} \quad \phi_k = \sum_{j=1}^{p} \phi_{kj}, \quad k = 1, \ldots, p\,.$$

Then

$$\pi_{jk} = \frac{\phi_{kj}}{\phi_k}, \quad k, j = 1, \ldots, p\,.$$

If for a harmful η_j the row vector $\boldsymbol{\pi}'_{[j]}$ contains *two or more* high variance decomposition proportions (greater than 0.5 say), then the corresponding independent variables are involved in near linear dependencies.

The matrix $\boldsymbol{\Pi}$ consisting of the rows $\boldsymbol{\pi}'_{[j]}$ is also called the matrix of variance-decomposition proportions. Each row is related to a condition index, while each column is related to an independent variable. One may also represent this matrix as in Table 6.4.

Table 6.4. Matrix of variance-decomposition proportions

Condition Index	Proportions of $\mathrm{Var}(\widehat{\beta}_1)$	$\mathrm{Var}(\widehat{\beta}_2)$	$\cdots$	$\mathrm{Var}(\widehat{\beta}_p)$
η_1	π_{11}	π_{12}	$\cdots$	π_{1p}
η_2	π_{21}	π_{22}	$\cdots$	π_{2p}
$\vdots$	$\vdots$	$\vdots$		$\vdots$
η_p	π_{p1}	π_{p2}	$\cdots$	π_{pp}
	x_1	x_2	$\cdots$	x_p
	Variable			

The smallest scaled condition index is always equal to 1, while the largest scaled condition index equals the scaled condition number $\widetilde{\kappa}(\boldsymbol{X})$. It is convenient to order the condition indexes is Table 6.4 in increasing order, so that

$\eta_1 = 1$ and $\eta_p = \widetilde{\kappa}(\boldsymbol{X})$. In addition, each condition index may be rounded off to an integer, neglecting the rather unimportant digits after the decimal point.

Example 6.4. Let us continue Example 6.3, dealing with the cement data from Table 1.1, p. 4. The $\boldsymbol{\Pi}$ matrix of variance-decomposition proportions for the model $\{\imath, x_1, x_2, x_3, x_4\}$ is given in Table 6.5, showing that there is only one harmful near dependency, involving all variables (including the intercept). When we consider the same model without intercept, then the highest condition index shows that there is no harmful collinearity in the model arising from possible near dependencies, see Table 6.6.

Table 6.5. Matrix $\boldsymbol{\Pi}$ for the model with variables $\{\imath, x_1, x_2, x_3, x_4\}$ in Example 6.4

Condition	Variable				
Index	$\imath$	x_1	x_2	x_3	x_4
1	0.0000	0.0004	0.0000	0.0002	0.0000
3	0.0000	0.0100	0.0000	0.0027	0.0001
4	0.0000	0.0006	0.0003	0.0016	0.0017
10	0.0001	0.0574	0.0028	0.0457	0.0009
250	0.9999	0.9316	0.9969	0.9498	0.9973

Table 6.6. Matrix $\boldsymbol{\Pi}$ for the model with variables $\{x_1, x_2, x_3, x_4\}$ in Example 6.4

Condition	Variable			
Index	x_1	x_2	x_3	x_4
1	0.0103	0.0071	0.0080	0.0182
2	0.1556	0.0025	0.0598	0.0308
3	0.0081	0.0739	0.0375	0.4706
9	0.8261	0.9165	0.8947	0.4804

When we assess a model via the AIC criterion, then for the model with variables $\{x_1, x_2, x_3, x_4\}$ we compute $\text{AIC}_p = 26.17$, while for the model with variables $\{\imath, x_1, x_2, x_4\}$ we have $\text{AIC}_p = 24.97$. Nonetheless, the latter one still implies collinearity problems, see Table 6.7.

Our choice of model also depends on our goals. If it is strongly desired to incorporate all four variables x_1, x_2, x_3, and x_4 into the model, then it might be better to exclude the intercept to avoid collinearity problems. In addition the model with variables $\{\imath, x_1, x_2, x_3, x_4\}$ has $\text{AIC}_p = 26.94$, being greater than $\text{AIC}_p = 26.17$ for the same model but without intercept. On

Table 6.7. Matrix $\boldsymbol{\Pi}$ for the model with variables $\{\imath, x_1, x_2, x_4\}$ in Example 6.4

Condition	Variable			
Index	$\imath$	x_1	x_2	x_4
1	0.0002	0.0227	0.0004	0.0008
3	0.0001	0.3387	0.0007	0.0160
4	0.0007	0.6278	0.0091	0.0077
52	0.9990	0.0108	0.9898	0.9756

the other hand, if we are interested in a *simple* model with rather small AIC and no harmful collinearity, then we may confine to the model with variables $\{\imath, x_1, x_2\}$. □

Remark 6.5. When collinearity is present, then one might be tempted to seek for independent variables which can be deleted from the considered regression model. This should, however, be regarded with caution. Collinearity is a problem with the data and not necessarily with the model. If *relevant* variables are omitted, then the model is misspecified, usually having two implications: First, the variances of the elements of the ordinary least squares estimator in the reduced model, as estimators for the remaining variables, will be smaller than the variances of the corresponding elements of the ordinary least squares estimator in the full model. Hence, it may appear that the negative effect of collinearity (large variances) is is fact reduced. Second, the ordinary least squares estimator in the reduced model will be biased. This shows that by omitting relevant variables one only barters lower variances for greater biases.

On the other hand, it is of course possible that the model is misspecified in the sense that it contains some *irrelevant* variables, having no other meaningful effect, than producing collinear data. If it is possible to identify those variables, then their deletion can also help to reduce collinearity problems.

6.4 Inspecting Residuals

The inspection of residuals is an important aid in finding out whether a specified linear regression model

$$y_i = \beta_1 x_{i,1} + \cdots + \beta_p x_{i,p} + \varepsilon_i = \boldsymbol{x}'_{[i]}\boldsymbol{\beta}, \quad i = 1, \ldots, n ,$$

together with the assumptions (i) to (iv) is plausible.

At first, we have to clarify what is usually meant by 'residual'. When we consider the vector $\widehat{\boldsymbol{\varepsilon}} = \boldsymbol{y} - \boldsymbol{X}\widehat{\boldsymbol{\beta}}$, then this may be seen as a substitute for the realization of the unobservable error vector $\boldsymbol{\varepsilon}$, thus comprising clues for adequacies/inadequancies of the model assumptions related to the distribution of $\boldsymbol{\varepsilon}$. However, when $\boldsymbol{\varepsilon} \sim (\boldsymbol{0}, \sigma^2 \boldsymbol{I}_n)$, then $\widehat{\boldsymbol{\varepsilon}} \sim (\boldsymbol{0}, \sigma^2(\boldsymbol{I}_n - \boldsymbol{P}))$

with $\boldsymbol{P} = \boldsymbol{X}(\boldsymbol{X}'\boldsymbol{X})^{-1}\boldsymbol{X}'$, implying that different elements of $\widehat{\boldsymbol{\varepsilon}}$ are neither uncorrelated nor homoscedastic.

To overcome difficulties arising from different variances of the elements $\widehat{\varepsilon}_i$, for diagnostic purposes it would be better to consider

$$\widetilde{\varepsilon}_i = \frac{\widehat{\varepsilon}_i}{\sigma\sqrt{1-p_{ii}}}, \qquad i = 1, \ldots, n ,$$

where p_{ii} is the i-th main diagonal element of $\boldsymbol{P}$. Clearly $\mathrm{E}(\widetilde{\varepsilon}_i) = 0$ and $\mathrm{E}(\widetilde{\varepsilon}_i^2) = 1$. Since σ is not known, it may be replaced by the (positive) root $\widehat{\sigma}$ of least squares variance estimate $\widehat{\sigma}^2 = \frac{1}{n-p}\widehat{\boldsymbol{\varepsilon}}'\widehat{\boldsymbol{\varepsilon}}$. This gives the *i-th standardized residual*

$$r_i = \frac{\widehat{\varepsilon}_i}{\widehat{\sigma}\sqrt{1-p_{ii}}},$$

for $i = 1, \ldots, n$. When instead of $\widehat{\sigma}^2$ the estimate $\widehat{\sigma}^2_{(i)}$ is applied, where $\widehat{\sigma}^2_{(i)}$ is the least squares variance estimate when the i-th observation $(y_i, x_{i1}, \ldots, x_{ip})$ is omitted, then

$$r_i^* = \frac{\widehat{\varepsilon}_i}{\widehat{\sigma}_{(i)}\sqrt{1-p_{ii}}}$$

is the *i-th studentized residual* for $i = 1, \ldots, n$.

Note that this terminology is not universally adopted. Studentized residuals are also called 'externally studentized', 'deleted studentized' or 'jackknifed' residuals in the literature, while standardized residuals are also called 'internally studentized' or simply 'studentized' residuals.

Usually, analysis of residuals is understood as analysis of r_i or r_i^*, where it can be shown that

$$r_i^* = r_i\sqrt{\frac{n-p-1}{n-p-r_i^2}} ,$$

see [20, Sect. 4.2.1] and Problem 6.5. The i-th studentized residual r_i^* has also the meaning of being the t-statistic for a test of the hypotheses that the i-th observation is not an outlier, see [20] and Sect. 6.6.6 for more details. In general, a large absolute value of r_i^* may be seen an indication for the i-th observation to be an outlier.

Definition 6.1. *In the linear regression model $y_i = \boldsymbol{x}'_{[i]}\boldsymbol{\beta} + \varepsilon_i$, the i-th observation $(y_i, \boldsymbol{x}'_{[i]})$ is defined to be a (vertical)* outlier, *if r_i^* is large compared to the studentized residuals of the other observations in the data set.*

Graphical methods involving studentized residuals (often preferred over standardized residuals) can give valuable information about the adequacy of the model. We do not intend to give a complete review on residual analysis here, but shortly discuss two very common residual plots in the following.

6.4.1 Normal Quantile Plot

In the linear regression model $\boldsymbol{y} = \boldsymbol{X}\boldsymbol{\beta} + \boldsymbol{\varepsilon}$, the assumption (v) of normality of the error vector $\boldsymbol{\varepsilon}$ cannot be checked, but one may find some evidence in favor or against (v).

Suppose that we have a sample of n independent observations $z_1, \ldots, z_n$ from some known distribution, which have been increasingly ordered as $z_{(1)}, \ldots, z_{(n)}$. Moreover, consider n probabilities $\gamma_1, \ldots, \gamma_n$ being equidistant and increasingly ordered. When we compute γ_i-quantiles Q_{γ_i} for $i = 1, \ldots, n$ from the distribution of the z's, then the points $(Q_{\gamma_i}, z_{(i)})$ will be almost on a straight line.

Now, the normal quantile plot of the studentized residuals is simply a plot of the points

$$(u_{\gamma_i}, r^*_{(i)}), \quad i = 1, \ldots, n ,$$

where u_{γ_i} denotes the γ_i-quantile of $N(0,1)$. The function `ppoints` in R generates the γ_i as

$$\gamma_i = \begin{cases} \dfrac{i - \frac{1}{2}}{n} & \text{for } n > 10 \\[2ex] \dfrac{i - \frac{3}{8}}{n + \frac{1}{4}} & \text{otherwise} \end{cases} .$$

For a better verification of the straight line property one may in addition plot a line through the points $(u_{0.25}, r^*_{0.25})$ and $(u_{0.75}, r^*_{0.75})$, where r^*_α is the empirical α-quantile of the observations $r^*_1, \ldots, r^*_n$. This can be done in R with the function `qqline`. Studentized residuals can easily be computed with the R function `rstudent`.

Figure 6.1 shows four normal quantile plots, each one being drawn for $n = 30$ generated realizations of independent $N(0,1)$ variables. As can be seen, even if a data vector stems from the $N(\boldsymbol{0}, \boldsymbol{I}_n)$ distribution, the plotted points do (of course) not exactly lie on a straight line. Usually, some deviances occur at the margins, which, however, should not be too pronounced, otherwise indicating outliers and/or a heavy-tail distribution.

Remark 6.6. Usually standardized and studentitzed residuals are plotted versus the quantiles of the $N(0,1)$ distribution. One may, however, argue that studentized residuals r^*_i may also be plotted versus the quantiles of the t_{n-p-1} distribution, since under the assumptions (i) to (iv) and (v), $r^*_i \sim t_{n-p-1}$, see e.g. Theorem 4.2 in [20]. Such a t quantile plot is done by the R function `qq.plot` from the package `car` when applied to a linear model, see also [40].

6.4.2 Residuals Versus Fitted Values Plot

If the linear regression model $\boldsymbol{y} = \boldsymbol{X}\boldsymbol{\beta} + \boldsymbol{\varepsilon}$ is adequate and the assumptions (i) to (iv) are satisfied, then $\mathrm{Cov}(\widehat{\boldsymbol{y}}, \widehat{\boldsymbol{\varepsilon}}) = \mathrm{Cov}(\boldsymbol{P}\boldsymbol{y}, (\boldsymbol{I}_n - \boldsymbol{P})\boldsymbol{y}) = \sigma^2 \boldsymbol{P}(\boldsymbol{I}_n - \boldsymbol{P}) =$

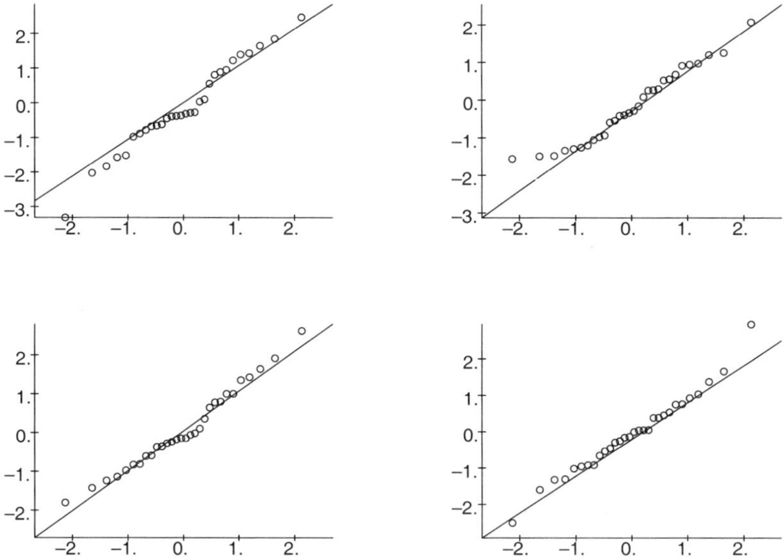

Fig. 6.1. Normal quantile plots of four samples of sizes $n = 30$, generated from the $N(0, 1)$ distribution

$\mathbf{0}$, and the plotted points $(\widehat{y}_i, \widehat{\varepsilon}_i)$ should scatter around the horizontal zero line without revealing any pattern.

We still prefer the analysis of studentized residuals, in which case a residual versus fitted values plot is simply a plot of the points

$$(\widehat{y}_i, r_i^*), \quad i = 1, \ldots, n .$$

When the plot reveals some pattern, then this may be seen as an indication for *violation of linearity* (curvature) or as an indication for *heteroscedasticity* (horn shape). The following two simple examples illustrate this.

Example 6.5. Consider $n = 30$ data pairs (y_i, x_i) with $x_i = 2i/n$ for $i = 1, \ldots, 30$, and y_i generated by $y_i = 10 + (1 + x_i)^3 + u_i$, where the u_i are independently $N(0, 1)$ distributed. When we apply a linear model

$$y_i = \beta_1 + x_i \beta_2 + \varepsilon_i, \quad i = 1, \ldots, n ,$$

then the residual versus fitted values plot reveals a curvature as in Fig. 6.2. (The adjusted coefficient of determination is $\overline{R}^2 = 0.9286$ for the generated data producing this specific plot.) □

Example 6.6. Consider $n = 30$ data pairs (y_i, x_i) with $x_i = i/n$ for $i = 1, \ldots, 30$, and y_i generated by $y_i = 10 + 10x_i + u_i$, where the u_i are independently $N(0, i)$ distributed. When we apply a linear model

$$y_i = \beta_1 + x_i\beta_2 + \varepsilon_i, \quad i = 1, \ldots, n\ ,$$

then the residual versus fitted values plot reveals a horn shape as in Fig. 6.3. (The adjusted coefficient of determination is $\overline{R}^2 = 0.8046$ for the generated data producing this plot.) □

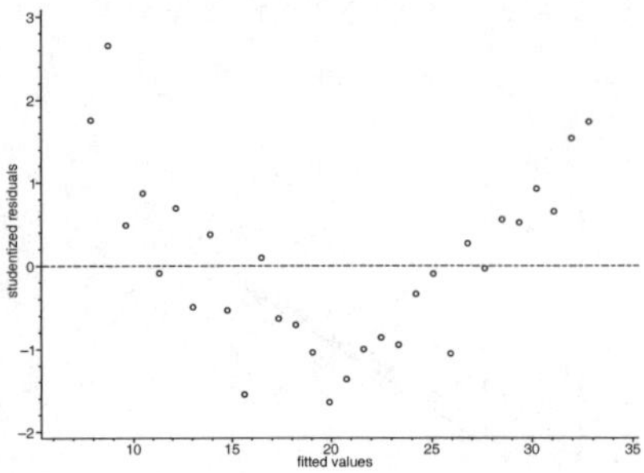

Fig. 6.2. Residual versus fitted values plot in Example 6.5

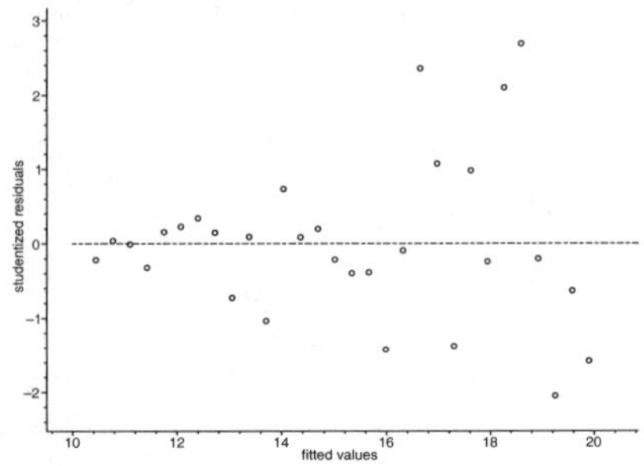

Fig. 6.3. Residual versus fitted values plot in Example 6.6

Note that although a systematic pattern in the residual versus fitted values plot is an indication for inadequacy of the considered model, it is not always clear whether this pattern is related to non-linearity or to heteroscedasticity. Cook and Weisberg [25, Sect. 14.1.5] give an example for a horn shaped residual plot with data from a model $y_i = f(x_{i1}, x_{12}) + \varepsilon_i$, being caused by wrongly considering $f(\cdot)$ to be linear.

6.4.3 Further Residual Plots

There are a number of further plots involving residuals. For example one may simply plot a *histogram* or a *box blot* of the studentized residuals r_i^* in order to get some impression of the distribution and possible peculiarities.

When the observations indexes in a linear regression model stand for subsequent time points, then it may also be useful to plot the *residuals versus the index*. If a pattern similar as in Fig. 5.5 or Fig. 5.6 (p. 282) shows up, then this may be an indication for a correctly specified model with autocorrelated errors. On the other hand, it may as well be an indication for the omission of relevant variables, since the latter may also cause autocorrelation among the errors.

Often, it is suggested to plot *residuals versus predictors* (independent variables in the model). If some systematic pattern is revealed, then this may imply that some transformation, e.g. x_j^2, of the variable x_j should also be incorporated within the model.

Other plots are *added variable (partial regression) plots* and *component plus residual (partial residual) plots*, see e.g. [20, 23, 24, 25, 40] for details.

6.5 Finding Influential Observations

Consider the linear regression model $\boldsymbol{y} = \boldsymbol{X\beta}+\boldsymbol{\varepsilon}$ with individual observations $(y_i, \boldsymbol{x}'_{[i]})$, where $\boldsymbol{x}'_{[i]}$ is the i-th row of the matrix $\boldsymbol{X}$. Then, the data may contain one or more observations with a rather strong influence on the results of the regression analysis as compared to other observations, and it is of specific interest to identify these observations.

6.5.1 Leverage

Usually, a slight change in the value y_i will not produce an according change in the value $\widehat{y}_i$, since $\widehat{y}_i = \boldsymbol{x}'_{[i]}\widehat{\boldsymbol{\beta}} = \boldsymbol{x}'_{[i]}(\boldsymbol{X}'\boldsymbol{X})^{-1}\boldsymbol{X}'\boldsymbol{y}$ depends on the whole observation matrix $(\boldsymbol{y}, \boldsymbol{X})$, and thus will be rather insensitive to small changes in the individual y_i.

The situation is different, however, when $\boldsymbol{x}'_{[i]}$ is far away from the rest of the data. Then $\widehat{y}_i$ will be close to y_i and small changes in y_i produce according changes in $\widehat{y}_i$. Hence one may take $\partial\widehat{y}_i/\partial y_i$ as a measure for the *leverage* of $\boldsymbol{x}'_{[i]}$. From the identity $\widehat{\boldsymbol{y}} = \boldsymbol{Py}$ with $\boldsymbol{P} = \boldsymbol{X}(\boldsymbol{X}'\boldsymbol{X})^{-1}\boldsymbol{X}'$ it follows that

$$\frac{\partial\widehat{y}_i}{\partial y_i} = p_{ii} \;,$$

where $p_{ii} = \boldsymbol{x}'_{[i]}(\boldsymbol{X}'\boldsymbol{X})^{-1}\boldsymbol{x}_{[i]}$ is the i-th main diagonal element of the *hat matrix* $\boldsymbol{P}$ and thus is also called the *i-th hat value.* (The orthogonal projector $\boldsymbol{P}$ puts a hat on $\boldsymbol{y}$ via the transformation $\boldsymbol{Py}$.) Since we have $\mathbf{0} \leq_{\mathrm{L}} \boldsymbol{P} \leq_{\mathrm{L}} \boldsymbol{I}_n$, it follows

$$0 \leq p_{ii} \leq 1 \,.$$

When the model contains an intercept, then we even have $(1/n) \leq p_{ii} \leq 1$, see Problem 6.3.

Definition 6.2. *In the linear regression model $y_i = \boldsymbol{x}'_{[i]}\boldsymbol{\beta}+\varepsilon_i$, the i-th observation $(y_i, \boldsymbol{x}'_{[i]})$ is defined to be a* high-leverage point, *if p_{ii} is large compared to the hat values of the other observations in the data.*

From this definition it is also clear that the concept of leverage is solely related to the given matrix $\boldsymbol{X}$ and not to the realized $\boldsymbol{y}$, although we have used the relationship $\boldsymbol{y} = \boldsymbol{Py}$ to motivate the application of the measure p_{ii}. Alternatively, one may show that the measure p_{ii} is equivalent to the Mahalanobis distance, see [20, Sect. 4.3.3.2].

Since $\sum_{i=1}^n p_{ii} = \mathrm{tr}(\boldsymbol{P}) = \mathrm{rk}(\boldsymbol{P}) = p$, it follows that the average of the diagonal elements of $\boldsymbol{P}$ is p/n. A value p_{ii} which is two times greater than the average, i.e.

$$p_{ii} > \frac{2p}{n} \,,$$

is often seen as an indication for a high-leverage point. (The R function `influence.measures` uses $3p/n$ as a critical value.) A further way to examine the hat values is to plot them versus the index. The critical value can be included as a horizontal line.

Example 6.7. In Example 3.33, a linear regression model $y_i = \beta_1 + \beta_2 x_i + \varepsilon_i$, $i = 1, \ldots, n$, with $n = 9$ observations from Table 3.7, p. 199, is considered. The hat values versus index plot is given in Fig. 6.4. As already noted in Example 3.33, the fourth observation is clearly a high-leverage point.

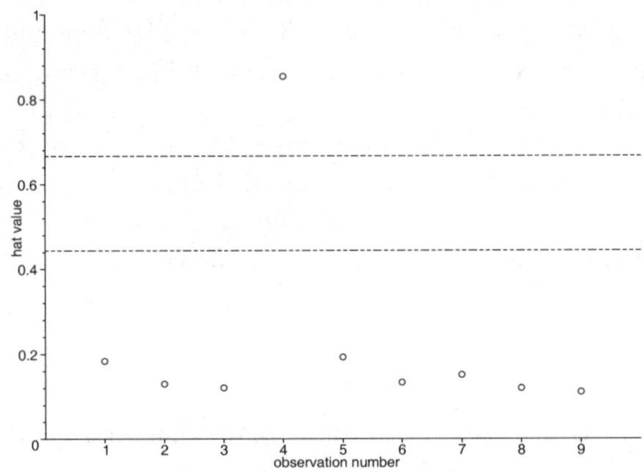

Fig. 6.4. Hat values versus index plot in Example 6.7 with critical values $2p/n$ and $3p/n$

6.5.2 Influential Observations

An individual observation $(y_i, \boldsymbol{x}'_{[i]})$ is said to be influential, if its presence has a strong influence on some or all regression results obtained from a model with observation matrix $(\boldsymbol{y}, \boldsymbol{X})$. If not present, the regression results for the rest of the data would look quite different. The detection of influential observations is thus of specific interest, making it for example possible to carry out an analysis for the bulk of the data separately. This does, however, not imply that influential observations should be ignored, but, to the contrary, should deserve special attention in a complete data analysis. It is of course possible that an influential observation can be explained by an error in the data collection process, but it may also be the case that it gives valuable information about some unusual but nonetheless existing behavior.

As shown in [20], neither outliers nor high-leverage points need to be influential. On the other hand, influential observations are either outliers or high-leverage points or both. Thus, influential measures should consider r_i^* (or r_i) and p_{ii} together. In the following we introduce four different measures for influence, see also [20].

Cook's Distance. A popular measure for the influence of the i-th observation is Cook's distance [22], given as

$$D_i = \frac{(\widehat{\boldsymbol{\beta}} - \widehat{\boldsymbol{\beta}}_{(i)})' \boldsymbol{X}' \boldsymbol{X} (\widehat{\boldsymbol{\beta}} - \widehat{\boldsymbol{\beta}}_{(i)})}{p \, \widehat{\sigma}^2} \,,$$

where $\widehat{\boldsymbol{\beta}}_{(i)}$ is the ordinary least squares estimator of $\boldsymbol{\beta}$ when the i-th observation is omitted. It can be shown, that

$$D_i = \frac{1}{p} \frac{p_{ii}}{1 - p_{ii}} r_i^2 \,,$$

where r_i is the i-th standardized residual. Thus, Cook's distance combines the two measures r_i (for outliers) and p_{ii} for leverage. High values of D_i indicate that the i-th observation is influential. The D_i may also be inspected via a Cook's distance versus index plot. Although, even under normality, D_i is not F distributed, values

$$D_i > F_{p,(n-p),0.5}$$

are often recommended as an indication for influence. (This is also the critical value used by the R function `influence.measures`.) One may include $F_{p,(n-p),0.5}$ as a horizontal line in the Cook's distance versus index plot.

Welsch-Kuh's Distance. An alternative measure is given by considering the difference of the the two predictions of y_i based on $\widehat{\boldsymbol{\beta}}$ and $\widehat{\boldsymbol{\beta}}_{(i)}$, respectively, relative to the standard error of $\widehat{y}_i$, see [10]. This gives

$$\text{DFFITS}_i = \frac{|\boldsymbol{x}'_{[i]} \widehat{\boldsymbol{\beta}} - \boldsymbol{x}'_{[i]} \widehat{\boldsymbol{\beta}}_{(i)}|}{\sigma^2_{(i)} \sqrt{p_{ii}}} \,,$$

where $\widehat{\sigma}^2_{(i)}$ is the least squares variance estimator for σ^2 when the i-th observation is omitted. It can be shown that

$$\text{DFFITS}_i = |r_i^*| \sqrt{\frac{p_{ii}}{1 - p_{ii}}} \,,$$

where r_i^* is the i-th studentized residual. If

$$\text{DFFITS}_i > 2 \sqrt{\frac{p}{n - p}} \,,$$

then this may be seen as an indication for the influence of the i-th observation. (The R function `influence.measures` uses $3\sqrt{\frac{p}{n-p}}$ as a critical value.)

Covariance Ratio. In order to find some indication for influence, one may also compare the estimated covariance matrices $\widehat{\sigma}^2(\boldsymbol{X}'\boldsymbol{X})^{-1}$ of $\widehat{\boldsymbol{\beta}}$ and $\widehat{\sigma}^2_{(i)}(\boldsymbol{X}'_{(i)}\boldsymbol{X}_{(i)})^{-1}$ of $\widehat{\boldsymbol{\beta}}_{(i)}$, where $\boldsymbol{X}_{(i)}$ is obtained from $\boldsymbol{X}$ by deleting its i-th row. This can be done by considering

$$\text{COVRATIO}_i = \frac{\det(\widehat{\sigma}^2_{(i)}(\boldsymbol{X}'_{(i)}\boldsymbol{X}_{(i)})^{-1})}{\det(\widehat{\sigma}^2(\boldsymbol{X}'\boldsymbol{X})^{-1})} \,,$$

see [10]. It can be shown that

$$\text{COVRATIO}_i = \left(\frac{n-p-r_i^2}{n-p-1}\right)^p \frac{1}{1-p_{ii}} \,.$$

If

$$|\text{COVRATIO}_i - 1| > \frac{3p}{n-p} \,,$$

then this is seen as an indiction for influence. (This is also the critical value used by the R function `influence.measures`.)

Influence on Results Concerning Individual Parameters. The above three measures do not give special emphasize on the influence of the i-th observation $(y_i, \boldsymbol{x}'_{[i]}\boldsymbol{\beta})$ on regression results concerning the j-th parameter β_j. Such a measure, see [10], is

$$\text{DFBETAS}_{ij} = \frac{\widehat{\beta}_j - \widehat{\beta}_{j(i)}}{\widehat{\sigma}_j} \,,$$

where $\widehat{\beta}_j$ and $\widehat{\beta}_{j(i)}$ is the j-th element of $\widehat{\boldsymbol{\beta}}$ and $\widehat{\boldsymbol{\beta}}_{(i)}$, respectively, and $\widehat{\sigma}_j$ is an estimator for the square root of $\sigma_j^2 = \text{Var}(\widehat{\beta}_j)$. As in Sect. 6.3.1, we have

$$\text{Var}(\widehat{\beta}_j) = \frac{\sigma^2}{\boldsymbol{x}'_j\boldsymbol{M}_j\boldsymbol{x}_j}, \quad \boldsymbol{M}_j = \boldsymbol{I}_n - \boldsymbol{X}_j(\boldsymbol{X}'_j\boldsymbol{X}_j)^{-1}\boldsymbol{X}_j \,,$$

where $\boldsymbol{x}_j$ is the j-th column of $\boldsymbol{X}$ and $\boldsymbol{X}_j$ is obtained from $\boldsymbol{X}$ by deleting the j-th column. By using $\sqrt{\widehat{\sigma}^2_{(i)}/\boldsymbol{x}'_j\boldsymbol{M}_j\boldsymbol{x}_j}$ as an estimator for σ_j, it can be shown that

$$\text{DFBETAS}_{ij} = r_i^* \frac{\boldsymbol{e}'_i\boldsymbol{M}_j\boldsymbol{x}_j}{\boldsymbol{x}'_j\boldsymbol{M}_j\boldsymbol{x}_j} \frac{1}{\sqrt{1-p_{ii}}} \,,$$

where $\boldsymbol{e}_i$ is the i-th $n \times 1$ unit vector. If

$$|\text{DFBETAS}_{ij}| > \frac{2}{\sqrt{n}} \,,$$

then this may be seen as an indication for influence (The R function `influence.measures` nominates absolute DFBETAS greater than 1 for special attention.)

6.5.3 Collinearity-Influential Observations

If some row $\boldsymbol{x}'_{[i]}$ of $\boldsymbol{X}$ has high leverage, then it may also have a strong influence on the eigenvalues of $\boldsymbol{X}'\boldsymbol{X}$, and thus on the presence of collinearity, although this is not necessarily the case.

As for example shown in [20, Sect. 4.2.9.2], it is possible that individual observations either create or hide collinearity present in the rest of the data. Such observations are called collinearity-influential observations and deserve special attention. See [20, Sect. 4.2.9] and [11, Sect. 8.2] for a detailed discussion.

6.6 Testing Model Assumptions

Significance tests in linear regression models can also be seen as tools for diagnosing inadequacies arising from the consideration of a given linear model with the usual standard assumptions as being appropriate. In this section only a short notion of testing model assumptions is given. A detailed review of the application of significance tests in linear regression models is provided by Krämer and Sonnberger [65]. Several tests can be carried out in R using the package `lmtest`, see also [128].

6.6.1 Preliminary Considerations

If it is felt that a given linear regression model is not adequately specified, then a significance test of the hypothesis

H_0: 'the given model is adequate with respect to the claimed inadequacy in H_1'

versus the alternative hypothesis

H_1: 'the given model is inadequate in one or more specific aspects concerning the model assumptions'

can be carried out. A considerable number of different tests is available in the literature, being created with respect to various possible inadequacies formulated in H_1.

Two Groups of Alternatives. In general, the possible inadequacies in H_1 may be divided into two groups, being

H_1: '$\boldsymbol{X\beta}$ is not an adequate substitute for $\mathrm{E}(\boldsymbol{y})$'

and

H_1: '$\sigma^2\boldsymbol{I}_n$ is not an adequate substitute for $\mathrm{Cov}(\boldsymbol{y})$'.

The first group concerns any possible misspecification with respect to the undisturbed functional relationship $y = f(x_1, \ldots, x_p)$ between the dependent variable y and the independent variables $x_1, \ldots, x_p$. This may concern a possible *non-linearity* of $f(\cdot)$, but also a possible instability in the regression coefficients, meaning that the undisturbed functional relationship may indeed be adequate, but related to different parameters for different groups of the data (*structural change* in the parameters).

The second group concerns possible misspecifications with respect to the disturbance of the error variable ε. This may be related to possible *heteroscedasticity* or to possible *autocorrelation*. A further inadequacy concerning the error vector $\boldsymbol{\varepsilon}$ (but not necessarily its covariance matrix) is its possible *non-normality*.

As pointed out in [65], misspecifications related to the first group usually have a more harmful influence on the analysis than misspecifications related to the second group.

Rejecting and Accepting H_0. Significance tests are designed in such a way that, provided all required assumptions are satisfied, the probability for rejecting H_0 when nonetheless H_0 is true (*Type I error*) is not greater than a number α, being specified by the user prior to the test. Usually one chooses a small number $\alpha = 0.05$ or $\alpha = 0.1$. Thus, a Type I error had rarely occurred if the same test would have been performed on a number of data sets independently obtained under the same conditions, giving some confidence that H_1 will be true when H_0 is rejected.

The situation is different when H_0 is not rejected. Then the probability for accepting H_0 when it is in fact true (*power* of the test) is usually not known, but should of course be quite high. (This is equivalent to the claim that the probability for accepting H_0 when nonetheless H_0 is false (*Type II error*) should be quite low.) Usually the descriptions of significance tests are accompanied by some remarks concerning their power, but, in general, the actual magnitude of the power remains unknown.

Assessing the Hypothesis. Instead of rejecting or accepting the hypothesis, one may consider the smallest significance level that would have led to rejection of H_0 (the *p-value*). This value is usually provided by implemented tests in computer packages and can be seen as a measure for how strong the data supports the hypothesis H_0. The smaller the p-value, the weaker the support of H_0 and the stronger the indication for some misspecification as formulated in H_1.

6.6.2 Testing for Heteroscedasticity

In the linear regression model, a test for heteroscedasticity is a test of the hypothesis

$$H_0 : \operatorname{Var}(\varepsilon_i) = \sigma^2 \quad \text{for } i = 1, \ldots, n$$

versus H_1 : notH_0. This alternative hypothesis is rather unspecified. If no further prior knowledge about the possible source of heteroscedasticity is available, then one may apply e.g. *White's general test* for heteroscedasticity, see [65, 126] and [46, Sect. 11.4.1]. If the possible source of heteroscedasticity can be narrowed, then White's general test will be inferior to tests accounting for a more specific formulation of H_1.

Goldfeld-Quandt [45] Test. This test applies to situations where the alternative hypothesis H_1 can be expressed as a decrease/increase of $\text{Var}(y_i)$ with increasing values $x_{i,j}$ of *one* of the independent variables x_j. The observations $(y_i, x_{i,1}, \ldots, x_{i,p})$ are reordered based on x_j, and then divided into two groups $(\boldsymbol{y}_1, \boldsymbol{X}_1)$ and $(\boldsymbol{y}_2, \boldsymbol{X}_2)$ with sizes n_1 and n_2, respectively, where $n = n_1 + n_2$. If H_0 is valid, then the variances of the error vectors in the two corresponding linear regression models are the same.

Now, the idea of the test is simply to compare the least squares variance estimators under both models (provided they are obtainable, of course). If the estimate from the first model is considerably larger/smaller than the estimate from the second model, the hypothesis is rejected. When the two residual vectors are $\widehat{\varepsilon}_1$ and $\widehat{\varepsilon}_2$, respectively, the test statistic is given as

$$R = \frac{\widehat{\varepsilon}_1'\widehat{\varepsilon}_1/(n_1 - p)}{\widehat{\varepsilon}_2'\widehat{\varepsilon}_2/(n_2 - p)} ,$$

being $F_{(n_1-p),(n_2-p),1-\alpha}$ distributed under H_0 when the errors are normally distributed. The hypothesis H_0 is rejected at level α if $R > F_{(n_1-p),(n_2-p),1-\alpha}$. This procedure tests for a decreasing variance. For a test of an increasing variance the subscripts in the test statistic should be reversed. The Goldfeld-Quandt test can be carried out with the R function `gqtest` from the package `lmtest`.

To increase the power of this test, it is sometimes suggested to exclude no more than $n/3$ observations out of the middle of the ordered observations before building the two groups with the remaining observations.

The Goldfeld-Quandt test may also be applied when the indexes i stand for subsequent time-points and it is suspected that $\text{Var}(y_i)$ decreases/increases over time. Then no reordering of the observations is required. Further, the Goldfeld-Quandt test may be applied in case of grouped heteroscedasticity, when it is suspected that the variances in the first groups are substantially smaller/greater than the variances in the last groups. In the latter situation one may also apply the *Bartlett test* for equality of variances, see also [62, p. 298]. The test can be carried out with the R function `bartlett.test` from the package `ctest`.

Breush-Pagan [18] Test. The Goldfeld-Quandt test allows only one independent variable as the basic explanation for the possible heteroscedasticity. There may, however, be more variables thought to influence heteroscedasticity. The alternative H_1 may then be expressed as

$$H_1 : \mathrm{Var}(\varepsilon_i) = h(\alpha_0 + \boldsymbol{z}'_{[i]}\boldsymbol{\alpha}) ,$$

where $\boldsymbol{z}_{[i]}$, $i = 1, \ldots, n$, is a $q \times 1$ vector with observations of variables $z_1, \ldots, z_q$ as elements and $\boldsymbol{\alpha}$ is $q \times 1$ unknown parameter vector. The function $h(\cdot)$ need not to be specified. The variables $z_1, \ldots, z_q$ may be independent variables from the model but may also be other variables.

When H_1 is specified in this manner, then the errors are homoscedastic if $\boldsymbol{\alpha} = \boldsymbol{0}$ and thus, a reformulation of H_0 according to H_1 is

$$H_0 : \alpha_1 = \alpha_2 = \cdots = \alpha_q .$$

To obtain some indication for the validity of $\mathrm{Var}(\varepsilon_i) = h(\alpha_0 + \boldsymbol{z}'_{[i]}\boldsymbol{\alpha})$, scaled squares of ordinary least squares residuals

$$g_i = \frac{n\widehat{\varepsilon}_i^2}{\widehat{\boldsymbol{\varepsilon}}'\widehat{\boldsymbol{\varepsilon}}}, \quad \widehat{\boldsymbol{\varepsilon}} = \boldsymbol{y} - \boldsymbol{X}\widehat{\boldsymbol{\beta}} ,$$

are regressed on $\alpha_0 + \boldsymbol{z}'_{[i]}\boldsymbol{\alpha}$, i.e. the model

$$g_i = \alpha_0 + \boldsymbol{z}'_{[i]}\boldsymbol{\alpha} + u_i, \quad i = 1, \ldots, n ,$$

is considered. The vector of fitted values $\widehat{\boldsymbol{g}} = \boldsymbol{Z}(\boldsymbol{Z}'\boldsymbol{Z})^{-1}\boldsymbol{Z}'\boldsymbol{g}$ will have considerably greater length if $\boldsymbol{\alpha} \neq \boldsymbol{0}$ than otherwise. The test statistic

$$Q = \frac{1}{2}\widehat{\boldsymbol{g}}'\widehat{\boldsymbol{g}}$$

is asymptotically $\chi^2_{(q)}$ distributed under H_0. The hypothesis H_0 is rejected at level α if $Q > \chi^2_{(q),1-\alpha}$. The Breush-Pagan test can be carried out with the R function `bptest` from the package `lmtest`.

6.6.3 Testing for Autocorrelation

Suppose the errors in the linear regression model are suspected to follow an AR(1) process

$$\varepsilon_i = \varrho\varepsilon_{i-1} + u_i, \quad |\varrho| < 1 ,$$

where the u_i are independently $N(0, \sigma_u^2)$ distributed. Then a test for autocorrelation is a test of the hypothesis

$$H_0 : \varrho = 0$$

versus the alternative $H_1 : \varrho \neq 0$. The alternative hypothesis may further be narrowed by testing for positive autocorrelation, i.e $H_1 : \varrho > 0$, or negative autocorrelation, i.e. $H_1 : \varrho < 0$. It is quite natural to expect that an appropriate test statistic is based on the estimator

$$\widehat{\varrho} = \frac{\sum_{i=2}^n \widehat{\varepsilon}_{i-1}\widehat{\varepsilon}_i}{\sum_{i=2}^n \widehat{\varepsilon}_{i-1}^2}$$

for ϱ, where $\widehat{\varepsilon}_i$ are the elements of the ordinary least squares residual vector $\widehat{\boldsymbol{\varepsilon}} = \boldsymbol{y} - \boldsymbol{X}\widehat{\boldsymbol{\beta}}$.

Durbin-Watson [31, 32, 33] Test. Clearly, either negative, no, or positive autocorrelation may be indicated by values $\widehat{\varrho}$ near to -1, 0, or 1. Alternatively, if one considers $d^* = 2 - 2\widehat{\varrho}$, then $d^* \in [0, 4]$ and small values of d^* indicate positive autocorrelation, values near 2 indicate no autocorrelation, and values near 4 indicate negative autocorrelation. The Durbin-Watson test statistic

$$d = \frac{\sum_{i=2}^{n} (\widehat{\varepsilon}_i - \widehat{\varepsilon}_{i-1})^2}{\sum_{i=1}^{n} \widehat{\varepsilon}_i^2}$$

satisfies $d \approx 2 - 2\widehat{\varrho}$. As a matter of fact, putting

$$\sum_{i=2}^{n} \widehat{\varepsilon}_i^2 \approx \sum_{i=2}^{n} \widehat{\varepsilon}_{i-1}^2 \approx \sum_{i=1}^{n} \widehat{\varepsilon}_i^2$$

gives

$$d = \underbrace{\frac{\sum_{i=2}^{n} \widehat{\varepsilon}_i^2 + \sum_{i=2}^{n} \widehat{\varepsilon}_{i-1}^2}{\sum_{i=1}^{n} \widehat{\varepsilon}_i^2}}_{\approx 2} - 2 \underbrace{\frac{\sum_{i=2}^{n} \widehat{\varepsilon}_i \widehat{\varepsilon}_{i-1}}{\sum_{i=1}^{n} \widehat{\varepsilon}_i^2}}_{\approx \widehat{\varrho}} .$$

For a given matrix $\boldsymbol{X}$, the exact distribution of d under H_0 can be obtained and thus a critical value $d(\alpha)$ for testing H_0 versus $H_1 : \varrho > 0$ can be computed. Then, H_0 is rejected at level α if $d < d(\alpha)$. A test of H_0 versus $H_1 : \varrho < 0$ rejects H_0 at level α if $d > 4 - d(\alpha)$, and a test of H_0 versus $H_1 : \varrho \neq 0$ rejects H_0 at level α if $d < d(\alpha)$ or $d > 4 - d(\alpha)$.

This procedure requires the computation of critical values $d(\alpha)$ whenever a new linear regression model with a specific model matrix $\boldsymbol{X}$ is considered. To avoid this, Durbin and Watson have considered a critical interval $[d_{\mathrm{L}}(\alpha), d_{\mathrm{U}}(\alpha)]$, only depending on n, p and α but not on the specific $\boldsymbol{X}$, such that it can be assumed that $d(\alpha) \in [d_{\mathrm{L}}(\alpha), d_{\mathrm{U}}(\alpha)]$. Tables of values $d_{\mathrm{L}}(\alpha)$ and $d_{\mathrm{U}}(\alpha)$ for different n, p and α can be found in most textbooks on econometrics. (Note that for these values it is assumed that the linear regression model contains an intercept variable.)

Now, for testing H_0 versus $H_1 : \varrho > 0$, the hypothesis is rejected if $d < d_{\mathrm{L}}(\alpha)$ and not rejected if $d > d_{\mathrm{U}}(\alpha)$, ensuring $d < d(\alpha)$ and $d > d(\alpha)$ respectively. However, when $d \in [d_{\mathrm{L}}(\alpha), d_{\mathrm{U}}(\alpha)]$, no conclusion upon rejection can be drawn, simply because it is not known whether $d < d(\alpha)$ or $d > d(\alpha)$. Therefore, the interval $[d_{\mathrm{L}}(\alpha), d_{\mathrm{U}}(\alpha)]$ is called the *inconclusive region* of the Durbin-Watson test, see Fig. 6.5.

Tests of H_0 versus $H_1 : \varrho < 0$ or $H_1 : \varrho \neq 0$ are carried out as above with $d(\alpha)$ replaced by $[d_{\mathrm{L}}(\alpha), d_{\mathrm{U}}(\alpha)]$ and $4 - d(\alpha)$ replaced by $[4 - d_{\mathrm{U}}(\alpha), 4 - d_{\mathrm{L}}(\alpha)]$.

The inconclusive region must be seen as a disadvantage of the Durbin-Watson test, since possible rejections of H_0 are given away. The application of computers, however, allows for avoiding the inconclusive region and even provides methods for computing p-values, see e.g. the R function `dwtest` from the package `lmtest` or the function `durbin.watson` from the package `car`.

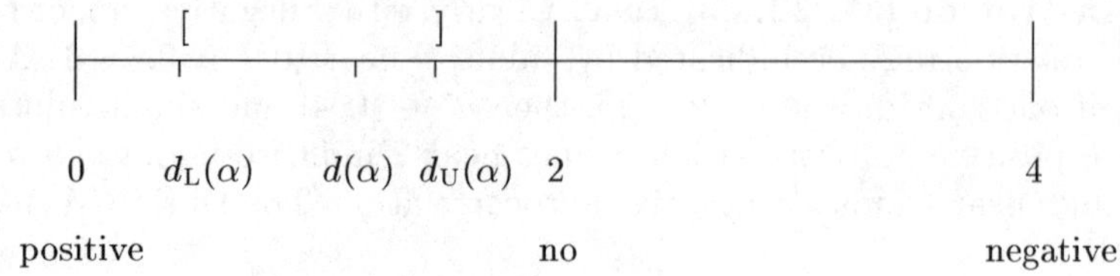

Fig. 6.5. Realization interval of the Durbin-Watson statistic d

6.6.4 Testing for Non-Normality

A well known test for the hypothesis H_0 of normality is the *Shapiro-Wilk* [100] test. It is available in R as the function `shapiro.test` from the package `ctest`. The function `shapiro.test` may be applied to the studentized residuals from a linear regression model to find some indication for non-normality of the errors.

In general, however, the Shapiro-Wilk test will not reveal more but only less than an inspection of the corresponding normal quantile plot. For example, the function `shapiro.test` yields a p-value of 0.677 for the $n = 30$ data points corresponding to the normal quantile plot in Fig 6.6. Thus, the hypothesis of normality is rather undoubted. The normal quantile plot supports this conclusion to a great extent, but in addition shows one residual being quite far away from the rest. For the data corresponding to the normal quantile plot in Fig 6.7, the function `shapiro.test` yields a p-value of 0.01477, indicating some departure from normality.

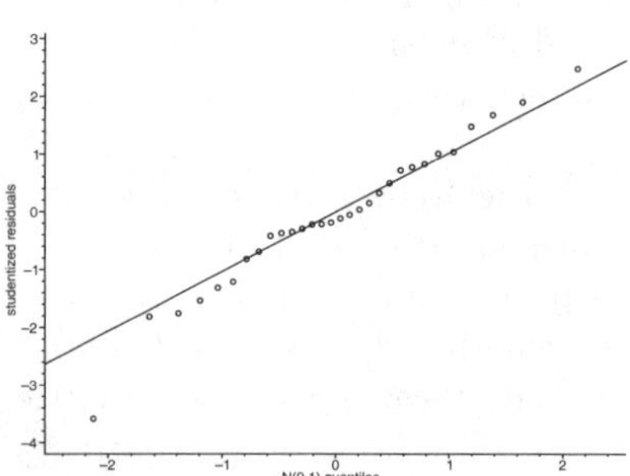

Fig. 6.6. Normal quantile plot from $n = 30$ points

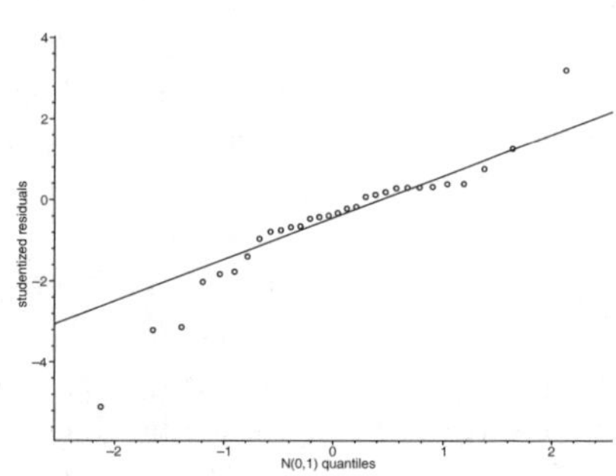

Fig. 6.7. Normal quantile plot from $n = 30$ points

Like any other conventional significance test, the Shapiro-Wilk test reduces the complete data to a single number, while the normal quantile plot still allows an inspection of n points and thus may reveal a lot more about the given data. Although graphical methods are advantageous over test procedures on that score, significant test results can nonetheless provide useful clues for further analysis.

6.6.5 Testing for Non-Linearity

When the undisturbed functional relationship

$$y = f(x_1, \ldots, x_p)$$

between the dependent variable y and the independent variables $x_1, \ldots, x_n$ is not linear, i.e. $f(\cdot)$ is not

$$f(x_1, \ldots, x_p) = \beta_1 x_1 + \beta_2 x_2 + \cdots + \beta_p x_p$$

for some parameters $\beta_1, \ldots, \beta_p$, then an analysis assuming linearity will yield incorrect conclusions. Thus, it is convenient to find some evidence in favor or against linearity.

Regression Specification Error Test (RESET). Suppose that the undisturbed relationship between a variable y and a variable x is given by

$$y = f(x) = \delta_1 + \delta_2 e^{\delta_3 x}$$

for parameters δ_1, δ_2 and δ_3. From

$$e^{\delta_3 x} = 1 + \delta_3 x + \frac{\delta_3^2 x^2}{2} + \frac{\delta_3^3 x^3}{6} + \cdots$$

one may conclude that the function $f(\cdot)$ can be approximated by

$$\begin{aligned} f(x) &\approx \delta_1 + \delta_2 + \delta_2\delta_3 x + \frac{\delta_2\delta_3^2 x^2}{2} + \frac{\delta_2\delta_3^3 x^3}{6} \\ &= \beta_1 + \beta_2 x + \gamma_1 x^2 + \gamma_2 x^3 \end{aligned}$$

for appropriate choices β_1, β_2 and γ_1, γ_2. Now, suppose that we have n observations of y and x available and we wish to examine whether the undisturbed relationship between y and x can be assumed as being linear. Then we may consider the linear regression model

$$y_i = \beta_1 + \beta_2 x_i + \gamma_1 x_i^2 + \gamma_2 x_i^3 + \varepsilon_i, \quad i = 1, \ldots, n ,$$

under which we examine the relevance of the variables x^2 and x^3 by testing the hypothesis $H_0 : \gamma_1 = \gamma_2 = 0$.

In this situation, we expect that the test result will give indication for the relevance of the variables x^2 and x^3, thus contradiction the hypothesis H_0 of linearity. By assuming that many other possible non-linearities in the specified relationship between y and $x_1, \ldots, x_n$ can similarly be approximated, we obtain the regression specification error test due to Ramsey [86], being simply the conventional F-test of the hypothesis

$$H_0 : \boldsymbol{\gamma} = \mathbf{0} \text{ in the model } \boldsymbol{y} = \boldsymbol{X\beta} + \boldsymbol{Z\gamma} + \boldsymbol{\varepsilon}$$

versus the alternative hypothesis

$$H_1 : \boldsymbol{\gamma} \neq \mathbf{0} \text{ in the model } \boldsymbol{y} = \boldsymbol{X\beta} + \boldsymbol{Z\gamma} + \boldsymbol{\varepsilon} \,,$$

where it is assumed that under the model $\boldsymbol{y} = \boldsymbol{X\beta} + \boldsymbol{Z\gamma} + \boldsymbol{\varepsilon}$ the assumptions (i) to (iv) and (v) are satisfied. If $\mathrm{RSS}(\boldsymbol{\beta}, \boldsymbol{\gamma})$ denotes the residual sum of squares obtained in the model $\boldsymbol{y} = \boldsymbol{X\beta} + \boldsymbol{Z\gamma} + \boldsymbol{\varepsilon}$ and $\mathrm{RSS}(\boldsymbol{\beta}, \boldsymbol{\gamma} = \mathbf{0})$ denotes the residual sum of squares obtained in the model $\boldsymbol{y} = \boldsymbol{X\beta} + \boldsymbol{\varepsilon}$, then the test statistic is

$$F = \frac{\mathrm{RSS}(\boldsymbol{\beta}, \boldsymbol{\gamma} = \mathbf{0}) - \mathrm{RSS}(\boldsymbol{\beta}, \boldsymbol{\gamma})}{\mathrm{RSS}(\boldsymbol{\beta}, \boldsymbol{\gamma})} \, \frac{n - p - q}{q} \,,$$

where q is the number of elements of the vector $\boldsymbol{\gamma}$. Under H_0, the statistic F is $F_{q,(n-p-q)}$ distributed and the hypothesis is rejected at level α if $F > F_{q,(n-p-q),1-\alpha}$. Rejection of H_0 indicates some misspecification in the functional form of the considered model $\boldsymbol{y} = \boldsymbol{X\beta} + \boldsymbol{\varepsilon}$.

For an actual application of RESET it is required to specify the matrix $\boldsymbol{Z}$, which comprises values of variables $z_1, \ldots, z_q$ being possibly responsible for the functional form of misspecification. In absence of any such prior knowledge, it is often proposed to use powers $\widehat{y}^2$, $\widehat{y}^3$, $\widehat{y}^4$ of the fitted values from the ordinary least squares regression, or powers x_j^2, x_j^3, x_j^4, $j = 1, \ldots, p$, of the independent variables. RESET can be carried out under R with the function `reset` from the package `lmtest`.

Rainbow Test. The main idea behind the rainbow test due to Utts [121] is, that under linearity the residual sum of squares obtained from a model with the complete observation matrix $(\boldsymbol{y}, \boldsymbol{X})$ does not differ considerably from the residual sum of squares obtained from a model with a subset $(\boldsymbol{y}_1, \boldsymbol{X}_1)$ of observations. On the other hand, a clear distinction in the two residual sum of squares may be seen as an indication for non-linearity, see Fig. 6.8 for an illustration.

An appropriate test can be derived as RESET when the artificial model

$$\begin{pmatrix} \boldsymbol{y}_1 \\ \boldsymbol{y}_2 \end{pmatrix} = \begin{pmatrix} \boldsymbol{X}_1 \\ \boldsymbol{X}_2 \end{pmatrix} \boldsymbol{\beta} + \begin{pmatrix} \mathbf{0} \\ \boldsymbol{I}_{n-n_1} \end{pmatrix} \boldsymbol{\gamma} + \begin{pmatrix} \boldsymbol{\varepsilon}_1 \\ \boldsymbol{\varepsilon}_2 \end{pmatrix}$$

with assumptions (i) to (iv) and (v) is considered, where $(\boldsymbol{y}_1, \boldsymbol{X}_1)$ contains the n_1 observations from the considered subset and $(\boldsymbol{y}_2, \boldsymbol{X}_2)$ contains the remaining $n - n_1$ observations. Then the alternative hypothesis of non-linearity may be reformulated as $\boldsymbol{\gamma} \neq \mathbf{0}$.

Now, the residual sum of squares obtained from the model with complete observation matrix $(\boldsymbol{y}, \boldsymbol{X})$ is $\mathrm{RSS}(\boldsymbol{\beta}, \boldsymbol{\gamma} = \mathbf{0})$, while it can be shown that the

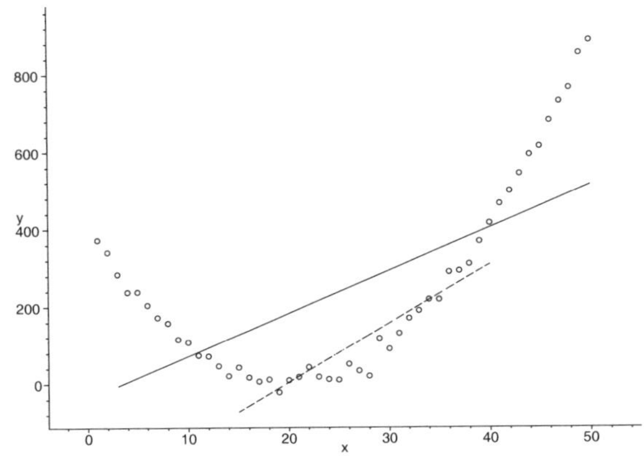

Fig. 6.8. Regression lines from a model $y_i = \beta_1 + \beta_2 x_i + \varepsilon_i$ with complete data (solid line) and with a subset (corresponding to x values between 15 and 40) of the data (dotted line)

residual sum of squares obtained from the model with the subset $(\boldsymbol{y}_1, \boldsymbol{X}_1)$ of observations is identical to $\mathrm{RSS}(\boldsymbol{\beta}, \boldsymbol{\gamma})$, see Problem 6.4. Then an appropriate test statistic is

$$F = \frac{\mathrm{RSS}(\boldsymbol{\beta}, \boldsymbol{\gamma} = \mathbf{0}) - \mathrm{RSS}(\boldsymbol{\beta}, \boldsymbol{\gamma})}{\mathrm{RSS}(\boldsymbol{\beta}, \boldsymbol{\gamma})} \frac{n_1 - p}{n - n_1},$$

being $F_{(n-n_1),(n_1-p)}$ distributed under H_0.

For an actual application of the rainbow test, it is necessary to specify the subset $(\boldsymbol{y}_1, \boldsymbol{X}_1)$ of observations along with the number n_1. The usual recommendation is to choose about one half out of the middle of the observations. This can be achieved by taking the $n_1 = n/2$ observations with the smallest hat values p_{ii} for the subset. The rainbow test can be carried out in R with the function `raintest` from the package `lmtest`.

When it is known that the functional form in some independent variable is convex or concave, then the *Harvey-Collier test* might be advantageous over the rainbow test, see e.g. [65] for details.

Structural Change. As a different form of non-linearity one may also see the occurrence of ***structural change*** in the parameters. This means that the functional form of the considered model might be correct, but the parameters are different for different groups of the data, see Fig 6.9 for an illustration.

The *Chow test* for structural change is identical to the rainbow test with the distinction that the criterion for choosing the appropriate subset is different. For the Chow test, the data points are arranged according to time and then the chosen subset consists of the first n_1 data points. Since the Chow test requires knowledge about the possible shift point, one may prefer

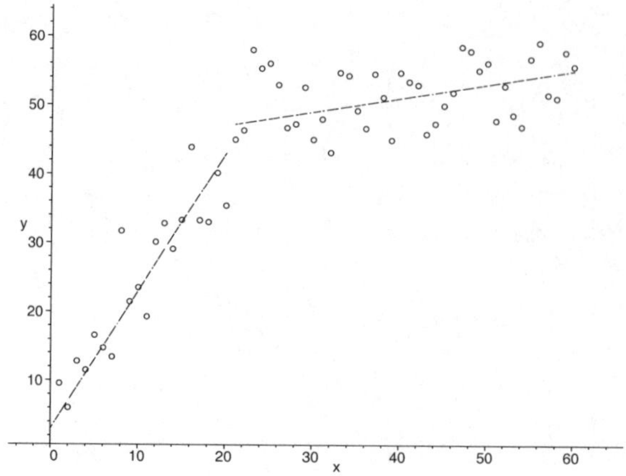

Fig. 6.9. Data points from a model $y_i = \beta_1 + \beta_2 i + \varepsilon_i$ with different parameters for the two groups $\{(i, y_i) : i = 1, \ldots, 20\}$ and $\{(i, y_i) : i = 21, \ldots, 60\}$

alternatives like the *CUSUM test.* See [65] for details and [127, 129] for descriptions of the R package `strucchange`, delivering advanced test methods for structural change.

6.6.6 Testing for Outlier

When the i-th observation is suspected to be an outlier, then the so called *mean-shift outlier model*

$$\boldsymbol{y} = \boldsymbol{X\beta} + \boldsymbol{e}_i\gamma + \boldsymbol{\varepsilon}$$

may be considered, where $\boldsymbol{e}_i$ denotes the i-th $n \times 1$ unit vector. The alternative hypothesis that the i-th observation is an outlier may then be formulated as $H_1 : \gamma \neq 0$ in the mean shift outlier model. The test may also be seen as a special case of RESET with the test statistic reducing to the i-th squared studentized residual $(r_i^*)^2$. Since r_i^* is t_{n-p-1} distributed, see e.g. [20], one may alternatively use the equivalent t-test, where the hypothesis $H_0 : \gamma = 0$ is rejected at level α if $|r_i^*| > t_{n-p-1,\alpha/2}$.

This test requires prior knowledge about the concrete observation being the possible outlier. If such knowledge is not available, then one may weaken the alternative hypothesis to 'there is at least one outlier in the model'. As an appropriate test statistic one may take $\max_{1 \le i \le n} |r_i^*|$ and reject the hypothesis H_0 : not H_1 at level α if

$$\max_{1 \le i \le n} |r_i^*| > t_{n-p-1,\alpha/2n} \,.$$

This is equivalent to testing each individual observation and rejecting the hypothesis if at least one test is significant, where, however, the level for each test is chosen as α/n, in order to ensure that the probability of the Type I error for the overall test is bounded by α (Bonferroni method). This test can be performed in R with the function `outlier.test` from the package `car`.

6.7 Problems

6.1. Consider the comparison of two linear regression models with different independent variables but the same observation $\boldsymbol{y}$. Show that each of inequalities $C_p(1) \le C_p(2)$ (for a fixed $\widehat{\sigma}$), $\mathrm{AIC}_p(1) \le \mathrm{AIC}_p(2)$, $\widehat{\sigma}^2(1) \le \widehat{\sigma}^2(2)$, $\overline{R}^2(2) \le \overline{R}^2(1)$, and $R_c^2(2) \le R_c^2(1)$ is equivalent to $\mathrm{RSS}_p(1) \le \mathrm{RSS}_p(2)$.

6.2. Consider the linear regression model $\boldsymbol{y} = \boldsymbol{X\beta} + \boldsymbol{\varepsilon}$ with assumptions (i) to (iv), and let the $n \times 1$ vector of ones $\boldsymbol{1}_n$ belong to the column space of $\boldsymbol{X}$. Let

$$R = \frac{\sum_{i=1}^n (y_i - \overline{y})(\widehat{y}_i - \overline{\widehat{y}})}{\left(\sum_{i=1}^n (y_i - \overline{y})^2 \sum_{i=1}^n (\widehat{y}_i - \overline{\widehat{y}})^2\right)^{1/2}}$$

be the *sample multiple correlation coefficient* between y and the prediction $\widehat{y}$.

(a) Show that $\boldsymbol{1}_n'\widehat{\boldsymbol{y}} = \boldsymbol{1}_n'\boldsymbol{y}$ for $\widehat{\boldsymbol{y}} = \boldsymbol{P}\boldsymbol{y}$, $\boldsymbol{P} = \boldsymbol{X}(\boldsymbol{X}'\boldsymbol{X})^{-1}\boldsymbol{X}'$.
(b) Show that R can be written as

$$R = \frac{\boldsymbol{y}'(\boldsymbol{P} - \boldsymbol{P}_1)\boldsymbol{y}}{\left[\boldsymbol{y}'(\boldsymbol{I}_n - \boldsymbol{P}_1)\boldsymbol{y}\boldsymbol{y}'(\boldsymbol{P} - \boldsymbol{P}_1)\boldsymbol{y}\right]^{1/2}},$$

where $\boldsymbol{P}_1 = (1/n)\boldsymbol{1}_n\boldsymbol{1}_n'$.
(c) Conclude that the centered coefficient of determination R_c^2 is the square of R.

6.3. Consider an $n \times p$ matrix $\boldsymbol{X}$ of full column rank such $\mathcal{C}(\boldsymbol{1}_n) \subseteq \mathcal{C}(\boldsymbol{X})$. Show that $(1/n) \le p_{ii}$ for $i = 1, \ldots, n$, where p_{ii} is the i-th main diagonal element of $\boldsymbol{P} = \boldsymbol{X}(\boldsymbol{X}'\boldsymbol{X})^{-1}\boldsymbol{X}'$.

6.4. Consider the linear regression model

$$\begin{pmatrix} \boldsymbol{y}_1 \\ \boldsymbol{y}_2 \end{pmatrix} = \begin{pmatrix} \boldsymbol{X}_1 & \boldsymbol{0}_{n_1 \times (n-n_1)} \\ \boldsymbol{X}_2 & \boldsymbol{I}_{n-n_1} \end{pmatrix} \begin{pmatrix} \boldsymbol{\beta} \\ \boldsymbol{\gamma} \end{pmatrix} + \begin{pmatrix} \boldsymbol{\varepsilon}_1 \\ \boldsymbol{\varepsilon}_2 \end{pmatrix}$$

with assumptions (i) to (iv), where $\boldsymbol{X}_1$ in an $n_1 \times p$ matrix of full column rank ($n_1 > p$).

(a) Show that

$$\begin{pmatrix} \boldsymbol{X}_1'\boldsymbol{X}_1 + \boldsymbol{X}_2'\boldsymbol{X}_2 & \boldsymbol{X}_2' \\ \boldsymbol{X}_2 & \boldsymbol{I}_{n-n_1} \end{pmatrix}^{-1} = \begin{pmatrix} (\boldsymbol{X}_1'\boldsymbol{X}_1)^{-1} & -(\boldsymbol{X}_1'\boldsymbol{X}_1)^{-1}\boldsymbol{X}_2' \\ -\boldsymbol{X}_2(\boldsymbol{X}_1'\boldsymbol{X}_1)^{-1} & \boldsymbol{I}_{n-n_1} + \boldsymbol{X}_2(\boldsymbol{X}_1'\boldsymbol{X}_1)^{-1}\boldsymbol{X}_2' \end{pmatrix}$$

holds true.

(b) Show that the ordinary least squares estimator for the vector $(\beta', \gamma')'$ is given as

$$\begin{pmatrix} \widehat{\beta} \\ \widehat{\gamma} \end{pmatrix} = \begin{pmatrix} (\boldsymbol{X}_1'\boldsymbol{X}_1)^{-1}\boldsymbol{X}_1'\boldsymbol{y}_1 \\ \boldsymbol{y}_2 - \boldsymbol{X}_2\widehat{\beta} \end{pmatrix} .$$

(c) Show that the residual sum of squares obtained in the above model is identical to

$$(\boldsymbol{y}_1 - \boldsymbol{X}_1\widehat{\beta})'(\boldsymbol{y}_1 - \boldsymbol{X}_1\widehat{\beta}) ,$$

being the residual sum of squares obtained under the model $\boldsymbol{y}_1 = \boldsymbol{X}_1\beta + \varepsilon_1$.

6.5. Consider the mean-shift outlier model

$$\boldsymbol{y} = \boldsymbol{X}\beta + \boldsymbol{e}_i\gamma + \varepsilon$$

with assumptions (i) to (iv), where $\boldsymbol{X}$ in an $n \times p$ matrix of full column rank.

(a) Show that the residual sum of squares obtained under the mean shift outlier model is identical to

$$\boldsymbol{y}_{(i)}'(\boldsymbol{I}_{n-1} - \boldsymbol{X}_{(i)}(\boldsymbol{X}_{(i)}'\boldsymbol{X}_{(i)})^{-1}\boldsymbol{X}_{i}')\boldsymbol{y}_{(i)} ,$$

being the residual sum of squares obtained under the model $\boldsymbol{y}_{(i)} = \boldsymbol{X}_{(i)}\beta + \varepsilon_{(i)}$, where $(\boldsymbol{y}_{(i)}, \boldsymbol{X}_{(i)})$ is obtained from $(\boldsymbol{y}, \boldsymbol{X})$ by deleting the i-th row.

(b) The orthogonal projector onto the column space of $(\boldsymbol{X}, \boldsymbol{e}_i)$ can be written as

$$\boldsymbol{P} + \frac{1}{1 - p_{ii}}(\boldsymbol{I}_n - \boldsymbol{P})\boldsymbol{e}_i\boldsymbol{e}_i'(\boldsymbol{I}_n - \boldsymbol{P}) ,$$

where $\boldsymbol{P} = \boldsymbol{X}(\boldsymbol{X}'\boldsymbol{X})^{-1}\boldsymbol{X}'$, see e.g. Property 2.4 in [20]. Show that the residual sum of squares obtained under the mean shift outlier model can be written as

$$\boldsymbol{y}'(\boldsymbol{I}_n - \boldsymbol{P})\boldsymbol{y} - \frac{\widehat{\varepsilon}_i^2}{1 - p_{ii}} .$$

(c) Conclude from (a) and (b) that

$$\widehat{\sigma}_{(i)}^2 = \widehat{\sigma}^2 \frac{n - p - r_i^2}{n - p - 1} ,$$

where r_i denotes the i-th standardized residual.

(d) Show that the i-th studentized residual can be written as

$$r_i^* = r_i\sqrt{\frac{n - p - 1}{n - p - r_i^2}} .$$

Table 6.8. Trees data

No.	Girth	Height	Volume	No.	Girth	Height	Volume
1	8.3	70	10.3	17	12.9	85	33.8
2	8.6	65	10.3	18	13.3	86	27.4
3	8.8	63	10.2	19	13.7	71	25.7
4	10.5	72	16.4	20	13.8	64	24.9
5	10.7	81	18.8	21	14.0	78	34.5
6	10.8	83	19.7	22	14.2	80	31.7
7	11.0	66	15.6	23	14.5	74	36.3
8	11.0	75	18.2	24	16.0	72	38.3
9	11.1	80	22.6	25	16.3	77	42.6
10	11.2	75	19.9	26	17.3	81	55.4
11	11.3	79	24.2	27	17.5	82	55.7
12	11.4	76	21.0	28	17.9	80	58.3
13	11.4	76	21.4	29	18.0	80	51.5
14	11.7	69	21.3	30	18.0	80	51.0
15	12.0	75	19.1	31	20.6	87	77.0
16	12.9	74	22.2				

6.6. Consider the data given in Table 6.8, providing measurements of the girth, height and volume of timber in 31 felled black cherry trees, see [93]. (This data set is available in R via `data(trees)`.)

(a) Consider a linear regression model with 'Volume' as dependent variable and 'Girth' and 'Height' as independent variables. A pairwise plot of the three variables may indicate that the postulated relationship is not linear. Compute the p-values for RESET (with squares and cubes of fitted values as z variables) and the rainbow test and conclude that the hypothesis of linearity is only weakly supported.
(b) Consider a regression model

$$\ln(\text{Volume})_i = \beta_1 \ln(\text{Girth})_i + \beta_2 \ln(\text{Height})_i + \varepsilon_i$$

and compute again p-values for RESET and the rainbow test. Conclude that the p-values do not act against the hypothesis of linearity.
(c) Compute the adjusted coefficient of determination for the model considered in (b) and conclude that the linear fit is quite high.
(d) Check whether stepwise regression will suggest an alternative model to (b) with less independent variables, and conclude that this is not the case.
(e) Check whether there are important outliers or high-leverage points in the model (b), and conclude that this is not the case.
(f) Compute the matrix $\boldsymbol{\Pi}$ of variance-decomposition proportions in the model (b) and conclude that there is one harmful near linear dependency in the model, involving the intercept variable and ln(Height).

A Matrix Algebra

This chapter presents basic and more far-reaching definitions and theorems in matrix algebra, being useful for the understanding of the results in this book. Proofs are omitted here. More comprehensive treatments of this topic with an emphasize on a statistical background are given in [50, 88, 91]. In addition, [59, 78, 133] provide beneficial presentations of the theory on matrices. A collection of useful results from different areas of matrix algebra is given in [73].

A.1 Preliminaries

This section gives basic definitions and elementary results on matrices and vectors.

A.1.1 Matrices and Vectors

The consideration of arrays of numbers as mathematical entities allows the treatment of problems in linear algebra in an elegant and clear fashion.

Definition A.1. *A rectangular arrangement*

$$\boldsymbol{A} = \begin{pmatrix} a_{11} & a_{12} & \cdots & a_{1n} \\ a_{21} & a_{22} & \cdots & a_{2n} \\ \vdots & \vdots & & \vdots \\ a_{m1} & a_{m2} & \cdots & a_{mn} \end{pmatrix}$$

of elements a_{ij}, $i = 1, \ldots, m$, $j = 1, \ldots, n$, in m rows and n columns is called an $m \times n$ matrix *$\boldsymbol{A}$.*

A matrix $\boldsymbol{A}$ is called *real-valued* if all its elements a_{ij} belong to the set $\mathbb{R}$ of real numbers. We only consider real-valued matrices here.

Definition A.2. *If every element of an $m \times n$ matrix is equal to zero, then this matrix is called* zero matrix. *A zero matrix is denoted by $\mathbf{0}$, or sometimes by $\mathbf{0}_{m \times n}$ to give additional information about the number of rows and columns when necessary.*

Definition A.3. *An $m \times n$ matrix $\boldsymbol{A}$ is called a* square matrix, *if $m = n$.*

In view of the above definition we may call an $m \times n$ matrix $\boldsymbol{A}$ *rectangular.* We may use this notion for $m \times n$ matrices with possibly different numbers of row and columns, not excluding the special case $m = n$.

Definition A.4. *The* transpose *of an $m \times n$ matrix $\boldsymbol{A}$ is the $n \times m$ matrix $\boldsymbol{A}'$, which is obtained from $\boldsymbol{A}$ by writing the rows of $\boldsymbol{A}$ as the columns of $\boldsymbol{A}'$.*

Of course, one may also obtain the transpose of $\boldsymbol{A}$ by writing its columns as rows.

Remark A.1. The (i, j)-th element of $\boldsymbol{A}'$ is the (j, i)-th element of $\boldsymbol{A}$.

Theorem A.1. *For an $m \times n$ matrix $\boldsymbol{A}$ the identity $\boldsymbol{A} = (\boldsymbol{A}')'$ holds true.*

Definition A.5. *A square matrix $\boldsymbol{A}$ is called* symmetric, *if $\boldsymbol{A} = \boldsymbol{A}'$.*

Definition A.6. *A* column vector *is a matrix with one column. A* row vector *is a matrix with one row.*

Assumption A.1. *A vector is understood as a column vector. A $1 \times n$ row vector is always represented as the transpose $\boldsymbol{a}'$ of an $n \times 1$ column vector $\boldsymbol{a}$.*

Definition A.7. *The $m \times 1$ vector having 1 as its i-th element and 0 as all other elements is called the i-th $m \times 1$* unit vector. *It is denoted by $\boldsymbol{e}_i$.*

A.1.2 Elementary Operations

Sum and product can be specified as elementary operations for matrices.

Definition A.8. *The sum and the product of matrices are defined as follows:*

(a) *An $m \times n$ matrix $\boldsymbol{A}$ and a $p \times q$ matrix $\boldsymbol{B}$ can only be added if $m = p$ and $n = q$. In this case, the sum $\boldsymbol{A} + \boldsymbol{B}$ is the matrix whose (i, j)th element equals $a_{ij} + b_{ij}$.*
(b) *An $m \times n$ matrix $\boldsymbol{A}$ can only be multiplied from the right by a $p \times q$ matrix $\boldsymbol{B}$ if $p = n$. In this case the product $\boldsymbol{AB}$ is the matrix whose (i, j)th element equals $\sum_{k=1}^{n} a_{ik} b_{kj}$.*

Remark A.2. If α is a real number and $\boldsymbol{A}$ is an $m \times n$ matrix, then $\alpha \boldsymbol{A}$ is the matrix whose (i, j)-th element equals αa_{ij}.

Remark A.3. The difference of two $m \times n$ matrices $\boldsymbol{A}$ and $\boldsymbol{B}$ can naturally be specified by $\boldsymbol{A} - \boldsymbol{B} = \boldsymbol{A} + (-1)\boldsymbol{B}$.

Remark A.4. Similarly to part (b) of Definition A.8, we can define the multiplication from the left. If we consider the product $\boldsymbol{AB}$, then we may either say that $\boldsymbol{A}$ is multiplied from the right by $\boldsymbol{B}$, or $\boldsymbol{B}$ is multiplied from the left by $\boldsymbol{A}$. The multiplication from one side is emphasized here, since the multiplication of matrices is *not* commutative. If the product $\boldsymbol{AB}$ of an $m \times n$ matrix $\boldsymbol{A}$ and an $n \times q$ matrix $\boldsymbol{B}$ exists, then this is not the case for $\boldsymbol{BA}$ unless $m = q$. If $m = q$, then $\boldsymbol{AB}$ is an $m \times m$ matrix while $\boldsymbol{BA}$ is an $n \times n$ matrix. Even if $m = q = n$, the identity $\boldsymbol{AB} = \boldsymbol{BA}$ is usually *not* true. If it is nonetheless satisfied, then $\boldsymbol{A}$ and $\boldsymbol{B}$ are said to be *commuting* matrices

Remark A.5. There exist further matrix products like 'Kronecker product', 'Hadamard product' or 'Khatri-Rao product', each one having specific properties.

For the transpose of the product of two matrices the reverse order law is valid.

Theorem A.2. *Let $\boldsymbol{A}$ be an $m \times n$ and $\boldsymbol{B}$ be an $n \times q$ matrix. Then $(\boldsymbol{AB})' = \boldsymbol{B}'\boldsymbol{A}'$.*

A.1.3 Rank of a Matrix

A characteristic number of any $m \times n$ matrix is its rank.

Definition A.9. *Let $\boldsymbol{A}$ be an $m \times n$ matrix.*

(a) *The* row rank *of $\boldsymbol{A}$ is the maximal number of linearly independent rows of $\boldsymbol{A}$ (where each row is considered as a row vector). If all rows of $\boldsymbol{A}$ are linearly independent, then $\boldsymbol{A}$ is called a matrix of* full row rank.
(b) *The* column rank *of $\boldsymbol{A}$ is the maximal number of linearly independent columns of $\boldsymbol{A}$ (where each column is considered as a column vector). If all columns of $\boldsymbol{A}$ are linearly independent, then $\boldsymbol{A}$ is called a matrix of* full column rank.

Theorem A.3. *For an $m \times n$ matrix $\boldsymbol{A}$ its row rank and its column rank coincide.*

Definition A.10. *The row rank (being identical to the column rank) of an $m \times n$ matrix $\boldsymbol{A}$ is simply called the* rank *of $\boldsymbol{A}$ and it is denoted by* $\mathrm{rk}(\boldsymbol{A})$.

Definition A.11. *If an $m \times m$ matrix $\boldsymbol{A}$ has rank m, then $\boldsymbol{A}$ is called* non-singular. *If* $\mathrm{rk}(\boldsymbol{A}) < m$, *then $\boldsymbol{A}$ is called* singular.

Theorem A.4. *For an $m \times n$ matrix $\boldsymbol{A}$ the identities*

$$\mathrm{rk}(\boldsymbol{A}) = \mathrm{rk}(\boldsymbol{A}') = \mathrm{rk}(\boldsymbol{AA}') = \mathrm{rk}(\boldsymbol{A}'\boldsymbol{A})$$

hold true.

Theorem A.5. *Let* $\boldsymbol{A}$ *be an* $m \times n$ *matrix. If* $\boldsymbol{C}$ *is a nonsingular* $m \times m$ *matrix and* $\boldsymbol{B}$ *is a nonsingular* $n \times n$ *matrix, then the identities*

$$\mathrm{rk}(\boldsymbol{CAB}) = \mathrm{rk}(\boldsymbol{AB}) = \mathrm{rk}(\boldsymbol{CA}) = \mathrm{rk}(\boldsymbol{A})$$

hold true.

The above theorem states that the multiplication of a matrix $\boldsymbol{A}$ with some nonsingular matrix from the right or the left gives always a matrix with the same rank as $\boldsymbol{A}$.

For the rank of the product and the sum of two matrices the following results are valid.

Theorem A.6. *For an* $m \times n$ *matrix* $\boldsymbol{A}$ *and an* $n \times q$ *matrix* $\boldsymbol{B}$ *the inequalities*

$$\mathrm{rk}(\boldsymbol{A}) + \mathrm{rk}(\boldsymbol{B}) - n \leq \mathrm{rk}(\boldsymbol{AB}) \leq \min\{\mathrm{rk}(\boldsymbol{A}), \mathrm{rk}(\boldsymbol{B})\}$$

hold true.

Theorem A.7. *For two* $m \times n$ *matrices* $\boldsymbol{A}$ *and* $\boldsymbol{B}$ *the inequality*

$$\mathrm{rk}(\boldsymbol{A} + \boldsymbol{B}) \leq \mathrm{rk}(\boldsymbol{A}) + \mathrm{rk}(\boldsymbol{B})$$

holds true.

A.1.4 Subspaces and Matrices

The columns of any $m \times n$ matrix generate a subspace of $\mathbb{R}^m$.

Definition A.12. *Let* $\boldsymbol{A}$ *be an* $m \times n$ *matrix.*

(a) *The* column space *of* $\boldsymbol{A}$ *is the set of all* $m \times 1$ *vectors* $\boldsymbol{y}$ *satisfying* $\boldsymbol{y} = \boldsymbol{Ax}$ *for some* $n \times 1$ *vector* $\boldsymbol{x}$. *The column space of* $\boldsymbol{A}$ *is denoted by* $\mathcal{C}(\boldsymbol{A})$.
(b) *The* null space *of* $\boldsymbol{A}$ *is the set of all* $n \times 1$ *vectors* $\boldsymbol{y}$ *satisfying* $\boldsymbol{Ay} = \boldsymbol{0}$. *The null space of* $\boldsymbol{A}$ *is denoted by* $\mathcal{N}(\boldsymbol{A})$.

The column space of an $m \times n$ matrix $\boldsymbol{A}$ is a subspace of $\mathbb{R}^m$. Its dimension equals the rank of the matrix $\boldsymbol{A}$. The null space of $\boldsymbol{A}$ is a subspace of $\mathbb{R}^n$. Its dimension equals $n - \mathrm{rk}(\boldsymbol{A})$ and is also called the *nullity* of $\boldsymbol{A}$. Hence, we have the *rank plus nullity theorem*

$$\dim[\mathcal{C}(\boldsymbol{A})] + \dim[\mathcal{N}(\boldsymbol{A})] = n$$

for any $m \times n$ matrix $\boldsymbol{A}$.

Theorem A.8. *For an* $m \times n$ *matrix* $\boldsymbol{A}$ *the identities*

$$\mathcal{C}(\boldsymbol{A}) = \mathcal{C}(\boldsymbol{AA}') \quad \text{and} \quad \mathcal{N}(\boldsymbol{A}) = \mathcal{N}(\boldsymbol{A}'\boldsymbol{A})$$

hold true.

If $\mathcal{U}$ and V are two subspaces, then the sum

$$\mathcal{U} + \mathcal{V} = \{\boldsymbol{u} + \boldsymbol{v} \ : \boldsymbol{u} \in \mathcal{U}, \, \boldsymbol{v} \in \mathcal{V}\}$$

is again a subspace. Moreover, the intersection $\mathcal{U} \cap \mathcal{V}$ is a subspace, while this is *not* necessarily the case for the union $\mathcal{U} \cup \mathcal{V}$. The intersection $\mathcal{U} \cap \mathcal{V}$ always contains the zero vector as an element. If $\mathcal{U} \cap \mathcal{V} = \{\boldsymbol{0}\}$, then $\mathcal{U}$ and $\mathcal{V}$ are also called *disjoint*. The sum of $\mathcal{U}$ and $\mathcal{V}$ is written as $\mathcal{U} \oplus \mathcal{V}$ (*direct sum*) whenever $\mathcal{U}$ and $\mathcal{V}$ are disjoint.

Theorem A.9. *For an $m \times n$ matrix $\boldsymbol{A}$ and an $n \times q$ matrix $\boldsymbol{B}$ the identity*

$$\mathrm{rk}(\boldsymbol{AB}) = \mathrm{rk}(\boldsymbol{B}) - \dim[\mathcal{N}(\boldsymbol{A}) \cap \mathcal{C}(\boldsymbol{B})]$$

holds true.

For any subspace $\mathcal{U}$ of $\mathrm{I\!R}^k$, there exists an *orthogonal complement* $\mathcal{U}^{\perp}$, given as

$$\mathcal{U}^{\perp} = \{\boldsymbol{v} \in \mathrm{I\!R}^k \,|\, \boldsymbol{v}'\boldsymbol{u} = 0 \, \forall \boldsymbol{u} \in \mathcal{U}\} \, .$$

Then $\mathrm{I\!R}^k = \mathcal{U} \oplus \mathcal{U}^{\perp}$.

Theorem A.10. *For an $m \times n$ matrix $\boldsymbol{A}$ the identities $\mathcal{C}(\boldsymbol{A})^{\perp} = \mathcal{N}(\boldsymbol{A}')$ and $\mathcal{N}(\boldsymbol{A})^{\perp} = \mathcal{C}(\boldsymbol{A}')$ hold true.*

A.1.5 Partitioned Matrices

Any matrix which is obtained form an $m \times n$ matrix $\boldsymbol{A}$ by deleting rows and/or columns is called a *submatrix* of $\boldsymbol{A}$.

Definition A.13. *An $m \times n$ matrix*

$$\boldsymbol{A} = \begin{pmatrix} \boldsymbol{A}_{11} & \boldsymbol{A}_{12} & \cdots & \boldsymbol{A}_{1t} \\ \boldsymbol{A}_{21} & \boldsymbol{A}_{22} & \cdots & \boldsymbol{A}_{2t} \\ \vdots & \vdots & & \vdots \\ \boldsymbol{A}_{s1} & \boldsymbol{A}_{s2} & \cdots & \boldsymbol{A}_{st} \end{pmatrix} ,$$

consisting of $m_i \times n_j$ submatrices $\boldsymbol{A}_{ij}$ for $i = 1, \ldots, s$ and $j = 1, \ldots, t$, is called partitioned matrix.

The matrix $\boldsymbol{A}_{ij}$ is often called the (i,j)-th *block* of the partitioned matrix $\boldsymbol{A}$. If an $m \times n$ matrix $\boldsymbol{A}$ and an $n \times q$ matrix $\boldsymbol{B}$ are partitioned as

$$\boldsymbol{A} = \begin{pmatrix} \boldsymbol{A}_{11} & \boldsymbol{A}_{12} & \cdots & \boldsymbol{A}_{1t} \\ \boldsymbol{A}_{21} & \boldsymbol{A}_{22} & \cdots & \boldsymbol{A}_{2t} \\ \vdots & \vdots & & \vdots \\ \boldsymbol{A}_{s1} & \boldsymbol{A}_{s2} & \cdots & \boldsymbol{A}_{st} \end{pmatrix} , \quad \boldsymbol{B} = \begin{pmatrix} \boldsymbol{B}_{11} & \boldsymbol{B}_{12} & \cdots & \boldsymbol{B}_{1u} \\ \boldsymbol{A}_{21} & \boldsymbol{A}_{22} & \cdots & \boldsymbol{A}_{2u} \\ \vdots & \vdots & & \vdots \\ \boldsymbol{A}_{t1} & \boldsymbol{A}_{t2} & \cdots & \boldsymbol{A}_{tu} \end{pmatrix}$$

such that the product $\boldsymbol{A}_{ik}\boldsymbol{B}_{kj}$ exists for every possible i, j and k, then the product $\boldsymbol{AB}$ is again a partitioned matrix with

$$\boldsymbol{A}_{i1}\boldsymbol{B}_{1j} + \boldsymbol{A}_{i2}\boldsymbol{B}_{2j} + \cdots + \boldsymbol{A}_{it}\boldsymbol{B}_{tj}$$

as its (i, j)-th block. In this case the matrices $\boldsymbol{A}$ and $\boldsymbol{B}$ are said to be conformable partitioned for the product $\boldsymbol{AB}$. (One may also differently specify conformable partitioning for the addition of two $m \times n$ matrices $\boldsymbol{A}$ and $\boldsymbol{B}$.)

A.1.6 Kronecker Product

As noted before, one may define the product of two matrices in different ways. The Kronecker product is useful for a convenient handling of specific partitioned matrices.

Definition A.14. *For an $m \times n$ matrix $\boldsymbol{A}$ and a $p \times q$ matrix $\boldsymbol{B}$, the Kronecker product $\boldsymbol{A} \otimes \boldsymbol{B}$ is the matrix*

$$\boldsymbol{A} \otimes \boldsymbol{B} = \begin{pmatrix} a_{11}\boldsymbol{B} & a_{12}\boldsymbol{B} & \cdots & a_{1n}\boldsymbol{B} \\ a_{21}\boldsymbol{B} & a_{22}\boldsymbol{B} & \cdots & a_{2n}\boldsymbol{B} \\ \vdots & \vdots & & \vdots \\ a_{m1}\boldsymbol{B} & a_{m2}\boldsymbol{B} & \cdots & a_{mn}\boldsymbol{B} \end{pmatrix},$$

where a_{ij} denotes the (i, j)-th element of $\boldsymbol{A}$.

Unlike the ordinary matrix product, the Kronecker product exists for matrices $\boldsymbol{A}$ and $\boldsymbol{B}$ of arbitrary dimensions.

Theorem A.11. *Let $\boldsymbol{A}$ be an $m \times n$ matrix and let $\boldsymbol{B}$ be a $p \times q$ matrix. Then the identities*

$$(k\boldsymbol{A}) \otimes \boldsymbol{B} = \boldsymbol{A} \otimes (k\boldsymbol{B}) = k(\boldsymbol{A} \otimes \boldsymbol{B})$$

hold true for any number k.

Theorem A.12. *Let $\boldsymbol{A}$ be an $m \times n$ matrix and let $\boldsymbol{B}$ be a $p \times q$ matrix. Then the identity*

$$(\boldsymbol{A} \otimes \boldsymbol{B})' = \boldsymbol{A}' \otimes \boldsymbol{B}'$$

holds true.

Note that the reverse order law does not hold for the transpose of the Kronecker product of two matrices, as opposed to the case when the ordinary matrix product is regarded.

Theorem A.13. *Let $\boldsymbol{A}$ be an $m \times n$ matrix and let $\boldsymbol{B}$ be a $p \times q$ matrix. Then the identity*

$$\operatorname{rk}(\boldsymbol{A} \otimes \boldsymbol{B}) = \operatorname{rk}(\boldsymbol{A}) \operatorname{rk}(\boldsymbol{B})$$

holds true.

Theorem A.14. *Let* $\boldsymbol{A}$ *and* $\boldsymbol{B}$ *be two* $m \times n$ *matrices and let* $\boldsymbol{C}$ *be a* $p \times q$ *matrix. Then the identities*

$$(\boldsymbol{A} + \boldsymbol{B}) \oplus \boldsymbol{C} = (\boldsymbol{A} \otimes \boldsymbol{C}) + (\boldsymbol{B} \otimes \boldsymbol{C})$$

and

$$\boldsymbol{C} \otimes (\boldsymbol{A} + \boldsymbol{B}) = (\boldsymbol{C} \otimes \boldsymbol{A}) + (\boldsymbol{C} \otimes \boldsymbol{B})$$

hold true.

Theorem A.15. *Let* $\boldsymbol{A}$ *be an* $m \times n$ *matrix, let* $\boldsymbol{B}$ *be a* $p \times q$ *matrix, and let* $\boldsymbol{C}$ *be a* $u \times v$ *matrix. Then the identity*

$$\boldsymbol{A} \otimes (\boldsymbol{B} \otimes \boldsymbol{C}) = (\boldsymbol{A} \otimes \boldsymbol{B}) \otimes \boldsymbol{C}$$

holds true.

Theorem A.16. *Let* $\boldsymbol{A}$ *be an* $m \times n$ *matrix, let* $\boldsymbol{B}$ *be a* $p \times q$, *let* $\boldsymbol{C}$ *be an* $n \times u$ *matrix, and let* $\boldsymbol{D}$ *be a* $q \times v$ *matrix. Then the identity*

$$(\boldsymbol{A} \otimes \boldsymbol{B})(\boldsymbol{C} \otimes \boldsymbol{D}) = (\boldsymbol{AC}) \otimes (\boldsymbol{BC})$$

holds true.

A.1.7 Moore-Penrose Inverse

For any $m \times n$ matrix, there exists a uniquely determined matrix satisfying the so-called Moore-Penrose conditions.

Theorem A.17. *For an* $m \times n$ *matrix* $\boldsymbol{A}$, *there exists a uniquely determined* $n \times m$ *matrix* $\boldsymbol{A}^+$, *satisfying the four conditions*

$$\boldsymbol{A}\boldsymbol{A}^+\boldsymbol{A} = \boldsymbol{A},\ \boldsymbol{A}^+\boldsymbol{A}\boldsymbol{A}^+ = \boldsymbol{A}^+,\ \boldsymbol{A}\boldsymbol{A}^+ = (\boldsymbol{A}\boldsymbol{A}^+)',\ \boldsymbol{A}^+\boldsymbol{A} = (\boldsymbol{A}^+\boldsymbol{A})' .$$

The matrix $\boldsymbol{A}^+$ *is called* Moore-Penrose inverse *or* pseudo inverse *of* $\boldsymbol{A}$.

The matrix $\boldsymbol{A}^+$ is sometimes also called *generalized inverse* of $\boldsymbol{A}$. Usually, however, the generalized inverse of $\boldsymbol{A}$ is understood as any matrix $\boldsymbol{A}^-$ satisfying $\boldsymbol{A}\boldsymbol{A}^-\boldsymbol{A} = \boldsymbol{A}$. The Moore-Penrose inverse $\boldsymbol{A}^+$ is thus a specific generalized inverse of the matrix $\boldsymbol{A}$.

Theorem A.18. *Let* $\boldsymbol{A}$ *be an* $m \times n$ *matrix and* $\alpha \neq 0$ *a real number. Then the conditions*

$$(\boldsymbol{A}^+)^+ = \boldsymbol{A}, \quad (\boldsymbol{A}^+)' = (\boldsymbol{A}')^+, \quad \textit{and} \quad (\alpha\boldsymbol{A})^+ = \frac{1}{\alpha}\boldsymbol{A}^+$$

hold true.

Theorem A.19. *Let* $\boldsymbol{A}$ *be an* $m \times n$ *matrix. Then*

$$\boldsymbol{A}^+ = (\boldsymbol{A}'\boldsymbol{A})^+\boldsymbol{A}' = \boldsymbol{A}'(\boldsymbol{A}\boldsymbol{A}')^+$$

holds true.

Remark A.6. The use of a generalized inverse $\boldsymbol{A}^-$ or the unique Moore-Penrose inverse $\boldsymbol{A}^+$ of an $m \times n$ matrix $\boldsymbol{A}$ can be quite helpful in algebraic manipulations. Nonetheless, it implicitly suggests that the matrix $\boldsymbol{A}$ in question can have arbitrary numbers of rows m and columns n and is possibly neither of full row rank, nor of full column rank. When this implicit assumptions are not satisfied, then the application of generalized inverses may lower the understanding of what is going on. For example, in the linear regression model with assumptions (i) to (iv), the ordinary least squares estimator $\widehat{\boldsymbol{\beta}} = (\boldsymbol{X}'\boldsymbol{X})^{-1}\boldsymbol{X}'\boldsymbol{y}$ for $\boldsymbol{\beta}$ may also be written as $\widehat{\boldsymbol{\beta}} = \boldsymbol{X}^+\boldsymbol{y}$. The latter does even exist when $\boldsymbol{X}$ is not of full column rank, but then it is not an unbiased estimator, since in that case no linear unbiased estimator for $\boldsymbol{\beta}$ exists. Hence the representation $\widehat{\boldsymbol{\beta}} = \boldsymbol{X}^+\boldsymbol{y}$ may misleadingly suggest that the column singularity of the matrix $\boldsymbol{X}$ has not an effect here.

A.2 Common Pitfalls

Having introduced basic notions and results, the subsequent sections deal with square and symmetric matrices in more detail.

In connection with *eigenvalues* and *definiteness* of square matrices there exist some common misjudgments of the following kind:

- The eigenvalues of a real-valued matrix are real.
- For any eigenvalue of a real-valued matrix there exists a real-valued eigenvector.
- The eigenvalues of an orthogonal matrix are either 1 or -1.
- A matrix is nonnegative definite if and only if all its eigenvalues are nonnegative.
- Without loss of generality a nonnegative definite matrix can be assumed as being symmetric.
- The Löwner partial ordering can only be defined within the set of symmetric matrices.

These statements are all incorrect if the usual definitions of term like 'eigenvalue', 'definiteness', or 'Löwner partial ordering' are considered. As the reason for the possible upcoming of statements as above, one may see the fact that in mathematical literature these terms are specified for arbitrary square matrices, while in statistical literature they are often applied to certain subsets of square matrices. As a matter of fact, each of the above statements is either correct or redundant, if the matrix in question is a square *symmetric* matrix.

The assumption of symmetry is quite reasonable when definiteness is considered, since the corresponding results are easier then, and in most applications symmetry is satisfied. When eigenvalues are regarded, then, however, non-symmetric matrices are not rare. In these cases it is of some importance to acknowledge the fact that real-valued matrices must not necessarily have real-valued eigenvalues.

A.3 Square Matrices

This section gives results for square matrices. Specific subclasses of square matrices are introduced, trace and determinant are considered, eigenvalues and eigenvectors are investigated, matrix and vector norms are specified, and the definiteness of square matrices is regarded.

A.3.1 Specific Square Matrices

We consider diagonal and identity matrix, inverse of a nonsingular matrix, orthogonal matrix, and projector.

Diagonal and Identity Matrix

Presumably the simplest square matrices are diagonal matrices.

Definition A.15. *Let $\boldsymbol{A}$ be an $m \times m$ matrix.*

(a) *The matrix $\boldsymbol{A}$ is called* diagonal matrix *if $a_{ij} = 0$ for all $i \neq j$.*
(b) *The matrix $\boldsymbol{A}$ is called* identity matrix *if if $a_{ij} = 0$ for all $i \neq j$ and $a_{ii} = 1$ for $i = 1, \dots, m$. The $m \times m$ identity matrix is denoted by $\boldsymbol{I}_m$.*

Inverse of a Matrix

The $m \times m$ identity matrix can be seen as the identity element for multiplication, i.e. $\boldsymbol{I}_m \boldsymbol{A} = \boldsymbol{A}$ for an $m \times n$ matrix $\boldsymbol{A}$. For a square matrix $\boldsymbol{A}$ one might ask whether there exists a multiplicative inverse.

Definition A.16. *For an $m \times m$ matrix $\boldsymbol{A}$, the $m \times m$ matrix $\boldsymbol{A}^{-1}$ satisfying the conditions*

$$\boldsymbol{A}\boldsymbol{A}^{-1} = \boldsymbol{I}_m \quad \text{and} \quad \boldsymbol{A}^{-1}\boldsymbol{A} = \boldsymbol{I}_m$$

is called the inverse *of $\boldsymbol{A}$.*

In contrast to the Moore-Penrose inverse of a possibly rectangular $m \times n$ matrix, the above defined inverse of a square $m \times m$ matrix does not necessarily exist. Hence, an $m \times m$ matrix $\boldsymbol{A}$ always has a Moore Penrose inverse $\boldsymbol{A}^{+}$ but not necessarily an inverse $\boldsymbol{A}^{-1}$. The existence of $\boldsymbol{A}^{-1}$ depends on the rank of $\boldsymbol{A}$.

Theorem A.20. *An $m \times m$ matrix $\boldsymbol{A}$ has an inverse $\boldsymbol{A}^{-1}$ if and only if it is nonsingular. In this case the inverse is uniquely determined.*

Theorem A.21. *Let $\boldsymbol{A}$ be an $m \times m$ nonsingular matrix and $\alpha \neq 0$ a real number. Then the conditions*

$$(\boldsymbol{A}^{-1})^{-1} = \boldsymbol{A}, \quad (\boldsymbol{A}')^{-1} = (\boldsymbol{A}^{-1})', \quad (\alpha \boldsymbol{A})^{-1} = \frac{1}{\alpha}\boldsymbol{A}^{-1}.$$

hold true.

For the inverse of the product of two matrices, the reverse order law is valid.

Theorem A.22. *Let $\boldsymbol{A}$ and $\boldsymbol{B}$ be two nonsingular $m \times m$ matrices Then $(\boldsymbol{AB})^{-1} = \boldsymbol{B}^{-1}\boldsymbol{A}^{-1}$.*

Theorem A.23. *Let $\boldsymbol{A}$ be an $m \times m$ and let $\boldsymbol{C}$ be an $n \times n$ matrix, both being nonsingular. Let $\boldsymbol{B}$ be an $m \times n$ and $\boldsymbol{D}$ be an $n \times m$ matrix. If $\boldsymbol{A} + \boldsymbol{BCD}$ is nonsingular, then*

$$(\boldsymbol{A} + \boldsymbol{BCD})^{-1} = \boldsymbol{A}^{-1} - \boldsymbol{A}^{-1}\boldsymbol{B}(\boldsymbol{C}^{-1} + \boldsymbol{DA}^{-1}\boldsymbol{B})^{-1}\boldsymbol{DA}^{-1}$$

holds true.

Theorem A.24. *Let $\boldsymbol{A}$ be an $m \times m$ matrix, $\boldsymbol{B}$ be an $m \times n$ matrix, $\boldsymbol{C}$ be an $n \times m$ matrix, and $\boldsymbol{D}$ be an $m \times m$ nonsingular matrix. Then $\begin{pmatrix} \boldsymbol{A} & \boldsymbol{B} \\ \boldsymbol{C} & \boldsymbol{D} \end{pmatrix}$ is nonsingular if and only if $\boldsymbol{S} = \boldsymbol{A} - \boldsymbol{BD}^{-1}\boldsymbol{C}$ is nonsingular, in which case*

$$\begin{pmatrix} \boldsymbol{A} & \boldsymbol{B} \\ \boldsymbol{C} & \boldsymbol{D} \end{pmatrix}^{-1} = \begin{pmatrix} \boldsymbol{S}^{-1} & -\boldsymbol{S}^{-1}\boldsymbol{BD}^{-1} \\ -\boldsymbol{D}^{-1}\boldsymbol{CS}^{-1} & \boldsymbol{D}^{-1} + \boldsymbol{D}^{-1}\boldsymbol{CS}^{-1}\boldsymbol{BD}^{-1} \end{pmatrix}$$

holds true.

Theorem A.25. *Let $\boldsymbol{A}$ be an $m \times m$ nonsingular matrix, $\boldsymbol{B}$ be an $m \times n$ matrix, $\boldsymbol{C}$ be an $n \times m$ matrix, and $\boldsymbol{D}$ be an $m \times m$ matrix. Then $\begin{pmatrix} \boldsymbol{A} & \boldsymbol{B} \\ \boldsymbol{C} & \boldsymbol{D} \end{pmatrix}$ is nonsingular if and only if $\boldsymbol{T} = \boldsymbol{D} - \boldsymbol{CA}^{-1}\boldsymbol{B}$ is nonsingular, in which case*

$$\begin{pmatrix} \boldsymbol{A} & \boldsymbol{B} \\ \boldsymbol{C} & \boldsymbol{D} \end{pmatrix}^{-1} = \begin{pmatrix} \boldsymbol{A}^{-1} + \boldsymbol{A}^{-1}\boldsymbol{BT}^{-1}\boldsymbol{CA}^{-1} & -\boldsymbol{A}^{-1}\boldsymbol{BT}^{-1} \\ -\boldsymbol{T}^{-1}\boldsymbol{CA}^{-1} & \boldsymbol{T}^{-1} \end{pmatrix}$$

holds true.

Theorem A.26. *Let $\boldsymbol{A}$ be an $m \times m$ nonsingular matrix. Then $\boldsymbol{A}^{-1} = \boldsymbol{A}^{+}$.*

The above result does not only hold for the Moore-Penrose inverse $\boldsymbol{A}^+$ but also for an arbitrary generalized inverse $\boldsymbol{A}^-$. Recall that a generalized inverse $\boldsymbol{A}^-$ of $\boldsymbol{A}$ is any matrix satisfying $\boldsymbol{A}\boldsymbol{A}^-\boldsymbol{A}$. A generalized inverse does always exist, even if $\boldsymbol{A}$ is a rectangular $m \times n$ matrix. Usually, however, it is not unique, i.e. there is a (presumably infinite) number of generalized inverses for a given matrix $\boldsymbol{A}$. The Moore-Penrose inverse is one of them, being unique in the sense that it is the only matrix satisfying the four Moore-Penrose conditions specified in Theorem A.17. The situation is different when $\boldsymbol{A}$ is a square nonsingular matrix. Then there exists only one matrix $\boldsymbol{A}^-$ satisfying the identity $\boldsymbol{A}\boldsymbol{A}^-\boldsymbol{A}$, being the inverse $\boldsymbol{A}^{-1}$ of $\boldsymbol{A}$. This justifies the notion 'generalized inverse' for a matrix $\boldsymbol{A}^-$ in two respects. First, it specifies an inverse for a *singular* square matrix and second, it even specifies an inverse for a *rectangular* matrix.

Theorem A.27. *Let $\boldsymbol{A}$ be an $m \times m$ nonsingular matrix and let $\boldsymbol{B}$ be a $p \times p$ nonsingular matrix. Then $\boldsymbol{A} \otimes \boldsymbol{B}$ is nonsingular and the identity*

$$(\boldsymbol{A} \otimes \boldsymbol{B})^{-1} = \boldsymbol{A}^{-1} \otimes \boldsymbol{B}^{-1}$$

holds true.

Orthogonal Matrix

The inverse of a square matrix might coincide with its transpose.

Definition A.17. *A nonsingular $m \times m$ matrix $\boldsymbol{A}$ is called* orthogonal *if $\boldsymbol{A}' = \boldsymbol{A}^{-1}$.*

The notion 'orthogonal matrix' is justified since the columns of such a matrix are orthogonal to each other. If $\boldsymbol{a}_i$ and $\boldsymbol{a}_j$ denote the i-th and the j-th column of an orthogonal matrix $\boldsymbol{A}$ with $i \neq j$, then $\boldsymbol{a}_i'\boldsymbol{a}_j = 0$. Moreover, in case $i = j$ it follows that $\boldsymbol{a}_i'\boldsymbol{a}_i = 1$. Hence the columns of an $m \times m$ orthogonal matrix $\boldsymbol{A}$ constitute an *orthonormal* basis of $\mathbb{R}^m$.

Example A.1. The matrix $\boldsymbol{A} = \begin{pmatrix} 1/\sqrt{2} & 1/\sqrt{2} \\ -1/\sqrt{2} & 1/\sqrt{2} \end{pmatrix}$ is a 2×2 orthogonal matrix.

Projectors

A square matrix multiplied by itself might yield the same matrix.

Definition A.18. *An $m \times m$ matrix $\boldsymbol{A}$ is called idempotent if $\boldsymbol{A} = \boldsymbol{A}^2$.*

Example A.2. The matrix $\boldsymbol{A} = \begin{pmatrix} 1 & b \\ 0 & 1 \end{pmatrix}$ is a 2×2 idempotent matrix for any number b.

Any idempotent matrix can be identified with a projector onto some subspace $\mathcal{U}$ along (parallel to) some subspace $\mathcal{V}$. If $\mathcal{U}$ and V are two complementary subspaces of some vector space $\mathcal{X}$, then any vector $\boldsymbol{x} \in \mathcal{X}$ can uniquely be written as $\boldsymbol{x} = \boldsymbol{u} + \boldsymbol{v}$ for some $\boldsymbol{u} \in \mathcal{U}$ and $\boldsymbol{v} \in \mathcal{V}$. Then the unique linear operator $\boldsymbol{A}$ defined by $\boldsymbol{A}\boldsymbol{x} = \boldsymbol{u}$ is called the projector onto $\mathcal{U}$ along $\mathcal{V}$, see Sect. 5.9 in [78]. For such a projector $\boldsymbol{A}$, it follows that $\mathcal{C}(\boldsymbol{A}) = \mathcal{U}$ and $\mathcal{N}(\boldsymbol{A}) = \mathcal{V}$. Hence, a projector $\boldsymbol{A}$ projects a vector onto its column space along its null space. As long as $\mathcal{N}(\boldsymbol{A})$ does not coincide with $\mathcal{C}(\boldsymbol{A})^{\perp}$ the projector is called *oblique*, while otherwise the projector $\boldsymbol{A}$ is called *orthogonal.* Orthogonal projectors can be identified with idempotent and symmetric matrices.

Definition A.19. *An $m \times m$ matrix $\boldsymbol{A}$ is called* orthogonal projector *if $\boldsymbol{A}$ is idempotent and symmetric*

Example A.3. The matrix $\boldsymbol{A} = \begin{pmatrix} 1/2 & 1/2 \\ 1/2 & 1/2 \end{pmatrix}$ is a 2×2 orthogonal projector.

Note that orthogonal projectors are not to be confused with orthogonal matrices. The term 'orthogonal' in connection with a projector has the above described geometric meaning.

Example A.4. Consider the two idempotent matrices

$$\boldsymbol{P}_1 = \begin{pmatrix} 1/2 & 1/2 \\ 1/2 & 1/2 \end{pmatrix} \quad \text{and} \quad \boldsymbol{P}_2 = \begin{pmatrix} 1/4 & 3/4 \\ 1/4 & 3/4 \end{pmatrix},$$

having identical column space

$$\mathcal{C}(\boldsymbol{P}_1) = \mathcal{C}(\boldsymbol{P}_2) = \mathcal{U} := \{\alpha \begin{pmatrix} 1 \\ 1 \end{pmatrix} : \alpha \in \mathbb{R}\}.$$

This is the line $x = y$ in the xy-plane.

The matrix $\boldsymbol{P}_1$ is symmetric and thus projects a vector $\boldsymbol{y} = (y_1, y_2)'$ onto $\mathcal{U}$ parallel to the orthogonal complement

$$\mathcal{U}^{\perp} = \{\alpha \begin{pmatrix} 1 \\ -1 \end{pmatrix} : \alpha \in \mathbb{R}\},$$

while the non-symmetric matrix $\boldsymbol{P}_2$ projects parallel to

$$\mathcal{N}(\boldsymbol{P}_2) = \{\alpha \begin{pmatrix} 3 \\ -1 \end{pmatrix} : \alpha \in \mathbb{R}\}.$$

Figures A.1 and A.2 illustrate the orthogonal projection of the vector $\boldsymbol{y} = (2, 6)'$ onto $\mathcal{U}$ being $\boldsymbol{P}_1\boldsymbol{y} = (4, 4)'$, and the oblique projection of $\boldsymbol{y} = (2, 6)'$ onto $\mathcal{U}$ parallel to $\mathcal{N}(\boldsymbol{P}_2)$ being $\boldsymbol{P}_2\boldsymbol{y} = (5, 5)'$. □

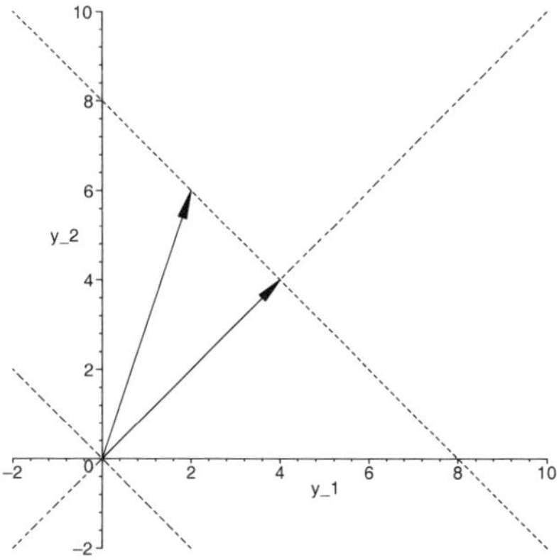

Fig. A.1. Orthogonal projection of the vector $(2,6)'$ onto the line $y_1 = y_2$

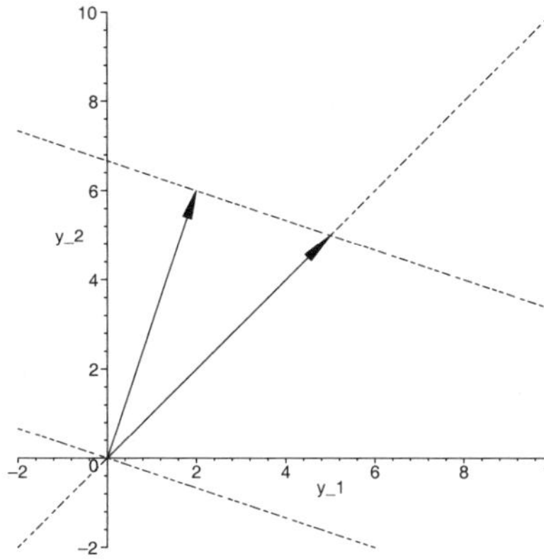

Fig. A.2. Oblique projection of the vector $(2,6)'$ onto the line $y_1 = y_2$

For a given matrix $\boldsymbol{B}$ the orthogonal projector onto $\mathcal{C}(\boldsymbol{B})$ can be computed solely from $\boldsymbol{B}$.

Theorem A.28. *Let $\boldsymbol{A}$ be an $m \times n$ matrix. Then the matrix*

$$\boldsymbol{P} = \boldsymbol{A}(\boldsymbol{A}'\boldsymbol{A})^+\boldsymbol{A}' = \boldsymbol{A}\boldsymbol{A}^+$$

is the orthogonal projector onto $\mathcal{C}(\boldsymbol{A}) = \mathcal{C}(\boldsymbol{P})$.

Two column spaces are uniquely determined by the orthogonal projectors onto them.

Theorem A.29. *Let $\boldsymbol{A}$ be an $m \times n$ and $\boldsymbol{B}$ be an $m \times p$ matrix. Then $\mathcal{C}(\boldsymbol{A}) = \mathcal{C}(\boldsymbol{B})$ if and only if the identity*

$$\boldsymbol{A}(\boldsymbol{A}'\boldsymbol{A})^+\boldsymbol{A}' = \boldsymbol{B}(\boldsymbol{B}'\boldsymbol{B})^+\boldsymbol{B}'$$

is satisfied.

Among all vectors from a specific column space, the orthogonal projection of a vector onto this column space has smallest distance to this vector.

Theorem A.30. *Let $\boldsymbol{A}$ be an $m \times m$ orthogonal projector and let $\boldsymbol{x}$ be an $m \times 1$ vector. Then the inequality*

$$||\boldsymbol{x} - \boldsymbol{A}\boldsymbol{x}|| \leq ||\boldsymbol{x} - \boldsymbol{y}||$$

holds true for every $m \times 1$ vector $\boldsymbol{y} \in \mathcal{C}(\boldsymbol{A})$.

The above theorem states that if $\boldsymbol{A}$ is an orthogonal projector, then the corresponding projection $\boldsymbol{A}\boldsymbol{x} \in \mathcal{C}(\boldsymbol{A})$ of a vector $\boldsymbol{x}$ onto $\mathcal{C}(\boldsymbol{A})$ has the smallest distance to $\boldsymbol{x}$ compared to any other vector $\boldsymbol{y} \in \mathcal{C}(\boldsymbol{A})$. Therefore, Theorem A.30 is also called the *closest point theorem.*

A.3.2 Trace and Determinant

Two characteristic numbers related to any square matrix are its *trace* and its *determinant.*

Trace

Presumably the most easily computed characteristic number of a square matrix is the sum of its diagonal elements.

Definition A.20. *The* trace *of an $m \times m$ matrix $\boldsymbol{A}$ is the number*

$$\operatorname{tr}(\boldsymbol{A}) = \sum_{i=1}^{m} a_{ii} ,$$

where a_{ii} denotes the i-th main diagonal element of $\boldsymbol{A}$.

Theorem A.31. *Let $\boldsymbol{A}$ and $\boldsymbol{B}$ be $m \times m$ matrices and let α be a real number. Then the identities*

$$\operatorname{tr}(\boldsymbol{A}) = \operatorname{tr}(\boldsymbol{A}'), \quad \operatorname{tr}(\alpha \boldsymbol{A}) = \alpha \operatorname{tr}(\boldsymbol{A}), \quad \operatorname{tr}(\boldsymbol{A}+\boldsymbol{B}) = \operatorname{tr}(\boldsymbol{A}) + \operatorname{tr}(\boldsymbol{B})$$

hold true.

The product of matrices can *cyclically* be permuted without altering its trace.

Theorem A.32. *Let $\boldsymbol{A}$ be an $m \times n$ matrix, $\boldsymbol{B}$ be an $n \times q$ matrix, and $\boldsymbol{C}$ be a $q \times m$ matrix. Then*

$$\operatorname{tr}(\boldsymbol{ABC}) = \operatorname{tr}(\boldsymbol{CAB}) = \operatorname{tr}(\boldsymbol{BCA})$$

is satisfied.

Theorem A.33. *Let $\boldsymbol{A}$ be an $m \times m$ matrix and let $\boldsymbol{B}$ be a $p \times p$ matrix. Then the identity*

$$\operatorname{tr}(\boldsymbol{A} \otimes \boldsymbol{B}) = \operatorname{tr}(\boldsymbol{A}) \operatorname{tr}(\boldsymbol{B})$$

holds true.

Determinant

A given permutation p of the numbers $(1, 2, \ldots, m)$ can only be rearranged to the natural order by either an even or an odd number of interchanges. Hence the *sign of a permutation* can be specified as

$$\sigma(p) = \begin{cases} +1 & \text{if } p \text{ can be restored to natural order} \\ & \text{by an } \textit{even} \text{ number of interchanges} \\ -1 & \text{if } p \text{ can be restored to natural order} \\ & \text{by an } \textit{odd} \text{ number of interchanges} \end{cases} ,$$

see also [78, Sect. 6.1].

Definition A.21. *The* determinant *of an $m \times m$ matrix $\boldsymbol{A}$ is the number*

$$\det(\boldsymbol{A}) = \sum_{p} \sigma(p)\, a_{1p_1} a_{2p_2} \cdots a_{mp_m} \,,$$

where the sum is taken over all permutations $p = (p_1, p_2, \ldots, p_m)$ of the numbers $(1, 2, \ldots, m)$.

Example A.5. Let

$$\boldsymbol{A} = \begin{pmatrix} a_{11} & a_{12} \\ a_{21} & a_{22} \end{pmatrix}$$

be a 2×2 matrix. Then $(1, 2)$ and $(2, 1)$ are all possible permutations of $(1, 2)$ with $\sigma(1, 2) = +1$ and $\sigma(2, 1) = -1$. Thus,

$$\det(\boldsymbol{A}) = \sigma(1, 2)a_{11}a_{22} + \sigma(2, 1)a_{12}a_{21} = a_{11}a_{22} - a_{12}a_{21}$$

is the determinant of $\boldsymbol{A}$. □

Definition A.22. *Let $\boldsymbol{A}$ be an $m \times m$ matrix.*

(a) *Let m_{ij} be the determinant of the matrix generated from $\boldsymbol{A}$ by deleting its i-th row and its j-th column. Then the number*

$$\ddot{a}_{ij} = (-1)^{(i+j)} m_{ij}$$

is called the cofactor *of $\boldsymbol{A}$ associated with the (i, j)th position. The matrix of cofactors is denoted by $\ddot{\boldsymbol{A}}$*

(b) *The transpose of $\ddot{\boldsymbol{A}}$ is called the* adjugate *of $\boldsymbol{A}$ and is denoted by* adj$(\boldsymbol{A})$.

Cofactors of a square matrix can be applied to compute its determinant.

Theorem A.34. *Let $\boldsymbol{A}$ be an $m \times m$ matrix. Then*

$$\det(\boldsymbol{A}) = \sum_{i=1}^{m} a_{ij}\ddot{a}_{ij}$$

for any $j \in \{1, \ldots, m\}$.

The above result is also called *cofactor expansion about the j-th column.* Similarly one may develop a cofactor expansion about a row, yielding

$$\det(\boldsymbol{A}) = \sum_{j=1}^{m} a_{ij}\ddot{a}_{ij}$$

for any $i \in \{1, \ldots, m\}$. The cofactor expansion about a column or a row is sometimes used as a definition for the determinant.

Any singular square matrix has a zero determinant.

Theorem A.35. *An $m \times m$ matrix $\boldsymbol{A}$ is singular if and only if* $\det(\boldsymbol{A}) = 0$. *If $\boldsymbol{A}$ is nonsingular, then the identity*

$$\boldsymbol{A}^{-1} = \frac{1}{\det \boldsymbol{A}} \operatorname{adj}(\boldsymbol{A})$$

holds true.

Theorem A.36. *Let $\boldsymbol{A}$ and $\boldsymbol{B}$ be $m \times m$ matrices and let α be a real number. Then the conditions*

$$\det(\boldsymbol{A}) = \det(\boldsymbol{A}'), \quad \det(\alpha \boldsymbol{A}) = \alpha^m \det(\boldsymbol{A}), \quad \det(\boldsymbol{A}\boldsymbol{B}) = \det(\boldsymbol{A}) \det(\boldsymbol{B})$$

hold true.

A.3.3 Eigenvalue and Eigenvector

Instead of reducing a square $m \times m$ matrix to a single characteristic number one may consider the m roots (eigenvalues) of the so-called characteristic polynomial.

The eigenvalue-eigenvector equation

Corresponding to each eigenvalue of a matrix there exists an infinite number of eigenvectors.

Definition A.23. *Let $\boldsymbol{A}$ be an $m \times m$ matrix.*

(a) *If a (possibly complex) number λ and a (possibly complex-valued) vector $\boldsymbol{x} \neq \boldsymbol{0}$ satisfy the equation*

$$\boldsymbol{A}\boldsymbol{x} = \lambda \boldsymbol{x}, \quad \boldsymbol{x} \neq \boldsymbol{0} ,$$

then λ is called eigenvalue *and $\boldsymbol{x}$ is called* eigenvector *of the matrix $\boldsymbol{A}$. Any such pair $(\lambda, \boldsymbol{x})$ is also called* eigenpair *of $\boldsymbol{A}$.*

(b) *The set of different eigenvalues of $\boldsymbol{A}$ is called the* spectrum *of $\boldsymbol{A}$ and is denoted by $\sigma(\boldsymbol{A})$.*

Note that the $m \times 1$ zero vector $\boldsymbol{0}$ trivially satisfies the equation $\boldsymbol{A}\boldsymbol{0} = \lambda \boldsymbol{0}$ for any number λ, but is by no means an eigenvector. For a fixed number λ, the set

$$\{\boldsymbol{x} \neq \boldsymbol{0} : \boldsymbol{A}\boldsymbol{x} = \lambda \boldsymbol{x}\} \cup \{\boldsymbol{0}\}$$

is called *eigenspace* of $\boldsymbol{A}$ associated with the eigenvalue λ. It is identical to

$$\mathcal{N}(\lambda \boldsymbol{I}_m - \boldsymbol{A}) ,$$

being a subspace of $\mathbb{C}^m$ and not of $\mathbb{R}^m$. The dimension of the eigenspace of $\boldsymbol{A}$ associated with the eigenvalue λ is called the *geometric multiplicity* of λ.

The characteristic polynomial

The eigenvalue-eigenvector equation $\boldsymbol{Ax} = \lambda \boldsymbol{x}$, $\boldsymbol{x} \neq \boldsymbol{0}$, can also be written as

$$(\lambda \boldsymbol{I}_m - \boldsymbol{A})\boldsymbol{x} = \boldsymbol{0}, \quad \boldsymbol{x} \neq \boldsymbol{0} .$$

Hence, the number λ is an eigenvalue of $\boldsymbol{A}$ if and only if $\lambda \boldsymbol{I}_m - \boldsymbol{A}$ is singular. This is satisfied if and only if $\det(\lambda \boldsymbol{I}_m - \boldsymbol{A}) = 0$.

Definition A.24. *Let $\boldsymbol{A}$ be an $m \times m$ matrix. Then $p_{\boldsymbol{A}}(\lambda) = \det(\lambda \boldsymbol{I}_m - \boldsymbol{A})$ is called the* characteristic polynomial *of $\boldsymbol{A}$ (in λ). The equation $p_{\boldsymbol{A}}(\lambda) = 0$ is called* characteristic equation *for $\boldsymbol{A}$.*

It is clear that any eigenvalue of the matrix $\boldsymbol{A}$ must be a solution to the characteristic equation $p_{\boldsymbol{A}}(\lambda) = 0$. Since from the *fundamental theorem of algebra* it follows that a polynomial of degree m has exactly m (not necessarily distinct) zeroes among the field of complex numbers, it follows that an $m \times m$ matrix $\boldsymbol{A}$ has exactly m (not necessarily distinct) eigenvalues.

Definition A.25. *The multiplicity of λ as a solution to the characteristic equation is called the* algebraic multiplicity *of λ.*

Assumption A.2. *The multiplicity of an eigenvalue is always understood as its algebraic multiplicity.*

Example A.6. The identity matrix $\boldsymbol{I}_m$ has m eigenvalues, each one being identical to 1. In other words, the eigenvalue 1 has multiplicity m. □

As demonstrated by the following example, a real-valued matrix does not necessarily have real eigenvalues.

Example A.7. The characteristic polynomial of the matrix $\boldsymbol{A} = \begin{pmatrix} 1 & -1 \\ 1 & 1 \end{pmatrix}$ is given by

$$p_{\boldsymbol{A}}(\lambda) = \det \begin{pmatrix} \lambda - 1 & 1 \\ -1 & \lambda - 1 \end{pmatrix} = (\lambda - 1)^2 + 1 = \lambda^2 - 2\lambda + 2 .$$

The two zeroes of $p_{\boldsymbol{A}}(\lambda)$ are the eigenvalues of $\boldsymbol{A}$ and given as $\lambda_1 = 1 + \mathrm{i}$ and $\lambda_2 = 1 - \mathrm{i}$, where $\mathrm{i} = \sqrt{-1}$ denotes the imaginary unit. □

In the above example the matrix $\boldsymbol{A}$ has two complex eigenvalues, each of them being the conjugate complex of the other. As a matter of fact, the strictly complex eigenvalues of a real-valued matrix always occur in conjugate pairs.

Theorem A.37. *Let $\boldsymbol{A}$ be an $m \times m$ (real-valued) matrix and let $\mathrm{i} = \sqrt{-1}$. If $\alpha + \mathrm{i}\beta$ with $\beta \neq 0$ is an eigenvalue of $\boldsymbol{A}$, then $\alpha - \mathrm{i}\beta$ is also an eigenvalue of $\boldsymbol{A}$.*

For an $m \times n$ matrix $\boldsymbol{A}$ and an $n \times m$ matrix $\boldsymbol{B}$ the two products $\boldsymbol{AB}$ and $\boldsymbol{BA}$ are well defined. Although they do not necessarily coincide, there exist the relationships

$$p_{\boldsymbol{BA}}(\lambda) = \lambda^{n-m} p_{\boldsymbol{AB}}(\lambda) \quad \text{when } m \leq n$$

and

$$p_{\boldsymbol{AB}}(\lambda) = \lambda^{m-n} p_{\boldsymbol{BA}}(\lambda) \quad \text{when } n \leq m\ .$$

From this, the following theorem can be deduced.

Theorem A.38. *Let $\boldsymbol{A}$ be an $m \times n$ matrix and let $\boldsymbol{B}$ be an $n \times m$ matrix. Then the nonzero eigenvalues of $\boldsymbol{AB}$ coincide with the nonzero eigenvalues of $\boldsymbol{BA}$, counting multiplicities.*

Eigenvectors

As noted before, any vector $\boldsymbol{x} \neq \boldsymbol{0}$ satisfying $\boldsymbol{Ax} = \lambda\boldsymbol{x}$ for some number λ is called *eigenvector* of $\boldsymbol{A}$ associated with the eigenvalue λ. (More precisely, any such vector $\boldsymbol{x}$ is a *right* eigenvector, since one may as well specify *left* eigenvectors, see e.g. [59, p.59]. The latter are not of importance here.)

Theorem A.39. *Let $\boldsymbol{A}$ be a (real-valued) $m \times m$ matrix and let λ be an eigenvalue of $\boldsymbol{A}$. Then:*

(a) *If λ is real, then there always exists a corresponding real-valued eigenvector.*
(b) *If λ is strictly complex, then there cannot exist a corresponding real-valued eigenvector.*

Example A.8. The real-valued matrix $\boldsymbol{A} = \begin{pmatrix} 1 & -1 \\ 1 & 1 \end{pmatrix}$ has the two complex eigenvalues $\lambda_1 = 1 + \mathrm{i}$ and $\lambda_2 = 1 - \mathrm{i}$. The eigenspace of λ_1 is

$$\mathcal{N}(\lambda_1 \boldsymbol{I}_2 - \boldsymbol{A}) = \{\alpha \begin{pmatrix} \mathrm{i} \\ 1 \end{pmatrix} \mid \alpha \in \mathbb{C}\}\ ,$$

showing that there cannot exist an eigenvector associated with λ_1 whose elements are all real. The same is true for the eigenvectors associated with λ_2. □

Theorem A.40. *If $\{(\lambda_1, \boldsymbol{x}_1), \ldots, (\lambda_k, \boldsymbol{x}_k)\}$ is a set of eigenpairs of an $m \times m$ matrix $\boldsymbol{A}$ such that $\lambda_1, \ldots, \lambda_k$ are distinct, then $\{\boldsymbol{x}_1, \ldots, \boldsymbol{x}_k\}$ is a set of linearly independent vectors.*

Trace and Determinant

The trace and the determinant of a square matrix can easily be obtained from its eigenvalues.

Theorem A.41. *For an $m \times m$ matrix $\boldsymbol{A}$ the identities*

$$\operatorname{tr}(\boldsymbol{A}) = \sum_{i=1}^{m} \lambda_i \quad \textit{and} \quad \det(\boldsymbol{A}) = \prod_{i=1}^{m} \lambda_i$$

hold true, where $\lambda_1, \ldots, \lambda_m$ are the m eigenvalues of $\boldsymbol{A}$.

Note that the trace and the determinant of a real-valued matrix are necessarily real. One may wonder whether the above theorem can hold true, although, as noted before, the eigenvalues of a real-valued matrix are not necessarily real. This apparent inconsistency clears up when notion is given to the fact that strictly complex eigenvalues of a real-valued matrix can only occur in conjugate pairs. Since the sum $(\alpha + i\beta) + (\alpha - i\beta) = 2\alpha$ as well as the product $(\alpha + i\beta)(\alpha - i\beta) = \alpha^2 + \beta^2$ are real numbers, it is obvious that no contradiction is implied by the above result.

Eigenvalues of Specific Matrices

Section A.3.1 considers square matrices with specific properties. It is clear that the eigenvalues of any diagonal matrix are its main diagonal elements. Moreover, the following holds.

Theorem A.42. *Let $\boldsymbol{A}$ be an $m \times m$ matrix.*

(a) *The matrix $\boldsymbol{A}$ is singular if and only if it has at least one zero eigenvalue.*
(b) *If $\boldsymbol{A}$ is idempotent, then $\boldsymbol{A}$ has only eigenvalues* 0 *or* 1.
(c) *If $\boldsymbol{A}$ is orthogonal, then all eigenvalues of $\boldsymbol{A}$ have absolute value* 1.

Statement (c) of the above theorem does not imply that the eigenvalues of an orthogonal matrix are either $+1$ or -1, since the eigenvalues of a (real-valued) orthogonal matrix can also be strictly complex.

Example A.9. The orthogonal matrix

$$\boldsymbol{A} = \begin{pmatrix} 1/\sqrt{2} & 1/\sqrt{2} \\ -1/\sqrt{2} & 1/\sqrt{2} \end{pmatrix}$$

has eigenvalues $\lambda_1 = \frac{1}{\sqrt{2}} + \mathrm{i}\frac{1}{\sqrt{2}}$ and $\lambda_2 = \frac{1}{\sqrt{2}} - \mathrm{i}\frac{1}{\sqrt{2}}$ with

$$|\lambda_1| = +\left((\frac{1}{\sqrt{2}})^2 + (\frac{1}{\sqrt{2}})^2 \right)^{1/2} = 1 = |\lambda_2| \,,$$

showing that the two strictly complex eigenvalues of $\boldsymbol{A}$ have absolute value 1. □

A.3.4 Vector and Matrix Norm

As a measure for the 'size' of a vector or a matrix one might consider its *norm*.

Definition A.26. *A vector norm $||\cdot||$ is a mapping from the set of $m \times 1$ vectors into* $\mathbb{R}$*, satisfying the conditions*

(1) $||\boldsymbol{x}|| \geq 0$,
(1a) $||\boldsymbol{x}|| = 0 \Leftrightarrow \boldsymbol{x} = \boldsymbol{0}$,
(2) $||c\boldsymbol{x}|| = |c|\,||\boldsymbol{x}||$,
(3) $||\boldsymbol{x} + \boldsymbol{y}|| \leq ||\boldsymbol{x}|| + ||\boldsymbol{y}||$

for any $m \times 1$ vectors $\boldsymbol{x}$ and $\boldsymbol{y}$ and any (real) number c.

If condition (1a) is not satisfied for every vector $\boldsymbol{x}$, i.e. there exists some nonzero $\boldsymbol{x}$ with $||\boldsymbol{x}|| = 0$, then the mapping is called a *vector seminorm.* There exist a number of different vector norms, the most common being the Euclidean norm specified by $||\boldsymbol{x}|| = \sqrt{\boldsymbol{x}'\boldsymbol{x}}$ which corresponds to the Euclidean inner product $\langle \boldsymbol{x}, \boldsymbol{y} \rangle = \boldsymbol{y}'\boldsymbol{x}$ of two vectors $\boldsymbol{x}$ and $\boldsymbol{y}$.

Theorem A.43. *For any two $m \times 1$ vectors $\boldsymbol{x}$ and $\boldsymbol{y}$ the inequality*

$$\langle \boldsymbol{x}, \boldsymbol{y} \rangle^2 \leq \langle \boldsymbol{x}, \boldsymbol{x} \rangle \langle \boldsymbol{y}, \boldsymbol{y} \rangle$$

is satisfied. Equality holds if and only if $\boldsymbol{x}$ and $\boldsymbol{y}$ are linearly dependent, i.e. $\boldsymbol{x} = \alpha \boldsymbol{y}$ for some number α.

The above inequality is known as the *Cauchy-Schwarz inequality.* Similarly to a vector norm, one may specify norms for matrices.

Definition A.27. *A matrix norm $||\cdot||$ is a mapping from the set of $m \times m$ matrices into* $\mathbb{R}$*, satisfying the conditions*

(1) $||\boldsymbol{A}|| \geq 0$,
(1a) $||\boldsymbol{A}|| = 0 \Leftrightarrow \boldsymbol{A} = \boldsymbol{0}$,
(2) $||c\boldsymbol{A}|| = |c|\,||\boldsymbol{A}||$,
(3) $||\boldsymbol{A} + \boldsymbol{B}|| \leq ||\boldsymbol{A}|| + ||\boldsymbol{B}||$,
(4) $||\boldsymbol{A}\boldsymbol{B}|| \leq ||\boldsymbol{A}||\,||\boldsymbol{B}||$

for any $m \times m$ matrices $\boldsymbol{A}$ and $\boldsymbol{B}$ and any (real) number c.

If condition (1a) is not satisfied for every $m \times m$ matrix $\boldsymbol{A}$, i.e. there exists some nonzero matrix $\boldsymbol{A}$ with $||\boldsymbol{A}|| = 0$, then the mapping is called a *matrix seminorm.* If condition (4) is not satisfied for every $m \times m$ matrices $\boldsymbol{A}$ and $\boldsymbol{B}$, then the mapping is also called *generalized matrix norm.* Well known matrix norms are the *Frobenius norm*

$$||\boldsymbol{A}|| = \sqrt{\mathrm{tr}(\boldsymbol{A}'\boldsymbol{A})}$$

(also called *Euclidean norm*, *Schur norm* or *Hilbert-Schmidt norm*) and the *spectral norm*

$$||\boldsymbol{A}|| = \max\{\sqrt{\lambda}\,|\,\lambda \text{ is an eigenvalue of } \boldsymbol{A}'\boldsymbol{A}\}\ .$$

A.3.5 Definiteness

In statistical applications, quadratic forms $\boldsymbol{x}'\boldsymbol{A}\boldsymbol{x}$ are often considered.

Definition A.28. *An $m \times m$ matrix $\boldsymbol{A}$ is called* nonnegative definite *if the inequality*

$$0 \leq \boldsymbol{x}'\boldsymbol{A}\boldsymbol{x}$$

is satisfied for every $m \times 1$ vector $\boldsymbol{x}$. A nonnegative definite matrix $\boldsymbol{A}$ is called positive definite *if the inequality*

$$0 < \boldsymbol{x}'\boldsymbol{A}\boldsymbol{x}$$

is satisfied for every $m \times 1$ vector $\boldsymbol{x} \neq \boldsymbol{0}$.

Usually, definiteness of a square matrix $\boldsymbol{A}$ is only specified for symmetric $\boldsymbol{A}$. As a reason for this one may see the fact that most applications of definiteness (especially in statistics) only involve symmetric matrices. Since the definiteness of a symmetric matrix is much easier to handle than the definiteness of an arbitrary square matrix, it is quite reasonable to confine to the set of symmetric matrices when regarding definiteness.

As a different reasoning one often finds the following. If one considers a quadratic form $\boldsymbol{x}'\boldsymbol{A}\boldsymbol{x}$ for some square matrix $\boldsymbol{A}$, then there always exists a symmetric matrix $\boldsymbol{B}$ such that

$$\boldsymbol{x}'\boldsymbol{A}\boldsymbol{x} = \boldsymbol{x}'\boldsymbol{B}\boldsymbol{x} \ ,$$

choose $\boldsymbol{B} = \frac{1}{2}(\boldsymbol{A} + \boldsymbol{A}')$. Hence, if we are solely interested in quadratic forms, then we can assume without loss of generality that the corresponding matrices are symmetric.

One should, however, acknowledge that often one is not interested in a quadratic form $\boldsymbol{x}'\boldsymbol{A}\boldsymbol{x}$ but in the matrix $\boldsymbol{A}$ itself. In that case the latter reasoning does not hold and as a matter of fact *it is* of importance whether the matrix $\boldsymbol{A}$ is symmetric or not.

Nonetheless, in view of the former reason given above, we only consider definiteness within the set of symmetric matrices, see Sect. A.4.4.

A.4 Symmetric Matrix

This section gives results for real-valued symmetric matrices.

A.4.1 Eigenvalues

Theorem A.44. *The m eigenvalues of a symmetric $m \times m$ matrix $\boldsymbol{A}$ are real.*

Clearly, there may also exist non-symmetric matrices whose eigenvalues are all real.

Example A.10. The matrix $\boldsymbol{A} = \begin{pmatrix} 1 & 1 \\ 0 & 1 \end{pmatrix}$ has the two real eigenvalues $\lambda_1 = \lambda_2 = 1$.

A.4.2 Spectral Decomposition

Any symmetric matrix can be diagonalized via an orthogonal matrix.

Theorem A.45. *If $\boldsymbol{A}$ is an $m \times m$ symmetric matrix, then $\boldsymbol{A}$ can be written as $\boldsymbol{A} = \boldsymbol{U}\boldsymbol{\Lambda}\boldsymbol{U}'$, where $\boldsymbol{U}$ is an orthogonal $m \times m$ matrix and $\boldsymbol{\Lambda}$ is an $m \times m$ diagonal matrix. The form $\boldsymbol{U}\boldsymbol{\Lambda}\boldsymbol{U}'$ is called a* spectral decomposition *of $\boldsymbol{A}$.*

Example A.11. The symmetric matrix $\boldsymbol{A} = \begin{pmatrix} 2 & 1 \\ 1 & 2 \end{pmatrix}$ can be written as $\boldsymbol{A} = \boldsymbol{U}_1\boldsymbol{\Lambda}_1\boldsymbol{U}_1'$, where

$$\boldsymbol{U}_1 = \begin{pmatrix} 1/\sqrt{2} & 1/\sqrt{2} \\ -1/\sqrt{2} & 1/\sqrt{2} \end{pmatrix} \quad \text{and} \quad \boldsymbol{\Lambda}_1 = \begin{pmatrix} 1 & 0 \\ 0 & 3 \end{pmatrix} .$$

The choices

$$\boldsymbol{U}_2 = \begin{pmatrix} -1/\sqrt{2} & -1/\sqrt{2} \\ -1/\sqrt{2} & 1/\sqrt{2} \end{pmatrix} \quad \text{and} \quad \boldsymbol{\Lambda}_2 = \begin{pmatrix} 3 & 0 \\ 0 & 1 \end{pmatrix}$$

give a further spectral decomposition $\boldsymbol{A} = \boldsymbol{U}_2\boldsymbol{\Lambda}_2\boldsymbol{U}_2'$ of $\boldsymbol{A}$. □

The above theorem ensures the existence of a spectral decomposition of a symmetric matrix. In each spectral decomposition of $\boldsymbol{A}$, the elements of the diagonal matrix are the eigenvalues of $\boldsymbol{A}$, while the columns of the orthogonal matrix are corresponding eigenvectors.

Theorem A.46. *If $\boldsymbol{U}\boldsymbol{\Lambda}\boldsymbol{U}'$ is a spectral decomposition of an $m\times m$ symmetric matrix $\boldsymbol{A}$, then the i-th diagonal element of $\boldsymbol{\Lambda}$ and the i-th column of $\boldsymbol{U}$ are an eigenpair of $\boldsymbol{A}$.*

Now, a symmetric $m \times m$ matrix $\boldsymbol{A}$ can be written as

$$\boldsymbol{A} = \boldsymbol{U}\boldsymbol{\Lambda}\boldsymbol{U}' = \sum_{i=1}^{m} \lambda_i \boldsymbol{u}_i \boldsymbol{u}_i' ,$$

where $\lambda_1, \ldots, \lambda_m$ are the m eigenvalues and $\boldsymbol{u}_1, \ldots, \boldsymbol{u}_m$ are corresponding orthonormal eigenvectors of $\boldsymbol{A}$. If $\mu_1, \ldots, \mu_k$ denote the k *distinct* eigenvalues of $\boldsymbol{A}$, then

$$\boldsymbol{A} = \sum_{i=1}^{k} \mu_i \boldsymbol{G}_i .$$

Here, $\boldsymbol{G}_i = \sum_j \boldsymbol{u}_j\boldsymbol{u}_j'$, where the sum is taken over all those j for which $\mu_i = \lambda_j$. This representation of $\boldsymbol{A}$ is also often called the spectral decomposition of $\boldsymbol{A}$.

Theorem A.47. *Let $\boldsymbol{A}$ and $\boldsymbol{B}$ be two symmetric $m \times m$ matrices satisfying $\boldsymbol{AB} = \boldsymbol{BA}$. Then there exists an $m \times m$ orthogonal matrix $\boldsymbol{U}$ such that*

$$\boldsymbol{A} = \boldsymbol{U\Lambda U}' \quad \text{and} \quad \boldsymbol{B} = \boldsymbol{U\Gamma U}' \ ,$$

where $\boldsymbol{\Lambda}$ and $\boldsymbol{\Gamma}$ are two $m \times m$ diagonal matrices.

The above result ensures that two commuting symmetric matrices are simultaneously diagonalizable via a single orthogonal transformation.

A.4.3 Rayleigh Ratio

For a symmetric $m \times m$ matrix $\boldsymbol{A}$ one may consider the *Rayleigh ratio*

$$\frac{\boldsymbol{x}'\boldsymbol{A}\boldsymbol{x}}{\boldsymbol{x}'\boldsymbol{x}} \ ,$$

where $\boldsymbol{x}$ is an $m \times 1$ vector. Then the following result holds true.

Theorem A.48. *Let $\lambda_1 \leq \cdots \leq \lambda_m$ be the eigenvalues of a symmetric $m \times m$ matrix $\boldsymbol{A}$. Then the inequalities*

$$\lambda_1 \leq \frac{\boldsymbol{x}'\boldsymbol{A}\boldsymbol{x}}{\boldsymbol{x}'\boldsymbol{x}} \leq \lambda_m$$

are satisfied for any $m \times 1$ vector $\boldsymbol{x} \neq \boldsymbol{0}$. The Rayleigh ratio attains its maximum, i.e. $\boldsymbol{x}'\boldsymbol{A}\boldsymbol{x}/\boldsymbol{x}'\boldsymbol{x} = \lambda_m$, when $\boldsymbol{x}$ is an eigenvector associated with λ_m, and it attains its minimum, i.e. $\boldsymbol{x}'\boldsymbol{A}\boldsymbol{x}/\boldsymbol{x}'\boldsymbol{x} = \lambda_1$, when $\boldsymbol{x}$ is an eigenvector associated with λ_1.

A.4.4 Definiteness

We now consider the definiteness of a square matrix $\boldsymbol{A}$ as specified in Definition A.28 when in addition $\boldsymbol{A}$ is assumed to be symmetric.

Characterization of Definiteness

Definiteness of symmetric matrices can be characterized in different ways.

Theorem A.49. *A symmetric $m \times m$ matrix $\boldsymbol{A}$ is nonnegative definite if and only if all eigenvalues of $\boldsymbol{A}$ are nonnegative. Moreover, $\boldsymbol{A}$ is positive definite if and only if all eigenvalues of $\boldsymbol{A}$ are positive.*

Theorem A.50. *An $m \times m$ matrix $\boldsymbol{A}$ is symmetric and nonnegative definite if and only if $\boldsymbol{A}$ can be written as $\boldsymbol{A} = \boldsymbol{F}\boldsymbol{F}'$ for some matrix $\boldsymbol{F}$ with m rows.*

Clearly, the matrix $\boldsymbol{F}$ must not necessarily be a square matrix to make $\boldsymbol{F}\boldsymbol{F}'$ symmetric nonnegative definite. On the other hand, it is easy to see from the spectral decomposition of symmetric matrices that for a symmetric nonnegative definite $m \times m$ matrix $\boldsymbol{A}$, one can always find some $m \times m$ matrix $\boldsymbol{F}$ with $\boldsymbol{A} = \boldsymbol{F}\boldsymbol{F}'$. Moreover for such a matrix $\boldsymbol{A}$ one can always find some matrix $\boldsymbol{F}$ with m rows and $\text{rk}(\boldsymbol{A})$ columns, but one can never find some matrix $\boldsymbol{F}$ with less than $\text{rk}(\boldsymbol{A})$ columns.

Corollary A.1. *If $\boldsymbol{A}$ is a symmetric nonnegative definite $m \times m$ matrix and $\boldsymbol{G}$ is a matrix with m columns, then $\boldsymbol{G}\boldsymbol{A}\boldsymbol{G}'$ is symmetric nonnegative definite.*

Theorem A.51. *An $m \times m$ matrix $\boldsymbol{A}$ is symmetric positive definite if and only if $\boldsymbol{A}$ can be written as $\boldsymbol{A} = \boldsymbol{F}\boldsymbol{F}'$ for some nonsingular $m \times m$ matrix $\boldsymbol{F}$.*

Theorem A.52. *Let $\boldsymbol{A}$ be an $m \times m$ matrix and let $\boldsymbol{G}$ be a full column rank matrix with m columns. Then $\boldsymbol{A}$ is symmetric nonnegative definite if and only if $\boldsymbol{G}\boldsymbol{A}\boldsymbol{G}'$ is symmetric nonnegative definite.*

Theorem A.53. *A symmetric nonnegative definite $m \times m$ matrix $\boldsymbol{A}$ is identical to the $m \times m$ zero matrix if and only if* $\text{tr}(\boldsymbol{A}) = 0$.

Theorem A.54. *A symmetric $m \times m$ matrix $\boldsymbol{A}$ is nonnegative definite if and only if* $\text{tr}(\boldsymbol{A}\boldsymbol{B}) \geq 0$ *for all symmetric nonnegative definite $m \times m$ matrices $\boldsymbol{B}$.*

The above result has been derived by Theobald [115]. An easy proof is as follows. Suppose the symmetric $m \times m$ matrix $\boldsymbol{A}$ satisfies $\text{tr}(\boldsymbol{A}\boldsymbol{B}) \geq 0$ for every symmetric nonnegative definite $m \times m$ matrix $\boldsymbol{B}$. Then also $\text{tr}(\boldsymbol{A}\boldsymbol{b}\boldsymbol{b}') \geq 0$ for every $m \times 1$ vector $\boldsymbol{b}$. In view of $\text{tr}(\boldsymbol{A}\boldsymbol{b}\boldsymbol{b}') = \text{tr}(\boldsymbol{b}'\boldsymbol{A}\boldsymbol{b}) = \boldsymbol{b}'\boldsymbol{A}\boldsymbol{b} \geq 0$ it follows that $\boldsymbol{A}$ must be nonnegative definite. Conversely, if $\boldsymbol{A}$ is symmetric nonnegative definite, then always $\text{tr}(\boldsymbol{A}\boldsymbol{B}) = \text{tr}(\boldsymbol{G}'\boldsymbol{A}\boldsymbol{G}) \geq 0$ for every symmetric nonnegative definite matrix $\boldsymbol{B} = \boldsymbol{G}\boldsymbol{G}'$.

Square Root of a Symmetric Nonnegative Definite Matrix

A square root of an $m \times m$ matrix $\boldsymbol{A}$ is any $m \times m$ matrix $\boldsymbol{B}$ such that $\boldsymbol{B}\boldsymbol{B} = \boldsymbol{A}$.

Theorem A.55. *Let $\boldsymbol{A}$ be a symmetric nonnegative definite $m \times m$ matrix. Then there exists a uniquely determined symmetric nonnegative definite $m \times m$ matrix $\boldsymbol{B}$ such that $\boldsymbol{B}\boldsymbol{B} = \boldsymbol{A}$. This matrix is denoted by $\boldsymbol{B} = \boldsymbol{A}^{1/2}$.*

In general, a square root of an $m \times m$ matrix $\boldsymbol{A}$ is not uniquely determined, meaning that there exist different $m \times m$ matrices $\boldsymbol{B}_1$ and $\boldsymbol{B}_2$ such that $\boldsymbol{B}_1\boldsymbol{B}_1 = \boldsymbol{B}_2\boldsymbol{B}_2 = \boldsymbol{A}$. Nonetheless, the above theorem claims that for a symmetric nonnegative definite $m \times m$ matrix $\boldsymbol{A}$ there can only exist *one* $m \times m$ matrix $\boldsymbol{B}$, being symmetric nonnegative definite and in addition satisfying $\boldsymbol{B}\boldsymbol{B} = \boldsymbol{A}$. One may compute the square root of a symmetric nonnegative definite matrix from one of its spectral decompositions.

Theorem A.56. *Let $\boldsymbol{A}$ be a symmetric nonnegative definite $m \times m$ matrix and let $\boldsymbol{U}\boldsymbol{\Lambda}\boldsymbol{U}'$ be a spectral decomposition of $\boldsymbol{A}$. Then $\boldsymbol{A}^{1/2} = \boldsymbol{U}\boldsymbol{\Lambda}^{1/2}\boldsymbol{U}'$, where $\boldsymbol{\Lambda}^{1/2}$ is the diagonal matrix whose i-th main diagonal element is the positive square root of the i-th main diagonal element of $\boldsymbol{\Lambda}$.*

Product of Symmetric Nonnegative Definite Matrices

Although the product of two symmetric nonnegative definite matrices must not be symmetric, its eigenvalues remain real nonnegative.

Theorem A.57. *Let $\boldsymbol{A}$ and $\boldsymbol{B}$ be two symmetric nonnegative definite $m \times m$ matrices. Then all eigenvalues of $\boldsymbol{A}\boldsymbol{B}$ are real and nonnegative.*

Theorem A.58. *Let $\boldsymbol{A}$ and $\boldsymbol{B}$ be two symmetric nonnegative definite $m \times m$ matrices. Then the inequalities*

$$0 \leq \operatorname{tr}(\boldsymbol{A}\boldsymbol{B}) \leq \operatorname{tr}(\boldsymbol{A}) \operatorname{tr}(\boldsymbol{B})$$

hold true.

Note that the above inequality $\operatorname{tr}(\boldsymbol{A}\boldsymbol{B}) \leq \operatorname{tr}(\boldsymbol{A}) \operatorname{tr}(\boldsymbol{B})$ can be deduced from $\|\boldsymbol{A}^{1/2}\boldsymbol{B}^{1/2}\| \leq \|\boldsymbol{A}^{1/2}\|\|\boldsymbol{B}^{1/2}\|$, where $\|\cdot\|$ denotes the Frobenius norm.

Kronecker Product of Symmetric Nonnegative Definite Matrices

If $\boldsymbol{A}$ is an $m \times m$ matrix and $\boldsymbol{B}$ is a $p \times p$ matrix, both being symmetric nonnegative definite, then $\boldsymbol{A}$ and $\boldsymbol{B}$ can be written as $\boldsymbol{A} = \boldsymbol{F}\boldsymbol{F}'$ and $\boldsymbol{B} = \boldsymbol{G}\boldsymbol{G}'$ for some $m \times m$ matrix $\boldsymbol{F}$ and some $p \times p$ matrix $\boldsymbol{G}$, respectively. Then

$$\boldsymbol{A} \otimes \boldsymbol{B} = (\boldsymbol{F}\boldsymbol{F}') \otimes (\boldsymbol{G}\boldsymbol{G}') = (\boldsymbol{F} \otimes \boldsymbol{G})(\boldsymbol{F}' \otimes \boldsymbol{G}') = (\boldsymbol{F} \otimes \boldsymbol{G})(\boldsymbol{F} \otimes \boldsymbol{G})' .$$

This shows that the Kronecker product of two symmetric nonnegative definite matrices is again symmetric nonnegative definite. Moreover, the identity $\operatorname{rk}(\boldsymbol{A} \otimes \boldsymbol{B}) = \operatorname{rk}(\boldsymbol{A}) \operatorname{rk}(\boldsymbol{B})$ shows that the Kronecker product of two symmetric positive definite matrices is again symmetric positive definite.

Theorem A.59. *Let $\boldsymbol{A}$ be an $m \times m$ matrix and let $\boldsymbol{B}$ be a $p \times p$ matrix, both being symmetric nonnegative definite. Then $\boldsymbol{A} \otimes \boldsymbol{B}$ is symmetric nonnegative definite. If in addition $\boldsymbol{A}$ and $\boldsymbol{B}$ both are positive definite, then $\boldsymbol{A} \otimes \boldsymbol{B}$ is positive definite.*

Rank 1 Modification

A symmetric nonnegative definite matrix of rank 1 can always be written as $\boldsymbol{a}\boldsymbol{a}'$. If such a matrix is subtracted from a positive definite matrix, then the resulting matrix is not necessarily nonnegative definite.

Theorem A.60. *Let $\boldsymbol{A}$ be a symmetric positive definite $m \times m$ matrix, $\boldsymbol{a}$ an $m \times 1$ vector, and α a positive number. Then $\alpha\boldsymbol{A} - \boldsymbol{a}\boldsymbol{a}'$ is nonnegative definite if and only if $\boldsymbol{a}'\boldsymbol{A}^{-1}\boldsymbol{a} \leq \alpha$ is satisfied.*

The above result can also be deduced from extremum properties of quadratic forms. As a matter of fact, $\alpha\boldsymbol{A} - \boldsymbol{a}\boldsymbol{a}'$ is nonnegative definite if and only if

$$\frac{(\boldsymbol{x}'\boldsymbol{a})^2}{\boldsymbol{x}'\boldsymbol{A}\boldsymbol{x}} \leq \alpha$$

is satisfied for all $m \times 1$ vectors $\boldsymbol{x} \neq \boldsymbol{0}$. Since the Cauchy-Schwarz inequality implies

$$\max_{\boldsymbol{x} \neq \boldsymbol{0}} \frac{(\boldsymbol{x}'\boldsymbol{a})^2}{\boldsymbol{x}'\boldsymbol{A}\boldsymbol{x}} = \boldsymbol{a}'\boldsymbol{A}^{-1}\boldsymbol{a} \,,$$

see e.g. [89, Sect. 1f.1]), the asserted equivalence follows. See also [36] for a similar but more statistically oriented proof. The following generalization of this result is given in [5].

Theorem A.61. *Let $\boldsymbol{A}$ be a symmetric $m \times m$ matrix, $\boldsymbol{a}$ an $m \times 1$ vector, and α a positive number. Then $\alpha\boldsymbol{A} - \boldsymbol{a}\boldsymbol{a}'$ is nonnegative definite if and only if the three conditions*

$$\boldsymbol{A} \text{ is nonnegative definite}, \quad \boldsymbol{a} \in \mathcal{C}(\boldsymbol{A}), \quad \boldsymbol{a}'\boldsymbol{A}^{+}\boldsymbol{a} \leq \alpha$$

are satisfied.

A.5 Löwner Partial Ordering

How can we order square matrices in the sense that we can say that one matrix is smaller or greater than the other? The most popular ordering is the so-called *Löwner partial ordering* $\leq_{\mathrm{L}}$.

Definition A.29. *Two $m \times m$ matrices $\boldsymbol{A}$ and $\boldsymbol{B}$ are related via $\boldsymbol{A} \leq_{\mathrm{L}} \boldsymbol{B}$ if the difference $\boldsymbol{B} - \boldsymbol{A}$ is symmetric nonnegative definite.*

Note that in the above definition it is not required that the matrices $\boldsymbol{A}$ or $\boldsymbol{B}$ must be symmetric, but only that the difference $\boldsymbol{B} - \boldsymbol{A}$ is symmetric.

Example A.12. Let $\boldsymbol{A} = \begin{pmatrix} 1 & 1 \\ 0 & 1 \end{pmatrix}$ and $\boldsymbol{B} = \begin{pmatrix} 2 & 1 \\ 0 & 2 \end{pmatrix}$. Then the difference

$$\boldsymbol{B} - \boldsymbol{A} = \begin{pmatrix} 1 & 0 \\ 0 & 1 \end{pmatrix}$$

is symmetric nonnegative definite, and hence $\boldsymbol{A} \leq_{\mathrm{L}} \boldsymbol{B}$. □

Theorem A.62. *The relation $\leq_{\mathrm{L}}$ is a partial ordering in the set of $m \times m$ matrices, meaning that*

(1) $\boldsymbol{A} \leq_{\mathrm{L}} \boldsymbol{A}$ *(reflexive),*
(2) $\boldsymbol{A} \leq_{\mathrm{L}} \boldsymbol{B} \wedge \boldsymbol{B} \leq_{\mathrm{L}} \boldsymbol{A} \Rightarrow \boldsymbol{A} = \boldsymbol{B}$ *(antisymmetric),*
(3) $\boldsymbol{A} \leq_{\mathrm{L}} \boldsymbol{B} \wedge \boldsymbol{B} \leq_{\mathrm{L}} \boldsymbol{C} \Rightarrow \boldsymbol{A} \leq_{\mathrm{L}} \boldsymbol{C}$ *(transitive)*

hold true for $m \times m$ matrices $\boldsymbol{A}$, $\boldsymbol{B}$ and $\boldsymbol{C}$.

Sometimes, in the literature, the symbol $\boldsymbol{A} < \boldsymbol{B}$ is used to denote a symmetric positive definite difference $\boldsymbol{B} - \boldsymbol{A}$. Such a relation $<$ does however not specify a partial ordering. Nonetheless, the use of $<$ may simplify the presentation of results, especially in those cases when $\leq_{\mathrm{L}}$ can be replaced by $<$ under certain conditions.

Quite often the Löwner partial ordering is considered within the set of symmetric nonnegative definite matrices.

Theorem A.63. *Let $\boldsymbol{A}$ and $\boldsymbol{B}$ be two symmetric nonnegative definite $m \times m$ matrices with $\boldsymbol{A} \leq_{\mathrm{L}} \boldsymbol{B}$. Then*

$$\lambda_i \leq \mu_i \quad \forall i \in \{1, \ldots, m\} ,$$

where $\lambda_1 \leq \cdots \leq \lambda_m$ are the ordered eigenvalues of $\boldsymbol{A}$ and $\mu_1 \leq \cdots \leq \mu_m$ are the ordered eigenvalues of $\boldsymbol{B}$.

Corollary A.2. *Let $\boldsymbol{A}$ and $\boldsymbol{B}$ be two symmetric nonnegative definite $m \times m$ matrices with $\boldsymbol{A} \leq_{\mathrm{L}} \boldsymbol{B}$. Then the inequalities*

$$\det(\boldsymbol{A}) \leq \det(\boldsymbol{B}) \quad \textit{and} \quad \operatorname{tr}(\boldsymbol{A}) \leq \operatorname{tr}(\boldsymbol{B})$$

hold true.

Theorem A.64. *Let $\boldsymbol{A}$ and $\boldsymbol{B}$ be two symmetric nonnegative definite $m \times m$ matrices with $\boldsymbol{A} \leq_{\mathrm{L}} \boldsymbol{B}$. Then $\boldsymbol{A}^{1/2} \leq_{\mathrm{L}} \boldsymbol{B}^{1/2}$.*

Note that the relation $\boldsymbol{A} \leq_{\mathrm{L}} \boldsymbol{B}$ for two symmetric nonnegative definite matrices $\boldsymbol{A}$ and $\boldsymbol{B}$ does *not* imply the relation $\boldsymbol{A}^2 \leq_{\mathrm{L}} \boldsymbol{B}^2$.

Example A.13. Consider the two symmetric nonnegative definite matrices $\boldsymbol{A} = \begin{pmatrix} 4.6 & 10.2 \\ 10.2 & 23.4 \end{pmatrix}$ and $\boldsymbol{B} = \begin{pmatrix} 5 & 11 \\ 11 & 25 \end{pmatrix}$. Then

$$\boldsymbol{B} - \boldsymbol{A} = \begin{pmatrix} 0.4 & 0.8 \\ 0.8 & 1.6 \end{pmatrix}$$

is nonnegative definite, while this is not true for

$$\boldsymbol{B}^2 - \boldsymbol{A}^2 = \begin{pmatrix} 20.8 & 44.4 \\ 44.4 & 94.4 \end{pmatrix},$$

see also [41, p. 608]. □

Theorem A.65. *Let* $\boldsymbol{A}$ *and* $\boldsymbol{B}$ *be two symmetric nonnegative definite* $m \times m$ *matrices. Then the relation* $\boldsymbol{A} \leq_{\mathrm{L}} \boldsymbol{B}$ *holds true if and only if the conditions*

$$\mathcal{C}(\boldsymbol{A}) \subseteq \mathcal{C}(\boldsymbol{B}) \quad \text{and} \quad \boldsymbol{A}\boldsymbol{B}^{+}\boldsymbol{A} \leq_{\mathrm{L}} \boldsymbol{A}$$

are satisfied.

Theorem A.66. *Let* $\boldsymbol{A}$ *and* $\boldsymbol{B}$ *be two symmetric nonnegative definite* $m \times m$ *matrices. Then the relation* $\boldsymbol{A} \leq_{\mathrm{L}} \boldsymbol{B}$ *holds true if and only if the conditions*

$$\mathcal{C}(\boldsymbol{A}) \subseteq \mathcal{C}(\boldsymbol{B}) \quad \text{and} \quad \lambda_{\max}(\boldsymbol{A}\boldsymbol{B}^{+}) \leq 1$$

are satisfied, where $\lambda_{\max}(\cdot)$ *denotes the largest eigenvalue of a matrix with only real eigenvalues.*

Theorem A.67. *Let* $\boldsymbol{A}$ *and* $\boldsymbol{B}$ *be two symmetric positive definite* $m \times m$ *matrices. Then* $\boldsymbol{A} \leq_{\mathrm{L}} \boldsymbol{B}$ *if and only if* $\boldsymbol{B}^{-1} \leq_{\mathrm{L}} \boldsymbol{A}^{-1}$.

Theorem A.68. *Let* $\boldsymbol{A}$ *and* $\boldsymbol{B}$ *be two symmetric nonnegative definite* $m \times m$ *matrices with* $\mathrm{rk}(\boldsymbol{A}) = \mathrm{rk}(\boldsymbol{B})$. *Then* $\boldsymbol{A} \leq_{\mathrm{L}} \boldsymbol{B}$ *if and only if* $\boldsymbol{B}^{+} \leq_{\mathrm{L}} \boldsymbol{A}^{+}$.

Theorem A.69. *Let* $\boldsymbol{A}$ *and* $\boldsymbol{B}$ *be two symmetric idempotent* $m \times m$ *matrices. Then* $\boldsymbol{A} \leq_{\mathrm{L}} \boldsymbol{B}$ *if and only if* $\mathcal{C}(\boldsymbol{A}) \subseteq \mathcal{C}(\boldsymbol{B})$.

B Stochastic Vectors

Results for vectors and matrices whose elements are random variables are given.

B.1 Expectation and Covariance

An $m \times n$ matrix $\boldsymbol{Z}$ is called *stochastic* if its elements are random variables. An $m \times 1$ stochastic matrix is called *stochastic vector* or *random vector*.

Expectation

The expectation of a stochastic matrix is the matrix of expectations of its elements.

Definition B.1. *Let $\boldsymbol{Z}$ be an $m \times n$ stochastic matrix with $\mathrm{E}(z_{ij}) = \mu_{ij}$ for $i = 1, \ldots, m$ and $j = 1, \ldots, n$. Then the expectation matrix $\mathrm{E}(\boldsymbol{Z})$ of $\boldsymbol{Z}$ is the matrix $\boldsymbol{M}$ whose (i, j)th element is μ_{ij}.*

The expectation of a random matrix is a linear operator.

Theorem B.1. *Let $\boldsymbol{Z}$ be an $m \times n$ stochastic matrix. Let $\boldsymbol{A}$ be a $k \times m$, $\boldsymbol{B}$ an $n \times p$, and $\boldsymbol{C}$ a $k \times p$ non-stochastic matrix. Then the identity*

$$\mathrm{E}(\boldsymbol{A}\boldsymbol{Z}\boldsymbol{B} + \boldsymbol{C}) = \boldsymbol{A}\,\mathrm{E}(\boldsymbol{Z})\boldsymbol{B} + \boldsymbol{C}$$

holds true.

If we consider a random vector $\boldsymbol{z}$, then $\mathrm{E}(\boldsymbol{z})$ may be seen as its first multivariate moment. As its second multivariate moment one may consider the matrix of covariances between its elements.

Covariance Matrix of a Random Vector

The covariance of a random vector is the matrix of covariances between its elements.

Definition B.2. *Let $\boldsymbol{z}$ be an $m \times 1$ random vector with* $\mathrm{E}(\boldsymbol{z}) = \boldsymbol{\mu}$. *Then the* covariance matrix $\mathrm{Cov}(\boldsymbol{z})$ *of $\boldsymbol{z}$ is the matrix*

$$\mathrm{Cov}(\boldsymbol{z}) = \mathrm{E}[(\boldsymbol{z}-\boldsymbol{\mu})(\boldsymbol{z}-\boldsymbol{\mu})'] = \begin{pmatrix} \sigma_{11} & \sigma_{12} & \cdots & \sigma_{1m} \\ \sigma_{21} & \sigma_{22} & \cdots & \sigma_{2m} \\ \vdots & \vdots & & \vdots \\ \sigma_{m1} & \sigma_{m2} & \cdots & \sigma_{mm} \end{pmatrix},$$

where $\sigma_{ij} = \sigma_{ji} = \mathrm{Cov}(z_i, z_j)$ for $i, j = 1, \dots, m$.

The covariance matrix of a random vector is also called *variance-covariance* or *dispersion* matrix.

Theorem B.2. *Let $\boldsymbol{z}$ be an $m \times 1$ random vector with* $\mathrm{E}(\boldsymbol{z}) = \boldsymbol{\mu}$ *and* $\mathrm{Cov}(\boldsymbol{z}) = \boldsymbol{V}$. *Let $\boldsymbol{A}$ be a non-stochastic $k \times m$ matrix and let $\boldsymbol{a}$ be a $k \times 1$ non-stochastic vector. Then the identities*

$$\mathrm{E}(\boldsymbol{Az}+\boldsymbol{a}) = \boldsymbol{A\mu}+\boldsymbol{a} \quad \text{and} \quad \mathrm{Cov}(\boldsymbol{Az}+\boldsymbol{a}) = \boldsymbol{AVA}'$$

hold true.

Theorem B.3. *The covariance matrix* $\mathrm{Cov}(\boldsymbol{z})$ *of an $m \times 1$ random vector $\boldsymbol{z}$ is symmetric nonnegative definite.*

Note that for any given symmetric nonnegative definite matrix $\boldsymbol{V}$ one may find a random vector $\boldsymbol{x}$ with $\mathrm{Cov}(\boldsymbol{x}) = \boldsymbol{V}$. Consider for example the vector $\boldsymbol{x} = \boldsymbol{V}^{1/2}\boldsymbol{z}$, where the elements z_i of the vector $\boldsymbol{z}$ are independent identically distributed random variables with $\mathrm{E}(z_i) = 0$ and $\mathrm{Var}(z_i) = 1$. In view of this, the set of symmetric nonnegative definite matrices can be identified with the set of covariance matrices.

Theorem B.4. *The covariance matrix* $\mathrm{Cov}(\boldsymbol{z})$ *of an $m \times 1$ random vector $\boldsymbol{z}$ is singular if and only if there exists a non-stochastic number α and a non-stochastic $m \times 1$ vector $\boldsymbol{b} \neq \boldsymbol{0}$ such that* $\mathrm{Prob}(\boldsymbol{b}'\boldsymbol{z} = \alpha) = 1$.

We may call a random vector *singular* or *nonsingular* depending on whether its covariance matrix is singular or nonsingular. Singular random vectors are not rare. If $\boldsymbol{A}$ is a matrix being not of full row rank, then the transformation $\boldsymbol{Az}+\boldsymbol{a}$ of a nonsingular random vector $\boldsymbol{z}$ is a singular random vector. Examples for singular random vectors in the linear regression model are the vector of least squares residuals $\widehat{\boldsymbol{\varepsilon}} = \boldsymbol{y} - \boldsymbol{X}\widehat{\boldsymbol{\beta}}$ and the estimator $\widehat{\boldsymbol{y}} = \boldsymbol{X}\widehat{\boldsymbol{\beta}}$ of $\mathrm{E}(\boldsymbol{y})$. Moreover, the restricted least squares estimator $\widehat{\boldsymbol{\beta}}_{\mathrm{R}}$ is a singular random vector.

Theorem B.5. *Let $\boldsymbol{z}$ be an $m \times 1$ random vector with* $\mathrm{E}(\boldsymbol{z}) = \boldsymbol{\mu}$ *and* $\mathrm{Cov}(\boldsymbol{z}) = \boldsymbol{V}$ *and let $\boldsymbol{A}$ be an $m \times m$ non-stochastic matrix. Then the identity*

$$\mathrm{E}(\boldsymbol{z}'\boldsymbol{Az}) = \mathrm{tr}(\boldsymbol{AV}) + \boldsymbol{\mu}'\boldsymbol{A\mu}$$

holds true.

Note that the above result holds true for any not necessarily symmetric square matrix $\boldsymbol{A}$.

Covariance Matrix Between Two Random Vectors

One may also specify the covariance between two random vectors.

Definition B.3. *Let $\boldsymbol{u}$ and $\boldsymbol{v}$ be two $m \times 1$ and $p \times 1$ random vectors. Then the* covariance matrix $\text{Cov}(\boldsymbol{u}, \boldsymbol{v})$ *between $\boldsymbol{u}$ and $\boldsymbol{v}$ is the matrix* $\text{Cov}(\boldsymbol{u}, \boldsymbol{v}) = \text{E}[(\boldsymbol{u} - \text{E}(\boldsymbol{u}))(\boldsymbol{v} - \text{E}(\boldsymbol{v}))']$.

The covariance matrix between $\boldsymbol{u}$ and $\boldsymbol{v}$ is an $m \times p$ matrix and thus it is not necessarily square. The covariance between $\boldsymbol{v}$ and $\boldsymbol{u}$ is the transpose of this matrix, i.e.

$$\text{Cov}(\boldsymbol{u}, \boldsymbol{v}) = \text{Cov}(\boldsymbol{v}, \boldsymbol{u})'.$$

If $\text{Cov}(\boldsymbol{u}, \boldsymbol{v}) = \boldsymbol{0}$ is satisfied, then $\boldsymbol{u}$ and $\boldsymbol{v}$ are called *uncorrelated.*

The covariance matrix between two vectors can be seen as the generalization of the covariance matrix of one vector, since

$$\text{Cov}(\boldsymbol{z}, \boldsymbol{z}) = \text{Cov}(\boldsymbol{z})$$

holds true for any random vector $\boldsymbol{z}$.

Theorem B.6. *Let $\boldsymbol{u}$ and $\boldsymbol{v}$ be two $m \times 1$ and $p \times 1$ random vectors. Let $\boldsymbol{A}$ and be $\boldsymbol{B}$ be two $h \times m$ and $k \times p$ non-stochastic matrices and let $\boldsymbol{a}$ and $\boldsymbol{b}$ be two $h \times 1$ and $k \times 1$ non-stochastic vectors. Then the identities*

$$\text{Cov}(\boldsymbol{A}\boldsymbol{u} + \boldsymbol{a}, \boldsymbol{B}\boldsymbol{v} + \boldsymbol{b}) = \text{Cov}(\boldsymbol{A}\boldsymbol{u}, \boldsymbol{B}\boldsymbol{v}) = \boldsymbol{A}\,\text{Cov}(\boldsymbol{u}, \boldsymbol{v})\boldsymbol{B}'$$

hold true.

Theorem B.7. *Let $\boldsymbol{z} = (\boldsymbol{z}_1', \boldsymbol{z}_2')'$, where $\boldsymbol{z}_1$ and $\boldsymbol{z}_2$ are two $m_1 \times 1$ and $m_2 \times 1$ random vectors. Then*

$$\text{Cov}(\boldsymbol{z}) = \begin{pmatrix} \text{Cov}(\boldsymbol{z}_1) & \text{Cov}(\boldsymbol{z}_1, \boldsymbol{z}_2) \\ \text{Cov}(\boldsymbol{z}_2, \boldsymbol{z}_1) & \text{Cov}(\boldsymbol{z}_2) \end{pmatrix},$$

where $\text{Cov}(\boldsymbol{z}_2, \boldsymbol{z}_1) = \text{Cov}(\boldsymbol{z}_1, \boldsymbol{z}_2)'$.

From the above two theorems the following corollary can be concluded.

Corollary B.1. *Let $\boldsymbol{u}$ and $\boldsymbol{v}$ be two $m \times 1$ and $p \times 1$ random vectors. Let $\boldsymbol{A}$ and $\boldsymbol{B}$ be two $h \times m$ and $h \times p$ non-stochastic matrices. Then the identity*

$$\begin{aligned} &\text{Cov}(\boldsymbol{A}\boldsymbol{u} + \boldsymbol{B}\boldsymbol{v}) \\ &= \boldsymbol{A}\,\text{Cov}(\boldsymbol{u})\boldsymbol{A}' + \boldsymbol{A}\,\text{Cov}(\boldsymbol{u}, \boldsymbol{v})\boldsymbol{B}' + \boldsymbol{B}\,\text{Cov}(\boldsymbol{v}, \boldsymbol{u})\boldsymbol{A}' + \boldsymbol{B}\,\text{Cov}(\boldsymbol{v})\boldsymbol{B}' \end{aligned}$$

holds true.

B.2 Multivariate Normal Distribution

One may generalize the normal distribution of a random variable to the multivariate normal distribution of a random vector.

Definition B.4. *Let* $\boldsymbol{x}$ *be a* $k \times 1$ *random vector whose elements* x_i *are independent identically* $N(0,1)$ *distributed. Then the distribution of* $\boldsymbol{x}$ *is called* k-variate standard normal distribution.

The multivariate standard normal distribution can be seen as the basis for the specification of the multivariate normal distribution.

Definition B.5. *Let* $\boldsymbol{z}$ *be an* $m \times 1$ *random vector with* $\mathrm{E}(\boldsymbol{z}) = \boldsymbol{\mu}$ *and* $\mathrm{Cov}(\boldsymbol{z}) = \boldsymbol{V}$ *with* $\mathrm{rk}(\boldsymbol{V}) = k$. *Let* $\boldsymbol{x}$ *have a* k*-variate standard normal distribution. Then the distribution of* $\boldsymbol{z}$ *is called* m-variate normal distribution, *denoted by* $\boldsymbol{z} \sim N_m(\boldsymbol{\mu}, \boldsymbol{V})$, *if there exists some* $m \times k$ *matrix* $\boldsymbol{A}$ *of full column rank satisfying* $\boldsymbol{V} = \boldsymbol{A}\boldsymbol{A}'$ *such that* $\boldsymbol{z}$ *and* $\boldsymbol{A}\boldsymbol{x} + \boldsymbol{\mu}$ *have the same distribution.*

As shown by the following example there might exist different $m \times k$ matrices $\boldsymbol{A}$ and $\boldsymbol{B}$ of full column rank with $\boldsymbol{V} = \boldsymbol{A}\boldsymbol{A}' = \boldsymbol{B}\boldsymbol{B}'$. Hence it is not clear whether the above definition specifies a well defined distribution, since one might suspect that $\boldsymbol{A}\boldsymbol{x} + \boldsymbol{\mu}$ and $\boldsymbol{B}\boldsymbol{x} + \boldsymbol{\mu}$ could have different distributions.

Example B.1. Let $\boldsymbol{V} = \begin{pmatrix} 1 & 1/2 \\ 1/2 & 1 \end{pmatrix}$. Then the two different full rank matrices

$$\boldsymbol{A} = \frac{1}{2}\begin{pmatrix} -1 & \sqrt{3} \\ 1 & \sqrt{3} \end{pmatrix} \quad \text{and} \quad \boldsymbol{B} = \frac{1}{2\sqrt{2}}\begin{pmatrix} -1+\sqrt{3} & 1+\sqrt{3} \\ 1+\sqrt{3} & -1+\sqrt{3} \end{pmatrix}$$

satisfy the identity $\boldsymbol{V} = \boldsymbol{A}\boldsymbol{A}' = \boldsymbol{B}\boldsymbol{B}'$.

Nonetheless, if one computes the characteristic function of $\boldsymbol{A}\boldsymbol{x} + \boldsymbol{\mu}$ and $\boldsymbol{B}\boldsymbol{x} + \boldsymbol{\mu}$ then it only depends on $\boldsymbol{\mu}$ and $\boldsymbol{V} = \boldsymbol{A}\boldsymbol{A}' = \boldsymbol{B}\boldsymbol{B}'$. Hence, the actual choice of the full column rank matrix $\boldsymbol{A}$ satisfying $\boldsymbol{V} = \boldsymbol{A}\boldsymbol{A}'$ is irrelevant for the distribution of $\boldsymbol{A}\boldsymbol{x} + \boldsymbol{\mu}$.

Theorem B.8. *Let* $\boldsymbol{z}$ *be an* $m \times 1$ *random vector with* $\mathrm{E}(\boldsymbol{z}) = \boldsymbol{\mu}$ *and* $\mathrm{Cov}(\boldsymbol{z}) = \boldsymbol{V}$. *Then* $\boldsymbol{z}$ *has an* m*-variate normal distribution if and only if the characteristic function of* $\boldsymbol{z}$ *is given as*

$$\phi_{\boldsymbol{z}}(\boldsymbol{t}) = \mathrm{e}^{\mathrm{i}\boldsymbol{t}'\boldsymbol{\mu} - \frac{1}{2}\boldsymbol{t}'\boldsymbol{V}\boldsymbol{t}}, \quad \boldsymbol{t} \in \mathrm{IR}^m ,$$

where $\mathrm{i} = \sqrt{-1}$ *denotes the imaginary unit.*

For the special case that the matrix $\boldsymbol{V}$ is symmetric positive definite, the density of $\boldsymbol{z} \sim N(\boldsymbol{\mu}, \boldsymbol{V})$ can be written as

$$f_{\boldsymbol{z}}(\boldsymbol{\zeta}) = \frac{1}{(2\pi)^{m/2}\det(\boldsymbol{V})^{1/2}} \exp\left[-\frac{1}{2}(\boldsymbol{\zeta} - \boldsymbol{\mu})'\boldsymbol{V}^{-1}(\boldsymbol{\xi} - \boldsymbol{\mu})\right], \quad \boldsymbol{\zeta} \in \mathrm{IR}^m .$$

If, however, $\boldsymbol{V}$ is singular, then no closed form formula for the density of $N_m(\boldsymbol{\mu}, \boldsymbol{V})$ exists.

Example B.2. Let $\boldsymbol{z} \sim N_2(\boldsymbol{0}, \boldsymbol{V})$ have a bivariate normal distribution with $\boldsymbol{V} = \begin{pmatrix} 1 & \varrho \\ \varrho & 1 \end{pmatrix}$, where $0 \le \varrho \le 1$. Figures B.1 and B.2 display the density for the two cases $\varrho = 0$ and $\varrho = 0.9$.

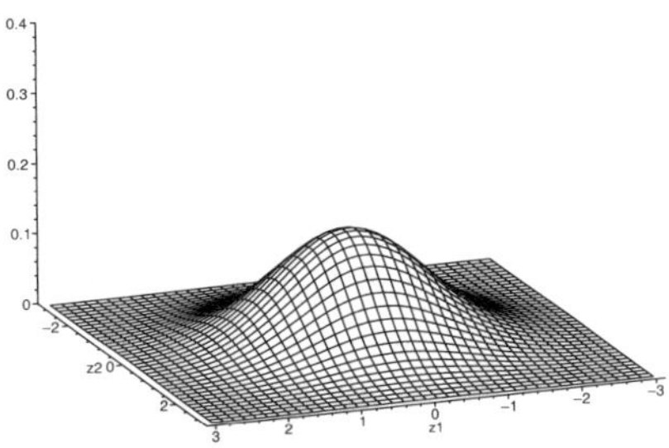

Fig. B.1. Density of $N_2(\boldsymbol{0}, \boldsymbol{V})$ with $\varrho = 0$ in Example B.2

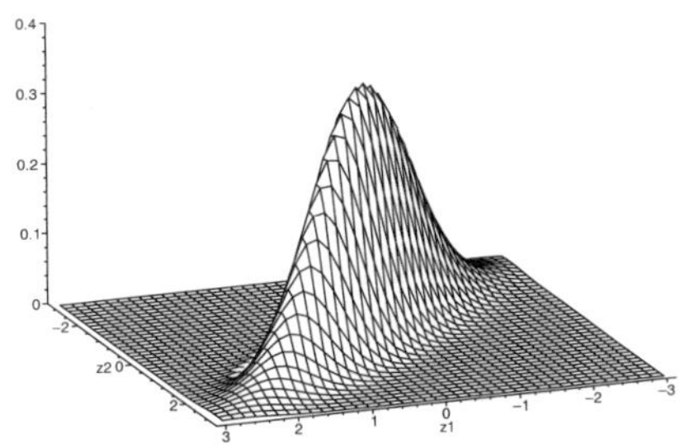

Fig. B.2. Density of $N_2(\boldsymbol{0}, \boldsymbol{V})$ with $\varrho = 0.9$ in Example B.2

As can be seen, greater values of ϱ press the density together and simultaneously raise its height. The graph for the extreme case $\varrho = 1$ (where $\boldsymbol{V}$ is singular) cannot be drawn, thus illustrating the fact that the density cannot be represented in closed form. □

Note that the distribution of a random vector is uniquely determined by its characteristic function. This fact can be of benefit in connection with the multivariate normal distribution, whose density does not necessarily exist. For example, the proof of the following theorem can be based on the characteristic function.

Theorem B.9. *Let $\boldsymbol{z} \sim N_m(\boldsymbol{\mu}, \boldsymbol{V})$. If $\boldsymbol{A}$ is a non-stochastic $h \times m$ matrix and $\boldsymbol{a}$ is a non-stochastic $h \times 1$ vector, then $\boldsymbol{A}\boldsymbol{z} + \boldsymbol{a} \sim N_h(\boldsymbol{A}\boldsymbol{\mu} + \boldsymbol{a}, \boldsymbol{A}\boldsymbol{V}\boldsymbol{A}')$.*

From the above theorem, any liner transformation of a multivariate normally distributed random vector has again a normal distribution. Since a part of a random vector is a specific linear transformation of this vector, any part of a multivariate normally distributed random vector has again a normal distribution.

Theorem B.10. *Let $\boldsymbol{z} \sim N_m(\boldsymbol{\mu}, \boldsymbol{V})$ with*

$$\boldsymbol{z} = \begin{pmatrix} \boldsymbol{z}_1 \\ \boldsymbol{z}_2 \end{pmatrix}, \quad \boldsymbol{\mu} = \begin{pmatrix} \boldsymbol{\mu}_1 \\ \boldsymbol{\mu}_2 \end{pmatrix}, \quad \boldsymbol{V} = \begin{pmatrix} \boldsymbol{V}_{11} & \boldsymbol{V}_{12} \\ \boldsymbol{V}_{21} & \boldsymbol{V}_{22} \end{pmatrix}.$$

Here $\boldsymbol{z}_1$ and $\boldsymbol{z}_2$ are $m_1 \times 1$ and $m_2 \times 1$ stochastic vectors with $m = m_1 + m_2$, and $\boldsymbol{\mu}$ and $\boldsymbol{V}$ are partitioned accordingly such that $\mathrm{Cov}(\boldsymbol{z}_1) = \boldsymbol{V}_{11}$, $\mathrm{Cov}(\boldsymbol{z}_2) = \boldsymbol{V}_{22}$, and $\mathrm{Cov}(\boldsymbol{z}_1, \boldsymbol{z}_2) = \boldsymbol{V}_{12}$. Then $\boldsymbol{z}_1 \sim N_{m_1}(\boldsymbol{\mu}_1, \boldsymbol{V}_{11})$ and $\boldsymbol{z}_2 \sim N_{m_2}(\boldsymbol{\mu}_2, \boldsymbol{V}_{22})$.

Stochastic Independence

The stochastic independence of two jointly multivariate normally distributed random vectors is equivalent to their uncorrelatedness.

Theorem B.11. *Let* $\boldsymbol{z} \sim N_m(\boldsymbol{\mu}, \boldsymbol{V})$ *with*

$$\boldsymbol{z} = \begin{pmatrix} \boldsymbol{z}_1 \\ \boldsymbol{z}_2 \end{pmatrix}, \quad \boldsymbol{\mu} = \begin{pmatrix} \boldsymbol{\mu}_1 \\ \boldsymbol{\mu}_2 \end{pmatrix}, \quad \boldsymbol{V} = \begin{pmatrix} \boldsymbol{V}_{11} & \boldsymbol{V}_{12} \\ \boldsymbol{V}_{21} & \boldsymbol{V}_{22} \end{pmatrix}.$$

Here $\boldsymbol{z}_1$ *and* $\boldsymbol{z}_2$ *are* $m_1 \times 1$ *and* $m_2 \times 1$ *stochastic vectors with* $m = m_1 + m_2$*, and* $\boldsymbol{\mu}$ *and* $\boldsymbol{V}$ *are partitioned accordingly such that* $\mathrm{Cov}(\boldsymbol{z}_1) = \boldsymbol{V}_{11}$*,* $\mathrm{Cov}(\boldsymbol{z}_2) = \boldsymbol{V}_{22}$*, and* $\mathrm{Cov}(\boldsymbol{z}_1, \boldsymbol{z}_2) = \boldsymbol{V}_{12}$*. Then* $\boldsymbol{z}_1$ *and* $\boldsymbol{z}_2$ *are stochastically independent if and only if* $\boldsymbol{V}_{12} = \boldsymbol{0}$*.*

In general, the stochastic independence of two random vectors implies their uncorrelatedness, but the converse is not necessarily true.

Theorem B.12. *Let* $\boldsymbol{z} \sim N_m(\boldsymbol{\mu}, \boldsymbol{I}_m)$*. Let* $\boldsymbol{A}$ *and* $\boldsymbol{B}$ *be two* $h \times m$ *and* $k \times m$ *non-stochastic matrices. If* $\boldsymbol{AB}' = \boldsymbol{0}$*, then* $\boldsymbol{Az}$ *and* $\boldsymbol{Bz}$ *are stochastically independent.*

Theorem B.13. *Let* $\boldsymbol{z} \sim N_m(\boldsymbol{\mu}, \boldsymbol{I}_m)$*. Let* $\boldsymbol{A}$ *be a symmetric non-stochastic* $m \times m$ *matrix and let* $\boldsymbol{B}$ *be a* $k \times m$ *non-stochastic matrix. If* $\boldsymbol{BA} = \boldsymbol{0}$*, then* $\boldsymbol{z}'\boldsymbol{Az}$ *and* $\boldsymbol{Bz}$ *are stochastically independent.*

Theorem B.14. *Let* $\boldsymbol{z} \sim N_m(\boldsymbol{\mu}, \boldsymbol{I}_m)$*. Let* $\boldsymbol{A}$ *and* $\boldsymbol{B}$ *be two symmetric non-stochastic* $m \times m$ *matrices. If* $\boldsymbol{BA} = \boldsymbol{0}$*, then* $\boldsymbol{z}'\boldsymbol{Az}$ *and* $\boldsymbol{z}'\boldsymbol{Bz}$ *are stochastically independent.*

Variance of Quadratic Form

Theorem B.5 states that for a random vector $\boldsymbol{z}$ with $\mathrm{E}(\boldsymbol{z}) = \boldsymbol{\mu}$ and $\mathrm{Cov}(\boldsymbol{z}) = \boldsymbol{V}$ the formula

$$\mathrm{E}(\boldsymbol{z}'\boldsymbol{Az}) = \mathrm{tr}(\boldsymbol{AV}) + \boldsymbol{\mu}'\boldsymbol{A\mu}$$

holds true, where $\boldsymbol{A}$ is a (possibly non-symmetric) square matrix. For the case that $\boldsymbol{z} \sim N_m(\boldsymbol{\mu}, \boldsymbol{V})$ and in addition $\boldsymbol{A} = \boldsymbol{A}'$, one may give a corresponding formula for the variance.

Theorem B.15. *Let* $\boldsymbol{z}$ *be an* $m \times 1$ *random vector with* $\boldsymbol{z} \sim N_m(\boldsymbol{\mu}, \boldsymbol{V})$ *and let* $\boldsymbol{A}$ *be an* $m \times m$ *non-stochastic symmetric matrix. Then the identity*

$$\mathrm{Var}(\boldsymbol{z}'\boldsymbol{Az}) = 2\,\mathrm{tr}[(\boldsymbol{AV})^2] + 4\boldsymbol{\mu}'\boldsymbol{AVA\mu}$$

holds true.

B.3 χ^2 Distribution

The distribution of the random variable $\boldsymbol{z}'\boldsymbol{z}$ with $\boldsymbol{z} \sim N_m(\boldsymbol{\mu}, \boldsymbol{I}_m)$ is called χ^2 distribution.

Definition B.6. *Let $\boldsymbol{z} \sim N_m(\boldsymbol{\mu}, \boldsymbol{I}_m)$. Then the distribution of $\boldsymbol{z}'\boldsymbol{z}$ is called* noncentral χ^2 distribution *with m degrees of freedom and noncentrality parameter $\lambda = \boldsymbol{\mu}'\boldsymbol{\mu}/2$, denoted by $\boldsymbol{z}'\boldsymbol{z} \sim \chi^2_{(m,\lambda)}$. In case $\lambda = 0$ the distribution is called* central χ^2 distribution *with m degrees of freedom, denoted by $\boldsymbol{z}'\boldsymbol{z} \sim \chi^2_{(m)}$.*

In the literature, the noncentrality parameter is sometimes defined as $\boldsymbol{\mu}'\boldsymbol{\mu}$. Hence, if for a given number $\lambda \geq 0$ we consider the $\chi^2_{(m,\lambda)}$ distribution in the sense of our definition, this distribution is identical to the $\chi^2_{(m,2\lambda)}$ distribution in the sense of the alternative definition.

Theorem B.16. *The density of $X \sim \chi^2_{(m,\lambda)}$ is given as*

$$f_X(x) = \begin{cases} \dfrac{e^{-\frac{x}{2}-\lambda}\, x^{\frac{m}{2}-1}}{2^{\frac{m}{2}}} \displaystyle\sum_{j=0}^{\infty} \frac{(\lambda x)^j}{2^j \Gamma\left(\frac{m}{2}+j\right) j!} & \textit{for } x > 0 \\ 0 & \textit{otherwise} \end{cases} .$$

Figure B.3 displays the densities of the χ^2 distribution with $m = 3$ degrees of freedom and noncentrality parameters 0, 1, 3, and 6.

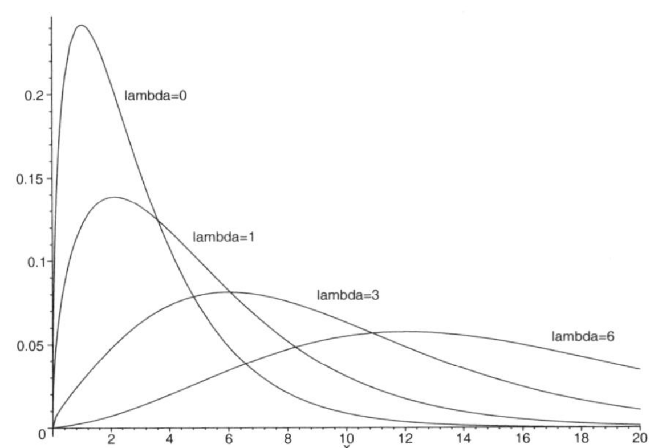

Fig. B.3. Densities of $\chi^2_{(3)}$, $\chi^2_{(3,1)}$, $\chi^2_{(3,3)}$, and $\chi^2_{(3,6)}$

Expectation and Variance of a $\chi^2_{(m,\lambda)}$ distributed random variable are given as follows.

Theorem B.17. *For $X \sim \chi^2_{(m,\lambda)}$ the identities*

$$\mathrm{E}(X) = m + 2\lambda \quad \text{and} \quad \mathrm{Var}(X) = 2(m + 4\lambda)$$

hold true.

Theorem B.18. *Let $\boldsymbol{z} \sim N_m(\boldsymbol{\mu}, \boldsymbol{I}_m)$ and let $\boldsymbol{A}$ be an $m \times m$ symmetric idempotent matrix. Then $\boldsymbol{z}'\boldsymbol{A}\boldsymbol{z} \sim \chi^2_{(a,\lambda)}$ with $a = \mathrm{rk}(\boldsymbol{A})$ and $\lambda = \boldsymbol{\mu}'\boldsymbol{A}\boldsymbol{\mu}/2$.*

Expectations of Certain Functions

We consider the expectation of certain functions in connection with the χ^2 distribution.

Assumption B.1. *The symbol $\chi^2_{(m,\lambda)}$ is also used to denote a random variable with a $\chi^2_{(m,\lambda)}$ distribution.*

The following results can also be found in [63, Appendix B].

Theorem B.19. *The expectation of the inverse of a $\chi^2_{(m,\lambda)}$ distributed random variable satisfies*

$$\mathrm{E}[1/\chi^2_{(m,\lambda)}] = \mathrm{e}^{-\lambda} \sum_{j=0}^{\infty} \frac{\lambda^j}{j!(m - 2 + 2j)} \,.$$

If $\lambda = 0$, then the identity

$$\mathrm{E}[1/\chi^2_{(m)}] = \frac{1}{m-2}$$

holds true for $m \geq 3$.

Theorem B.20. *Let $\boldsymbol{z} \sim N_m(\boldsymbol{\mu}, \boldsymbol{I}_m)$ and let $\phi(\cdot)$ be a Borel measurable function. Then*

$$\mathrm{E}[\phi(\boldsymbol{z}'\boldsymbol{z})\, \boldsymbol{z}] = \boldsymbol{\mu}\ \mathrm{E}[\phi(\chi^2_{(m+2,\lambda)})] \,,$$

where $\lambda = \boldsymbol{\mu}'\boldsymbol{\mu}/2$.

Theorem B.21. *Let $\boldsymbol{z} \sim N_m(\boldsymbol{\mu}, \boldsymbol{I}_m)$. Let $\boldsymbol{A}$ be an $m \times m$ symmetric non-negative definite matrix. Let $\phi(\cdot)$ be a Borel measurable function. Then*

$$\mathrm{E}[\phi(\boldsymbol{z}'\boldsymbol{z})\, \boldsymbol{z}'\boldsymbol{A}\boldsymbol{z}] = \mathrm{E}[\phi(\chi^2_{(m+2,\lambda)})]\ \mathrm{tr}(\boldsymbol{A}) + \mathrm{E}[\phi(\chi^2_{(m+4,\lambda)})]\ \boldsymbol{\mu}'\boldsymbol{A}\boldsymbol{\mu} \,,$$

where $\lambda = \boldsymbol{\mu}'\boldsymbol{\mu}/2$.

If in the above theorem $\boldsymbol{A} = \boldsymbol{I}_m$ and $\phi(x) = 1/x$, then the following result is obtained as a corollary.

Corollary B.2. *The identity*

$$1 = \mathrm{E}[1/\chi^2_{(m+2,\lambda)}]m + \mathrm{E}[1/\chi^2_{(m+4,\lambda)}]2\lambda$$

holds true for any integer $m \geq 1$ and any real number $\lambda \geq 0$.

B.4 *F* Distribution

The distribution of the ration of two independent χ^2 distributed random variables is called F distribution.

Definition B.7. *Let* $\chi^2_{(m,\lambda)}$ *and* $\chi^2_{(n,\delta)}$ *be stochastically independent. Then the distribution of*

$$X = \frac{\chi^2_{(m,\lambda)}/m}{\chi^2_{(n,\delta)}/n}$$

is called noncentral F distribution *with* (m,n) *degrees of freedom and noncentrality parameters* (λ,δ). *If* $\delta = \lambda = 0$, *then the distribution of* X *is called* central F distribution *with* m *and* n *degrees of freedom, denoted by* $X \sim F_{(m,n)}$.

In general, the density of noncentral F distributions cannot be expressed in closed form. For the density of the central F distribution the following holds true.

Theorem B.22. *The density of* $X \sim F_{(m,n)}$ *is given as*

$$f_X(x) = \begin{cases} \frac{\left(\frac{m}{n}\right)^{m/2} x^{\frac{m}{2}-1}}{B\left(\frac{m}{2},\frac{n}{2}\right)} \left(1+\frac{m}{n}x\right)^{-\frac{m+n}{2}} & \text{for } x > 0 \\ 0 & \text{otherwise} \end{cases},$$

where $B(\cdot,\cdot)$ *denotes the Beta function.*

Theorem B.23. *Let* $X \sim F_{(m,n)}$. *Then*

$$\mathrm{E}(X) = \frac{n}{n-2}$$

whenever $n > 2$, *and*

$$\mathrm{Var}(X) = \frac{2n^2(m+n-2)}{m(n-2)^2(n-4)}$$

whenever $n > 4$.

C An Example Analysis with R

In this chapter we describe a possible analysis of a linear regression model with the statistical-computing environment R, based on the S language. R is available through the Internet under the General Public License (GPL) and can be downloaded from one of the CRAN (Comprehensive R Archive Network) sites. See the main site

```
http://cran.r-project.org/
```

For textbooks presenting the application of R for the purpose of statistical analysis see e.g. [27, 40, 122].

When R is running, text following the prompt '>' will be interpreted as a command, usually producing some reaction. For example

```
> 2 + 2
[1] 4
```

directly computes the sum $2 + 2$, while

```
> a <- 2 + 2
```

assigns the sum $2 + 2$ to the variable a and can be made visible via

```
> a
[1] 4
```

C.1 Problem and Goal

Let us reconsider Example 1.1 with the data given in Table 1.1, p. 4. The variable y is the heat evolving during the hardening of a batch of cement. Observations are available for $n = 13$ batches. Each batch consists of four main ingredients whose amounts are denoted by x_1, x_2, x_3, and x_4, given as percentages.

Suppose that it is our aim to find an appropriate prediction formula for the heat y (dependent variable), given values of the amounts of ingredients x_1, x_2, x_3, and x_4 (independent variables).

C.2 The Data

The cement data is already available in the package `MASS`. It may be assigned to `cement` via the two commands

```
> library(MASS)
> data(cement)
```

Then,

```
> cement
   x1 x2 x3 x4     y
1   7 26  6 60  78.5
2   1 29 15 52  74.3
3  11 56  8 20 104.3
4  11 31  8 47  87.6
5   7 52  6 33  95.9
6  11 55  9 22 109.2
7   3 71 17  6 102.7
8   1 31 22 44  72.5
9   2 54 18 22  93.1
10 21 47  4 26 115.9
11  1 40 23 34  83.8
12 11 66  9 12 113.3
13 10 68  8 12 109.4

> summary(cement)
       x1               x2              x3
 Min.   : 1.000   Min.   :26.00   Min.   : 4.00
 1st Qu.: 2.000   1st Qu.:31.00   1st Qu.: 8.00
 Median : 7.000   Median :52.00   Median : 9.00
 Mean   : 7.462   Mean   :48.15   Mean   :11.77
 3rd Qu.:11.000   3rd Qu.:56.00   3rd Qu.:17.00
 Max.   :21.000   Max.   :71.00   Max.   :23.00
       x4              y
 Min.   : 6   Min.   : 72.50
 1st Qu.:20   1st Qu.: 83.80
 Median :26   Median : 95.90
 Mean   :30   Mean   : 95.42
 3rd Qu.:44   3rd Qu.:109.20
 Max.   :60   Max.   :115.90
```

The values of each variable are directly accessible as data vectors. For example,

```
> cement$y
 [1]  78.5  74.3 104.3  87.6  95.9 109.2 102.7  72.5  93.1
[10] 115.9  83.8 113.3 109.4
```

A first impression of the given data may be obtained from a pairwise scatter plot of the variables. This may reveal possible functional relationships, extreme observations, or other peculiarities. The command

```
> pairs(cement)
```

produces Fig. C.1. The pairwise relationships between y and x_1, x_2, and x_4, respectively, appear to be somewhat linear. The pairwise relationship between y and x_3 might indicate that the variance of y decreases with increasing values of x_3 (possible heteroscedasticity). There is also a rather strong linear relationship between x_2 and x_4, giving some indication for collinearity problems. When considering the plots involving x_1, there appears to be one outstanding observation.

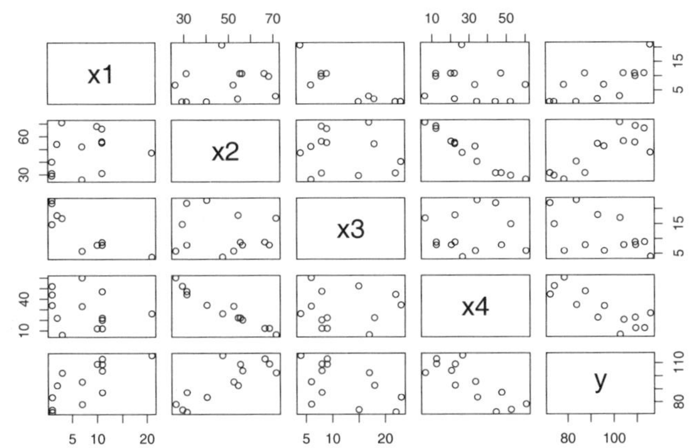

Fig. C.1. Scatter plots involving the variables in `cement`

C.3 The Choice of Variables

For the purpose of predicting the evolving heat in a batch of cement, we fit a linear model with the amounts of ingredients 1 to 4 as possible predictors (independent variables).

C.3.1 The Full Model

When building a linear model with y as dependent variable, we may start from the full model with x_1, x_2, x_3, x_4, and the intercept as independent variables, denoted by `fm1`.

```
> fm1 <- lm(y~x1+x2+x3+x4,data=cement)
> summary(fm1)
```

```
Call:
lm(formula = y ~ x1 + x2 + x3 + x4, data = cement)

Residuals:
    Min      1Q  Median      3Q     Max
-3.1750 -1.6709  0.2508  1.3783  3.9254

Coefficients:
            Estimate Std. Error t value Pr(>|t|)
(Intercept)  62.4054    70.0710   0.891   0.3991
x1            1.5511     0.7448   2.083   0.0708 .
x2            0.5102     0.7238   0.705   0.5009
x3            0.1019     0.7547   0.135   0.8959
x4           -0.1441     0.7091  -0.203   0.8441
---
Signif. codes:  0 '***' 0.001 '**' 0.01 '*' 0.05 '.' 0.1 ' ' 1

Residual standard error: 2.446 on 8 degrees of freedom
Multiple R-Squared: 0.9824,     Adjusted R-squared: 0.9736
F-statistic: 111.5 on 4 and 8 DF,  p-value: 4.756e-07
```

The estimated standard deviations (`Std. Error`) of the ordinary least squares estimates for the individual parameters (`Estimate`) are quite large, implying strong uncertainty about preciseness. The adjusted coefficient of determination $\overline{R}^2$ is 0.9736, revealing a good fit, but the p-values (`Pr(>|t|)`) of the t-statistics question (with different intensity) the relevance of all variables, except for x_1, in this model. (The t value is the statistic for testing the hypothesis that the parameter for the corresponding variable is zero. The F-statistic is for testing the hypothesis that all parameters corresponding to the independent variables except for the intercept are simultaneously zero. Hence, the small p-value associated with the F-statistic shows that the independent variables as a whole must be regarded as relevant.)

The `Pr(>|t|)` values should, of course, not lead to an exclusion of all variables except for x_1. As a matter of fact, when we drop just one variable, namely the intercept, the situation looks quite different. This can be seen from

```
> fm1a <- lm(y~-1+x1+x2+x3+x4,data=cement)
> summary(fm1a)
...

Coefficients:
   Estimate Std. Error t value Pr(>|t|)
x1  2.19305    0.18527  11.837 8.65e-07 ***
x2  1.15333    0.04794  24.057 1.77e-09 ***
x3  0.75851    0.15951   4.755  0.00104 **
x4  0.48632    0.04141  11.744 9.25e-07 ***
```

```
---
Signif. codes:  0 '***' 0.001 '**' 0.01 '*' 0.05 '.' 0.1 ' ' 1

Residual standard error: 2.418 on 9 degrees of freedom
Multiple R-Squared: 0.9996,     Adjusted R-squared: 0.9994
F-statistic:  5176 on 4 and 9 DF,  p-value: 4.083e-15
```

Further analysis will show that model fm1a has good properties and we may end up with this choice. On the other hand, we may also apply the stepwise regression procedure to find an appropriate model.

C.3.2 Stepwise Regression

We start with model fm1, but we want the intercept variable to be treated as any other dependent variable. This can be achieved by introducing a new variable Int

```
> n <- length(cement$y)
> Int <- rep(1,n)
> Int
 [1] 1 1 1 1 1 1 1 1 1 1 1 1 1
```

and excluding the usual intercept in the formula for the linear model, while at the same time including the new variable Int.

```
fm2 <- lm(y~-1+Int+x1+x2+x3+x4,data=cement)
```

Then, backward stepwise regression gives

```
> fm3 <- step(fm2)
Start:  AIC= 26.94
 y ~ Int + x1 + x2 + x3 + x4 + 0

        Df Sum of Sq    RSS    AIC
- x3     1     0.109 47.973 24.974
- x4     1     0.247 48.111 25.011
- x2     1     2.972 50.836 25.728
- Int    1     4.746 52.609 26.173
<none>               47.864 26.944
- x1     1    25.951 73.815 30.576

Step:  AIC= 24.97
 y ~ Int + x1 + x2 + x4 + 0

        Df Sum of Sq    RSS    AIC
<none>                47.97  24.97
- x4     1      9.93  57.90  25.42
- x2     1     26.79  74.76  28.74
- Int    1    136.81 184.78  40.51
- x1     1    820.91 868.88  60.63
```

The resulting model fm3 is the full model except for x_3. It has AIC of 24.97. Note that when in addition the variable x_4 is removed from this model, the AIC for the new model only slightly increases to 25.42.

C.3.3 Collinearity Diagnostics

We have already noted from the pairwise scatter plot in Fig. C.1 that there is a rather strong linear relationship between the variables x_2 and x_4. This gives the motivation for considering possible collinearity problems at this stage, since both variables are contained in model fm3. The matrix $\boldsymbol{\Pi}$ of variance-decomposition proportions may be obtained from the function

```
> PI.lm <- function(lm.obj) {
+ nam <- names(coef(lm.obj))
+ M <- model.matrix(lm.obj)
+ X <- M %*% diag((1/sqrt(diag(crossprod(M)))))
+ SV <- svd(X)
+ ev <- round(max(SV$d)/SV$d, digits=0)
+ evnam <- as.character(ev)
+ phi <- SV$v^2 %*% diag(1/(SV$d)^2)
+ PI <- round(t(phi) %*% diag(1/rowSums(phi)),digits=4)
+ colnames(PI) <- nam
+ rownames(PI) <- evnam
+ cat("Variance-Decomposition Proportions for\n")
+ cat("Scaled Condition Indexes:\n\n")
+ print(PI)
+ cat("\n")
+ }
```

Then, application of this function gives

```
> PI.lm(fm3)
Variance-Decomposition Proportions for
Scaled Condition Indexes:

      Int     x1     x2     x4
1  0.0002 0.0227 0.0004 0.0008
3  0.0001 0.3387 0.0007 0.0160
4  0.0007 0.6278 0.0091 0.0077
52 0.9990 0.0108 0.9898 0.9756
```

This shows that there is one harmful scaled condition index of 52. The variance-decomposition proportions reveal one near linear relationship involving the variables x_2, x_4 and the intercept. To avoid possible collinearity effects, we exclude the variable x_4. This can in addition be 'justified' by only a slight increase in AIC and a `Pr(>|t|)` value of 0.205395 for the variable x_4 in model fm3.

```
> fm4 <- lm(y~x1+x2,data=cement)
> summary(fm4)
```

```
...

Coefficients:
            Estimate Std. Error t value Pr(>|t|)
(Intercept) 52.57735    2.28617   23.00 5.46e-10 ***
x1           1.46831    0.12130   12.11 2.69e-07 ***
x2           0.66225    0.04585   14.44 5.03e-08 ***
---
Signif. codes:  0 '***' 0.001 '**' 0.01 '*' 0.05 '.' 0.1 ' ' 1

Residual standard error: 2.406 on 10 degrees of freedom
Multiple R-Squared: 0.9787,     Adjusted R-squared: 0.9744
F-statistic: 229.5 on 2 and 10 DF,  p-value: 4.407e-09
```

In model fm4, the standard errors are moderate, the fit is still good, and the considered variables are strongly relevant. In addition,

```
> PI.lm(fm4)
Variance-Decomposition Proportions for
Scaled Condition Indexes:

  (Intercept)     x1     x2
1      0.0109 0.0390 0.0107
3      0.0549 0.9598 0.0483
8      0.9342 0.0012 0.9410
```

reveals no harmful condition indexes, and thus no collinearity problems. We may thus confine to model fm4 in the following.

C.4 Further Diagnostics

Having chosen the independent variables, the linear regression model may further be analyzed by considering residuals and checking for influential observations.

C.4.1 Residuals

A plot of studentized residuals versus fitted values may be generated by

```
> plot(fitted(fm4),rstudent(fm4))
```

and is shown in Fig. C.2. It gives no indication for a possible non-linearity or other peculiarities.

The normal quantile plot of the studentized residuals may be obtained from

```
> qqnorm(rstudent(fm4))
> qqline(rstudent(fm4))
```

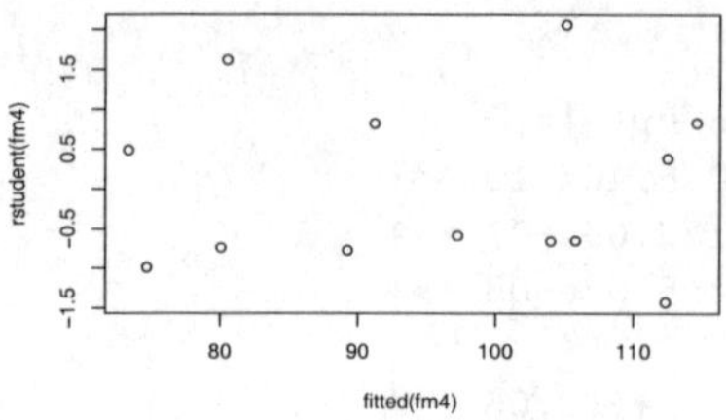

Fig. C.2. Residual versus fitted values plot for model `fm4`

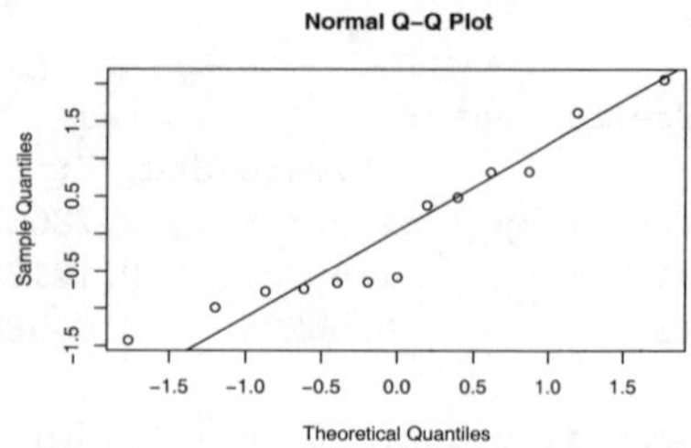

Fig. C.3. Normal quantile plot for model `fm4`

and is displayed in Fig. C.3. Figures C.2 and C.3 show no relevant outlier in the data. One may, however, suspect a moderate departure from normality. The p-value of the Shapiro-Wilk test

```
> library(ctest)
> shapiro.test(rstudent(fm4))

        Shapiro-Wilk normality test

data:  rstudent(fm4)
W = 0.9133, p-value = 0.2032
```

supports the hypothesis of normality to a certain extent.

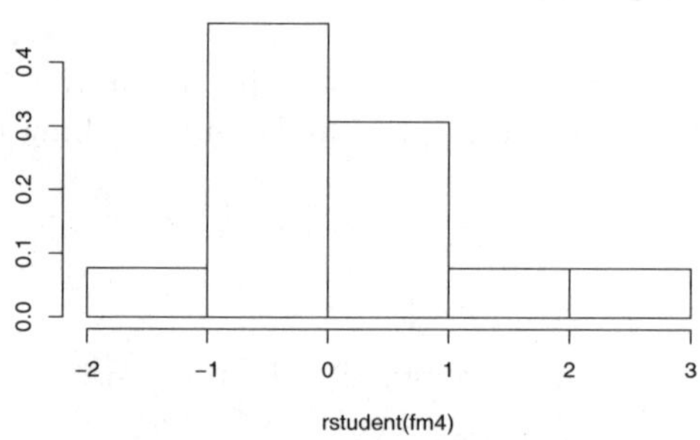

Fig. C.4. Histogram of studentized residuals for model `fm4`

When considering a histogram of the studentized residuals via

```
> truehist(rstudent(fm4),col=0)
```

one might conclude that the distribution is skewed to the right and hence not normal, see Fig. C.4. (The function `truehist` is available in the package `MASS`, which has already been loaded via `library(MASS)`.) All in all, the assertion of normality may be regarded with some caution.

C.4.2 Influential Observations

The pairwise scatter plot in Fig. C.1 had revealed an outstanding point in the plots of x_1 versus the other variables. We may consider this in more detail by plotting y versus x_1, and then identifying the outstanding point interactively with the mouse. This can be achieved by

```
> plot(cement$x1,cement$y)
> identify(cement$x1,cement$y)
[1] 10
```

When `identify` is running, a left-click near the point gives the index number of the observation. A right-click ends `identify`. The outstanding x_1 observation is identified as observation number 10, see Fig. C.5. Note also from the data in Table 1.1 that the 10-th batch contains an unusual high amount of ingredient 1.

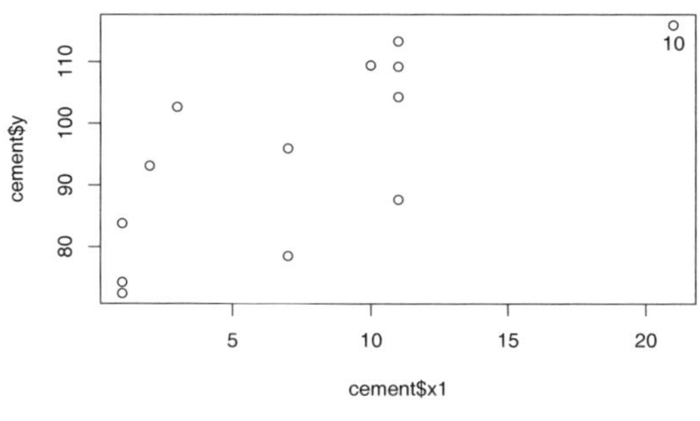

Fig. C.5. Outstanding x_1 observation in model `fm4`

When we seek for influential observations, then from

```
> summary(influence.measures(fm4))
Potentially influential observations of
         lm(formula = y ~ x1 + x2, data = cement) :

   dfb.1_ dfb.x1 dfb.x2 dffit cov.r    cook.d hat
10 -0.02    0.85  -0.22   0.92  2.44_*  0.29   0.55
```

the 10-th observation is in fact identified as influential, but only with respect to the COVRATIO criterion (indicated by the asterisk). Nonetheless, the hat-value also looks quite high. A hat-value versus index plot may be obtained by

```
> p <- length(coef(fm4))
> plot(hat(model.matrix(fm4)),ylim=c(0,1))
> abline(h=c(2,3)*p/n,lty=2)
```

The result is given in Fig. C.6. It shows that the hat-value of the 10-th observation is greater than $2p/n$ but smaller than $3p/n$.

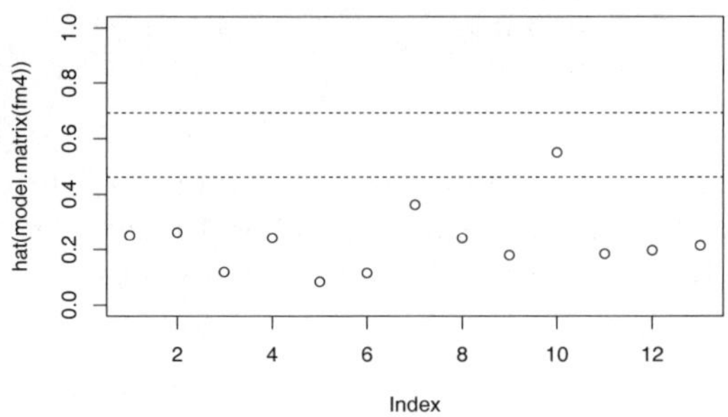

Fig. C.6. Hat-value versus index plot for model `fm4`

Now, should the analysis be carried out without the 10-th observation? As it seems, the 10-observation is not due to an error in the data collection process, since the values of x_1 to x_4 (percentages) sum up to nearly 100. In addition, Fig. C.5 suggests that the 10-th observation of x_1, although outstanding, is more or less in accordance with the rest of the observations. From this we may conclude that the 10-th observation rather supports than disturbs the relationship fitted to the rest of observations and thus should not be given away.

C.5 Prediction

Suppose we want to predict the heat y for $x_1 = 15$, $x_2 = 50$, $x_3 = 10$, and $x_4 = 25$ on the basis of model `fm4`. Then, only the first two values are relevant, and the prediction formula based on the ordinary least squares estimates (`Estimate` in the `summary` of model `fm4`) is

$$y = 52.5774 + 1.4683x_1 + 0.6623x_2 \ .$$

It may be computed by

```
> new <- data.frame(x1=15,x2=50,x3=10,x4=25)
> predict(fm4,new)
[1] 107.7145
```

When we choose model `fm1a` for prediction, then

```
> predict(fm1a,new)
[1] 110.3051
```

yielding a greater value than before.

We cannot definitely decide on the basis of the given data which of the two models `fm4` and `fm1a` is the better one, since both seem to be 'good' models. Model `fm4` has a smaller AIC but also a smaller $\overline{R}^2$ than model `fm1a`. (The complete analysis of model `fm1a` is left to the reader.)

In general, the choice of model will not only depend on the given data (and thus on criteria like AIC and $\overline{R}^2$), but also on the intention and the goal of the researcher. (What is relationship one wishes to investigate? Which variables should be included for specific (theoretical, logical) reasons? Can certain possible predictors (independent variables) be neglected, when it is costly to obtain their values? Etc.)

References

1. Aitken, A.C. (1935): On least squares and linear combination of observations. Proceedings of the Royal Society of Edinburgh, **55**, 42–48.
2. Akaike, H. (1974): A new look at statistical model identification. IEEE Transactions on Automatic Control, **AU-19**, 716–722.
3. Andrews, D.F., Bickel, J.F., Hampel, F.R., Huber, P.J., Rogers, W.H., Tukey, J.W. (1972): Robust Estimation of Location: Survey and Advances. Princeton University Press, Princeton, NJ.
4. Askin, R.G., Montgomery, D.C. (1980): Augmented robust estimators. Technometrics, **22**, 333–341.
5. Baksalary, J.K., Kala, R. (1983): Partial orderings between matrices one of which is of rank one. Bulletin of the Polish Academy of Science, Mathematics, **31**, 5–7.
6. Baksalary, J.K., Markiewicz, A. (1985): Admissible linear estimators in restricted linear models. Linear Algebra and Its Applications, **70**, 9–19.
7. Baksalary, J.K., Markiewicz, A. (1986): Characterization of admissible linear estimators in restricted linear models. Journal of Statistical Planning and Inference, **13**, 395–398.
8. Baksalary, J.K., Markiewicz, A. (1990): Admissible linear estimators of an arbitrary vector of parametric functions in the general Gauss-Markov model. Journal of Statistical Planning and Inference, **26**, 161–171.
9. Baye, M.R., Parker, D.F (1984): Combining ridge and principal components regression: a money demand illustration. Communications in Statistics, Theory and Methods, **13**, 197–205.
10. Belsley, D.A., Kuh, E., Welsch, R.E. (1980): Regression Diagnostics. Wiley, New York.
11. Belsley, D.A. (1991): Conditioning Diagnostics, Collinearity and Weak Data. Wiley, New York.
12. Berger, J. (1976): Admissible minimax estimation of a multivariate normal mean with arbitrary quadratic loss. Annals of Statistics, **4**, 223– 226.
13. Berger, J.O. (1985): Statistical Decision Theory and Bayesian Analysis (2nd ed.). Springer, New York.
14. Bibby, J. (1972): Minimum mean square error estimation, ridge regression, and some unanswered questions. Proceedings of the 9th European meeting of Statisticians, Budapest. Reprinted (1974): Gani, J., Sarkadi, K., Vincze, I. (ed) Progress is Statistics, Colloquia Mathematica, Janos Bolyai, 107–121.
15. Bickel, P.J., Doksum, K.A. (2001): Mathematical Statistics, Vol. 1. Second Edition. Prentice Hall, Uppers Saddle River, New Jersey.
16. Birkes, D., Dodge, Y. (1993): Alternative Methods of Regression. Wiley, New York.

17. Bock, M.E. (1975): Minimax estimators of the mean of a multivariate normal distribution. Annals of Statistics, **3**, 209–218.
18. Breusch, T.S., Pagan, A.R. (1979): A simple test for heteroscedasticity and random coefficient variation. Econometrica. **47**, 1287–1294.
19. Casella, G., Berger, R.L. (2002): Statistical Inference. Second Edition. Duxburry Press. Belmont, California.
20. Chatterjee, S., Hadi, A.S. (1988): Sensitivity Analysis in Linear Regression. Wiley, New York.
21. Christensen, R.C. (2002): Plane Answers to Complex Questions. The Theory of Linear Models. Third Edition. Springer, New York.
22. Cook, R.D. (1977): Detection of influential observations in linear regression. Technometrics, **19**, 15–18.
23. Cook, R.D. (1998): Regression Graphics: Ideas for Studying Regressions Through Graphics. Wiley, New York.
24. Cook, R.D., Weisberg, S. (1982): Residuals and Influence in Regression. Chapman and Hall, London.
25. Cook, R.D., Weisberg, S. (1999): Applied Regression Including Computing and Graphics. Wiley, New York.
26. Cohen, A. (1966): All admissible linear estimates of the mean vector. Annals of Mathematical Statistics, **37**, 458–463.
27. Dalgaard, P. (2002): Introductory Statistics with R. Springer, New York.
28. Davidson, R., MacKinnon, J.G. (1993): Estimation and Inference in Econometrics. Oxford University Press, New York.
29. Dey, D.K., Ghosh, M., Strawderman, W.E. (1999): On estimation with balanced loss function. Statistics & Probability Letters, **45**, 97–101.
30. Dodge, Y. (1997): LAD regression for detecting outliers in response and explanatory variables. Journal of Multivariate Analysis, **61**, 144–158.
31. Durbin, J., Watson, G.S. (1950): Testing for serial correlation in least squares regression–I. Biometrika, **37**, 409–428.
32. Durbin, J., Watson, G.S. (1951): Testing for serial correlation in least squares regression–II. Biometrika, **38**, 159–178.
33. Durbin, J., Watson, G.S. (1971): Testing for serial correlation in least squares regression–III. Biometrika, **58**, 1–42.
34. Escobar, L.A., Skarpness, B. (1984). A closed form solution for the least squares regression problem with linear inequality constraints. Communications in Statistics, Theory and Methods, **13**, 1127–1134.
35. Farebrother, R.W. (1975): The minimum mean square error linear estimator and ridge regression. Technometrics, **17**, 127–128.
36. Farebrother, R.W. (1976): Further results on the mean square error of ridge regression. Journal of the Royal Statistical Society, Ser. B, **38**, 248–250.
37. Farebrother, R.W. (1984): The restricted least squares estimator and ridge regression. Communications in Statistics, Theory and Methods, **13**, 191–196.
38. Ferguson, T.S. (1967): Mathematical Statistics. Academic Press, New York.
39. Firinguetti, L. (1987): Exact moments of Lawless and Wang's operational ridge regression estimator. Communications in Statistics, Theory and Methods, **16**, 731–745.
40. Fox, J. (2002): An R and S-PLUS Companion to Applied Regression. Sage Publications, Thousand Oaks.

41. Gaffke, N., Krafft, O. (1982): Matrix inequalities in the Löwner ordering. In: Korte, B. (ed) Modern Applied Mathematics. North-Holland, Amsterdam, 595–622.
42. Galton, F. (1886): Regression towards mediocrity in hereditary stature. Journal of the Anthropological Institute, **15**, 246–263.
43. George, E.I., Oman, S.D. (1996): Multiple-shrinkage principal component regression. The Statistician, **45**, 111–124.
44. Goldberger, A.S. (1991): A Course in Econometrics. Harvard University Press, Cambridge, Massachusetts.
45. Goldfeld, S.M., Quandt, R.E. (1965): Some tests for homoscedasticity. Journal of the American Statistical Association, **60**, 539–547.
46. Greene, W.H. (2002): Econometric Analysis. Fifth Edition. Prentice Hall, Upper Saddle River, New Jersey.
47. Gruber, M.H.J. (1998): Improving Efficiency by Shrinkage. The James-Stein and Ridge Regression Estimators. Marcel Dekker, New York.
48. Hadi, A.S., Ling, R.F. (1998): Some cautionary notes on the use of principal components regression. The American Statistician, **52**, 15–19.
49. Hald, A. (1952): Statistical Theory with Engineering Applications. Wiley, New York.
50. Harville, D.A. (1997): Matrix Algebra From a Statistician's Perspective. Springer, New York.
51. Hoerl, A.E., Kennard, R.W. (1970): Ridge regression: biased estimation for nonorthogonal problems. Technometrics, **12**, 55–67.
52. Hoerl, Kennard (1976): Ridge regression iterative estimation of the biasing parameter. Communications in Statistics, Theory and Methods, **5**, 77–88.
53. Hoerl, A.E., Kennard, R.W., Baldwin, K.F. (1975): Ridge regression: some simulations. Communications in Statistics, Theory and Methods, **4**, 105–123.
54. Hoffmann, K. (1977): Admissibility of linear estimators with respect to restricted parameter sets. Mathematische Operationsforschung Statistik, Ser. Statistics, **8**, 425–438.
55. Hoffmann, K. (1992): Improved estimation of distribution parameters: Stein-type estimators. Teubner, Leipzig.
56. Hoffmann, K. (1995): A note on the Kuks-Olman estimator. Statistics, **26**, 185–187.
57. Hoffmann, K. (2000): Stein estimation – a review. Statistical Papers, **41**, 127–158.
58. Holland, P. (1973): Weighted ridge and robust regression methods. NBER Working Paper No. 11, Cambridge, MA.
59. Horn, R.A., Johnson, C.R. (1985): Matrix Analysis. Cambridge University Press, Cambridge.
60. Huber, P.J. (1964): Robust estimation of a location parameter. Annals of Mathematical Statistics, **35**, 73–101.
61. James, W., Stein, C. (1961): Estimation with quadratic loss. In: Proceedings of the Fourth Berkeley Symposium on Mathematical Statistics and Probability, University of California Press, Berkeley, CA), 361–379.
62. Johnston, J. (1984): Econometric Methods. Third Edition. McGraw-Hill, New York.
63. Judge, G.G., Bock, M.E. (1978): The Statistical Implications of Pre-Test and Stein-Rule Estimators in Econometrics. North-Holland Publishing Company, Amsterdam.

64. Keijan, L. (1993): A new class of biased estimate in linear regression. Communications in Statistics, Theory and Methods, **22**, 393–402.
65. Krämer, W., Sonnberger, H. (1986): The linear regression model under test. Physica-Verlag, Heidelberg.
66. Kuks, J., Olman, W. (1971): A minimax estimator of regression coefficients (in Russian). Iswestija Akademija Nauk Estonskoj SSR, **20**, 480–482.
67. Kuks, J., Olman, W. (1972): A minimax estimator of regression coefficients II (in Russian). Iswestija Akademija Nauk Estonskoj SSR, **21**, 66–72.
68. Lawless, J.F., Wang, P. (1976): A simulation study of ridge and other regression estimators. Communications in Statistics, Theory and Methods, **5**, 307–323.
69. Läuter, H. (1975): A minimax linear estimator for linear parameters under restrictions in form of inequalities. Mathematische Operationsforschung Statistik, Ser. Statistics, **6**, 680–695.
70. Lindley, D.V. (1962): Discussion of Professor Stein's paper. Journal of the Royal Statistical Society, Ser. B, **24**, 285–287.
71. Lindley, D.V., Smith, A.F.M. (1972): Bayes estimate for the linear model (with discussion) part 1. Journal of the Royal Statistical Society, Ser. B, **34**, 1–41.
72. Long, J.S, Ervin, L.H. (2000): Using heteroscedasity consistent standard errors in the linear regression model. American Statistician, **54**, 217–224.
73. Lütkepohl, H. (1996): Handbook of Matrices. Wiley , Chichester.
74. Mallows, C.L. (1973): Some comments on C_p. Technometrics, **15**, 661–675.
75. Marquardt, D.W. (1970): Generalized inverses, ridge regression, biased linear estimation and nonlinear estimation. Technometrics, **12**, 591–612.
76. Marquardt, D.W., Snee, R.D. (1975): Ridge Regression in practice. The American Statistician, **29**, 3–20.
77. Mayer, L.S., Willke, T.A. (1973): On biased estimation in linear models. Technometrics, **15**, 497–508.
78. Meyer, C.D. (2000): Matrix Analysis and Applied Linear Algebra. SIAM, Philadelphia.
79. Mood, A.M., Graybill, F.A., Boes, D.C. (1974): Introduction to the Theory of Statistics. Third Edition. McGraw-Hill, Auckland.
80. Nomura, M. (1988): On the almost unbiased ridge regression estimator. Communications in Statistics, Simulation, **17**, 729–743.
81. Obenchain, R.L. (1978): Good and optimal ridge estimators. The Annals of Statistics, **6**, 1111–1121.
82. Ohtani, K. (1996): On an adjustment of degrees of freedom in the minimum mean squared error estimator. Communications in Statistics, Theory and Methods, **25**, 3049–3058.
83. Pilz, J. (1991): Bayesian Estimation and Experimental design in Linear Regression Models. Wiley, Chichester.
84. Puntanen, S. and Styan, G.P.H. (1989): The equality of the ordinary least squares estimator and the best linear unbiased estimator with discussion. The American Statistician, **43**, 153–164.
85. Quenouille, M. (1956): Notes on bias in estimation. Biometrika, **43**, 353–360.
86. Ramsey, J.B. (1969): Tests for specification error in classical linear least squares regression analysis. Journal of the Royal Statistical Society, Ser. B, **31**, 350–371.

87. Ramsay, J.O. (1977). A comparative study of several robust estimates of slope, intercept and scale in linear regression. Journal of the American Statistical Association, **72**, 608–615.
88. Rao, A.R., Bhimasankaram, P. (1992): Linear Algebra. Tata McGraw–Hill, New Delhi.
89. Rao, C.R. (1973). Linear Statistical Inference and Its Applications. Second Edition. Wiley, New York.
90. Rao, C.R. (1976). Estimation of parameters in a linear model. The Annals of Statistics, **4**, 1023–1037. Corrigenda: The Annals of Statistics, **7**, 696.
91. Rao, C.R., Rao, M.B. (1998): Matrix Algebra and Its Applications to Statistics and Econometrics. World Scientific, Singapore.
92. Rousseeuw, P.J., Leroy, A.J. (1987): Robust Regression and Outlier Detection. Wiley, New York.
93. Ryan, T.A., Joiner, B.L., Ryan, B.F. (1976): The Minitab Student Handbook. Duxbury Press.
94. Sarkar, N. (1992): A new estimator combining the ridge regression and the restricted least squares method of estimation. Communications in Statistics, Theory and Methods, **21**, 1987–2000.
95. Sarkar, N. (1996): Mean square error matrix comparison of some estimators in linear regressions with multicollinearity. Statistics & Probability Letters, **30**, 133–138.
96. Schipp, B. (1990): Minimax-Schätzer im simultanen Gleichungsmodell bei vollständiger und partieller Vorinformation. A. Hain, Frankfurt/M.
97. Schipp, B., Trenkler, G., Stahlecker, P. (1988): Minimax estimation with additional linear restrictions — a simulations study —. Communications in Statistics, Ser. B, **17**, 393–406.
98. Sclove, S.L. (1968): Improved estimators for coefficients in linear regression. Journal of the American Statistical Association, **63**, 596–606.
99. Seber, G.A.F. (1977): Linear Regression Analysis. Wiley, New York.
100. Shapiro, S.S., Wilk, M.B. (1965): An analysis of variance test for normality (complete samples). Biometrika, **52**, 591–611.
101. Shao, R.Y.-S., Strawderman, W.E. (1994): Improving on the James-Stein positive-part estimator. The Annals of Statistics, **22**, 1517–1538.
102. Silvapulle (1991): Robust ridge regression based on an M-estimator. Australian Journal of Statistics, **33**, 319–333.
103. Simpson, J.R., Montgomery, D.C. (1996): A biased-robust regression technique for the combined outlier-multicollinearity problem. Journal of Statistical Computation and Simulation, **56**, 1–22.
104. Simpson, J.R., Montgomery, D.C. (1998): The development and evaluation of alternative generalized M-estimation techniques. Communications in Statistics, Simulation, **27**, 999–1018.
105. Singh, B., Chaubey, Y.P., Dwivedi, T.D. (1986): An almost unbiased ridge estimator. Sankhyā, Ser., **48**, 342–346.
106. Sommers, R.W. (1964): Sound application of regression analysis in chemical engineering. Presented during the American Institute of Chemical Engineers Symposium on Avoiding Pitfalls in Engineering Applications of Statistical Methods, Memphis, Tennessee.
107. Srivastava, A.K., Shukla, P. (1996): Minimax estimation in linear regression model. Journal of Statistical Planning and Inference, **50**, 77–89.

108. Srivastava, V.K., Giles, D.E.A. (1987): Seemingly unrelated regression equations models. Marcel Dekker, New York.
109. Stahlecker, P. (1987): A priori Information und Minimax-Schätzung im Linearen Regressionsmodell. Athenäum, Frankfurt/M.
110. Stahlecker , P., Trenkler, G. (1987): Quasi minimax estimation in the linear regression model. Statistics, **18**, 219–226.
111. Stein, C. (1956): Inadmissibility of the usual estimator for the mean of a multivariate normal distribution. In: Proceedings of the Third Berkeley Symposium on Mathematics, Statistics and Probability, University of California Press, Berkeley), 197–206.
112. Swindel, B.F. (1976): Good estimators based on prior information. Communications in Statistics, Theory and Methods, **5**, 1065–1075.
113. Theil, H. (1971): Principles of Econometrics. John Wiley & Sons, New York.
114. Theil, H., Goldberger, A.S. (1961): On pure and mixed statistical estimation in economics. International Economic Review, **2**, 65–78.
115. Theobald, C.M. (1974): Generalizations of mean square error applied to ridge regression. Journal of the Royal Statistical Society, Ser., **36**, 103–106.
116. Toro-Vizcarrondo, C., Wallace, T.D. (1968): A test of the mean square error criterion for restrictions in linear regression. Journal of the American Statistical Association, **63**, 558–572.
117. Toutenburg, H. (1982): Prior Information in Linear Models. Wiley, Chichester.
118. Toutenburg, H., Roeder, B. (1978): Minimax-linear and Theil estimator for restrained regression coefficients. Mathematische Operationsforschung Statistik, Ser. Statistics, **9**, 499–505.
119. Trenkler, G. (1978): An iteration estimator for the linear model. In: COMPSTAT 1978, Physica-Verlag, 125–131.
120. Trenkler, D., Trenkler, G. (1995): An objective stability criterion for selecting the biasing parameter from the ridge trace. Industrial Mathematics, **45**, 93–104.
121. Utts, J. (1982): The rainbow test for lack of fit in regression. Communications in Statistics, Theory and Methods, **11**, 1801–1815.
122. Venables, W.N., Ripley, B.D. (2002): Modern Applied Statistics with S. Fourth Edition. Springer, New York.
123. Vinod, H.D. and Ullah, A. (1981): Recent Advances in Regression Methods. Marcel Dekker, New York.
124. Wald, A. (1950): Statistical Decision Functions. Wiley, New York.
125. Werner, H.J. (1990): On inequality constrained generalized least squares estimation. Linear Algebra and Its Applications, **127**, 379–392.
126. White, H. (1980): A heteroscedasticity-consistent covariance matrix estimator and a direct test for heteroscedasticity. Econometrica. **50**, 483–500.
127. Zeileis, A. (2001): strucchange: Testing for structural change in linear regression relationships. R News, **1/3**, 8-11 (http://cran.R–project.org/doc/Rnews/).
128. Zeileis, A., Hothorn, T. (2002). Diagnostic Checking in Regression Relationships. R News, **2/3**, 7–10 (http://cran.R–project.org/doc/Rnews/).
129. Zeileis, A., Leisch, F., Hornik, K., Kleiber, C. (2002): strucchange: An R package for testing for structural change in linear regression models. Journal of Statistical Software, **7**, 1–38 (http://www.jstatsoft.org/v07/i02/).

130. Zellner, A. (1962): An efficient method of estimating seemingly unrelated regressions and tests for aggregation bias. Journal of the American Statistical Association. **57**, 348–368.
131. Zellner, A. (1986): Bayesian estimation and prediction using asymmetric loss functions. Journal of the American Statistical Association, **81**, 446–451.
132. Zellner, A. (1994): Bayesian and non-Bayesian estimation using balanced loss functions. In: Gupta, S.S., Berger, J.O. (ed) Statistical Decision Theory and Related Topics, Springer New York, 371–390.
133. Zhang, F. (1999): Matrix Theory. Springer, New York.

Index

Lecture Notes in Statistics

121: Constantine Gatsonis, James S. Hodges, Robert E. Kass, Robert McCulloch, Peter Rossi, and Nozer D. Singpurwalla (Editors), Case Studies in Bayesian Statistics, Volume III. xvi, 487 pp., 1997.

122: Timothy G. Gregoire, David R. Brillinger, Peter J. Diggle, Estelle Russek-Cohen, William G. Warren, and Russell D. Wolfinger (Editors), Modeling Longitudinal and Spatially Correlated Data. x, 402 pp., 1997.

123: D. Y. Lin and T. R. Fleming (Editors), Proceedings of the First Seattle Symposium in Biostatistics: Survival Analysis. xiii, 308 pp., 1997.

124: Christine H. Müller, Robust Planning and Analysis of Experiments. x, 234 pp., 1997.

125: Valerii V. Fedorov and Peter Hackl, Model-Oriented Design of Experiments. viii, 117 pp., 1997.

126: Geert Verbeke and Geert Molenberghs, Linear Mixed Models in Practice: A SAS-Oriented Approach. xiii, 306 pp., 1997.

127: Harald Niederreiter, Peter Hellekalek, Gerhard Larcher, and Peter Zinterhof (Editors), Monte Carlo and Quasi-Monte Carlo Methods 1996. xii, 448 pp., 1997.

128: L. Accardi and C.C. Heyde (Editors), Probability Towards 2000. x, 356 pp., 1998.

129: Wolfgang Härdle, Gerard Kerkyacharian, Dominique Picard, and Alexander Tsybakov, Wavelets, Approximation, and Statistical Applications. xvi, 265 pp., 1998.

130: Bo-Cheng Wei, Exponential Family Nonlinear Models. ix, 240 pp., 1998.

131: Joel L. Horowitz, Semiparametric Methods in Econometrics. ix, 204 pp., 1998.

132: Douglas Nychka, Walter W. Piegorsch, and Lawrence H. Cox (Editors), Case Studies in Environmental Statistics. viii, 200 pp., 1998.

133: Dipak Dey, Peter Müller, and Debajyoti Sinha (Editors), Practical Nonparametric and Semiparametric Bayesian Statistics. xv, 408 pp., 1998.

134: Yu. A. Kutoyants, Statistical Inference For Spatial Poisson Processes. vii, 284 pp., 1998.

135: Christian P. Robert, Discretization and MCMC Convergence Assessment. x, 192 pp., 1998.

136: Gregory C. Reinsel, Raja P. Velu, Multivariate Reduced-Rank Regression. xiii, 272 pp., 1998.

137: V. Seshadri, The Inverse Gaussian Distribution: Statistical Theory and Applications. xii, 360 pp., 1998.

138: Peter Hellekalek and Gerhard Larcher (Editors), Random and Quasi-Random Point Sets. xi, 352 pp., 1998.

139: Roger B. Nelsen, An Introduction to Copulas. xi, 232 pp., 1999.

140: Constantine Gatsonis, Robert E. Kass, Bradley Carlin, Alicia Carriquiry, Andrew Gelman, Isabella Verdinelli, and Mike West (Editors), Case Studies in Bayesian Statistics, Volume IV. xvi, 456 pp., 1999.

141: Peter Müller and Brani Vidakovic (Editors), Bayesian Inference in Wavelet Based Models. xiii, 394 pp., 1999.

142: György Terdik, Bilinear Stochastic Models and Related Problems of Nonlinear Time Series Analysis: A Frequency Domain Approach. xi, 258 pp., 1999.

143: Russell Barton, Graphical Methods for the Design of Experiments. x, 208 pp., 1999.

144: L. Mark Berliner, Douglas Nychka, and Timothy Hoar (Editors), Case Studies in Statistics and the Atmospheric Sciences. x, 208 pp., 2000.

145: James H. Matis and Thomas R. Kiffe, Stochastic Population Models. viii, 220 pp., 2000.

146: Wim Schoutens, Stochastic Processes and Orthogonal Polynomials. xiv, 163 pp., 2000.

147: Jürgen Franke, Wolfgang Härdle, and Gerhard Stahl, Measuring Risk in Complex Stochastic Systems. xvi, 272 pp., 2000.

148: S.E. Ahmed and Nancy Reid, Empirical Bayes and Likelihood Inference. x, 200 pp., 2000.

149: D. Bosq, Linear Processes in Function Spaces: Theory and Applications. xv, 296 pp., 2000.

150: Tadeusz Caliński and Sanpei Kageyama, Block Designs: A Randomization Approach, Volume I: Analysis. ix, 313 pp., 2000.

151: Håkan Andersson and Tom Britton, Stochastic Epidemic Models and Their Statistical Analysis. ix, 152 pp., 2000.

152: David Ríos Insua and Fabrizio Ruggeri, Robust Bayesian Analysis. xiii, 435 pp., 2000.

153: Parimal Mukhopadhyay, Topics in Survey Sampling. x, 303 pp., 2000.

154: Regina Kaiser and Agustín Maravall, Measuring Business Cycles in Economic Time Series. vi, 190 pp., 2000.

155: Leon Willenborg and Ton de Waal, Elements of Statistical Disclosure Control. xvii, 289 pp., 2000.

156: Gordon Willmot and X. Sheldon Lin, Lundberg Approximations for Compound Distributions with Insurance Applications. xi, 272 pp., 2000.

157: Anne Boomsma, Marijtje A.J. van Duijn, and Tom A.B. Snijders (Editors), Essays on Item Response Theory. xv, 448 pp., 2000.

158: Dominique Ladiray and Benoît Quenneville, Seasonal Adjustment with the X-11 Method. xxii, 220 pp., 2001.

159: Marc Moore (Editor), Spatial Statistics: Methodological Aspects and Some Applications. xvi, 282 pp., 2001.

160: Tomasz Rychlik, Projecting Statistical Functionals. viii, 184 pp., 2001.

161: Maarten Jansen, Noise Reduction by Wavelet Thresholding. xxii, 224 pp., 2001.

162: Constantine Gatsonis, Bradley Carlin, Alicia Carriquiry, Andrew Gelman, Robert E. Kass Isabella Verdinelli, and Mike West (Editors), Case Studies in Bayesian Statistics, Volume V. xiv, 448 pp., 2001.

163: Erkki P. Liski, Nripes K. Mandal, Kirti R. Shah, and Bikas K. Sinha, Topics in Optimal Design. xii, 164 pp., 2002.

164: Peter Goos, The Optimal Design of Blocked and Split-Plot Experiments. xiv, 244 pp., 2002.

165: Karl Mosler, Multivariate Dispersion, Central Regions and Depth: The Lift Zonoid Approach. xii, 280 pp., 2002.

166: Hira L. Koul, Weighted Empirical Processes in Dynamic Nonlinear Models, Second Edition. xiii, 425 pp., 2002.

167: Constantine Gatsonis, Alicia Carriquiry, Andrew Gelman, David Higdon, Robert E. Kass, Donna Pauler, and Isabella Verdinelli (Editors), Case Studies in Bayesian Statistics, Volume VI. xiv, 376 pp., 2002.

168: Susanne Rässler, Statistical Matching: A Frequentist Theory, Practical Applications and Alternative Bayesian Approaches. xviii, 238 pp., 2002.

169: Yu. I. Ingster and Irina A. Suslina, Nonparametric Goodness-of-Fit Testing Under Gaussian Models. xiv, 453 pp., 2003.

170: Tadeusz Caliński and Sanpei Kageyama, Block Designs: A Randomization Approach, Volume II: Design. xii, 351 pp., 2003.

171: D.D. Denison, M.H. Hansen, C.C. Holmes, B. Mallick, B. Yu (Editors) Nonlinear Estimation and Classification. x, 474 pp., 2002.

172: Sneh Gulati, William J. Padgett, Parametric and Nonparametric Inference from Record-Breaking Data. ix, 112 pp., 2002.

173: Jesper Møller (Editor), Spatial Statistics and Computational Methods. xi, 214 pp., 2002.

174: Yasuko Chikuse, Statistics on Special Manifolds. xi, 418 pp., 2002.

175: Jürgen Gross, Linear Regression. xiv, 394 pp., 2003.

Statistics on Special Manifolds
Yasuko Chikuse
This book is concerned with statistical analysis on the two special manifolds, the Stiefel manifold and the Grassmann manifold, treated as statistical sample spaces consisting of matrices. Starting with fundamental material of the special manifolds and some knowledge in multivariate analysis, the book discusses population distributions (especially the matrix Langevin distributions that are used for the most of the statistical analyses in this book), decompositions of the special manifolds, sampling distributions, and statistical inference on the parameters (estimation and tests for hypotheses). Asymptotic theory in sampling distributions and statistical inference is developed for large sample size, for large concentration and for high dimension. This book is designed as a reference book for both theoretical and applied statisticians.
2002/418 PAGES/SOFTCOVER/ISBN 0-387-00160-3
LECTURE NOTES IN STATISTICS, VOL. 174

Spatial Statistics and Computational Methods
Jesper Møller
Spatial statistics and Markov Chain Monte Carlo (MCMC) techniques have each undergone major developments in the last decade. Also, these two areas are mutually reinforcing, because MCMC methods are often necessary for the practical implementation of spatial statistical inference, while new spatial stochastic models in turn motivate the development of improved MCMC algorithms. This volume shows how sophisticated spatial statistical and computational methods apply to a range of problems of increasing importance for applications in science and technology.
2002/214 PAGES/SOFTCOVER/ ISBN 0-387-00136-0
LECTURE NOTES IN STATISTICS, VOL. 173

Parametric and Nonparametric Inference from Record-Breaking Data
Sneh Gulati and William J. Padgett
This book provides a comprehensive look at statistical inference from record-breaking data in both parametric and nonparametric settings, including Bayesian inference. A unique feature is that it treats the area of nonparametric function estimation from such data in detail, gathering results on this topic to date in one accessible volume. Previous books on records have focused mainly on the probabilistic behavior of records, prediction of future records, and characterizations of the distributions of record values, addressing some inference methods only briefly.
2002/112 PAGES/SOFTCOVER/ ISBN 0-387-00138-7
LECTURE NOTES IN STATISTICS, VOL. 172

Druck und Bindung: Strauss Offsetdruck GmbH